AF344426

THE NATURE OF LIFE

THE NATURE OF LIFE

DERRY D. KOOB
Utah State University

WILLIAM E. BOGGS

ADDISON-WESLEY
PUBLISHING COMPANY
Reading, Massachusetts
Menlo Park, California • London
Don Mills, Ontario

This book is in the
ADDISON-WESLEY SERIES IN LIFE SCIENCE

John W. Hopkins, III
Consulting Editor

To P.W.R. and H.K.B.

PREFACE

The philosopher Blaise Pascal once apologized to a colleague for writing him a long letter because he did not have time to write a short one. In biology, too, perhaps the harder task is to write a short book. We wrote ours because we felt the need for a book that would present the ideas of the new biology in language intelligible to the layman.

We have organized the book around the three great unifying ideas of modern biology: cell theory, genetics, and evolution. We have made no attempt at a complete survey of the plant and animal kingdoms; we leave this to later, more concentrated courses. We introduce organisms which have proved useful in solving some basic biological question, or which present some interesting unsolved problem. This has led to a certain emphasis on one-celled organisms, the simplest and in many ways the most versatile living systems. We defend this first, because much of the important research of recent decades has focused on such organisms; second, because these organisms illustrate all the basic functions of life; and third, because there is no point in using a complex model when a simple one will do.

The argument of the book can be summarized in a few paragraphs. The earth is probably not unique in the universe in sustaining life. In quest of a concept that will apply to all forms of life, we view life as a fourth state of matter, a state characterized by a remarkably high level of organization. We illustrate this idea with the fictitious organism described by astrophysicist Fred Hoyle in his novel *The Black Cloud*. According to the Second Law of Thermodynamics, an input of energy is required to maintain this organization. (We include a brief chapter on the basic ideas of thermodynamics and biochemistry.) On earth, this energy ultimately comes from the sun.

The continuity of life is maintained by the transfer of information encoded in giant molecules. All the hereditary information in these molecules directly or indirectly concerns the synthesis of proteins. Among proteins the most important are enzymes; they are the master control that determines which of thousands of possible reactions actually will occur within the cell. The cell is the basic unit of life on earth. Information and energy enable it to organize the substances in the environment into biological structures.

The cell is faced with a constant quest for energy and raw materials, and the diversity of organisms past and present reflects the spectrum of possible solutions to this problem. Life on earth began a few billion years ago, and we explore several theories of the origin of life. Once life began, the process of mutation tended to invent new life processes and organisms. Natural selection sorts the adaptive from the nonadaptive, and the environment tends to incorporate organisms into ecosystems. In man, natural selection has produced a dominant and sometimes blundering animal that threatens the survival of many species, including his own. The study of the behavior patterns of other species provides a fresh perspective for studying the nature of man.

We wrote the book for brief courses for liberal arts students, but reviewers of the manuscript suggested several additional ways in which the book might be used. Some reviewers suggested that the book might also be useful in the first half of a full-year course for science students, of which the second half might consist of a semester on higher plants and animals. Others suggested the book as the nucleus of a highly flexible full-year course for nonscience majors supplemented by a variety of lecture topics and readings. Many possible branching points are suggested by the bibliographies. Of course no textbook is meant to replace the teacher or even to lead him by the hand. It merely presents a point of view with which the teacher may agree at times and disagree at others. Both are inherent in the process of education.

We believe that all authors should warn the reader of their backgrounds and biases and we shall promptly confess ours. One of us is a biologist whose research involves studies of the growth and reproduction of bacteria, single-celled plants, and microscopic animals. This background leads to an emphasis on verifiable experimental work and an awareness of the non-permanence of concepts. The other author is a science editor, formerly with *Scientific American*. As a journalist he tends to emphasize recent topics in research and believes that no idea is too complex for a layman to understand if it is presented logically and lucidly. We both believe that ours is the last generation of Americans that will be encumbered by the use of English units. We use metric units in the book, sometimes indicating the English equivalent in parentheses, but where possible indicating a common object for comparison (for example, one square centimeter is roughly the area of a man's fingernail). We share a low opinion of end-of-chapter problems and have not included any. Because the book focuses on many ideas that have drawn wide attention in the press and on television, readers gifted with curiosity may want to explore some topics in more detail. For these students we have compiled an extensive chapter-by-chapter list of further readings that includes a brief "thumbnail" review of each work.

We wish to acknowledge the help of Professor Paul Mundinger of Rockefeller University, who drafted the chapter on behavior; and of Professor Peter Marler of the same institution, who reviewed the final version of that material.

September, 1971   D.D.K.
                                                                    W.E.B.

CONTENTS

CHAPTER 1   EXOTIC LIFE, THE CELL, THE GAME OF SCIENCE   1

Beyond the solar system   7
What is life?   10
The black cloud   11
Life probes   14
The cell   16
The logic of science   21
A medieval experiment   23
A fictitious experiment   23

CHAPTER 2   LIFE ON EARTH   27

Classification simplified   29
Scientific names   29
Monera   32
Protista   35
True plants (Metaphyta)   42
Plants in research   42
True animals (Metazoa)   43
A catalogue of beasts   44
Animals in research   52
The fourth-state model   55
Energy   56
Temperature   57
Moisture   59
Pressure   62
Chemicals   64
Time   65
Cause and effect   68

CHAPTER 3   THE PHYSICS AND CHEMISTRY OF LIFE   73

The first law of thermodynamics   76
The second law of thermodynamics   77
Atoms   78
Molecules and compounds   80
Biochemistry   81
Carbohydrates   82
Lipids   86
Proteins   90
Nucleic acids   92
Adenosine phosphates   94

CHAPTER 4   THE STRUCTURE OF CELLS   97

The eukaryotic cell   101
The cell membrane   101
Diffusion   102
Active transport   105
The nucleus   107
Cytoplasm   111
Mitochondria   113
Chloroplasts   114
Lysosomes   115
Ribosomes   116
Cell diversity   116
Multicellularity   121
Plants   122
Animals   122
Levels of organization   123
The new biology   125

CHAPTER 5   THE CELL NUCLEUS   129

The role of the nucleus   130
Chromosomes   134
Structure of chromosomes   137
Mitosis and cell division   139
Nucleic acids   144
Genetic transformation   144
Viral nucleic acids   148
Tracer experiments   150
Transduction   151
Cytoplasmic inheritance   153

CHAPTER 6   THE DOUBLE HELIX   159

The Watson-Crick model   159
Replication of DNA   160
Mutants   165
Protein synthesis   166
The genetic code   173
Levels of protein structure   175
Methemoglobinemia   181

CHAPTER 7   REPRODUCTION   185

Asexual reproduction   186
Bacteria and blue-green algae   187
The higher plants   188
Multicellular animals   191
Sexual reproduction   192
Conjugation   192
Fertilization   194

Sex and chromosome number 196
Meiosis 197
Green algae 203
Ferns 203
Flowering plants 206
Animals 212
Haploidy versus diploidy 215
Polyploid animals 218
Haploid syndrome 218

CHAPTER 8 CLASSICAL GENETICS 221

Classical genetics 222
Mendel 222
Genes 223
Genotype 224
Phenotype 227
Genetic prediction 230
Determination of genotypes 234
Dihybrids 237
Trihybrids and more 240
Incomplete dominance 240
Multiple alleles 243
Polygenic inheritance 244
Linkage 244
Mutation 250

CHAPTER 9 CONTROL MECHANISMS 253

Control of development 254
Totipotency 255

Induction and repression 257
Organizers 263
Hormonal control 263
Insect metamorphosis 266
Plant hormones 267
Gibberellins 269
Cytokinins 269
Plant development 269
Nerve control 271
The nerve impulse 275
The reflex arc 276
Other control functions 280
Biological versus artificial systems 281

CHAPTER 10 PHOTOSYNTHESIS 287

Solar energy 288
Generalized equations of photosynthesis 290
A refinement of the equations 292
Photochemistry 294
Bacterial photosynthesis 297
Photosynthesis in higher plants 300
A model of the chloroplast 304

CHAPTER 11 NUTRITION AND METABOLISM 309

Nutrition 310
Digestion 314
Fermentation and respiration 320
The respiratory chain 326
The mitochondrion 328

Biosynthesis 330
The assembly of a cell 335
The complexity of life 336

CHAPTER 12 THE ORIGIN OF LIFE 339

The geologic timetable 340
The birth of the planet 342
The primitive atmosphere 343
Temperature and climate 344
Small organic molecules 345
Giant molecules 346
The primordial soup 347
Self-duplicating systems 348
Primitive cellular organisms 350
Photosynthesis and the oxygen revolution 353
Life from nonlife 354

CHAPTER 13 THE RISE OF MODERN ORGANISMS 357

Lamarck 360
Darwin 360
Darwin's finches 367
Darwin's theory 367
Natural selection 369
The modern theory 372
The Hardy-Weinberg law 374
The mechanism of evolution 374
The peppered moth 375
The origin of species 377

Biology as history   381
The Paleozoic era   386
The Mesozoic era   391
The Cenozoic era   396
Future evolution   399

CHAPTER 14   THE BIOLOGY OF BEHAVIOR   403

Inherited behavior   404
Learning   406
Development   408
The evolution of behavior   410
The evolution of displays   411
Communication   414
Tactile signals   415
Visual signals   416
Acoustical signals   417
Chemical signals   419
Signal and language   420
Orientation   423
Sun compass   424
Navigation   424
Smell   429
Aggression   429

CHAPTER 15   ECOLOGY AND MAN   435

Human ecology   436
Community and ecosystems   437
The oceans   443
Energy transfer   446

An Arctic energy pyramid 449
The recycling of matter 450
The human impact 454
Farming the jungle 455
The war with the insect 456
The gypsy moth 456
The ring bark borer 457
Control by sterilization 458
Man versus nature 459

CHAPTER 16   TOWARD THE TWENTY-FIRST CENTURY   463

Population and the green revolution 465
City and suburb 467
Energy and electricity 470
Chemical pollution 474
Corporation versus public 476
An end to individualism 478

ACKNOWLEDGMENTS 483
INDEX 487

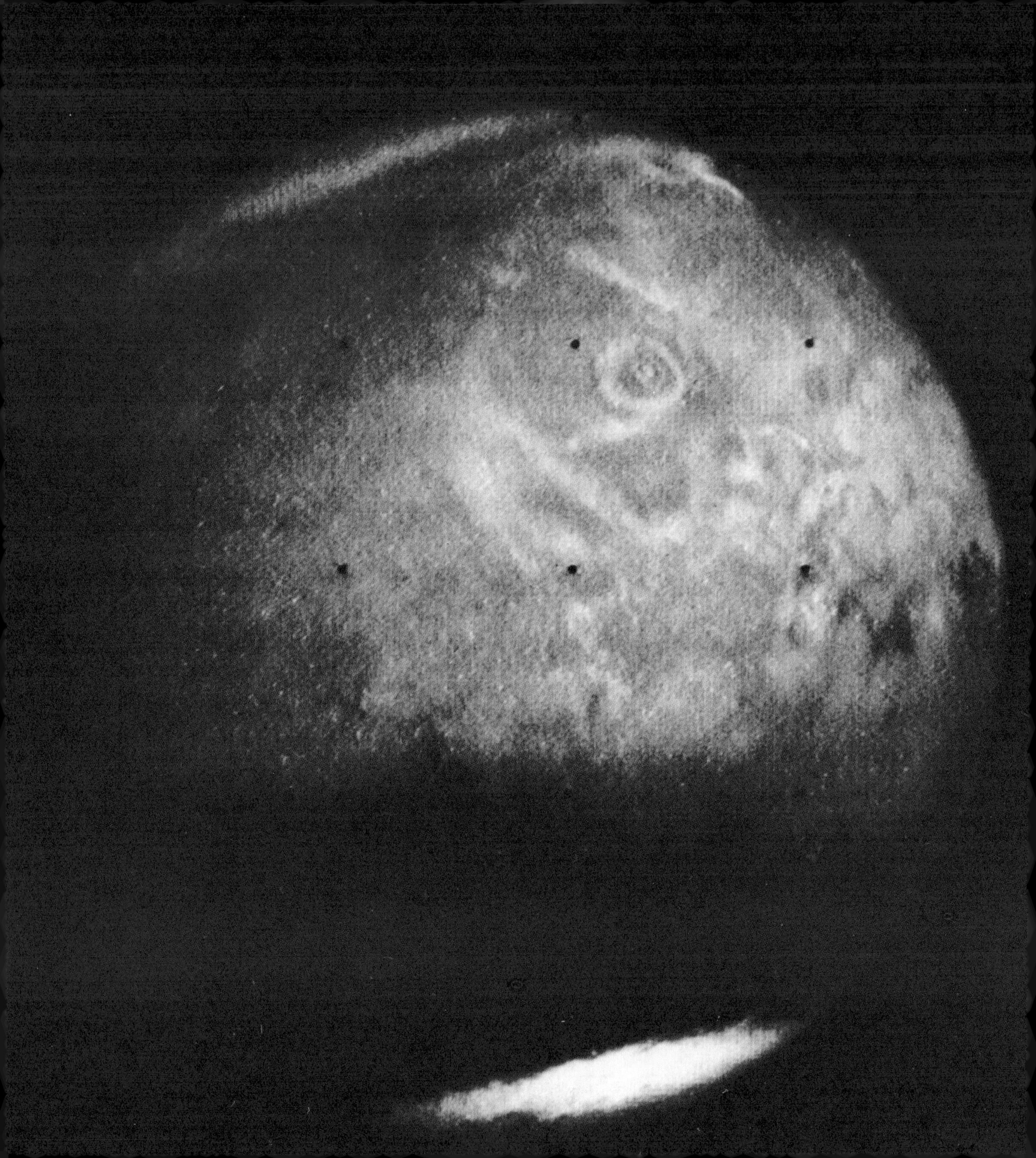

1. EXOTIC LIFE, THE CELL, THE GAME OF SCIENCE

There is a place with four suns in the sky—red, white, blue and yellow; two of them are so close together that they touch, and star-stuff flows between them. I know of a world with a million moons. I know of a sun the size of the earth—and made of diamond. There are atomic nuclei a few miles across which rotate thirty times a second. There are tiny grains between the stars, with the size and atomic composition of bacteria. There are stars leaving the Milky Way, and immense gas clouds falling into it. There are turbulent plasmas writhing with X- and gamma rays and mighty stellar explosions. There are, perhaps, places which are outside our universe. The universe is vast and awesome, and for the first time we are becoming a part of it.*

Are we alone in the universe? This is an ancient, almost instinctive question. For centuries it has troubled mankind, and the inability to answer it has powerfully influenced religion, philosophy, and science. Two or three decades ago the answer seemed clear; biologists did not take seriously the possibility of life on the other planets in our solar system. Mercury and

---

* Carl Sagan, *Planetary Exploration.* University of Oregon Press. Eugene, Oregon, 1970. Reprinted by permission.

◀**The planet Mars** was photographed by the space probe *Mariner 7*. This frame was made at a distance of about 435,000 kilometers, somewhat more than the distance from the earth to the moon. It shows the south polar cap (bottom) and the bright bullseye crater called *Nix Olympica* (right center). Dark area at bottom center is the *Mare Sirenum*. Photographs reproduced in Figs. 2 and 3 depict regions that lie beyond the eastern horizon (right).

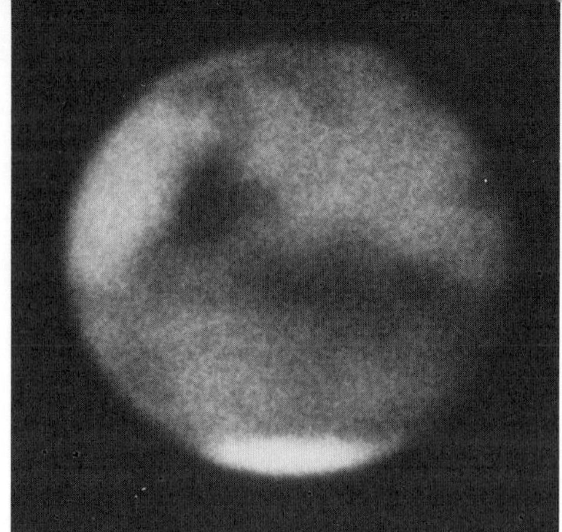

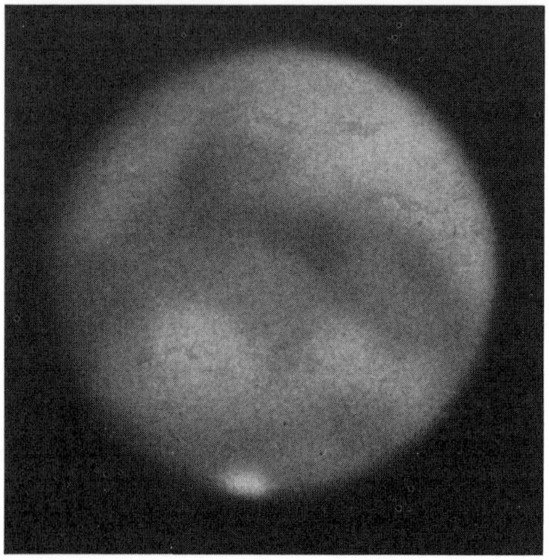

1

Seasonal darkening of the surface of Mars is evident in these two photographs made at the Lowell Observatory. In the Martian summer (bottom) the white polar cap is smaller and the contrast between the light and dark areas of the planet decreases. Many observers have ascribed this effect to the presence of vegetation.

2

Surface of Mars appears barren and moonlike in this photograph made by *Mariner 7*. Cratered landscape lies in region known as *Noachis,* near the southern polar cap. This wide-angle view shows about 225,000 square kilometers, roughly the area of New Mexico. No sign of life is visible.

Venus seemed too hot. Mars had lost most of its atmosphere and moisture. The giant planets, Jupiter, Saturn, Uranus, and Neptune, seemed too cold and were blanketed with poisonous atmospheres. Pluto, the most distant planet, was too far away for precise study. And speculation about life beyond the solar system belonged in the pages of science fiction.

The intellectual climate has changed since the launching of the first *Sputnik*. The radio beeps from a metal sphere in orbit meant that man and his instruments were no longer earthbound. The ancient question was no longer unanswerable.

So far the evidence for extraterrestrial life is shaky but tantalizing. If we study the spectrum of Mars from observatories on earth, we detect the presence on Mars of organic molecules (carbon compounds) characteristic of living organisms on earth. The darkening colors associated with the change of seasons on Mars have long suggested the presence of vegetation (Fig. 1). Although photographs of the planet made by the space vehicles *Mariner 6* and *Mariner 7* reveal a desolate terrain with no sign of life (Figs. 2 and 3), the photographs are not conclusive. Each vehicle carried two cameras, the more powerful of which could only reveal structures more than 300 meters in diameter. Comparable photographs of the earth, taken by orbiting satellites with lenses that resolve objects as small

3
Closeup of Mars taken by *Mariner* 6 reveals
a desolate moonlike terrain devoid of any
evidence of life. The resolution of the optical
system is about 300 meters. The small deep
crater at center is about four kilometers
across, a little wider than Manhattan. This
photograph shows an area just south of the
Martian equator.

     Chapter 1. Exotic life, the cell, the game of science

◄**Closeup of earth** was made by the crew of *Apollo 9*. The Imperial Valley of southern California lies at center, with the Salton Sea at top center and the Colorado River barely visible at far right. In this oblique view the resolution is of the same order as in Fig. 3, although the area shown here is larger. For scale, note that the Salton Sea is about five times the width of the small crater in Fig. 3. The only evidence of biological activity is the regular geometry of cultivated farms (small squares). At this resolution, no individual organisms are visible.

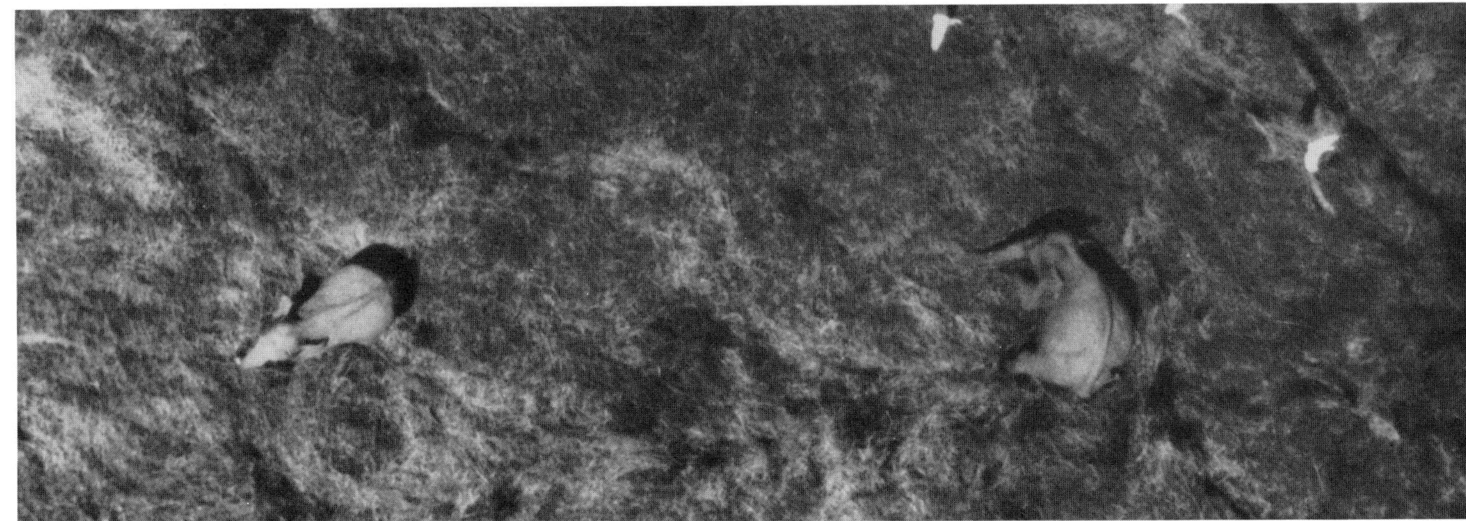

as 100 meters across (about the diameter of a small stadium), show scant evidence of life. In fact one must look closely to observe the large-scale works of man, even when one knows they are there (Fig. 4). Obvious evidence of living organisms on earth shows up in aerial photographs taken with special lenses that can resolve objects from one to ten meters in diameter (Fig. 5).

The earth is steadily bombarded with showers of meteors, presumably the debris of shattered asteroids, and some of this material contains molecules characteristic of living systems. In meteors of the type known as carbonaceous chondrites, about one part in 10,000 consists of organic molecules. Recently a team of scientists from the National Aeronautics and Space Administration (NASA) discovered amino acids, the chemical building blocks of proteins, in a small meteor that fell in Australia. Perhaps the organic molecules are a result of contamination after the meteor struck the earth. But some of the amino acids analyzed by the NASA team were of a type very rarely found in organisms on earth. Coupled with astronomical observations that suggest the presence of organic compounds in

5
Evidence of life on earth appears in photographs that reveal objects as small as one meter across. This aerial view of the grasslands of Africa shows a few large organisms: elephant and flying cranes.

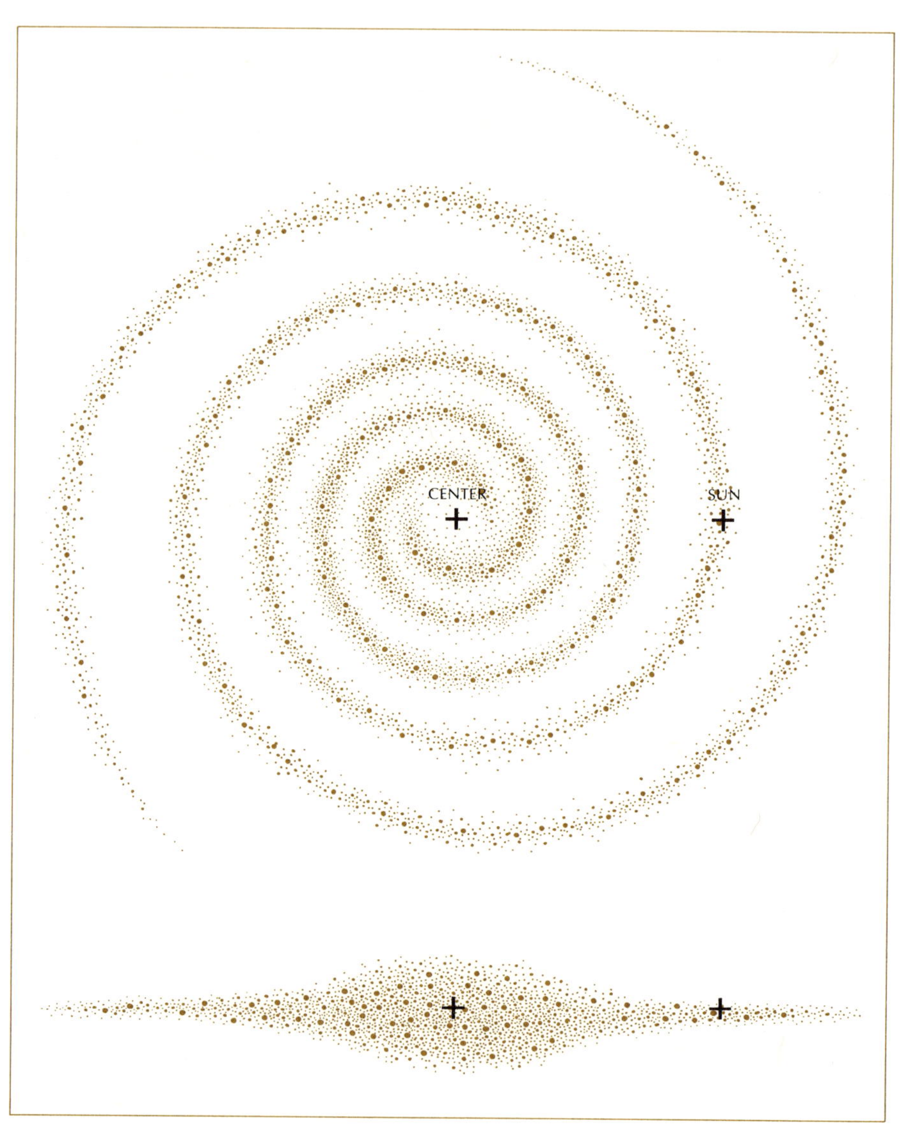

CENTER
SUN

◄**Our galaxy** is a flattened disk of stars, gas, and cosmic dust. Its five spiral arms contain some two billion stars. This diagram depicts it from above (top) and edge on (bottom). Our sun lies in one of the outer arms, some 30,000 light-years from the center of the galaxy.

distant clouds of interstellar gas, such evidence indicates that the earliest stages in the emergence of life—the evolution of small organic molecules— can proceed not only on earth, but even in space. Life may not be a rare phenomenon in the universe.

In the search for environments that might harbor organisms roughly similar to some of those found on earth, biologists have taken a fresh look at the solar system. The hot clouds of Venus, the subsoil of Mercury, the moons of Jupiter have all been suggested as possible outposts of life by scientists with excellent credentials.

BEYOND THE SOLAR SYSTEM

The possibility of life outside the solar system is more difficult to assess. Our sun is an undistinguished star that lies toward the outer edge of a flattened, pinwheel-shaped galaxy that contains roughly 200 billion stars (Fig. 6). In the night sky we see our galaxy edge on, stretched across the sky in a luminous band that we call the Milky Way (Fig. 7). There is good evidence that a few of these stars have planets, and various theories suggest the existence of many planetary systems. The distances between stars are measured in terms of the distance that an object moving at the speed of light (300,000 kilometers per second) could travel in a year. One light-year equals about 10^{13} kilometers (six billion miles). The nearest star beyond the sun is Alpha Centauri, about 4.3 light-years away.

It is unlikely that within the next few centuries man can develop the technology to send a space vehicle to the neighborhood of another star. But we can already eavesdrop on radio activity at interstellar distances. The giant antennas and amplifiers of radio telescopes such as those at Jodrell Bank in Great Britain and at the National Radio Astronomy Observatory at Green Bank, West Virginia, are designed to detect very weak and distant

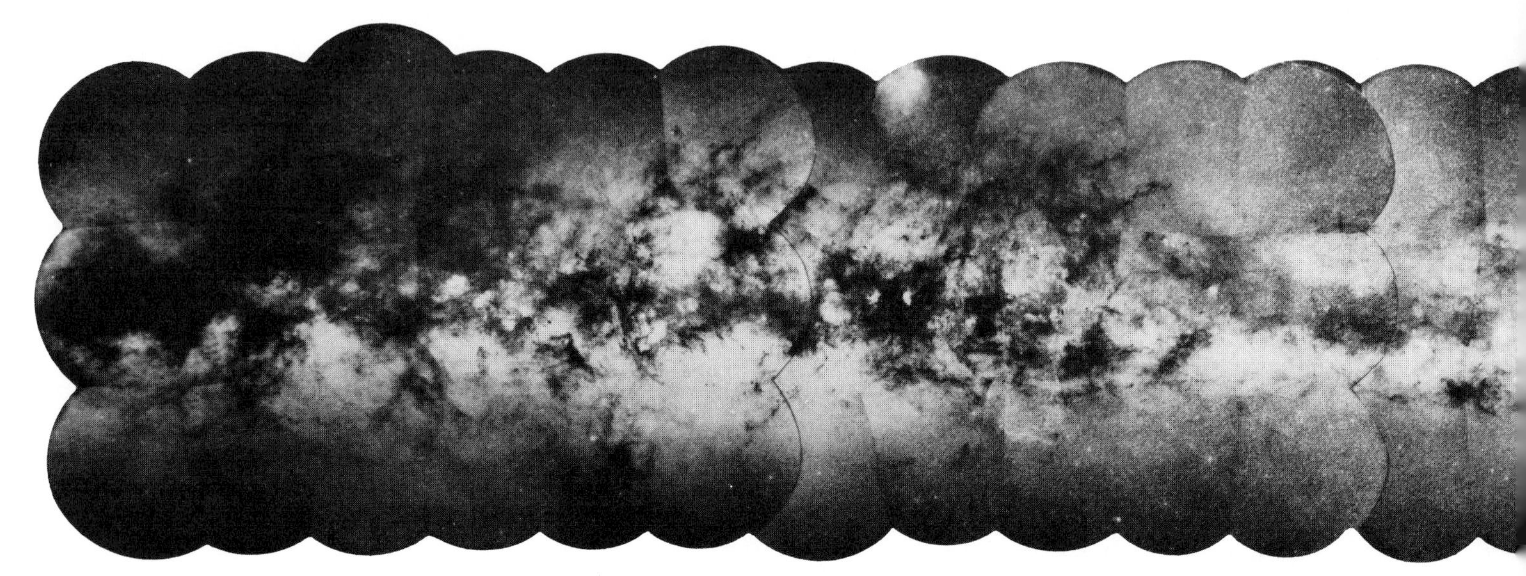

    Chapter 1. Exotic life, the cell, the game of science

Milky Way is our galaxy, seen here edge on. The dark wisps along the plane of the galaxy are clouds of cosmic dust that absorb light from stars lying beyond them. Such clouds hide the center of the galaxy, which lies in the direction of the constellation Sagittarius (left). This composite photograph was assembled from scores of smaller ones made at the Mt. Stromlo Observatory in Australia. Nearest star beyond the sun, Alpha Centauri, appears as a bright disk at center.

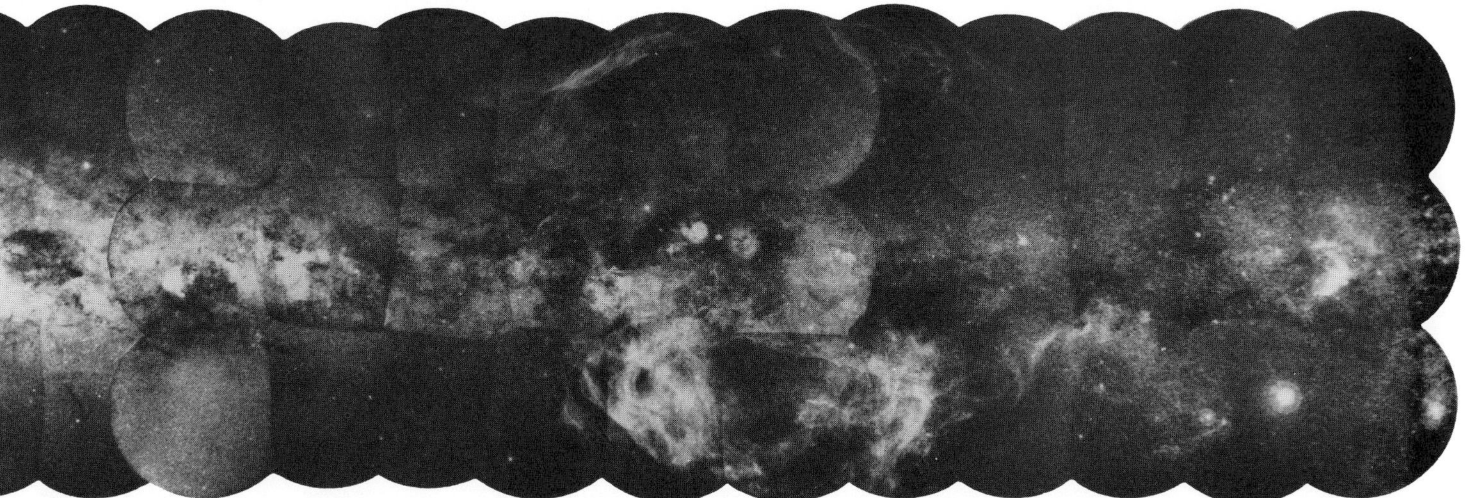

signals (Fig. 8). These powerful instruments could detect the present radio transmissions on earth from distances of up to 1000 light-years away. Within this range there are 10 million stars; among them there is a reasonable probability of planetary systems occupied by one or more detectable civilizations. In fact, a two-month search for such signals was made in 1960 at Green Bank, and another is now in progress at the Sternberg Observatory in Moscow. Properly done, such searches must be regarded as longterm ventures. Many astronomers in the United States have urged the funding of a radio search of our part of the galaxy; they estimate that it would take roughly 30 years and cost perhaps 50 million dollars, a very

8

◄**Radio telescope** at Green Bank, West Virginia, detects weak radio emissions from matter in space. Curved dish covered with wire mesh focuses incoming signals on the antenna mounted in front of it. Dish is almost 100 meters in diameter, as can be seen by comparing its size with that of the automobile at far left.

small fraction of the billions spent to send a man to the moon. For many scientists, the question now is not whether we shall discover extraterrestrial life, but when.

WHAT IS LIFE?

The possibility of discovering some exotic form of life has forced biologists to reexamine their ideas about the nature of life. Many still doubt that there exists, at least in the solar system, "life as we know it." This is a cautious phrase, but what does it mean? One way to define life is in terms of the characteristics common to all things that live on earth: growth, reproduction, continuous chemical activity, and responsiveness to stimuli.

Taken separately, these characteristics can also be found in nonliving systems. All living things grow, but so do salt crystals, hailstones, and flames. Most living organisms can reproduce, but flames can reproduce by means of sparks, and carry on continuous chemical activity, converting energy from one form to another. A flask of nitroglycerin is remarkably responsive to stimuli, especially to heat and shock.

To a classical biologist, the coexistence of all these characteristics in a single system makes it "alive." Some research workers in the new biology—the study of living systems at the molecular level—prefer to define life in terms of genetics. To them, a living system is one that can reproduce and evolve. To evolve, a genetic system must be capable of undergoing mutations—small random errors in the chemical structure of a gene or in the copying of genes from one generation to the next—and the mutant genes must themselves be inheritable generation after generation. Any system that can reproduce, mutate, and reproduce its mutant genes will evolve. It will develop the genetic potential to adapt to changes in its environment and under the competitive pressures of nature it will, over the course of centuries or millenia, become a progressively more effective "machine for living." Mutation "invents" an assortment of genetic traits that can be incorporated in different organisms. Environmental conditions such as climate, predators, and food supply favor some of these inventions and discourage others. The favored organisms survive and multiply; the others become extinct. In retrospect, the process seems to have a strategy and a purpose; actually, the diversity and adaptability of modern organisms are the result of roughly four billion years of interaction between life and the earth.

Even the genetic definition of life has its shortcomings. In fact, both the classical and the modern definitions exclude some organisms that are unquestionably alive but sexually sterile, such as the mule, the offspring of a mare and a male donkey.

Recent research in biochemistry has led many workers to conclude that the real "secret" of life lies in its intricate organization. Microbes, plants, and animals are complex systems whose chemical architecture is much more complicated than any system in the nonliving world. Within a single cell the molecules are organized into pumps, communication systems, control mechanisms, chemical processing units, and analog computers. And each cell contains in its giant molecules the complete blueprint for duplicating the entire system. The most complicated industrial plant, when compared with a single living cell, is a relatively crude system.

Compared with the human brain, the most elaborate digital computer is a simple robot. Computers of the future will display a much higher level of organization and performance. No one who saw Stanley Kubrick's film *2001* will forget the powers of HAL, the onboard computer that controlled the mission to Jupiter. Designed with logic circuits that could make 10 million decisions a second, and with a memory that could easily engulf the contents of *The Encyclopaedia Britannica,* HAL could outperform the human brain in everything but imagination. Linked to television and sound systems, HAL could see, hear, and speak in human language. Computers like HAL are not a fantasy, but a foreseeable reality. Will these sophisticated electronic organisms be alive? It depends on how one defines life.

Each of these approaches is relevant, but the results are disappointing to anyone who wants a short definition of life. Biologists cannot agree on what they mean by "life as we know it." It is even more difficult to construct a definition that will fit all life, including undiscovered organisms that may stretch the limits of our imaginations.

THE BLACK CLOUD

Astrophysicist Fred Hoyle of Cambridge University describes such an organism in his novel *The Black Cloud.* Terror grips the earth as an enormous

Cloud of ionized gas, millions of kilometers in diameter, sweeps into the solar system from outer space and engulfs the sun. All attempts to predict the movements of the Cloud according to astronomical laws fail. Then one astronomer proposes an incredible idea:

> All our mistakes have a certain hallmark about them. They're just the sort of mistake that it'd be natural to make if instead of the Cloud being inanimate, it were alive.*

His astonished colleagues ask him to explain what he means by the word "alive." From a biologist's viewpoint, his reply is brilliant:

> Well, John, you know better than I do that the distinction between animate and inanimate is more a matter of verbal convenience than anything else. By and large, inanimate matter has a simple structure and comparatively simple properties. Animate or living matter, on the other hand, has a highly complicated structure and is capable of very involved behavior. When I said the Cloud may be alive I meant that the material inside it may be organized in an intricate fashion, so that its behaviour and consequently the behaviour of the whole Cloud is far more complex than we previously supposed. . . . In short, I think the Cloud has a brain.

The scientists attempt to communicate with the Cloud by transmitting a beam of radio pulses. After 33 hours of transmission, the lab picks up the first signal from the Cloud. Using computer techniques, they translate this message:

> Your first transmission came as a surprise, for it is most unusual to find animals with technical skills inhabiting planets, which are in the nature of extreme outposts of life.

Asked why, the Cloud replies:

> For two quite simple reasons. Living on the surface of a solid body you are exposed to a strong gravitational force. This greatly limits the size to which your animals can grow and hence limits the scope

---

* Harper & Row, Publishers, Inc. New York, 1957. William Heinemann, Ltd., British publishers. This and the following excerpts reprinted by permission.

of your neurological activity. . . . By and large, one only expects
intelligent life to exist in a diffuse gaseous medium, not on planets
at all.

The second unfavorable factor is your extreme lack of basic
chemical foods. For the building of chemical foods on a large
scale, starlight is necessary. Your planet, however, absorbs only a
very minute fraction of the light from the Sun. At this very moment I myself am building basic chemicals at about 10 billion
times the rate at which building is occurring on the entire surface of
your planet.

This shortage of food chemicals leads to a tooth and claw existence
in which it is difficult for the first glimmerings of intellect to gain
a foothold in competition with bone and muscle. Of course once
intelligence becomes firmly established, competition with sheer
bone and muscle becomes easy, but the first steps along the road
are excessively difficult—so much so that your own case is a rarity
among planetary life forms.

Professor Hoyle's fantasy carries the idea of biological organization to its
logical limit, and in the process he challenges some of our smug, almost
unconscious assumptions about the nature of life.

A few more of these assumptions are exposed by other creators of
fictitious organisms. In his novel *Childhood's End,* Arthur C. Clarke describes the intelligent life on an imaginary planet caught in the conflicting
gravities of six colored suns, a planet which sees no day and night, no
seasons and years, only a continuous change of light as it spins along its
unbelievably complex orbit:

The great, many-faceted crystals stood grouped in intricate geo-
metrical patterns, motionless in the eras of cold, growing slowly
along veins of mineral when the world was warm again. No matter
if it took a thousand years for them to complete a thought. The
universe was still young, and time stretched endlessly before
them. . . .*

---

*   Ballantine Books. New York, 1953.

In *The Sirens of Titan,* Kurt Vonnegut, Jr. describes imaginary creatures called harmoniums that live in caves beneath the planet Mercury. Shaped like small blue kites, they cling to the phosphorescent yellow walls in bright lozenge patterns and feed on the mechanical energy of the planet's vibrations. Our science, too, might overlook the intelligence of Clarke's crystals or the existence of organisms like Vonnegut's harmoniums.

LIFE PROBES

The search for extraterrestrial life is one of the prime objectives of the space programs of both the U.S. and the U.S.S.R. During the present decade they will probably land packages of instruments on Mars and send flyby missions or instrumented orbiters to Jupiter. At first the search will concentrate on finding organisms similar to those on earth, particularly bacteria. This seems to be the most promising approach because for hundreds of years biologists have refined the techniques for studying earthly life. No one would know how to go about designing equipment to detect life forms as alien as those described by Hoyle and Clarke.

On earth it is difficult to find a spot not inhabited by bacteria. They inhabit the desert, survive in the frozen soil of Antarctica, and even float freely in the atmosphere. Abundance is important because the landing site for the first life probe will be selected primarily on the basis of favorable landing characteristics. A vehicle submerged in soft surface material or wrecked on a boulder field can transmit no useful information.

The first life probes have already been designed as part of NASA's Project Viking. In 1975, according to the current schedule, two Viking spacecraft will be launched toward Mars. From orbits above Mars they will each send a smaller vehicle for a soft landing on the surface (Fig. 9). The Viking landers will carry several types of sensors, including at least one designed to detect the presence of living organisms.

Several types of life sensors have been proposed. One would scoop up a soil sample and dump it into a tube of nutrient culture medium. If any organisms are present in the soil, and if they can grow in the culture broth— two big if's—they will begin to multiply. An increase in the population of cells would make the broth cloudy, and the resulting change in transparency could be detected by a photocell. A miniature transmitter can relay this information back to earth.

9
Unmanned landing on Mars is the goal of NASA's Project Viking. Viking spacecraft in orbit above Mars (top) will send instrument package for soft landing on
surface (bottom). Sampling devices on lander will test Martian soil for evidence
of living organisms, particularly bacteria. Results will be transmitted back to
earth, either directly or by relay from the orbiter, which will also transmit photographs of the landing site.

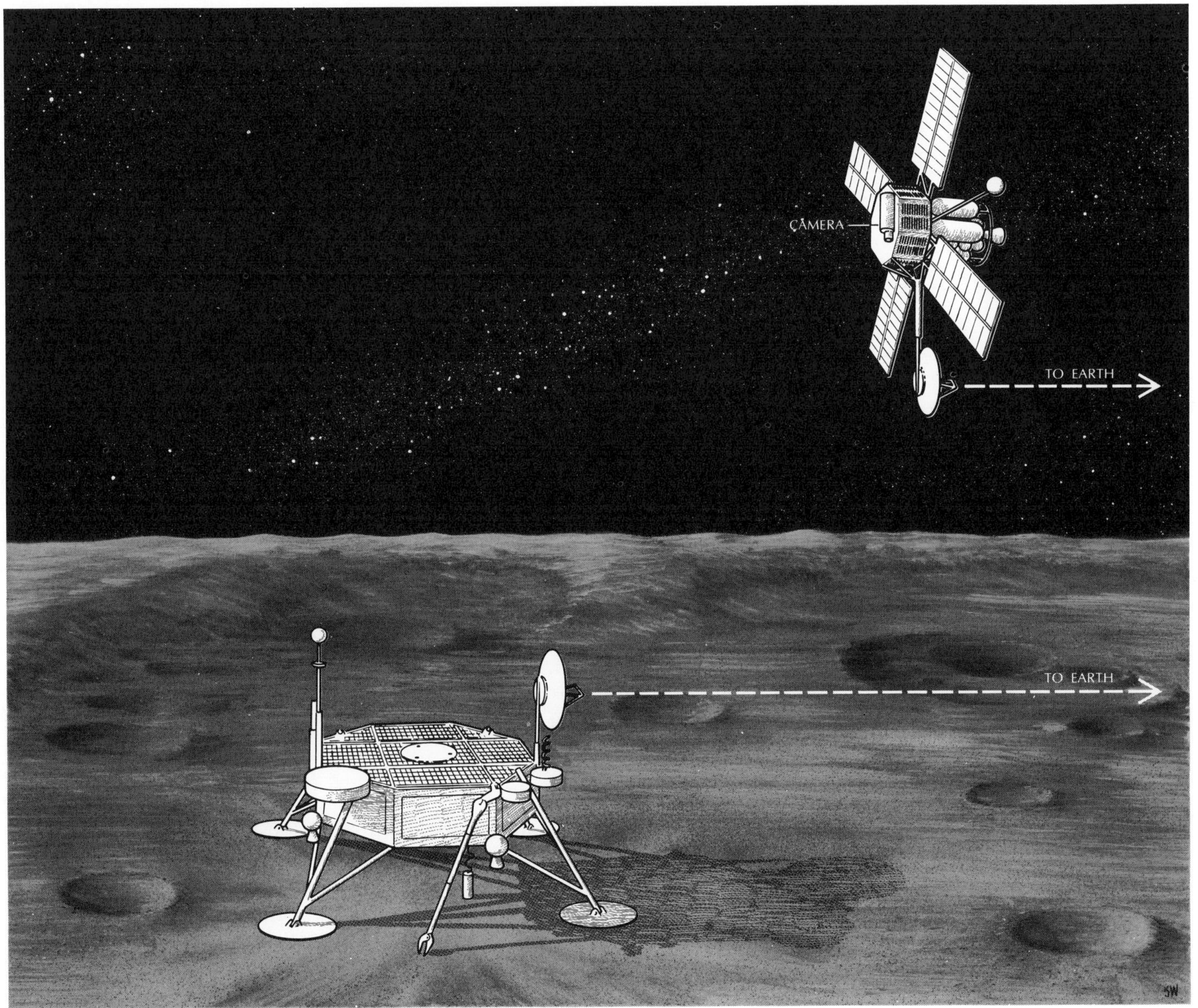

When earthly bacteria grow and multiply in liquid culture they change the acidity of the broth they grow in. This too can easily be detected in the culture medium of the lander. If the transmitter notifies earth that the acidity of the culture medium is changing, we shall have additional evidence that something is growing in it, something that might be called "alive."

Another life probe has been designed assuming that life on another planet will involve some of the biochemical processes carried out by microorganisms on earth. For example, as bacteria grow they use sugars as food and produce carbon dioxide as a waste product. The culture medium would contain sugars "labeled" with radioactive carbon. If soil placed in this medium contains organisms similar to our bacteria, they will use the radioactive sugar and produce radioactive carbon dioxide. The presence of this radioactive gas will be detected by a Geiger counter. Without life, no gas will appear.

These are only two of the ingenious devices designed to discover whether life exists on planets like Mars and Venus. If the first reports from these probes are negative, no one will be surprised. The chance of landing at the right spot at the right time and providing the right growing conditions for unknown organisms is certainly a long shot. If the first mission fails, we shall certainly try again, and again, and again.

In a very real sense these probes will search the solar system for something that we cannot define. We can discuss life, but only very crudely. We can say that somehow a living system is more than the sum of its parts, in much the same way as music, superbly played, is more than notes on a staff. Like musicians, we proceed from a study of the notes to an interpretation of the music. In this way we can hope to understand some of the principles that govern life on earth and throughout the universe.

THE CELL

The cell is the basic unit of life on earth. Most introductory biology students have already memorized the classic definition: A cell is the microscopic unit of which all plants and animals are composed. It contains a nucleus, cytoplasm, and a cell membrane. This was a pretty good definition in 1839 when it was postulated as "The Cell Theory" by the German biologists Matthias Schleiden and Theodore Schwann. But that was more than a century ago. The information known in science at any given time is only

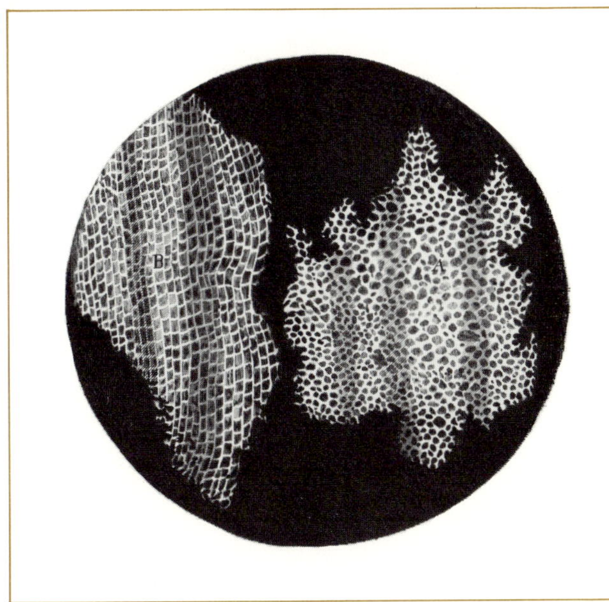

10
Cork cells observed in the 17th century by Robert Hooke with his primitive microscope were not cells at all, but only cell walls. He published these engravings of them in his famous *Micrographia*.

the basis for further knowledge. Research is constantly yielding new data. So, too, the concept of the cell continues to change.

Cells have been on earth for at least three billion years, but their existence was not discovered until the invention of a device to magnify very small structures. Like so many other breakthroughs, this basic biological discovery depended on the development of a technical tool of the physical sciences, the microscope. A primitive early microscope showed that pieces of cork were composed of empty cubicles or "cells" (Fig. 10). No such units were found in animal tissues examined at that time. The structural organization of plants seemed to differ sharply from that of animals. Years later, with the development of techniques for rapidly killing, staining, and slicing cells, biologists discovered that animal tissue was composed of cells quite similar to those found in plants. Moreover, they discovered that the cells were not empty. Each cell contained a round body, or nucleus, surrounded by a region of jellylike material that was named cytoplasm.

This was part of the knowledge that led Schleiden and Schwann to stress the importance of the cell as the basic structural unit of living organisms. The persistence of this definition is testimony to its validity. By now biologists have examined innumerable additional organisms and used more powerful tools, such as the electron microscope, to explore the microstructure of cells. This work has produced a long list of exceptions to the simple classic definition.

First, all cells are not microscopic in size. The marine alga *Acetabularia* is a single cell, but it is readily visible to the eye of an alert skindiver (see

11

Acetabularia, a seaweed, is a single giant cell. Found in salt-water, it grows to almost the size shown here.

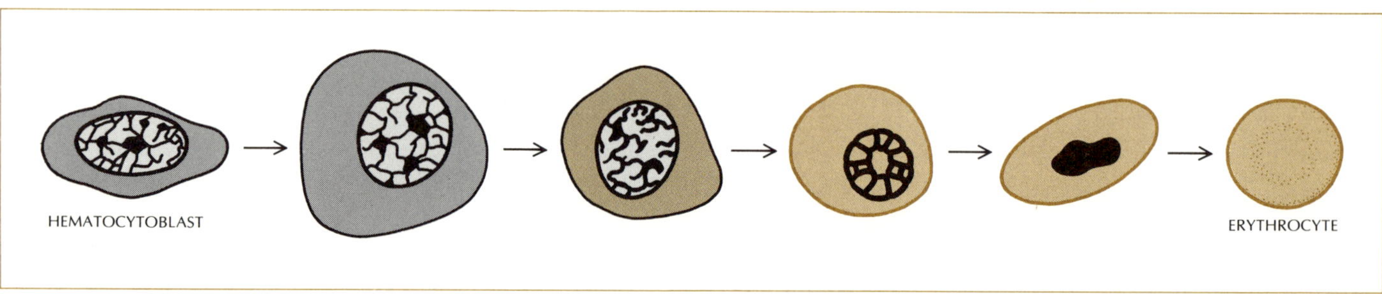

12

Human red blood cells (erythrocytes) are atypical cells in that they contain no nuclei. Erythrocytes develop from nucleated forms (left), but the nucleus disappears when the cells mature (right).

Fig. 11). The nerve cells of the giant squid grow to six feet or more in length. A chicken's egg is probably the most familiar example of a single giant cell.

Second, not all cells contain nuclei. Mature human red blood cells lack nuclei but they continue to live and function for a time, although they cannot divide. It is true that each red blood cell has a nucleus when it is formed, and that the nucleus disintegrates later in the cell's life (Fig. 12). But some cells never contain nuclei. Bacteria and closely related organisms called blue-green algae have no nuclei (see Fig. 13); they contain nuclear material, but no structurally distinct nucleus is ever present. Conversely, some cells contain more than one nucleus. A stage in the life cycle of many fungi is characterized by the presence of two distinct nuclei in each cell (Fig. 14).

Third, some organisms simply are not made up of discrete units called cells. Certain seaweeds, like the Merman's Shaving Brush (*Penicillus*), grow

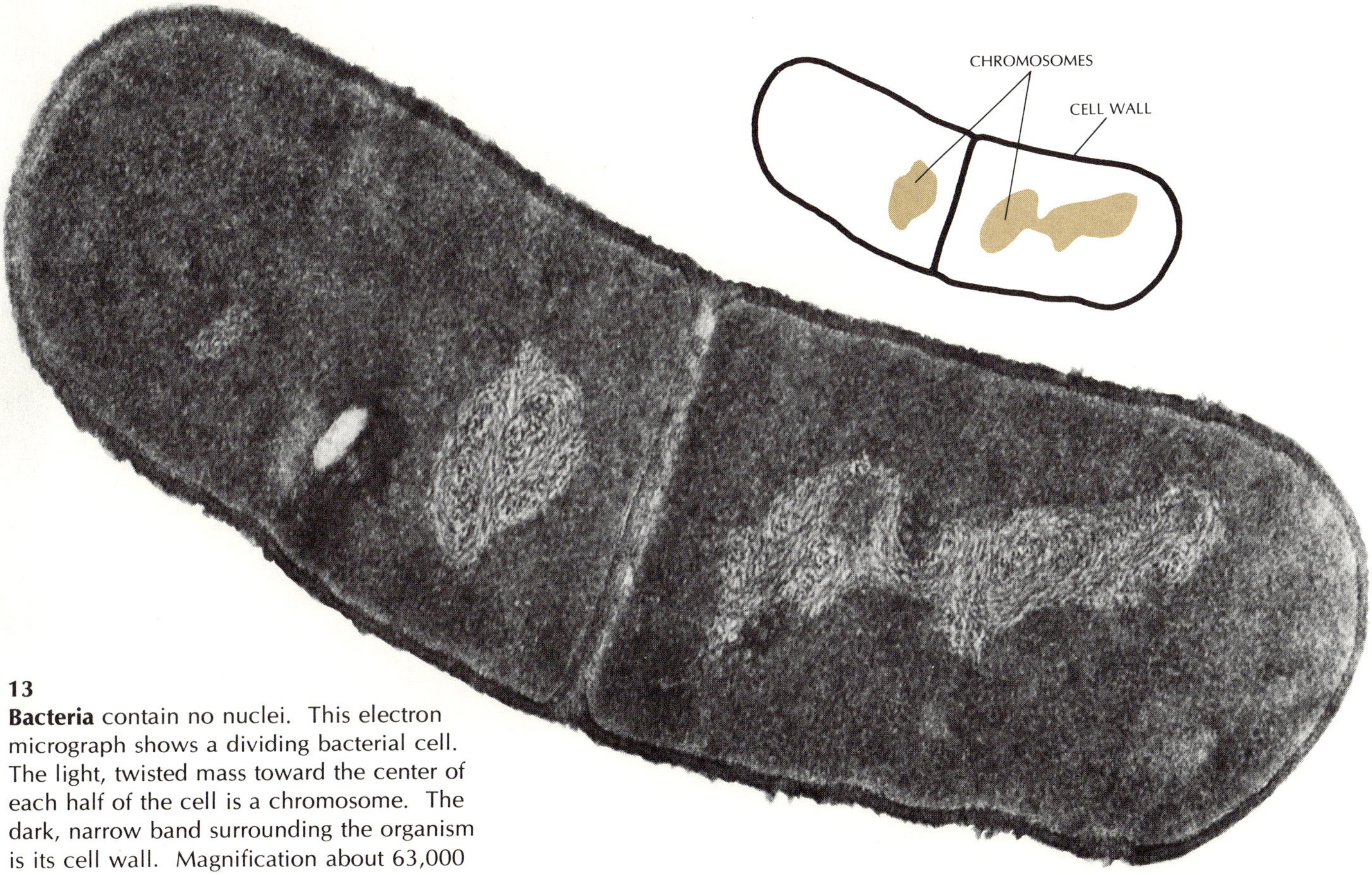

13
Bacteria contain no nuclei. This electron micrograph shows a dividing bacterial cell. The light, twisted mass toward the center of each half of the cell is a chromosome. The dark, narrow band surrounding the organism is its cell wall. Magnification about 63,000 diameters.

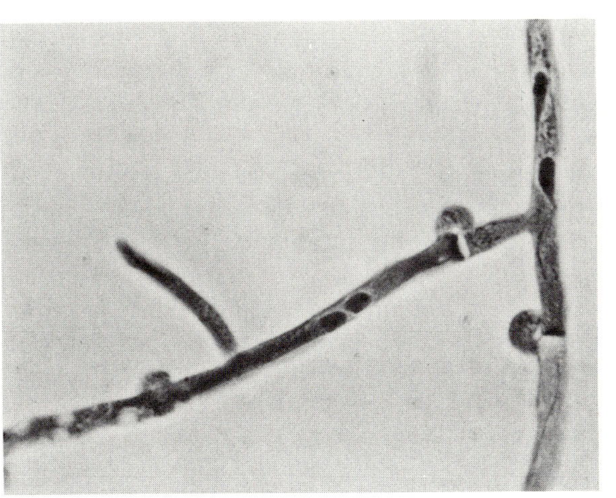

14
Cells with two nuclei occur in some fungi. When the cell divides, the two nuclei divide simultaneously, producing two cells each with two nuclei. Later the nuclei in each cell may fuse, resulting in the mixing of their genetic material. Magnification about 1100 ×.

15

Merman's shaving brush is the common name of the seaweed *Penicillus*. Shown here almost actual size, this organism is not composed of cells, but contains many nuclei embedded in its branching, brushlike tubules.

relatively large and yet are not composed of cells. Many nuclei are distributed throughout the branching series of tubes that comprise this organism (see Fig. 15).

To add to the confusion, biologists have discovered that structure is only one basis for defining a cell. Of equal importance is function. Yet all cells do not do the same things. In addition, there are types of cell—even in well-studied organisms like humans—whose functions still are only partly known.

In short, the original definition of a cell is simply not complete. To define a cell critically in terms of structure and function would take a rather thick book. The concept of the cell, like the concept of life, illustrates the problem with definitions in general. In this book we shall sometimes discuss a subject without giving a concise definition of it. By now, the reason for this should be clear: it is better to develop slowly an understanding of a subject than to memorize a definition. Biology deals with material not easily defined, and true knowledge comes slowly.

The inability to come up with concise definitions of life and the cell should not lead one to conclude that biology is a futile game. In a way, any science is a contest with nature, an attempt to decode the subtle riddles of the universe. But nature's resistance to being decoded is entirely passive. She does not change the rules to prevent our discovering them. The late mathematician Norbert Wiener, in his book *Cybernetics and Society,* compared the tasks that confront a research scientist with those faced by a chess player. He pointed out that the chess player, faced with an alert opponent, cannot make a single mistake without risking defeat; his fate is governed mainly by his worst moments. The scientist, on the other hand, has all the time in the world to carry out his inquiry, and thus the progress of his work is governed by his best moments—the moments of insight and discovery—rather than by his mistakes. Before we begin the study of the new biology, let us consider briefly the rules of the game called science.

THE LOGIC OF SCIENCE

Each science has its own language, and every scientist, being an individual, tends to work more or less in his own way. Some biologists spend most of their working days simply theorizing; some spend long and laborious hours in the laboratory; and some are administrators, the executives of science who organize large projects, obtain funds, and hire personnel. Most biologists are a combination of these types, and the emphasis varies from one man to another. Their approaches to the investigation of the unknown follow a variety of pathways.

In reality, a biologist's most important tool is not a microscope, a scalpel, or a rack of electrical sensors: it is his mind. Each biologist enters every investigation with his own preconceived ideas, biases, and hopes, as well as with his own mental store of facts, including relevant and irrelevant knowledge alike. There is no single way in which a biologist works. His approach to each problem is intensely personal. He utilizes a subtle mixture of imagination, intuition, and rigid logic.

Perhaps first and foremost, a biologist is a critical observer. In observing the world around him, he must learn to see things as they are. This is an art which requires trained judgment. In his novel, *Stranger in a Strange Land,* Robert A. Heinlein describes a future society whose legal system depends on the testimony of rigorously trained professionals known as Fair Witnesses. To test the powers of observation of one Witness, an attorney asks her the color of a distant house. Her reply: "This side is white." A layman would have said the house was white, but the Witness refused to infer that the rest of the house was painted the same color as the side facing her. Unsuspected bias and a lack of training lead most of us to see what we want to see or what we think we should see. Even experienced scientists sometimes make inaccurate observations which invalidate their later work. For this reason, science does not accept an observation at face value until several other scientists have repeated it and reported the same result.

Because of its insistence that observations be reproducible, science is an inadequate tool for the study of onetime events, such as the wondrous manifestations that religious men call miracles. Faced with reports of such events, the scientist will most likely remain skeptical but open-minded.

An observation often raises questions such as How? What? Why? The scientist defining a problem for research frames his questions more carefully, of course, than this. To be scientifically pertinent, his question must be testable by experiment. "Why does man exist?" cannot be answered by experiment, and so falls between the realms of science and philosophy. Asking significant questions is as difficult as making accurate observations, and just as crucial to science.

The biologist can usually make an educated guess at the answer to his question. Among scientists this is known as *postulating an hypothesis*. For simplicity, let us call this educated guess an *idea*. In everyday life many of us make observations, ask questions, and guess at the answers. If the answer sounds logical, that is as far as we go. But the scientist cannot stop here; if he does, he will never know whether he is right.

His next step is to design an experiment to test his idea. This is the hardest step in science, because each experiment presents its own set of intellectual and technical demands. The results of the experiment may confirm the hypothesis, or they may not. If they do, the scientist publishes his idea, and the evidence for it, in a scientific journal. Some ideas are more sweeping, more elegant than others, but even the smallest may form a link in a chain of reasoning that leads to a powerful theory, like the theory of evolution. If the results do not confirm the hypothesis, the scientist will probably reexamine his work and try again, perhaps using a new approach.

Science is a self-correcting body of knowledge, but the process of discovery and correction is far from straightforward. In practice, science proceeds by trial and error, insight and accident; it often takes decades or even centuries to solve a problem. Sometimes the difficulty is the lack of the right idea, sometimes the lack of the right tool, and sometimes the lack of the right technique. Two anecdotes will serve to illustrate some of the things that can go wrong.

A MEDIEVAL EXPERIMENT

Historians tell a story, perhaps apocryphal, about the 12th-century Holy Roman Emperor, Frederick Barbarossa. One of the most enlightened men of the day, Frederick observed that the children of German-speaking parents grew up speaking German, the children of French-speaking parents speaking French, and so on. This quite sound observation led Barbarossa to the question: What language is spoken in heaven? Latin? Hebrew? German? He guessed at the answer, reasoning that since the souls of infants come directly from heaven, the language spoken in heaven would be the language spoken by infants who were not exposed to human speech. He designed an experiment to test this hypothesis. In a special nursery in the palace, a group of infants was raised by deaf mutes. The experiment was cut short by Frederick's departure for the Third Crusade. At that time the children had passed their twelfth birthday without speaking a word.

Frederick's experimental approach was far ahead of his time, but his question fell outside the realm of science. The results were unfortunate, but if Frederick had lived to continue his experiment (he died in the Crusade), he was certainly intelligent enough to have concluded that something was wrong.

A FICTITIOUS EXPERIMENT

The other anecdote describes a fictitious experiment on the poinsettia, a tropical plant that grows well in gardens in Florida and southern California. It has been adopted as one of the symbols of Christmas because it blooms in December. Observing this, a fictitious botanist in Wisconsin wondered why this plant flowers in the winter and not in the summer. He noticed that it was colder outside in December than it was in June, and reasoned that poinsettias bloom in December because of the low temperature. He has made critical observations, has asked a pertinent question, and has made an educated guess or hypothesis.

To test his hypothesis he had some plants shipped from Florida to Wisconsin where he transplanted them to a greenhouse. As December approached, he began to lower the temperature in the greenhouse. Not too much, just about enough to equal the natural decrease in temperature in

Florida. Sure enough, in mid-December the plants bloomed. He concluded that his hypothesis was correct: lowering the temperature led to blossoming.

But one of his assistants had carelessly set a few of the poinsettias in another greenhouse, where the temperature was kept constant all year round. In mid-December the poinsettias there began to flower, too. The assistant concluded that temperature was not the critical factor; the plants would bloom in any reasonable temperature.

The question posed by the botanist was a good one, but the experimental method that he used to test his hypothesis was poor. He lowered the temperature in the greenhouse, but he had no way of knowing what the plants might have done if he hadn't. He did not have a control group of plants, and so his work was actually a nonexperiment.

For each group of plants exposed to the lower temperature, he should have set aside a similar group that was not exposed to it. By comparing the results of the two groups, he would have had some evidence as to whether flowering was really due to low temperature. In this case, the plants set aside by his assistant accidentally provided a control group.

Once the botanist, like his assistant, recognized that temperature was not the crucial factor, he sought a new hypothesis. He observed that the period of daylight is shorter in winter than in summer, and postulated that poinsettias have some kind of internal mechanism that causes them to bloom when the days get shorter. To test this new hypothesis he designed a new experiment. He kept two groups of poinsettias under identical conditions, except that he exposed one group to artificially short days by growing them in a greenhouse that he covered every day a few hours before sunset to shut off all light. The botanist found that poinsettias raised this way would bloom in say, August, while those in the control group would not. This time he concluded, rightly, that his new hypothesis was correct.

The experimental approach is one of the most powerful tools ever invented by man to explain and control the forces of the natural world. But it does not give us answers to questions involving morals, values, and politics. It cannot tell us whether it is just, or even logical, to devastate an enemy's crops with herbicides; whether Bach was a greater composer than Verdi; or whether capitalism is a better system of economics than socialism. Nevertheless, it has opened to human understanding the patterns of structure and activity that distinguish the living world from the nonliving. This book traces some of those patterns, and maps the avenues of research that will lead to the biology of the future.

READINGS

Clarke, A. C., *Childhood's End*. Ballantine Books, New York, 1969 (paper)
    A mind-expanding novel about man's first contact with extraterrestrial intelligence, by the coauthor of the screenplay for the film *2001*.

Conklin, G. (ed.), *Great Science Fiction by Scientists*. Collier Books, New York, 1962 (paper)
    Sixteen short stories written by leading scientists between 1926 and 1962 with an emphasis on the sometimes bizarre implications of scientific discovery for human society.

Hoyle, F., *The Black Cloud*. New American Library, New York, 1968 (paper)
    A novel about an organism that engulfs the sun, by a famous British astrophysicist who is also a superb storyteller.

Margenau, H., and D. Bergamini, *The Scientist*. Life Science Library, Time-Life Books, New York, 1964
    Scientists in history and in modern times. Stresses their contributions to society and the problems implicit in the profession.

National Aeronautics and Space Administration, *Ecological Surveys from Space*. U.S. Government Printing Office, Washington, D.C., 1970 (paper)
    A glance at the spectacular color photographs in this thin volume will demonstrate just how hard it is to detect evidence of life from a satellite.

Sagan, C., *Planetary Exploration*. University of Oregon Press, Eugene, Oregon, 1970 (paper)
    A well-illustrated publication of the 1968 Condon lectures by an astronomer who now heads the Laboratory for Planetary Studies at Cornell University. Colorfully written.

Sagan, C., and J. Leonard, *The Planets*. Life Science Library, Time-Life Books, New York, 1966
    This richly illustrated summary of information about the planets in our solar system contains a brief chapter on the possibility of life on other planets.

Savage, M., *In Vivo*. Fawcett World Library, New York, 1964 (paper)
    An engrossing novel that depicts the intuitive and human sides of scientific investigation.

Shklovskii, I. S., and C. Sagan. *Intelligent Life in the Universe*. Dell, New York, 1966 (paper)
    A speculative, thoroughly interesting work born of a collaboration between two astronomers, one Russian, one American.

Straum nes
Almungar
Adabrig
Langabak
Iokultfiord
Grumavig
Issiord
Mundarsfiord
Sugandersfiord
Dyrafiord
Reikuig
Serfialla strand
Arnarfiord
Selardal
Talknafiord
Patra fiord
Kaluig
Brudeuig
alo tuo
Bard
Rauga sand
Baridi strang
Valti
BLOE.
Skagin
Stegzstrand
Steigzfiord
Malan ey
Kolbens dalur
HOLA altera episcopalis, cum scola
Drang ey
Huala Dal
Sub eodem tecto homines, canes, et oues
Romæ sed clo ster
VUESTFIORDVNG
Reykiafiord
Berdi leisa
Strungrumsfiord
Butrufiord
Kollafiord
Buk fiord
Glama
Perpetuæ niues
Krosfiord
Gilsfiord
Getlands Iokul.
Bald Iokul.
Fons cereuisialis, qui ali quando ob domini sui san sedem mutauit.
Skialbre
Kluka
Blak høfdi
Flatey.
Bunc eyar
Pelle strand
Huams suert
Huams fiord
Staphoolt.
Hiatus terræ fœtentes
Lundur
Reykia Dallur
Finguol lur
Thermo
Breydafiordur
Hrunsfiord
Kolgusfiord
Grissta fiord
Albafiord
Regkia phu.
Megin sueit.
Sgomsfiord
Melafadur
Borgerfiord
Aranes
Hallarsfiord
Kumbrum vig
Sneuels Iokul.
Staps
Stadarstad
Hersey
Hualfiord
Mospelt2 sneit
Fontes ferui dissimi
Skerisfiord
AVLVOS.
Brimnes
Ondvertnes
C.
D.
Londranga
Jokul
Haffiorderey
Besrafiord
Kongard
Hafnarsfiore
Sehuopa
Hafna
E.
G.
Bormalanes
Kefiavig
Klin vig
Grunda
Fons communis albai in nigra
Eldey
F.
Rykianes
Fon
Geie fuelasker
Tangi
Geie eiar
has communes la nas nigras in albas
H.
I.

2. LIFE ON EARTH

Legends and primitive religions have fashioned a menagerie of fantastic organisms: the minotaur of Crete, Ganesha of India, the Sphinx and the animal-gods of Egypt. From the rain forests of Central America sprang Quetzalcoatl, the plumed serpent of the Mayas and Aztecs. A medieval Latin bestiary described gorgons and manticores, griffins and basilisks, but dismissed centaurs and mermaids as fictions. These imaginary creatures betray a certain poverty of the human imagination; they are but composites assembled from various parts of real organisms (see Fig. 1 and illustration on opposite page).

The natural world has outstripped human inspiration in fashioning incredible animals and plants. Without knowing they exist, who could have invented a bacterium, a honeybee, a peacock, a firefly, or an electric eel, plants with the properties of hashish, or others whose roots have digested the stones of the medieval Cambodian temples at Angkor Wat?

Ancient and medieval natural philosophy attempted to impose some sort of order on the astonishing diversity of living things. Only in the last hundred years has the task become manageable. The pioneering work of the 18th-century Swedish naturalist Carl Linnaeus, and the emergence of the theory of evolution in the 19th century, provided the intellectual scaffolding that enabled biologists to name and classify well over a million species of animals and plants. Another few million species may still remain unclassified.

◀**Sea monsters** on this early map of Iceland combine anatomical features of marine and terrestrial animals: claws, hog snouts, hawk beaks, reptilian bodies. Map appears in supplement to *Theatri Orbis Terrarum,* an atlas by Abraham Ortelius published in 1573. Volume is now in the collection of the American Geographical Society, New York.

1

Griffin was described by medieval writers as having the body of a lion and the head and wings of an eagle. It was taken quite seriously, as is evident from this fierce and very literal woodcut by the German master Albrecht Dürer. This is a detail from his giant *Triumphal Arch,* completed in 1515.

At the molecular level, a physical chemist might envision all living organisms as comprising a thin film of highly organized matter that envelops the surface of the planet. The extent of this film is limited by certain physical and chemical factors, particularly solar energy. The downward spread of life is limited by the absence of sunlight; the upward spread is limited by gravity, because a heavy organism must expend considerable energy to remain aloft in the atmosphere or, in the case of a tall plant, must have a structurally massive base to support the soaring branches. Life is most abundant at the interfaces between land and sky, land and water, water and sky.

In this chapter we shall outline the major groups of organisms and illustrate a few species, with emphasis on organisms of interest in biological research. We shall sketch the range of the conditions under which life is possible on this planet, and from this perspective shape a few general conclusions about the properties of life.

CLASSIFICATION SIMPLIFIED

The classification upon which all modern schemes are based was set up by Linnaeus. The scheme has since been modified, many new categories added, and many old ones discarded. An important recent modification was the creation of two new categories: the Monera (bacteria and blue-green algae) and the Protista (fungi, slime molds, algae, and protozoans). The separation of these organisms from the true plant and animal groups reflects the slow recognition by many biologists that some organisms possess characteristics of both plants and animals and therefore do not fit neatly into the old system.

Although a knowledge of classification schemes is essential to the serious biologist, a student taking an introductory course need understand only three main ideas: the use of scientific names, the hierarchy of classification, and the concept of a species.

SCIENTIFIC NAMES

The biologist uses a Latin name to denote a specific kind of organism because its common name may vary from place to place. An oak tree in English becomes *chêne* in French, *Eiche* in German, and *roble* in Spanish.

But to all botanists, this type of plant is denoted by the Latin word *Quercus*. People in various regions of the United States use the words cougar, mountain lion, panther, painter, puma, and catamount to identify a single kind of animal. Zoologists call it *Felis concolor*.

Sometimes the Latin and the common name of an organism are the same. Almost any woman recognizes the name *Begonia* or *Philodendron*. In this book we shall use either common names or scientific names, depending on which is most convenient.

The main schemes of classification in use today are based on the structure of organisms and on their evolutionary history. These attributes enable taxonomists to sort organisms into progressively smaller groups: kingdom, phylum (or division in the plant kingdom), class, order, family, genus, and species. Classification schemes for two organisms, the American white oak and the mountain lion, appear in Table I.

The classification system is much more than an empty naming scheme. From the table we can infer that the oak evolved from an ancestor common to all flowering plants, and that the mountain lion traces its descent from an ancestor common to all animals with backbones.

The scientific Latin name identifies an organism by genus and species. The mountain lion is a member of the genus *Felis,* which includes all cats; the species name *concolor* identifies the particular kind of cat. Members of the same species are closely enough related to interbreed or exchange genetic material. A leopard, *Felis pardus,* does not interbreed with a mountain lion. The concept of species as a breeding group is particularly important in understanding modern biology, with its emphasis on genetics.

Four major taxonomic groupings are now recognized by biologists: (1) Monera, (2) Protista, (3) true plants, and (4) true animals (Table II). Biological interest in each of these groups has waxed and waned over the years, with most of the attention focused on organisms that impinge on man. Early biological studies centered around experiments with food plants and animals. Only later did the discovery of the role of bacteria in disease bring the Monera to the forefront of research. Lately biologists have recognized that protists are also excellent research material. These simple organisms can often provide answers to specific questions in a fraction of the time needed to get the same answer from experiments on ''higher'' plants and animals.

Classification scheme of a plant and an animal. Biologists usually refer to an organism merely by genus and species: the white oak as *Quercus alba;* the mountain lion as *Felis concolor.*

	White Oak	Mountain Lion
kingdom	*Plantae*	*Animalia*
division/phylum	*Anthophyta*	*Chordata*
class	*Dicotyledonae*	*Mammalia*
order	*Fagales*	*Carnivora*
family	*Fagaceae*	*Felidae*
genus	*Quercus*	*Felis*
species	*alba*	*concolor*

II

Four main kingdoms of living organisms are listed here, together with a few of the best known groups in each kingdom. Metaphyta are true plants, Metazoa true animals.

Monera	bacteria blue-green algae
Protista	fungi slime molds green, brown, golden-brown, and red algae protozoans
Metaphyta	mosses liverworts ferns conifers flowering plants
Metazoa	sponges coelenterates flatworms sac worms segmented worms molluscs arthropods echinoderms chordates

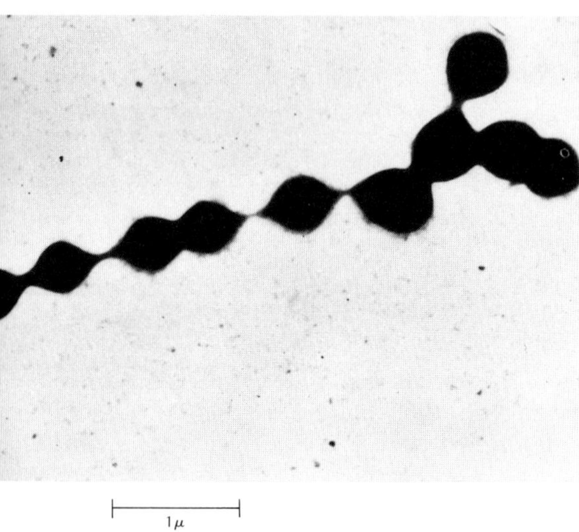

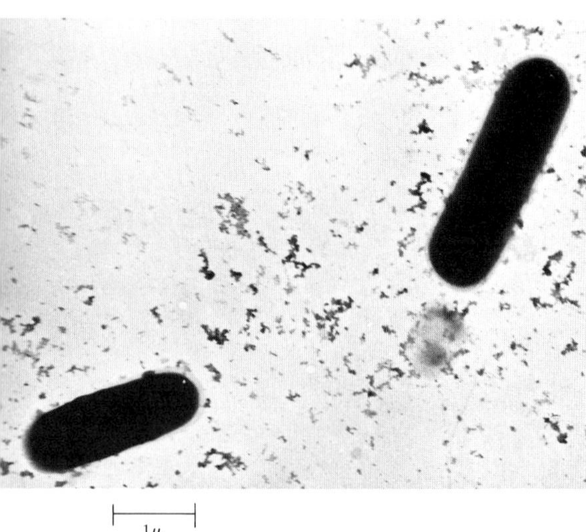

2
Bacterial cells under the microscope mostly appear as either spheres (cocci) or rods (bacilli). The chain of spheres (top) is *Streptococcus pyogenes*, an inhabitant of the human mouth. The rods (bottom) are *Clostridium botulinum*, probably the most toxic bacterium on earth. It causes the deadly form of food poisoning known as botulism.

MONERA

The organisms in this group contain the simplest cells in the living world. Of all living organisms, they are believed to bear the closest resemblance to the first cells on earth. They lack a true nucleus and the well-formed internal structures characteristic of the cells of all other organisms, yet they carry out many of the same life processes. They grow to maturity rapidly: whereas a human reaches reproductive age in about 15 years, a bacterium takes only 20 minutes. Because of their rapid growth rate and the ease with which they can be cultured, bacteria are excellent experimental organisms. It is possible to study basic processes in many generations of these organisms in a few days. Bacteriological research has led to major advances in the study of genetics, nutrition, and metabolism.

To most laymen, the word "bacteria" automatically means disease. A journalist once wrote that the complete elimination of the 2000 existing species of bacteria would be the salvation of mankind. True, there would be no syphilis, tetanus, rabies, leprosy, or meningitis. There would also be no vinegar, buttermilk, sauerkraut, butter, or nearly as many types of cheese. More important, our entire food supply would quickly be exhausted. Bacteria (and blue-green algae) are the only known organisms that can convert atmospheric nitrogen into a form utilizable by plants and animals. Without this continuing renewal of usable nitrogen compounds, crops would fail and famine sweep the earth. In addition, we would slowly be buried in a midden of dead organisms because bacteria, with the help of fungi, are the organisms responsible for the process of decay: the breakdown of dead or dying material. Without decay the bodies of dead organisms would accumulate until the planet became a heap of corpses, and the human race would be part of the pile. Eliminate the bacteria and you eliminate humans. The beneficial activities of bacteria far outweight their mischief. Although as a group bacteria display enormous anatomical diversity—some grow

Box A

The Size of Microorganisms

Under an ordinary light microscope individual bacteria become distinguishable at magnifications that approach 1000 diameters (1000×). At this magnification a man's fingernail would cover the floor of a classroom 22 feet square; and on that floor the rod-shaped bacterium in Fig. 2 would appear smaller than an uncooked grain of rice. A human blood cell about 8 microns in diameter would look about the size of the head of a tack.

An electron microscope can provide much higher magnification: in special cases, more than 2,000,000×, although 5,000 to 100,000× is a more useful range for biological work. Magnified 100,000×, the bacterium in Fig. 2 would appear to be about the length of a sausage, and a red blood cell would more than cover this page.

Although photomicrographs made with either type of instrument can be further enlarged photographically, the measure of the performance of an optical system is not its magnification but its resolving power: its ability to record fine detail. The vastly higher magnification of the electron microscope (compared with the light microscope) would be worthless without a corresponding increase in resolving power.

In this book, where possible, we include a micron "measuring stick" beside photomicrographs to indicate scale. If we do not know the precise size of the structure depicted, we mention in the caption the magnification indicated on the print submitted to us, which should give a vague but approximate idea of the real size of the object.

on stalks, some aggregate into strands, clusters, or elaborate networks—most single bacterial cells are either spherical or rod shaped (Fig. 2). Most bacteria are about one or two microns long. A micron, abbreviated by the Greek letter mu (μ), is a convenient unit in microbiology. It is 1/10,000 of a centimeter (0.0001 cm), a unit so small that it tends to be meaningless to the nonscientist (see Box A).

Study of the blue-green algae (about 2500 species) has lagged. With the exception of their ability to fix nitrogen, their only economically im-

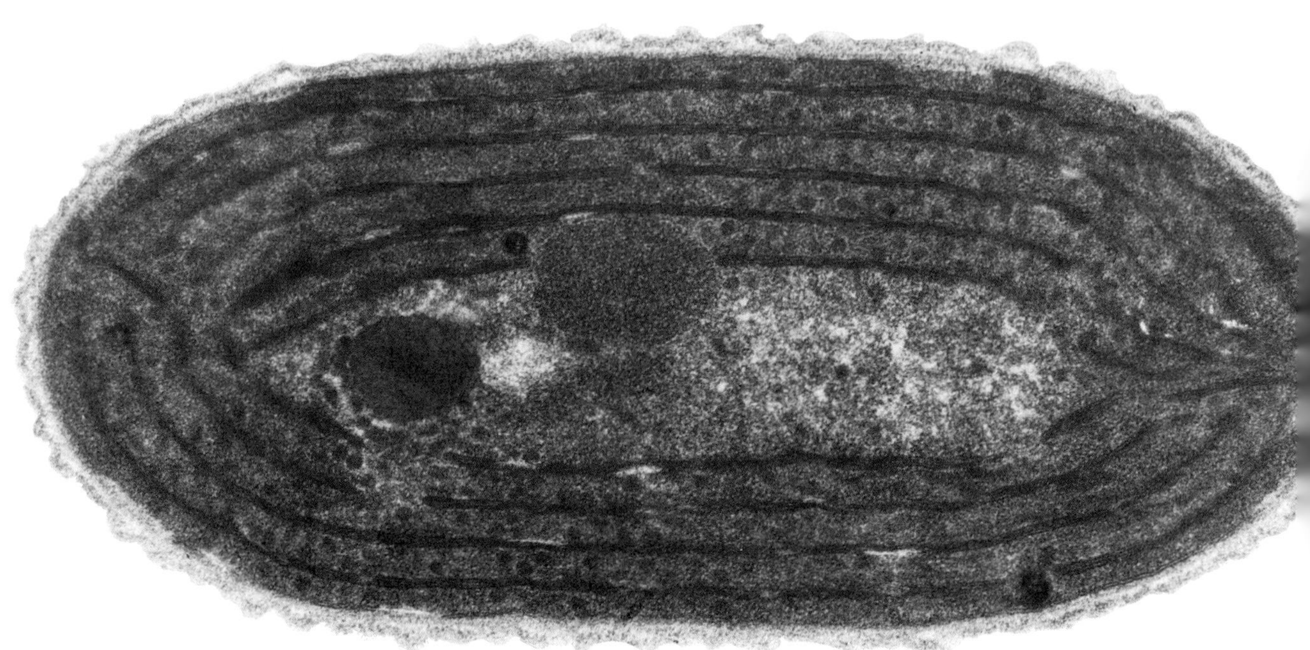

portant attribute seems to be their activity in soil formation. Undoubtedly because of their extremely simple cellular organization (Fig. 3), they are able to inhabit some of the harshest environments on earth. They are often the first organisms to invade devastated areas: they were the first forms of life to return to the barren slopes of Krakatoa after the volcano erupted in 1883 killing everything on the island. From water, the minerals in rocks, and the gases of the air, blue-green algae can synthesize organic matter which in turn can support populations of plants and animals. Without these algae, many areas now covered with rich soil and lush vegetation would be lifeless wastelands. In India blue-green algae have been introduced into saline and alkaline soils and even into rice paddies to improve their content of nitrogen and organic matter.

The role of soil algae has just begun to be investigated extensively in this country. In the future, as man colonizes the more rigorous environments of the world, knowledge of the beneficial activities of blue-green algae will become essential.

3
◄**Blue-green algae,** like bacteria, lack a true nucleus. This electron micrograph
shows an alga from the warm springs of Yellowstone Park, reproduced at a
magnification of about 68,000 ×. The nuclear region (nucleoplasm) at the
center of the cell is ringed by four or five layers of photosynthetic membranes
(thylakoids). The cell membrane (plasmalemma) is surrounded by a cell wall.

PROTISTA

Protists are a strange collection of organisms: algae (except blue-greens),
fungi, slime molds, and protozoa. There are about 123,000 species of
protists; some can make their own food by photosynthesis; others must
absorb or ingest food from the external environment. Some are no more
than a micron in diameter; others are visible to the unaided eye. Some are
motile and others are sedentary. The one characteristic that unites these
diverse life forms is that the organisms are constructed from only a small
number of cell types. In general, plantlike protists lack the complicated
cell types found in true plants, and the animallike protists lack the tissues
and organs characteristic of true animals. The cells of Protista differ from
those of Monera in that their genetic material is segregated from the rest
of the cell in a nuclear region bounded by a membrane. In this respect,
protistan cells resemble the cells of higher organisms.

The protists were originally studied for their beauty. To a neophyte
biologist, one of the greatest revelations is his first glimpse of a drop of pond
water under the microscope: a miniature world alive with organisms of

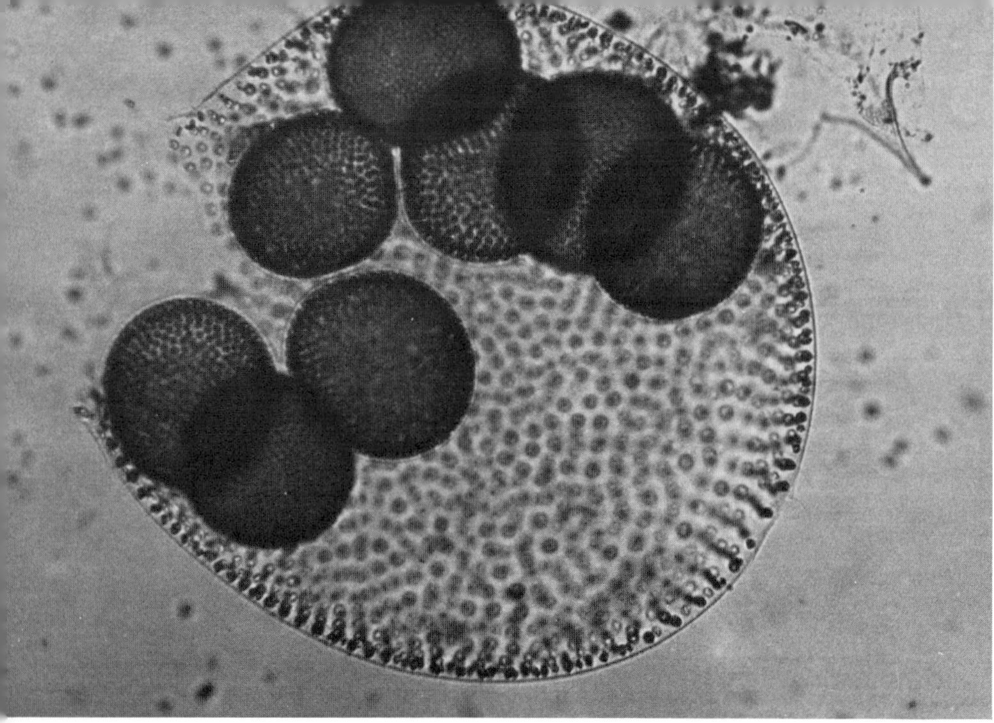

4
Freshwater protists include the green alga *Volvox* (top) and the protozoan *Actino-spherium* (bottom). *Volvox* is a spherical colony of plantlike cells. It can be seen in ponds as a green sphere about the size of a pinhead. *Actinospherium* is more like an animal. This one just engulfed some diatoms (dark mass at center). These organisms are hundreds of times larger than bacteria.

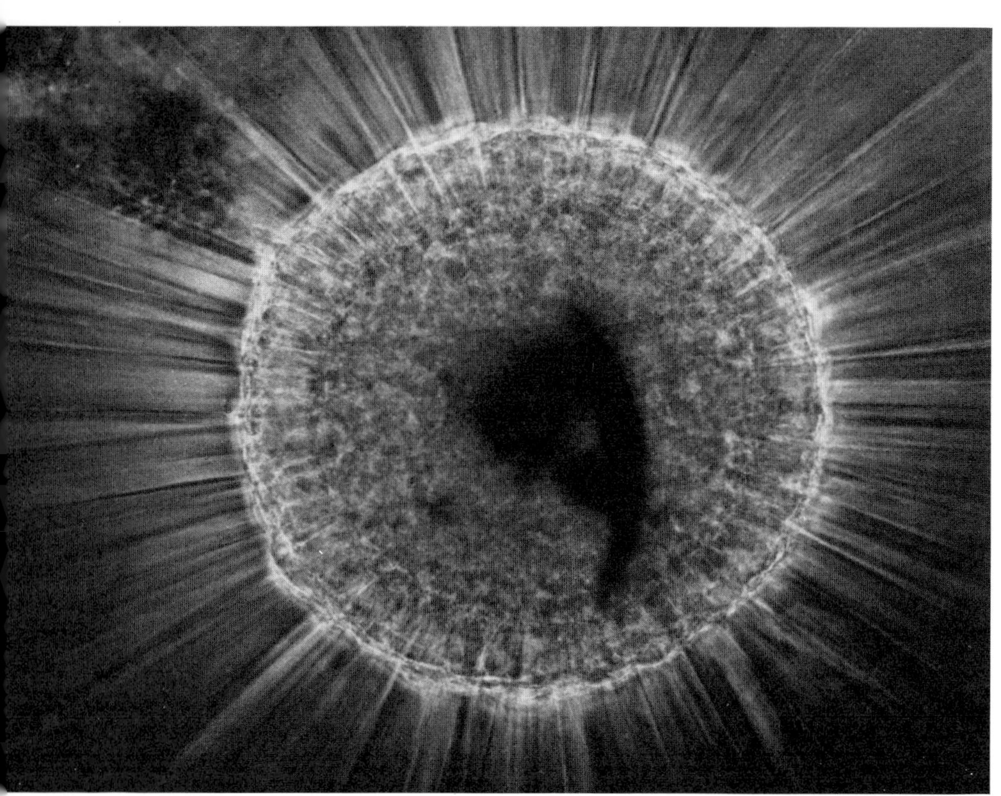

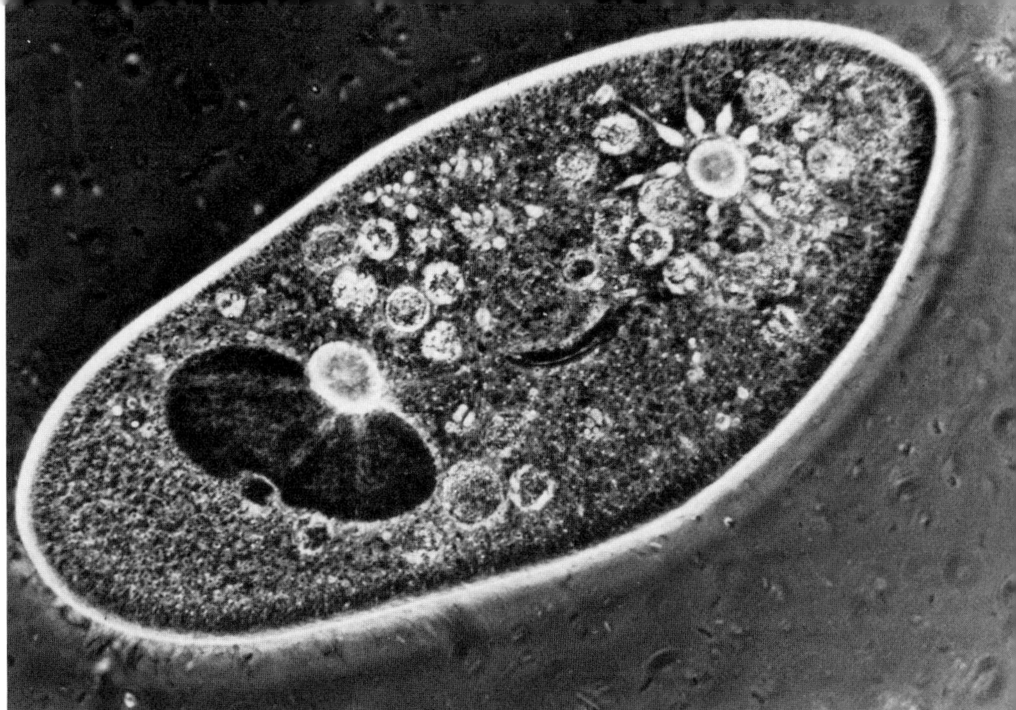

5
Paramecium is a complex one-celled
animal. Large dark nucleus is visible at
front of cell and star-shaped contractile
vacuole at rear. Bright round objects
inside cell are vacuoles containing food.
Almost large enough to be seen with the
unaided eye, paramecia are giant cells
compared with bacteria.

exotic shapes and colors. Among the most numerous inhabitants of such
a community are the green, gold, and brown algae and the perpetually
swimming protozoa (Fig. 4).

Although most of the studies on the algae have been taxonomic, these
organisms have proved most useful in two fields of research: genetics and
physiology. The green one-celled alga *Chlorella* has revealed many of the
secrets of the food-making process common to all green plants. This laboratory "weed" grows so easily that it often contaminates bottles of distilled
water. Exploiting its small size and its phenomenal ability to reproduce
under virtually any conditions, scientists have cultured *Chlorella* to study
its physiological response to controlled environments. A small laboratory
at the University of California at Berkeley was able to house all the complex
equipment needed to perform sophisticated experiments on photosynthesis
in *Chlorella*. The result was a detailed understanding of the chemistry of
photosynthesis and a Nobel Prize for the investigator, Melvin Calvin.

Various small protozoa have been valuable in studies of biological
phenomena. The one-celled *Paramecium* has fascinated biologists since
the invention of the microscope. Paramecia contain tiny spherical structures called contractile vacuoles which pump excess water out of the cell
(Fig. 5). In 15 or 20 minutes the rhythmic contraction and expansion of

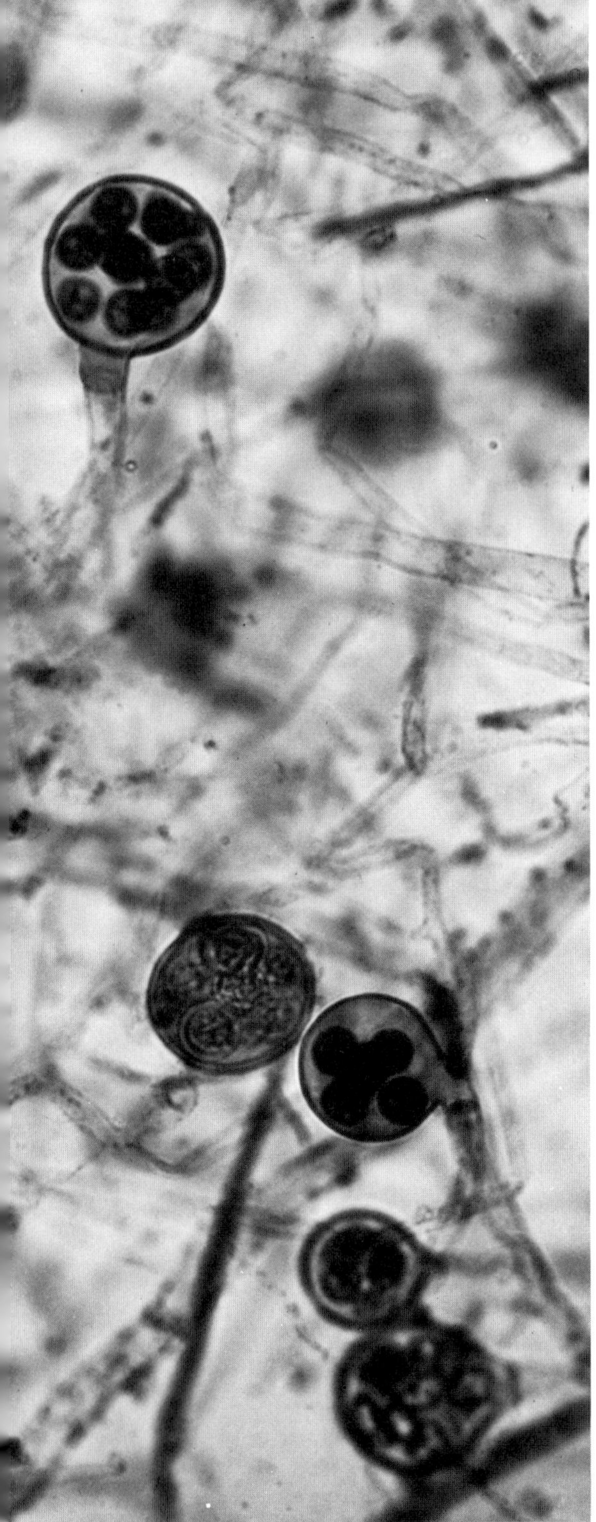

6
Aquatic fungi include *Saprolegnia,* shown here at
a magnification of about 200 ×. Microscopic
filaments support round egg sacs. These fungi
grow on many kinds of aquatic detritus or are
parasitic on fish eggs and young fish.

these vacuoles can pump a quantity of water equal to the volume of the
entire cell. When a paramecium is poisoned with cyanide, but not killed,
the functioning of the contractile vacuoles is impaired. The paramecium
continues to absorb water but cannot excrete it; eventually the cell swells
and bursts. A study of these structures has led to a better understanding of
the water balance in all organisms.

The fungi, another group within the Protista, are equally interesting
and useful. They include the well-known mushrooms, as well as the molds
that attack bread, fruit, and other foodstuffs. Less familiar are the aquatic
fungi, which are often microscopic in size (Fig. 6). Many are parasitic on
fish, water plants, algae, and other aquatic organisms.

The importance of fungi in causing disease, in producing various antibiotics, and in the process of decay is well known. But how many laymen
know that bread rises or puffs up because of carbon dioxide produced by
Saccharomyces cerevisiae, a simple fungus growing in the dough? Or that
a closely related fungus produces the alcohol in beer and wine? These
versatile fungi are better known as baker's and brewer's yeast. Few laymen
have ever seen a single yeast organism. Under the microscope it is a single
spherical or egg-shaped cell (Fig. 7). Growing from some of the cells may
be lobes or extrusions; these are buds or developing new cells. A single cell
can reproduce by budding up to 20 times. Each bud, when it separates
from its parent, becomes a new yeast organism capable of initiating another
reproductive cycle.

Yeasts are useful in basic genetic studies. Over 100 of their genes have
been studied and mapped. By detailed analysis geneticists have worked
out many of the steps in the complex pathways leading to the manufacture
of essential food products within the yeast cell. This type of information can
be applied to studies of similar reactions in other organisms ranging from
bacteria to man.

7
Yeast cells appear egg-shaped under the light
microscope. Some of the cells are budding. A
view of yeast cells under the electron microscope
is reproduced in Chapter 11. Magnification about
1000 ×.

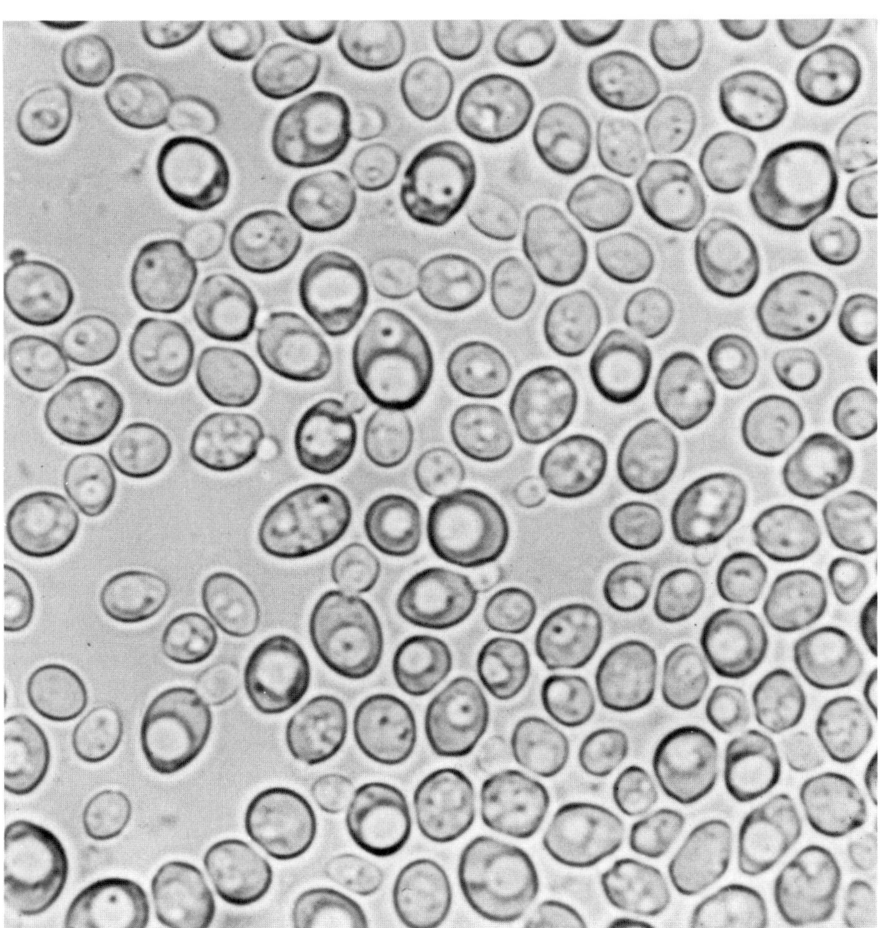

8

◄ **Slime molds** are aggregations of single-celled social amoebas. Photograph
depicts fruiting bodies: sacs of spores supported by stalks. About half an inch
high, each stalk is composed of thousands of individual amoebas.

9

Life cycle of slime mold is depicted schematically. Free-living amoebas (far left)
aggregate to form a slimy slug (left center). Slug moves about as if it were a
single organism, then settles down and produces a fruiting body (right). Eventually fruiting body releases spores, each of which develops into a single amoeba.

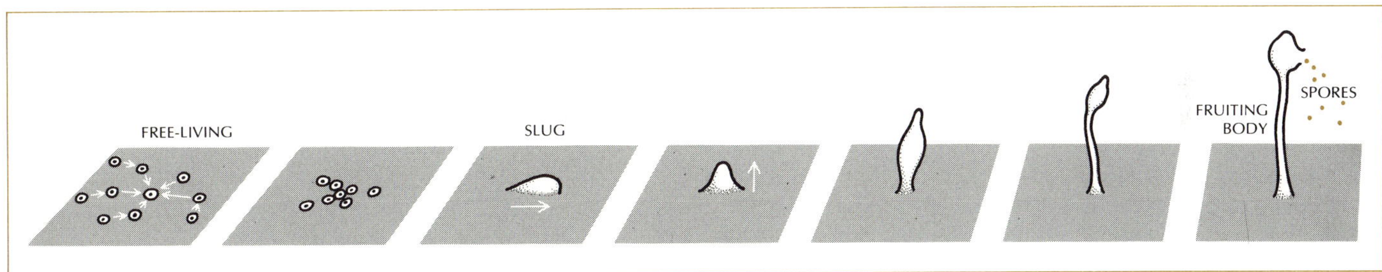

Slime molds are a group of part-time colonial organisms. At certain
stages of their life history, they behave like fungi, but at other stages they
behave like simple animals. Any organism that deviates from an expected
pattern of reproduction or development excites the curiosity and imagination of biologists. In this respect, slime molds are some of the most interesting organisms on earth.

One slime mold, *Dictyostelium discoideum,* has been studied for about
70 years. One might expect that by now everything would be known about
this organism, but that is hardly the case. The central problem is the peculiar
method by which it reproduces. During one stage of its life history *Dictyostelium* is a free-living single-celled organism that feeds on bacteria. But
at some critical period in the life cycle thousands of these ameboid cells
migrate to a single spot and fuse into a slimy, cylindrical mass of cells. This
new "organism," called a slug, moves around for 5 to 24 hours. Then it
settles down and changes shape. A long stalk forms and at its tip appear
numerous spores. When these spores are released, each develops into a
single one-celled organism. The entire cycle is depicted in Figs. 8 and 9.

What controls this curious series of events? The aggregation phase is initiated by a chemical produced by a small mass of cells; these cells attract all the others. Finding this simple answer required work by numerous investigators and the development of special culture techniques. The chemical itself, called acrasin, has been isolated, but no one knows the exact nature of the conditions that cause a few cells to start producing it. More than half a century of work has only scratched the surface of the mysteries surrounding the reproductive cycle of this organism.

TRUE PLANTS (METAPHYTA)

According to one scheme of classification the true plants, the Metaphyta, comprise eight divisions. One division, the *bryophytes*, or moss plants, includes about 25,000 species; the other seven divisions contain the vascular plants which number about 260,000 species. Except for insects, they comprise the largest group of organisms on earth, and one of the most diverse. These plants have vascular tissues, and usually have roots, stem, and leaves. The vascular plants include ferns, club mosses, horsetails, cycads, and flowering plants. Contrary to popular belief, most trees are flowering plants. So are all grasses, including wheat, corn, and even crabgrass. Most true plants are photosynthetic, and all are incapable of locomotion.

PLANTS IN RESEARCH

Two examples illustrate the use of plants in basic research. One involves the use of carrot cells to study the process of cell growth; the other, the use of squash plants to study the food transportation mechanism of vascular plants. The results of both projects have surprisingly practical applications.

For many years biologists stated flatly that the only cell capable of producing a mature functioning organism was the fertilized egg. The cells of many plants and animals, including those of man, could be cultured in test tubes for extended periods but they never formed an intact, fully developed organism. That was the state of affairs until F. C. Steward at Cornell University placed cells from a common carrot in a specially designed culture medium. The cells, which were originally taken from the root, grew and multiplied and ultimately gave rise to a complete plant: leaf, stem, root, and flowers. Later experiments with cells from tobacco plants added further knowledge of the process. These studies suggested the chemical processes that stimulate cell enlargement, initiate cell division, and trigger the formation of buds or the development of roots. Within limits, it became possible to control the plant's pattern of development. It is not surprising that these studies, and similar work on other flowering plants, were largely funded with money from the National Institutes of Health and the American Cancer Society. Only when we understand normal growth can we understand the abnormal growth processes of cancer cells.

The other example concerns one of the basic questions of plant physiology: How is the food produced in the leaves by photosynthesis transported to the roots for storage or for use in the growth of new roots? The question is central to the study of the growth of trees and other flowering plants. But trees are not practical experimental organisms; they are simply too big and grow too slowly. The common garden squash proved to be the perfect plant for this study. Its large leaves provide an abundance of photosynthetic products. More important, the squash has very long leaf stalks through which the movement of these products can be traced in minute detail. By injecting radioactive tracers into the leaf, Carroll A. Swanson and his colleagues at Ohio State University were able to discover the movement rates of naturally occurring foods and the site of the tissue through which they move. They were also able to establish the chemical form in which this food is transported. By freezing very small portions of the leaf stalk, they were able to study the mechanisms by which the food was forced through the plant. Not only was this information valuable in its own right, but it provided much knowledge applicable to agriculture. We now know much more about the movement of such substances as the sugary sap of maple trees and the way in which sugar beets accumulate and store reserve food.

TRUE ANIMALS (METAZOA)

The true animals are the Metazoa. Two fundamental differences between animals and plants arise from their pattern of nutrition. Animals are incapable of photosynthesis, and hence cannot manufacture their own food. They ingest food, digest it, and excrete the unusable remains. This process is called alimentation, and it is characteristic of virtually all animals.

Another characteristic of animals is locomotion. Most animals are relatively large organisms, and require sizable quantities of food. Given

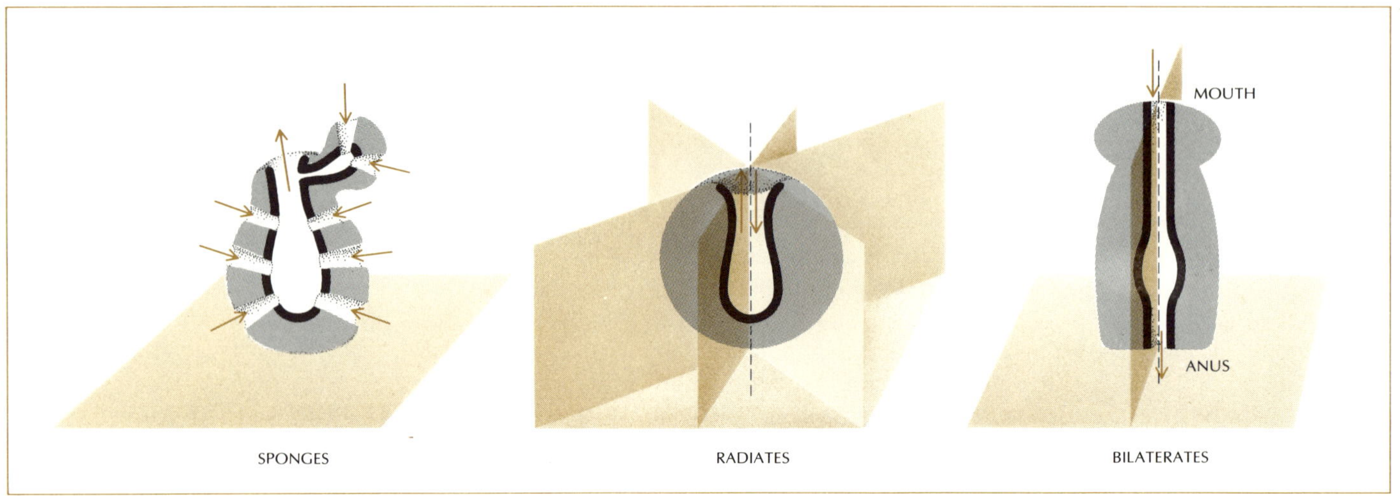

10

Planes of symmetry are the bases for sorting animals into three large categories.
Sponges show no anatomical symmetry in any plane (left). Radially symmetrical
animals like sea anemones are anatomically symmetrical around a central axis
(center). All other animals, including man, are bilaterally symmetrical (right);
that is, they are symmetrical side to side but not front to back. Arrows indicate
the flow of nutrients through the organisms.

enough sunlight and water, most plants can synthesize their own food. A
small, nonphotosynthetic organism like a bacterium or a protozoan, with
its specialized nutritional requirements, can rely passively on water currents
to bring food to it, or vice versa. Attached aquatic animals like sponges
and clams use movable specialized structures to propel food to them. But
most animals forage for food, and this usually requires some mechanism
for movement of the entire organism.

A CATALOGUE OF BEASTS

Animals have been classified into three comprehensive categories: sponges,
radiates, and bilaterates (Fig. 10).

Sponges are essentially living pumps, perforated by a network of channels. Cells along the channels absorb food from the water as it passes
through. This primitive system has survived the test of evolution. There are
about 15,000 living species of sponge. A few of them are depicted in Fig. 11.

Radiates, so named because their bodies are radially symmetrical, have
an alimentary (digestive) system that is essentially a one-holed sack. The
same opening serves as both mouth and anus for the roughly 10,000 species

Sponges are simple multicellular animals. The slimy sponge *Desmacidon* (left) belongs to the class that includes commercial sponges. *Leucosolenia* (right) is a bristly calcareous sponge, so named because its cells are supported by a "skeleton" composed of tiny needles of calcium carbonate. It is shown here about actual size.

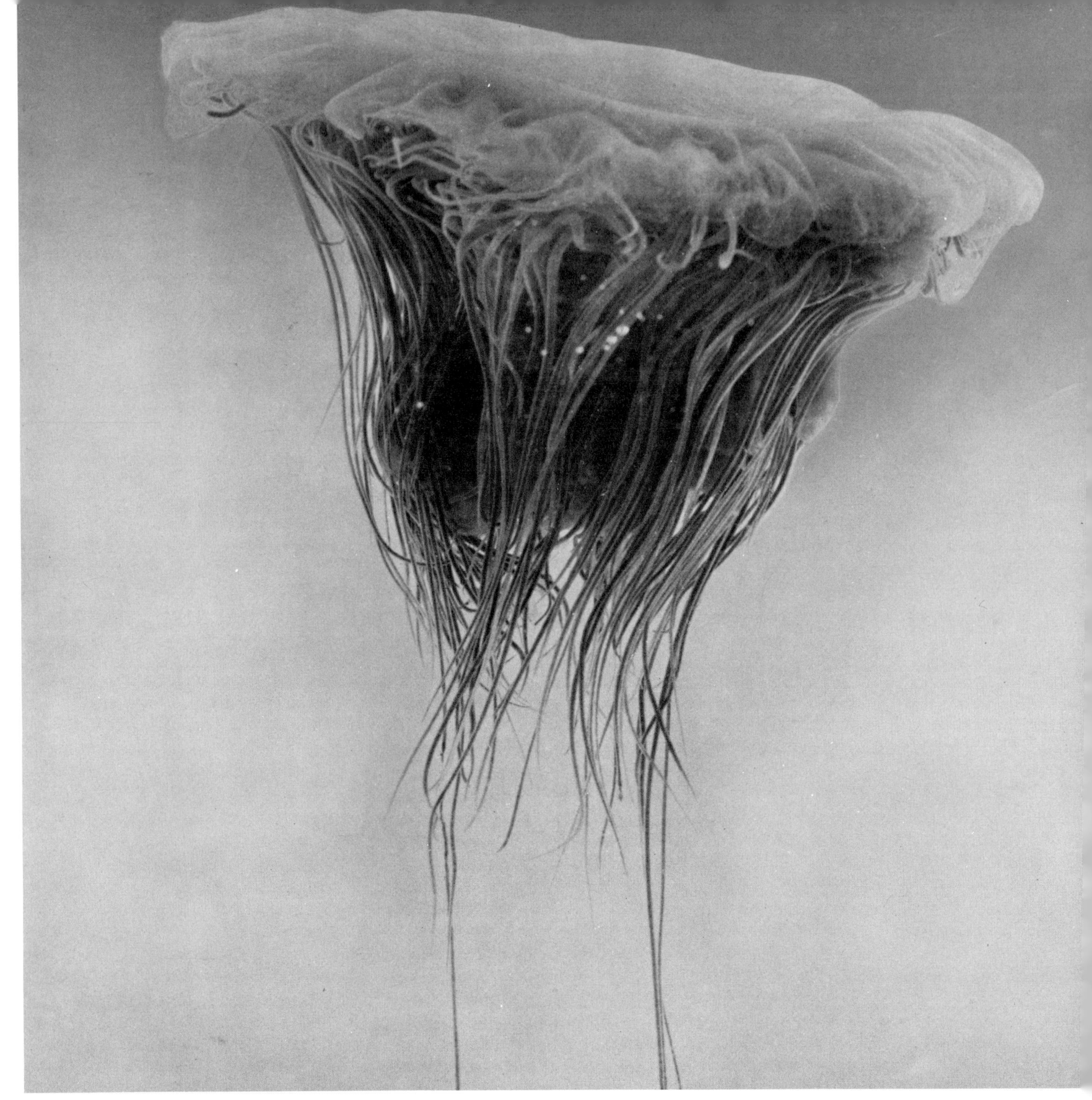

12
Radiates include simple marine animals like jellyfish (left) and anemones (above).
Jellyfish swim by a form of jet propulsion, alternately contracting and relaxing
their buoyant bells. Photo shows Lion's Mane jellyfish, a native of the North
Atlantic. Brightly colored sea anemone looks like a flower but actually is a
carnivorous animal that catches small fish with its delicate tentacles. Anemones
are commonly found on rocks in shallow coastal waters.

in this superphylum. Among them are jellyfish, sea anemones, corals, and
the familiar hydra (Fig. 12).

    All other animals are bilaterates, whose bodies, like man's, are symmet-
rical side to side but not front to back. Their alimentary system is a two-
holed tube: the mouth at one end and the anus at the other. This is the most
diverse of the three categories of animal organisms; among its many groups
it includes flatworms, roundworms, mollusks, arthropods, and chordates.

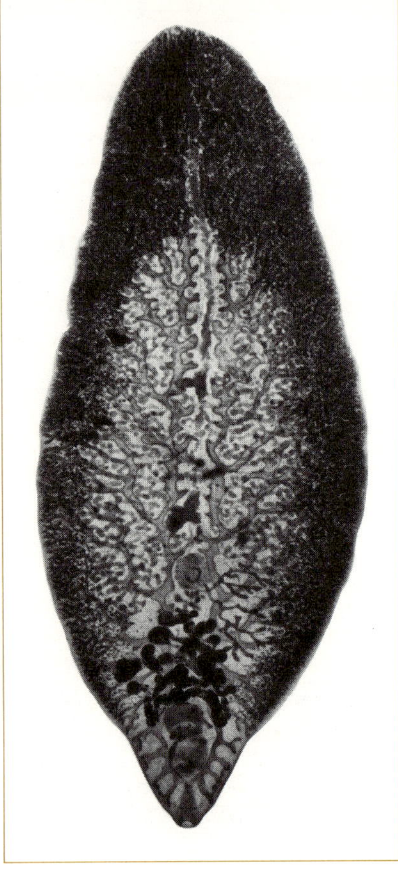

13
Flatworms such as the liver fluke are bilaterates. This particular fluke, about the length of a paper clip, infests the liver of sheep and cattle, often causing death. Specimen here has been stained to show internal structures.

Segmented worms include the familiar earthworm (left) and the tubeworm ▶ *Polymmia* (right), which lies embedded in the sea floor and feeds on tiny organisms swept toward it by the waving tentacles. Tube consists of tiny particles of sand and shell cemented together by a mucous secretion. In terms of complexity, segmented worms stand about midway between primitive protozoa and complex animals like insects and vertebrates.

A few of the organisms belonging to these groups are depicted in Figs. 13 through 15.

Arthropods comprise the largest phylum on earth; there are roughly a million known species, and some taxonomists estimate that the actual total (including animals still uncatalogued) may be many times this number. Arthropods include such creatures as insects, crabs, scorpions, spiders, centipedes, and millipedes (Fig. 16). Unquestionably this phylum represents a triumph of evolution, a triumph that derives directly from the versatility of the arthropod body plan (Fig. 17). The essential features of that plan are the armored exoskeleton and the segmented architecture of the body. The external skeleton, of chiton, protects the animal superbly. The segmented body has enhanced the development of specialized appendages: built-in tools for walking, flying, swimming, piercing, biting, sucking, cutting, boring, grasping, egg-laying, sensing, silk-spinning, and even music-making.

One of the most despised organisms on earth represents one of the most overwhelming evolutionary successes. The European cockroach, a familiar roommate to millions of Americans, is a living fossil. So well adapted are its structure and responses to the environment in which it lives that more than 200 million years of evolution have resulted in surprisingly few improvements. The cockroach scuttled across the earth before men evolved; it has adapted to man's rise to dominance, and because of its formidable resistance to radiation, it may survive after man has gone. In the event of a nuclear holocaust, the cockroach and its relatives could inherit the earth.

Where does man fit into this picture? We belong to the chordates, the group that includes all animals with a bony internal skeleton, particularly a backbone: fish, amphibians, reptiles, birds, and mammals.

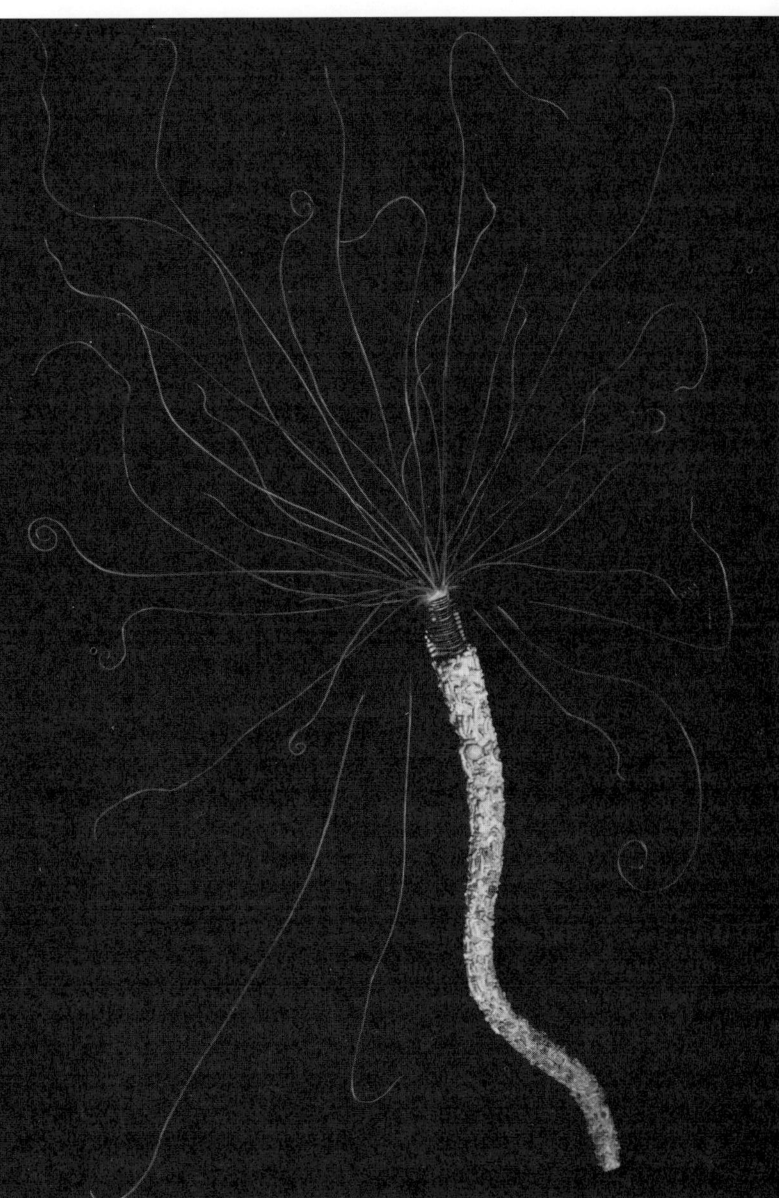

15
Mollusks are the second largest group of invertebrates (animals without backbones). Garden snail (top) is a typical mollusk in that it has a shell and a large fleshy foot. The coiling of the shell is an evolutionary modification that makes the animal more compact and mobile. Squid (bottom) has virtually lost the molluscan shell. Highly adapted as a swimmer, it seizes prey with two tentacles specially equipped with suckers. Squids not only represent one of the most highly developed invertebrates, but also the largest: the giant squid attains a length of 15 meters or more.

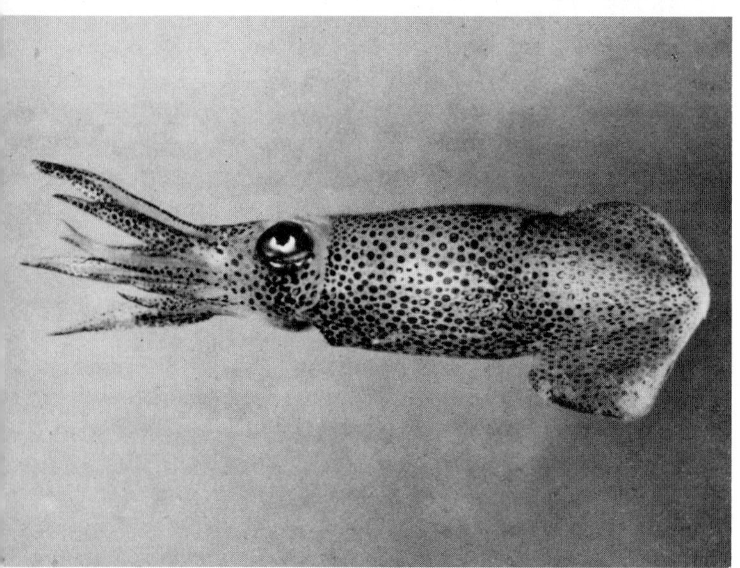

16

Arthropods (animals with jointed legs) are the largest group of invertebrates. The phylum includes not only insects (six legs) but a variety of other organisms such as millipedes (top) and scorpions (bottom). Although millipede means "thousand-legged," they actually have only two pairs of legs on each segment of the abdomen. Scorpion is a nocturnal animal that catches spiders and insects with its pincers, stings them to death with its tail, then sucks them dry. Its sting is painful but not dangerous to adults, although it is sometimes fatal to children.

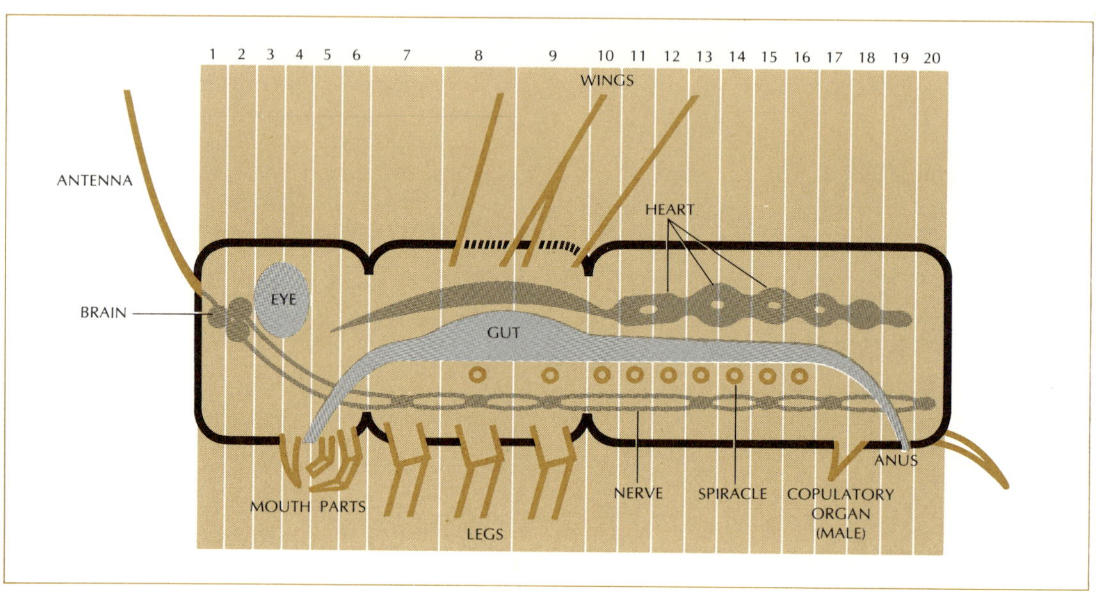

ANIMALS IN RESEARCH

Experiments with three animals—the sea urchin, the salmon, and the fruit fly—illustrate the ways that researchers have exploited the diversity of animals to gain insights into processes common to all organisms. The eggs of the sea urchin have been used since the late 19th century to study the process of embryo formation. The earliest workers thought that an embryo developed because prepackaged organ-forming substances were present in the fertilized egg; when distributed to certain cells of the developing embryo, these substances caused the cells to develop into specific organs. Later investigators found that if they artificially separated the two cells produced by the first division of the fertilized egg, the result was the development of two complete intact living organisms, not two half organisms. This evidence not only provided insight into the embryonic development of the sea urchin, but also suggested the method of origin of identical twins in humans. The eggs and embryos of sea urchins are now used to study

17

◄**Anatomy of arthropods** is responsible for their spectacular biological success. Diagram depicts the 20-segment body plan of a hypothetical insect. The appendages, protected by a rigid exoskeleton of chitin, have evolved into an assortment of structures specialized for a variety of chores, including walking, flying, swimming, biting, piercing, cutting, grasping, stinging, and reproducing. Insect shown here is a male; in female, egg-laying device (ovipositor) replaces copulatory organ. Some body segments are modified as sensory structures, notably the antennae. The only blood vessel extends the length of the body. Posteriorly it enlarges into a multisegmented heart. Spiracles are valves through which the insect breathes.

the way in which the hereditary material of the cell "switches on" crucial chemical processes at certain stages of embryo growth. Such investigations may lead to a better understanding of what goes wrong in the development of malformed human embryos.

Among the first experiments in the now active field of molecular biology were the attempts of a group of pioneering biochemists, late in the 19th century, to isolate the genetic material in cells and to identify it chemically. The problem was to obtain large quantities of the hereditary material in a relatively pure condition. They first tried to extract it from eggs. The egg was unsatisfactory because it contains large quantities of stored food and it proved extremely difficult to separate the genetic material from the food. Sperm, on the other hand, are relatively free of foodstuffs. The next step was to find an organism that could supply large quantities of sperm. The common salmon and the herring proved to be ideal. They shed enor-

mous quantities of sperm which could be obtained with little inconvenience to either the fish or the scientist. Experiments on chemical extracts of this material were one of the critical early steps toward the famed biological revolution of the 1960's.

Insects are excellent experimental animals. The biological activity of cockroaches has given us an insight into the intimate workings of biological clocks. Hormone studies with moths have led to the discovery of a juvenile hormone that controls the metamorphosis of a caterpillar into a mature winged moth. But perhaps the insect used most extensively in biological work is the common fruit fly, *Drosophila melanogaster*.

More is known about the genetics of this one insect than of any other known organism. Because its chromosomes are large and easy to see, they have been studied extensively for more than 50 years. Classical studies of the effect of X-rays on the mutation rate were done first with the fruit fly. The influence of the number and kinds of chromosomes on the expression of maleness and femaleness was studied in this organism. Experiments on the inheritance of eye color in *Drosophila* have led to an understanding of how genes control the expression of observable traits in all kinds of organisms. Studies on the mating behavior of fruit flies have given us clues as to how a new species may evolve: with the fruit fly a geneticist can study evolution in the laboratory. The analysis of experiments on populations of fruit flies has sparked the mathematical approach to the study of the inheritance of traits in large groups of individuals.

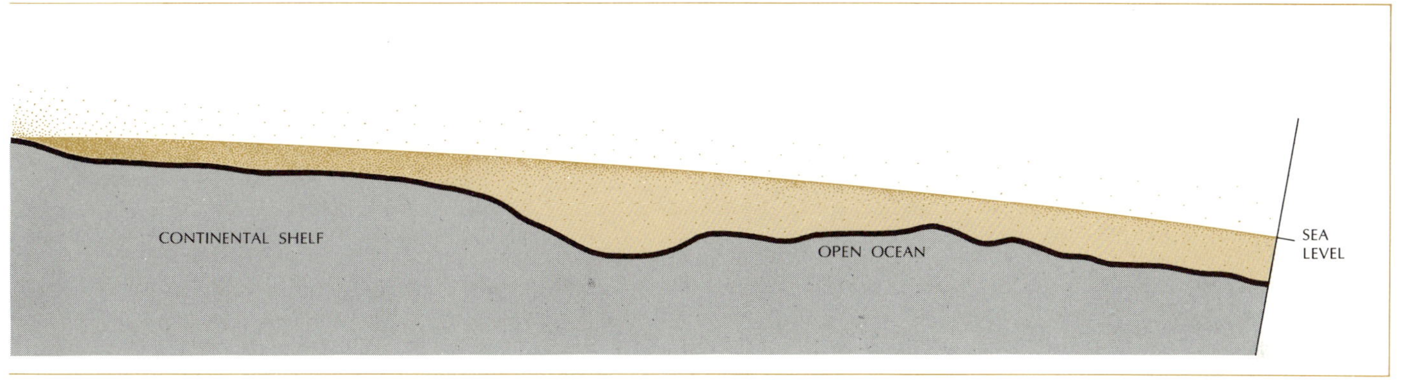

This brief illustrated survey can only sketch the incredible diversity of living things that inhabit this planet. The nature of their similarities and differences will become much clearer in later chapters, but even at this point a few principles about life on earth should begin to emerge.

1. All known organisms may be sorted into one of four categories: Monera, Protista, Metaphyta, and Metazoa.

2. Classification is not just a naming system; from it we can infer something about an organism's kinship to other living things and its evolutionary history.

3. Excluding man, in the course of evolution three overwhelmingly successful life forms have emerged: bacteria, flowering plants, and arthropods.

4. The diversity of living organisms has been exploited by modern research to elucidate mechanisms common to all forms of life.

18
Thin film of living matter (colored dots) spreads across the surface of the earth. Life is most abundant near the interfaces between land and air, land and sea, and sea and air. Few places on the earth's surface are completely sterile, but life does not penetrate far beneath the surface of the land nor is it abundant at extreme altitudes or in the depths of the open ocean.

THE FOURTH-STATE MODEL

If we look at the vast and diverse populations of organisms as a film of living matter, we are struck by the thinness of the film compared with the diameter of the earth. Life does not extend far into the sky nor penetrate deeply into the earth's crust. The film is only a few meters deep at its thinnest spots, and about a thousand at its thickest (Fig. 18).

We can think of this film as a fourth state of matter, a state marked by complex organization and complex properties. The idea of life as a fourth state of matter is what scientists call a thought model: a generalization that unifies and simplifies a complicated assortment of phenomena. Models are used often in science, and we shall refer to several of them in this book. Some are physical models, such as the chemist's ball-and-stick representations of molecules; some are abstract, such as Freud's model of the id, ego, and superego. It is important to remember that all models are constructs: representations of reality, not reality itself. They are useful only so long as their explanations correspond to observations of the real world. When they fail this purpose, they must be modified or discarded. The fourth-state model provides an abstract, wide-angle perspective for examining many of the important ideas of modern biology.

The solid, liquid, and gaseous states of matter vary with temperature and pressure. We recall that at atmospheric pressure, pure water is a solid at temperatures below 0°C, a liquid from 0° to 100°C, and a gas at higher temperatures. At a pressure of several atmospheres, the boiling point of water is considerably higher than 100°C, and in a near vacuum the boiling point is much lower. In the language of the physical chemist, temperature and pressure are the boundary conditions for these three states.

Among the boundary conditions that govern the living state the most important is solar energy, because directly or indirectly it drives the biochemical machinery of every living thing, providing the energy for the synthesis of the special chemical compounds required by living systems. Other boundary conditions include temperature, moisture, pressure, the availability of certain chemicals, and to a greater or lesser extent the presence of other organisms. The ultimate boundary condition is time. The film of living matter endures for eons, but the existence of each individual organism appears to be inherently temporary. As we shall see, there are some notable and thought-provoking exceptions.

ENERGY

Sunlight provides the energy for all life on earth—directly in plants and other photosynthetic organisms, indirectly in animals. Photosynthesis traps solar energy and converts it to chemical energy. Animals obtain these compounds by feeding on plants, or by feeding on other animals. Photosynthesizers are the foundation of the "food pyramid" of classical ecology (see Chapter 15).

In terms of our fourth-state model, we can say that solar energy sustains the high level of organization that characterizes the living state. From the pool of chemical compounds created by photosynthesis, individual organisms draw the energy necessary to convert small molecules into the giant macromolecules of life. By now, the thermonuclear fires of the sun have burned roughly half of its fuel. In a few billion years, as the fuel supply dwindles, those fires will diminish. On the earth, the organization of the fourth state will become ever more difficult to maintain. Finally, life on this planet will falter and disappear. Long after life has vanished, the barren planet will continue to circle a shrinking dwarf star that once was the sun. We shall have more to say about the energy economy of the fourth state in the next chapter, and about photosynthesis in Chapter 10.

TEMPERATURE

Much of the radiation from the sun that strikes the earth is converted into heat. The temperature of whole geographic regions is governed by the duration of exposure and the intensity of the sun's rays. Air temperatures near the surface of the earth range from the dry heat of the equatorial deserts (60°C) to the bone-cracking cold of the Antarctic continental plateau (−88°C).

At the molecular level, the thermal limitations of life are imposed by the physical properties of water and the chemical properties of proteins. All organisms contain water; none can survive having the water boiled out of it, nor can the chemistry of life proceed if the water is frozen. All organisms also contain proteins, giant molecules that are particularly sensitive to heat. Their structure and chemical properties, so crucial to the metabolism of living systems (see Chapter 6) are destroyed by even moderate heating. A familiar example is the transformation of egg white from a straw-colored, water-soluble protein to a white insoluble mass when an egg is heated.

Organisms vary in their ability to survive extremes in temperature, but in general monera and protists are more resistant than true plants and animals. The hot springs of Yellowstone National Park contain many kinds of algae which can grow and multiply at the extremely warm water temper-

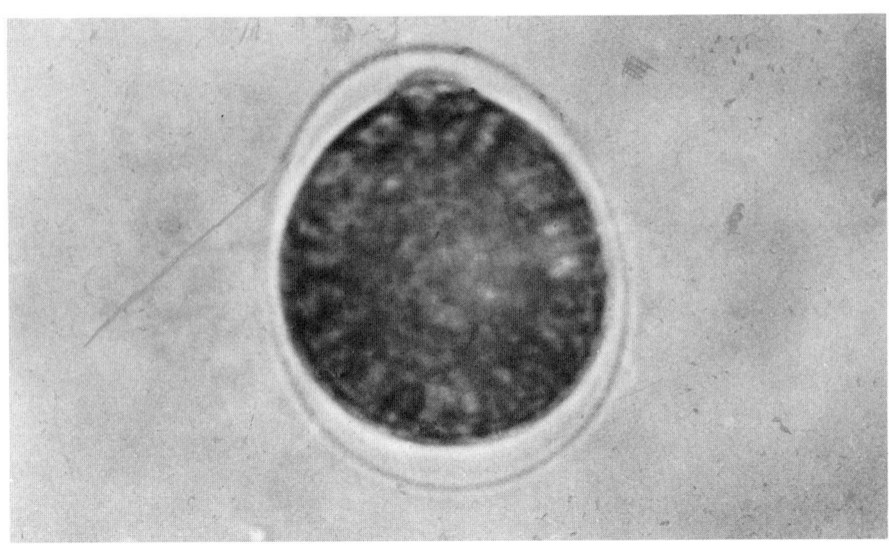

19
Unicellular alga that grows at freezing temperatures tints snowfields red in many parts of the world. *Chlamydomonas nivalis* is usually observed in an encysted resting stage that is bright red. Photomicrograph depicts the alga in its swimming vegetative form, which is about 50 microns in diameter.

ature of 70°C. At the other extreme there are different kinds of algae that grow directly on the surface of mountain snowfields. The most common type is responsible for the red snow found in many parts of the world (Fig. 19). The alga which gives the characteristic bloody color to the snow completes its life cycle at temperatures not far from 0°C. By careful laboratory studies the temperature range known to support active growth has been widened even further. Some fungi and bacteria can grow at −12°C; other types of bacteria are active at 104°C (Fig. 20). The mechanisms that permit survival under these extreme conditions are not yet understood.

Animals are more limited in their ability to resist temperature extremes. And yet one mammal has survived the coldest and the hottest climates on earth: man. His ability to inhabit any region of the planet depends on two factors. The first is the regulatory mechanism that maintains a constant temperature inside his body regardless of the temperature outside. Animals that have this ability are said to be homeothermous (*homeo* = same; *thermous* = temperature). Only mammals and birds are homeotherms. All other animals are poikilothermous (*poikilo* = variable); their internal temperature varies with that of the environment. The second factor is man's habit of carrying an artificial environment with him. He wears protective clothing, builds heated structures in the polar regions, and air-conditions

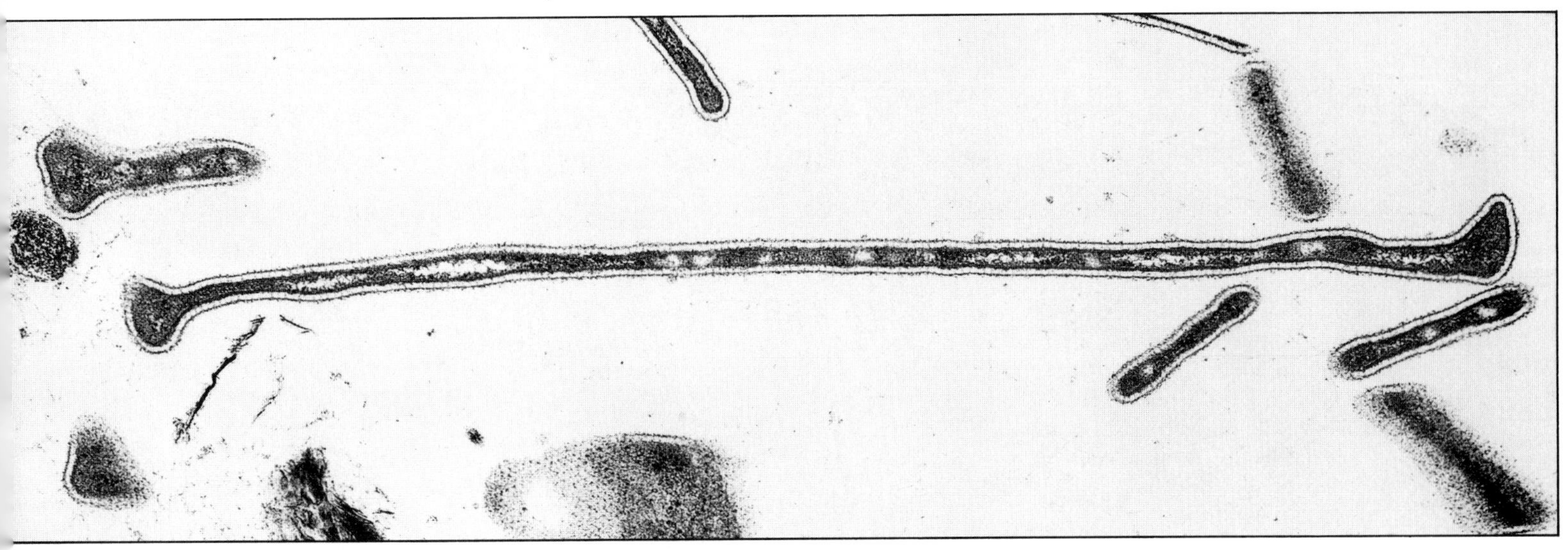

20
Thermophilic bacteria thrive at 92°C, a temperature close to the boiling point of water. Electron micrograph shows long filamentous type (center) and short rodlike type (right). Magnification about 36,000 ×.

buildings in the tropics. But even without his clothes, houses, furnaces, and air conditioners, man is still highly adaptable. In hot climates his ability to perspire provides a cooling mechanism and in cold climates his metabolic rate increases, compensating for the loss of heat.

MOISTURE

No organism can grow without water. Conversely, few organisms can grow in pure double-distilled water. In habitats between these extremes dwell millions of species of plants and animals. Plants inhabiting arid regions are structurally different from those growing in moister areas. Mesquite trees have roots up to 55 meters long, which enable them to reach water deep underground. Succulent plants such as the cactus store water during brief wet spells and use it during the long dry season. The leaves of cacti are reduced to spines whose small surface area minimizes the loss of water by evaporation. The leaves of other plants, such as the creosote bush, are covered by a thick waxy layer that restricts the amount of water lost by evaporation. Moreover, the leaves of the creosote bush and other desert shrubs contain a compound that inhibits the germination (sprouting) of seeds. When the leaves fall the inhibitor passes into the soil. No other

Flowering of desert occurs shortly after a heavy spring rain—if there is one—and lasts only a few days. This photograph was made near Ajo Mountain in the Sonoran Desert of southern Arizona.

22
Kangaroo rat, adapted to the desert, can
survive indefinitely on the water it obtains by
digesting food. Its urine contains so little
water that it solidifies a few moments after
excretion.

seeds can germinate in the vicinity, and there is less competition for available water. Other desert plants survive the dry season by effectively avoiding it. Their seeds germinate only when a minimum amount of rain has fallen; then they grow abruptly, flower, set seed, and die before the drought returns. The spring flowering of the desert is spectacular but brief, and it does not occur every year (Fig. 21).

A man stranded in the desert may lose a liter of water in an hour through perspiration. After one day without water he may die of dessication. Even by drinking several liters a day he cannot live in the desert more than a week. But some desert animals survive without ever drinking a drop of water. In all animals some water is produced chemically through the breakdown of food. The kangaroo rat, for example, is so frugal in its use of water that this small source is enough (Fig. 22). It excretes very little water because it has unusually effective kidneys, and confines its feeding to the cool hours of the evening and night. The camel, too large to burrow like the rat, is insulated by his wool and his fatty hump. His tissues can lose more than 25 percent of their water without ill effects. Moreover, the camel's body temperature is somewhat flexible; he will not sweat freely until his temperature reaches 40°C (105°F).

Aquatic animals would seem to have no moisture problems, but many ponds and streams dry up each year during the dry season. Some organisms survive these dry periods by going into a state of inactivity called aestivation. During this time the chemical machinery of the body slows down, to return to normal only when moisture again becomes available. Other species lay thick-walled eggs that can withstand long periods of dryness. Drought kills the adult, but new organisms hatch from the eggs at the beginning of the next wet season.

PRESSURE

During the 1968 Olympic Games in Mexico City much was written about the effect of altitude, with its diminished oxygen pressure, on the performance of athletes. However, the only environment where low atmospheric pressure poses serious obstacles to life lies at the crest of the highest mountain ranges, like the Himalayas. The teams of climbers who conquered Mt. Everest—8900 meters (more than 29,000 feet) above sea level—used oxygen masks during the last stage of their ascent.

Water pressure is quite another matter. A diver with scuba equipment cannot long endure the pressure at 100 meters: 10 kilograms per square centimeter (361 pounds per square inch). Not many years ago any suggestion of finding organisms in the abyss of the oceans was dismissed as absurd. Scientists reasoned that there was no light to support plants, and that the pressures would crush any imaginable animal. Nevertheless observers in deep-sea diving machines like the bathyscaphe and the bathysphere have photographed unknown types of fish and other organisms at depths of a few thousand meters (Fig. 23). Recent investigations by workers from the Woods Hole Oceanographic Institute in Massachusetts and Scripps Institute of Oceanography in California have discovered organisms at depths exceeding 10,000 meters, where the pressures exceed one ton per square inch. Most of these organisms die as they are brought to the surface, but deep-sea bacteria have been transferred under pressure to the laboratory, and have grown and multiplied in a chamber at pressures 1400 times greater than those at the surface of the sea (Fig. 24). The exact source of food in these regions of eternal darkness remains a mystery, but its ultimate origin must be the sunlit upper regions of the sea. Because of the difficulties involved in observing such organisms in their native surroundings, and the extensive problems involved in designing and conducting controlled experiments on them, the effects of pressure, temperature, and

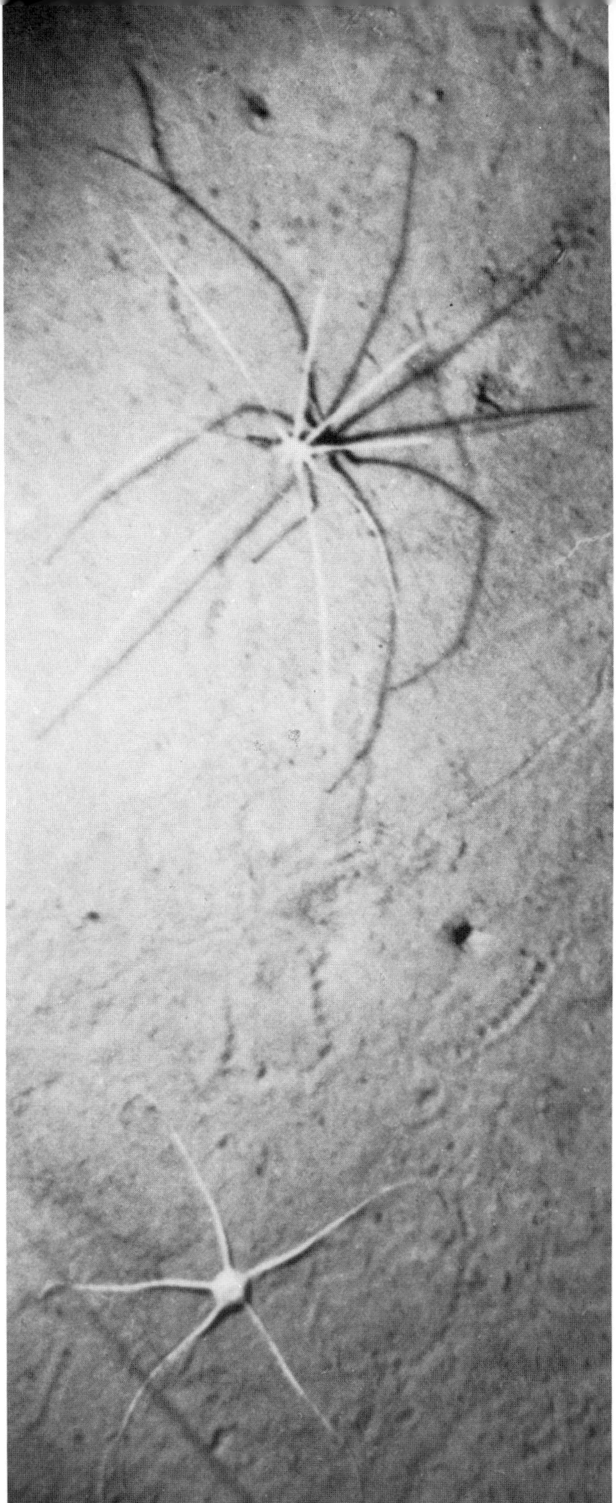

23
Deep-sea animals live in perpetual darkness thousands of meters beneath the surface. Sea spider and five-pointed starfish (left) belong to unknown species. Tentacled polyp about 1 meter tall (right) was photographed at a depth of 4850 meters (roughly three miles). It is related to sea anemones. Pressure on each square centimeter of its body surface is more than half a ton.

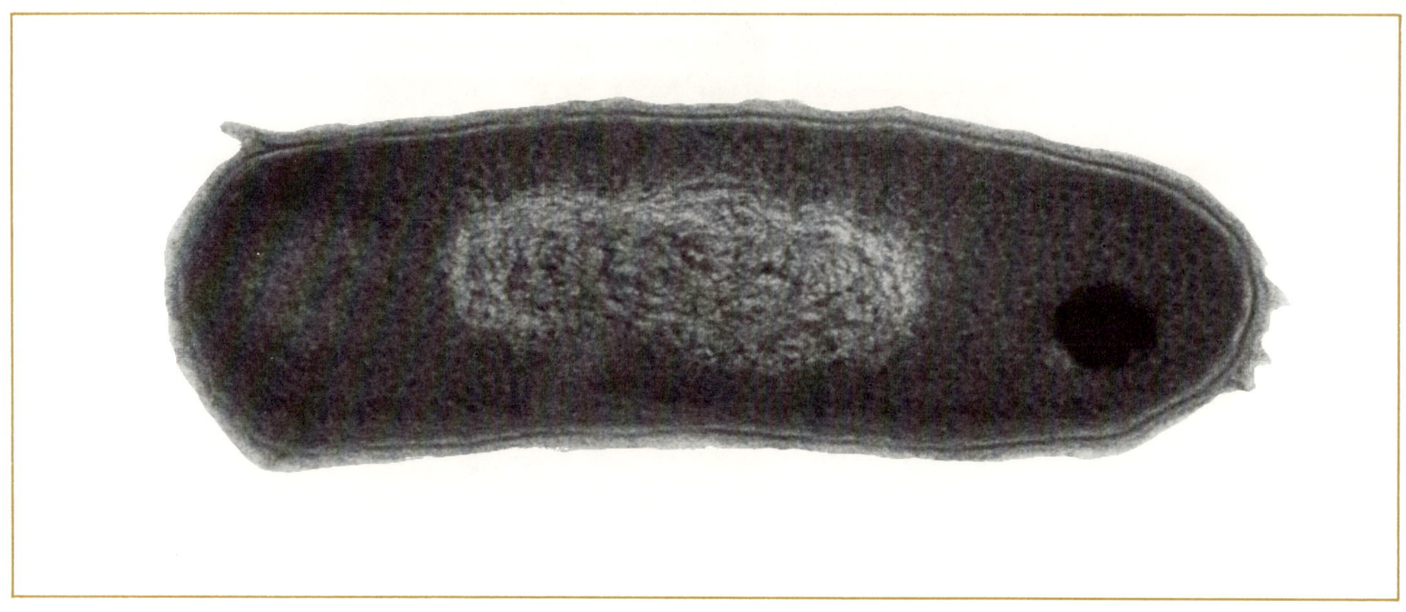

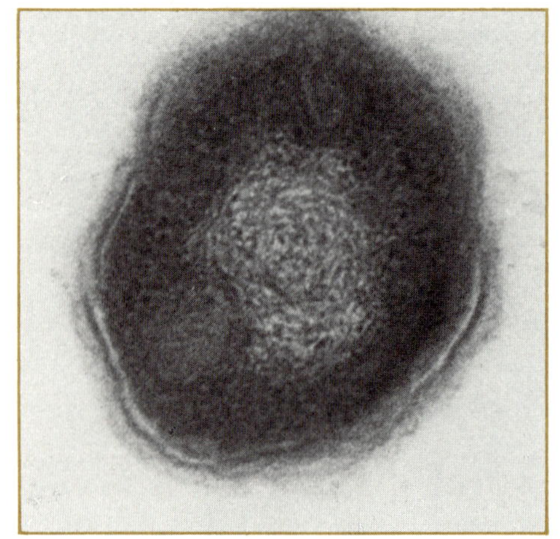

nutrient utilization are very poorly known. Even as the first life probes await launching toward the planets, the ocean depths remain a virgin frontier for the biological explorer.

CHEMICALS

All organisms require at least a few chemical substances as raw materials. The minimum requirements for life are found in certain species of photosynthetic bacteria. Given a supply of carbon dioxide, water, and a few minerals, they can synthesize all of the complicated organic molecules that comprise their cells. Most organisms, including man, rely on the environment to provide not only minerals, oxygen, and water, but also up to 30 or 40 preformed organic molecules, including vitamins, amino acids, carbohydrates and fats.

Despite enormous differences in structure and function, all living cells are made up of a strikingly similar assortment of molecules. The organism must obtain from the environment any of these compounds that its cells cannot synthesize for themselves. In terms of the ability to synthesize the molecules they need, many of the so-called lower organisms, such as bacteria and molds, are considerably more versatile than man. The ability

24

◄**Deep-sea bacterium** was found in sediment on ocean floor at depth of 1700 meters (about one mile). This organism is a *Corynebacterium,* a relative of the organism that causes diphtheria. One is shown in longitudinal section (above) and another in cross section (below). It exhibits the structure characteristic of prokaryotic cells. Note the absence of a membrane around the light-toned nuclear region.

of higher organisms to synthesize the compounds they need probably disappeared in the course of evolution as they began to obtain those molecules by digesting other organisms.

TIME

With the sun playing a crucial role in the maintenance of life, it is not surprising that the life cycles of many organisms are regulated by means of internal timing mechanisms set by the sun. In 1920 two scientists of the U.S. Department of Agriculture, working at the experimental station in Maryland discovered that photoperiod (the number of hours in daylight) somehow governs the flowering season in many kinds of plants. So-called long-night plants (like chrysanthemums and ragweed) bloom only when the night reaches a certain length. Temperature also plays a role: these plants do not bloom outdoors in the winter (when the nights are longest) because it is too cold. But in a heated greenhouse long-night plants can bloom during the winter. Long-night plants characteristically bloom during the late summer months. Hay fever sufferers can avoid ragweed pollen by moving north into Canada or Alaska. Although the ragweed plant can survive the lower temperatures of these regions, it does not bloom because

25

◄**Oldest living organisms** on earth are bristlecone pines, some of which have survived for almost 50 centuries. This one stands in the White Mountains of California. Men at lower left indicate its size.

the temperature is too low when the northern nights become long enough to initiate flowering. If there are no flowers there can be no pollen.

In some animals the daily light cycle influences the physiological changes associated with reproduction. Migratory and reproductive behavior in birds, territorial defense and mating in fish, and the pattern of development of many insects are all triggered by a biological clock set by the sun. In squirrels, the prevailing light conditions of autumn stimulate an increasing tendency to store food. The autumnal change in the coat color of the snowshoe hare from dark to light is also related to the changing length of the day. The survival of many fish, insects, birds, and mammals depends on reactions initiated by the photoperiod of the area in which they live.

The sun also serves as a navigation beacon for bees, birds, and other peripatetic animals. We shall discuss biological clocks and animal navigation in more detail in Chapter 14.

The great majority of organisms live only a few hours, days, or weeks. Of almost a million species of insect, relatively few can live into their second year. Virtually no organisms survive for more than a century: among animals, only marine turtles, elephants, and an occasional human; among plants, only a few species of tree. Probably the oldest organisms on earth are not the majestic redwoods and sequoias of California, but the gnarled, scraggly bristlecone pines found in the mountains of our western states (Fig. 25). Some of them are 4900 years old; they were saplings when the architects of Egypt's Third Dynasty fashioned the first pyramids. Eventually they too will be harvested by time.

Despite individual mortality, each living organism is the temporary custodian of life's permanent component: a heritage of genetic material that stretches in an unbroken chain across the abyss of time. Before it dies, the organism will bequeath this heritage to its offspring. Individuals come and go, but the species persists. Taking a much longer perspective, we see that species, too, come and go. Walrus and peacock, slime mold and mushroom, orchid and cactus, all have inherited the genes of primeval species not only long dead, but long extinct. In the long view, the custodians change, but life itself endures.

Aging and death are not an inevitable part of the life cycle of all organisms. When a bacterial cell divides, it yields two daughter cells, identical to each other and to the parent cell. Each cell is equally young, and a new life stretches before it. Somehow this method of reproduction carries with it the promise of self-renewal. Bacteria can be killed, of course, or can die for lack of nutrients, but in a sense each cell is potentially immortal. The mechanism of self-renewal in these simple organisms raises some profound questions about the nature of aging and death, questions that modern biology has only begun to study.

CAUSE AND EFFECT

The diversity of organisms on earth can be attributed to changes resulting from challenges posed by a variety of environments. At times it may seem that the onward surge of life has a direction and a purpose. In a way, this may be true. Unlike the other states of matter, life is not a passive one. Living matter tends to perpetuate its own existence, to disperse, and to organize the environment into its own image. In this sense, life has an aggressive strategy for survival and expansion. One might speculate that man's drive to colonize new continents, and his current thrusts toward the moon and the planets, are only the expression of a striving that all organisms carry in their genes.

The argument does not apply to individual species. For example, in discussing evolution, laymen often imply that the environment of an organism somehow causes the appearance of adaptive structures: a fish has gills *because* it lives in water. Applying the same kind of reasoning, is it not logical to suppose that if humans decided to live under water they also would grow gills? It is more accurate to say that a fish *can* live in water because it has gills.

The correct elucidation of cause and effect relationships is one of the most critical and essential parts of scientific thinking. Many of us were taught that a giraffe has a long neck because it feeds on the leaves at the tops of trees, but this statement simply reverses cause and effect. The statement should suggest that giraffes can nibble on treetops because they have long necks. Of course this doesn't answer the question of why the giraffe has a long neck, or why fish have gills, or why cacti have thorns, or why humans have two arms instead of three. The answers—when they are known—are never simple. The scientific heresy of ascribing a purpose to nature is called teleology, and it is an easy trap to fall into.

Let us take a not-so-simple example: Why can a bird inhabit the air? Because it has wings? But penguins and ostriches have wings and can't fly; pasting wings on a human will not give him flight. Wings alone are not enough. The entire anatomy of a bird is adapted to the requirements of flight. The bones are light and hollow. Fused into a rigid skeleton, they combine maximum strength with minimum weight. An enormous breastbone anchors the massive flight muscles. Feathers insulate the body without adding much weight. An examination of structure often elucidates the ability of an organism to exploit a particular environment.

The next logical question is why the bird possesses such structures. Why does a bird have hollow bones? The answer is that it inherited from its parents the genes (genetic potential) for the production of hollow bones. The genes initiate a sequence of chemical reactions in the developing embryo that leads to the production of hollow bones. But even this line of reasoning begs the ultimate question, "Why?", which cannot be answered by a description of developmental mechanisms. Ultimately the answer lies in the genetic basis of evolution. This explains the emphasis of modern biology on the study of genetics. In the following chapters we shall outline the relationships between the gene and the chemistry, structure, and function of the organism. With that background we shall attempt to reconstruct the origin and evolution of modern organisms.

READINGS

Allen, J. M. (ed.), *The Nature of Biological Diversity*. McGraw-Hill, New York, 1963

A collection of essays, many at an advanced level, on the principles that explain the existence of the wide variety of organisms on earth.

Bonner, J. T., *The Cellular Slime Molds*. Princeton University, Princeton, second edition, 1966

The biology of these peculiar organisms has baffled investigators for decades. This account is a description of recent research, much of it done by the author.

Borges, J. L., and M. Guerrero, *The Book of Imaginary Beings*. Dutton, New York, 1969

The zoology of fantasy, from both Eastern and Western myths. Borges is a modern Argentine writer who seems to have read everything. The first story, *A Bao A Qu,* is recommended to smokers.

Doyle, W. T., *Nonvascular Plants: Form and Function*. Wadsworth, Belmont, Calif., 1965 (paper)

A rapid survey of the biology of bacteria, viruses, algae, mosses, and liverworts.

Lewinsohn, R., *Animals, Men, and Myths*. Premier Books, Harper & Row, 1964 (paper)

A sociological and zoological look at prehistoric, modern, and futuristic animals. A masterly blend of mythology and fact concerning the interrelationships of man and his animal contemporaries.

Salisbury, F. B., and R. V. Parke, *Vascular Plants: Form and Function*. Wadsworth, Belmont, Calif., 1964 (paper)

A quick course in botany, with emphasis on the relationship between structure and function in flowering plants.

Vickerman, K., and F. E. G. Cox, *The Protozoa*. Houghton Mifflin, Boston, 1967 (paper)

A short and enjoyable introduction to the world of protozoa, with consistently good text, drawings, and photographs.

White, T. H., *The Bestiary*. Putnam, New York, 1954 (paper)

A translation of a 12th century Latin bestiary that is both informative and amusing. Illustrated.

The World of Nature. Doubleday, Garden City, N.Y.

> The following six volumes are all part of this interesting series. All are profusely illustrated with excellent color photographs. The texts are short and informative.

Buchsbaum, R., and L. J. Milne, *The Lower Animals: Living Invertebrates of the World.* 1960

Cochran, D. M , *Living Amphibians of the World.* 1961

Gillard, E. T., *Living Birds of the World.* 1958

Herald, E. S., *Living Fishes of the World.* 1961

Klots, A. B., and E. B. Klots, *Living Insects of the World.* 1959

Sanderson, I. T., *Living Mammals of the World.* 1955

3. THE PHYSICS AND CHEMISTRY OF LIFE

The universe and all living things in it are composed entirely of matter. According to our fourth-state model, living systems—the Black Cloud, a bacterium, a man—are all composed of matter in a highly organized state, a state established and maintained by the expenditure of energy.

This abstract philosophy is peculiarly modern. Less than a century ago, most scientists believed that the condition of being alive added something unique, something not explainable by the laws that apply to inanimate matter. No one was quite sure what this something was, so they called it the Vital Force. Scientists used the term to explain why the structure of living matter was so much more complex than that of nonliving matter, and why the behavior of living matter was so much less predictable. The philosophy of Vitalism assumed that living organisms were governed by laws different from those of chemistry and physics.

The quest for some insight into the nature of the Vital Force has led biologists to explore organisms in finer and finer detail. Gradually the emphasis of research has shifted from whole organisms to single cells, parts of cells, and ultimately to the molecules that comprise those parts.

Until the German chemist Friedrich Wöhler prepared urea (one of the components of urine) from ammonium cyanate in 1828, most scientists believed that certain compounds, called organic, could be assembled only

◄**Bundles of cellulose fibrils** strengthen the walls of plant cells. The wall contains two layers of bundles at right angles to each other. This electron micrograph shows the inside wall of an alga cell. One layer of cellulose bundles has been stripped away to reveal the other. Magnification about 60,000 ×.

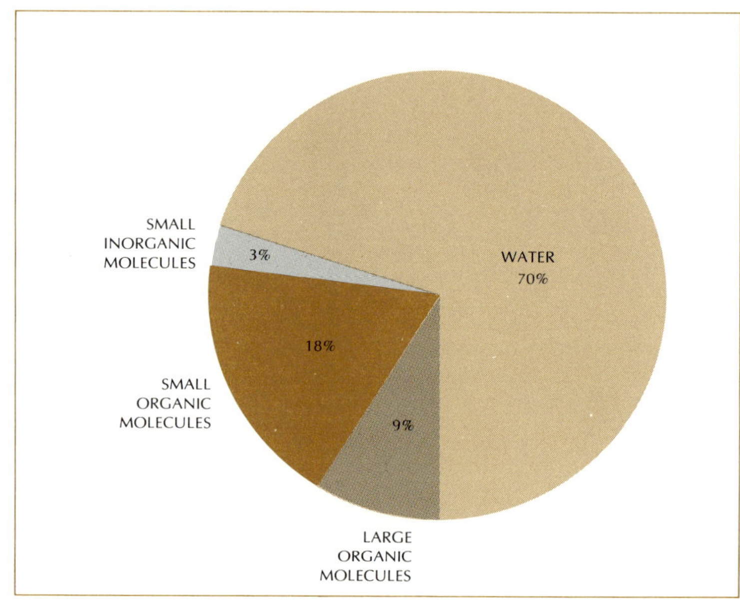

1

Chemical composition of a cell. Only about 9 percent of the total weight of a cell is large organic molecules, such as proteins and nucleic acids.

within living organisms. Chemical substances were classified as organic or inorganic according to whether or not they were characteristic of living systems. The words still persist in chemistry, but the distinction now is more accurate: organic compounds are the compounds of carbon; the rest are inorganic. Biochemists have since demonstrated that even proteins and nucleic acids, the two most complex types of compounds known to modern chemistry, can be synthesized in the laboratory. So far biologists have found that even at the molecular level, no living thing violates any known law of physics or chemistry. Many scientists now assert that physical and chemical laws will eventually explain all activities of plants and animals.

The chemistry of living organisms can be summarized in a few general statements. If we could analyze the chemical makeup of all cells on earth, we should find that the "average" cell consists of about 70 percent water, 18 percent small organic molecules (such as fats and carbohydrates), 3 percent small inorganic molecules (mostly mineral solids), and 9 percent large organic molecules (nucleic acids and proteins) (Fig. 1). These giant

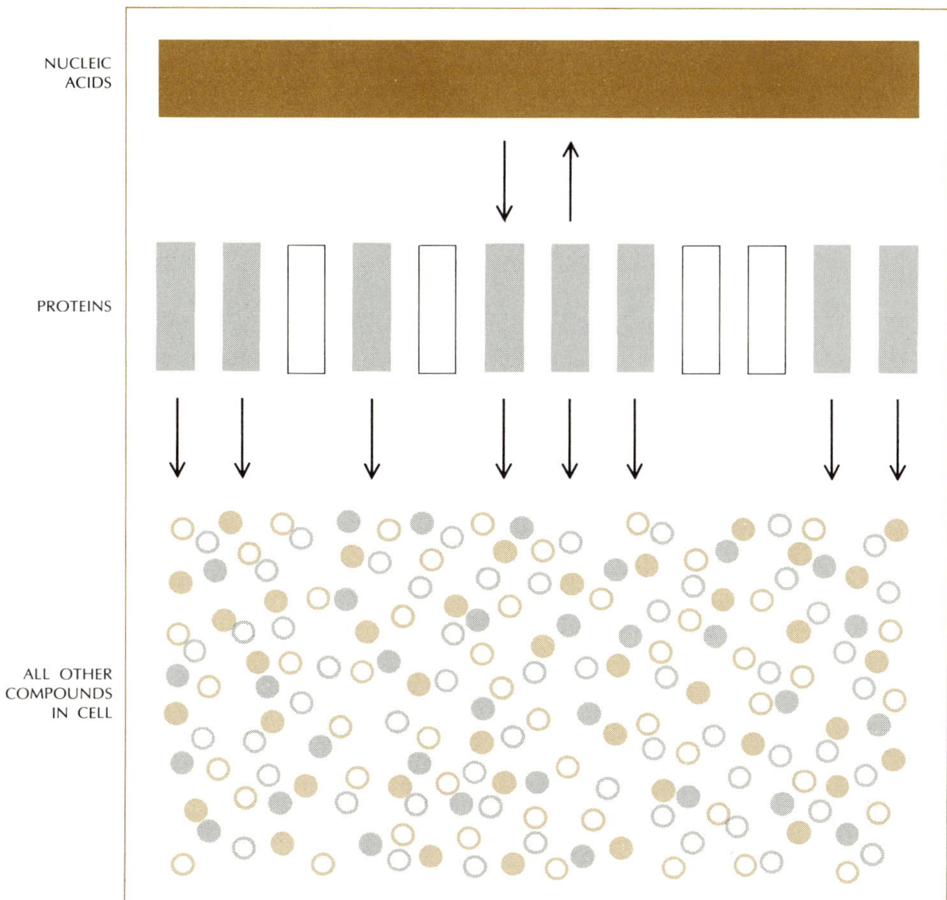

2
Nucleic acids and enzymes control the chemistry of life. Nucleic acids (top) contain the molecular instructions for making proteins, including enzymes (middle). Enzymes in turn facilitate the synthesis of all other biological compounds (bottom), including more protein and nucleic acid molecules.

molecules are the key to the chemistry of life. Encoded in the structure of nucleic acids is the blueprint for making a cell, particularly the "instructions" for making enzymes: special protein molecules that speed up chemical reactions. Enzymes can increase the rate of a reaction by as much as 10^8 times, which means the difference between a reaction time of one second and one of three years. Many cells carry out thousands of reactions per second, and the rate of each reaction is controlled by one specific enzyme. The number of different enzymes in a cell can run into the hundreds or even higher, and if even one of them is absent, the reaction that it controls will proceed so slowly as to be negligible. The activity of enzymes controls the reactions that synthesize all the components of the cell, including other proteins, from the small molecules that the cell obtains from its environment (Fig. 2).

Remodeling the architecture of molecules requires energy. Plants get this energy from sunlight, and all other organisms get it from food; but neither the energy of sunlight nor that of food is in a form that can be used directly by the cell, just as an automobile cannot use dynamite as fuel. All living systems convert energy into a form that they can gear directly to their biochemical machinery: specifically, into the chemical energy of two compounds designated in chemical shorthand as ADP and ATP.

With such broad strokes we can sketch only a crude and understated picture of the remarkably sophisticated mechanisms that enable an organism to survive, grow, and reproduce. It is as if we described a Ferrari roadster as being merely an adequate means of transportation from the campus to the corner drugstore. In later chapters the picture will take on greater depth and richness. A fuller understanding of the splendor of living systems requires at least a nodding familiarity with the language of thermodynamics and chemistry. Thermodynamics is a rigorous and mathematical discipline with profound implications for all science; its province is the flow of energy throughout the universe, and there are no exceptions to its basic laws.

THE FIRST LAW OF THERMODYNAMICS

In nonmathematical terms, the First Law of Thermodynamics states that "The sum of the energy and mass of the universe is a constant." Einstein's famous relativity equation, $e = mc^2$ (where e is energy, m is mass, and c is the speed of light) expresses concisely the interconvertibility of mass and energy. Because living systems convert only infinitesimal amounts of mass directly into energy (atomic bombs are far more efficient at this particular process), biologists use an older, less accurate statement of the First Law: "Matter and energy can be neither created nor destroyed."

Matter is substance and occupies space. It has mass. In big enough and compact enough pieces one can see it and feel it. Energy is less tangible. Its many forms include light, heat, nuclear, chemical, mechanical, and electrical energy. They are not static but may change from one form to another. For example, in a green plant such as grass, the process of photosynthesis converts light energy into the chemical energy of newly synthesized sugar molecules. A horse that eats the grass can convert some of the chemical energy of the sugar into the mechanical energy of muscular contraction. If the horse is hitched to a generator, some of his mechanical energy can be converted into electrical energy. In a kitchen toaster, the electrical energy can be changed into heat.

In none of these conversions is energy created or destroyed; it is changed only in form. Biological systems are completely dependent upon orderly conversions of energy from one form to another. If this flow ceases for any reason, the organism dies.

THE SECOND LAW OF THERMODYNAMICS

The Second Law of Thermodynamics deals with an abstraction called *entropy*. In a sense, entropy is a measure of disorder, or randomness. The more disorder, the higher the entropy; the greater the organization of the system, the lower its entropy. The law says that "The entropy of any isolated system tends to increase." Stated another way, the Second Law implies that any isolated system becomes more disorganized as time goes on. Both matter and energy can become disorganized: matter by becoming more randomly arranged, and energy by being degraded into heat. When a crystal of salt dissolves in a cup of soup the salt goes from a highly organized crystalline state to a state of random dispersion throughout the liquid. In the process, the entropy of the system increases. When you rub your hands together they get hot. In this case, some of the chemical energy of food-stuffs is converted into mechanical energy, and in the process some energy is lost as heat. When any form of energy is converted into heat, the entropy of the system increases.

If matter continually becomes more disorganized and energy continually becomes degraded into heat, then the universe must be running down. According to one theory, it is. Some scientists believe that in perhaps several billion years, the thermonuclear fires of all stars will have burnt out, and the universe will reach thermal equilibrium. This is what physicists mean when they speak of the "heat death" of the universe. As the mathematician Norbert Wiener put it, "In a very real sense, we are all shipwrecked travelers on a doomed planet."

But growing organisms seem to be "running up." Anyone who has watched a fertilized hen egg grow into an embryo and then into a small chick knows that complexity can be created from simple beginnings. During the incubation of the fertilized egg, the supply of food within the egg is incorporated into the structure of a developing, complex living organism. Is life an exception to the Second Law?

The answer becomes clear if we recall that entropy deals with changes in energy as well as changes in matter. All living organisms constantly convert chemical energy into heat energy that escapes into the environment. Growth requires energy. Some of the energy goes to create order, that is, to form complex living structures from simple food substances. But much of the energy is degraded into heat. On balance, the amount of energy degraded is greater than the amount of order created. Growth does not violate the Second Law.

In later chapters we shall emphasize the importance of the degradation of energy to the functioning of biological systems. In simple terms it means that no reaction—physical, chemical, or biological—is 100 percent efficient. Either energy is wasted as heat, or order is degraded into disorder. We shall see that the more advanced biological systems, refined by more than four billion years of evolution, are as efficient in their use of energy as the best power plants that engineers have designed.

ATOMS

The structure and properties of matter are the province of chemistry. To the relief of many professional biologists, one doesn't have to be a chemist to study biology, but even biologists must be willing to master a few of its basic ideas. As every undergraduate knows, all substances are composed of small particles called atoms. An atom consists of a nucleus—a dense central core containing one or more positively charged particles called protons—surrounded by a "cloud" of one or more negatively charged electrons. The nucleus may also contain noncharged particles called neutrons.

Figure 3 depicts the atomic structures of hydrogen and carbon. The hydrogen atom is essentially a proton associated with a single electron. There are no neutrons in a hydrogen atom. An atom of carbon is much more complex. In the central core there are six protons and six neutrons. The electron cloud contains six electrons.

It is certainly less important to memorize the number of protons, neutrons, and electrons in each of more than 100 known elements, than to learn three important generalizations that apply to all atoms:

1. Atoms are electrically neutral. The number of electrons equals the number of protons.

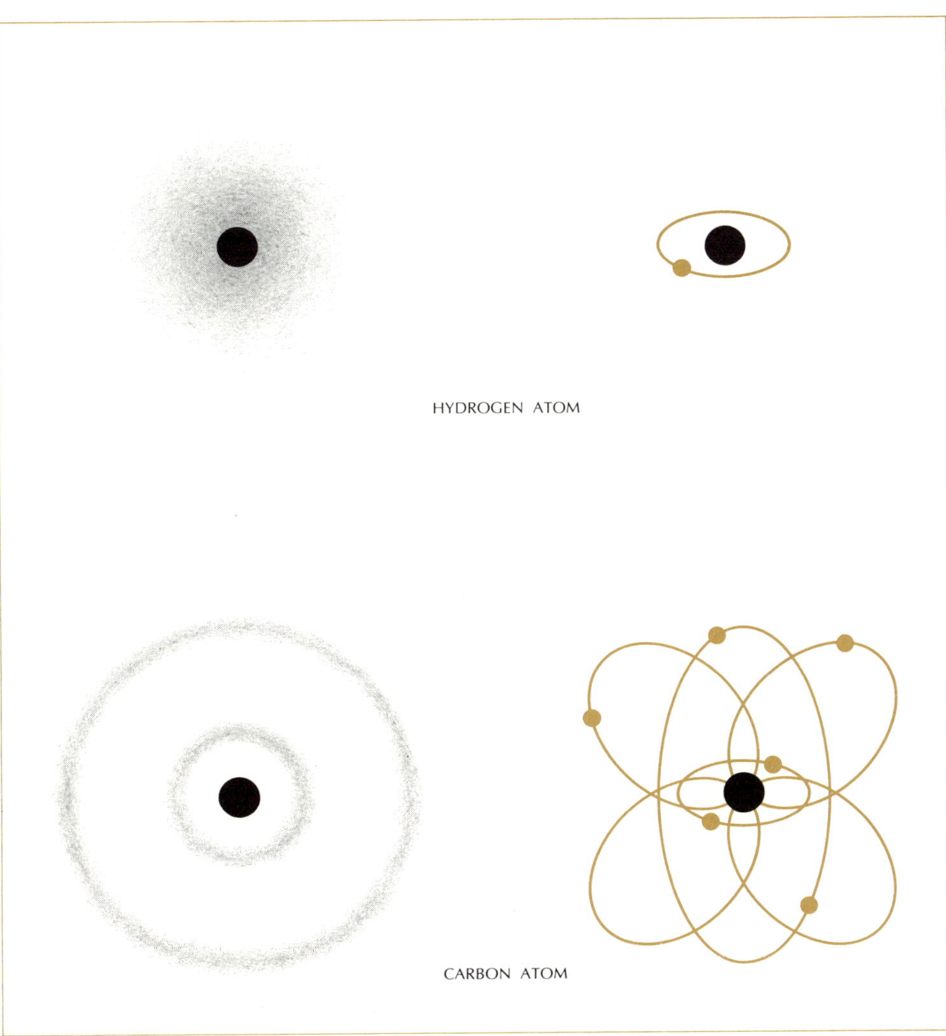

3
Atoms of hydrogen and carbon.
Diagrams at left depict the relative positions of the nuclei and the surrounding electron clouds. Diagrams at right show electrons "frozen" at one point along their orbital paths.

2. The mass number of an atom is calculated by adding the number of protons and neutrons in the nucleus. The mass of the electrons is ignored in the calculation of the atomic mass number because electrons are very light compared with the protons and neutrons. Therefore, the atomic mass number of hydrogen is 1 (one proton, no neutrons), and that of carbon is 12 (six protons, six neutrons).

3. The number and the arrangement of the electrons of an atom determines its combining or bonding potential. The bonding potential of an atom

Eight elements most commonly found in living organisms. The mass number indicates the total number of protons and neutrons in the nucleus of the atom. The bonding potential indicates the number of other atoms with which an atom of the element may combine at one time.

Element	Symbol	Mass Number	Bonding Potential
hydrogen	H	1	1
carbon	C	12	4
nitrogen	N	14	3
oxygen	O	16	2
sodium	Na	23	1
phosphorus	P	31	3
sulfur	S	32	2
potassium	K	39	1

indicates the number of other atoms with which it can combine at any one time. Table I presents the atomic mass numbers and the bonding potentials of the elements that are most important in biology.

MOLECULES AND COMPOUNDS

A molecule consists of two or more atoms chemically bonded together. A compound is composed of two or more *different kinds* of atoms chemically bonded together. H is the symbol for an atom of hydrogen; H_2 represents a molecule of hydrogen; and H_2O is the formula for a molecule of the chemical compound called water.

A chemical formula is merely a list of the kinds and numbers of atoms in a molecule or compound. Structural formulas show how the atoms are bonded together. CH_4 is the chemical formula of the compound methane, more commonly known as marsh gas. The structural formula (Fig. 4) sug-

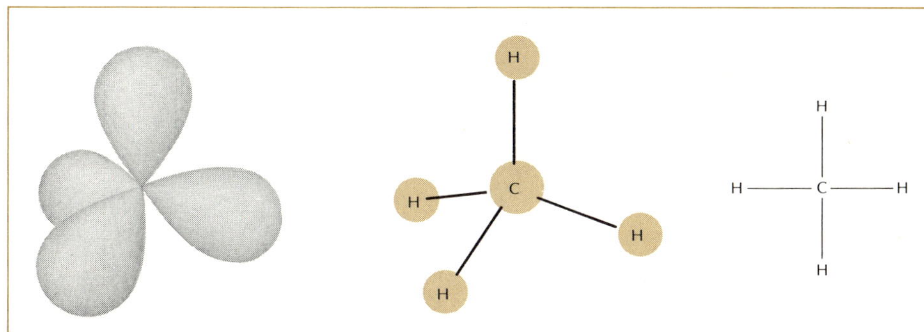

4
Three models of the methane molecule (CH_4). Orbital diagram (left) depicts the shape of electron clouds. Ball-and-stick model (center) indicates the geometry of the bonds. Structural formula (right), much easier to write, gives only a rough picture of the three-dimensional shape of the molecule.

gests the spatial arrangement of the four hydrogen atoms and the single carbon atom to which they are bonded. A chemical formula tells less about the molecule than a structural formula but it is easier to write. We shall use both kinds of formulas later as the need arises.

BIOCHEMISTRY

Many of the classes of compounds found in living systems are familiar to anyone who has ever glanced at a cookbook or a dieting manual. They include carbohydrates, fats, and proteins. Dieticians think of carbohydrates as quick-energy foods, fats as energy-storage compounds, and proteins as body-building material. In general these concepts are valid, but these food groups play many diverse roles in the functioning of living organisms. Other less familiar types of compounds are also essential to life. These include the nucleic acids and adenosine phosphates.

5

Simple sugars such as glucose and fructose can combine to form complex sugars such as sucrose. Combination involves the removal of hydrogen atom (color) from one sugar and an hydrogen and an oxygen atom (color) from the other sugar. The oxygen and the two hydrogens combine into a molecule of water (right).

CARBOHYDRATES

Carbohydrate is a class of organic compounds including sugars, starch, cellulose, and glycogen. The word carbohydrate derives from an old idea that these compounds contain equal ratios of carbon and water. The formula of glucose, $C_6H_{12}O_6$, can also be written as $6(C \cdot H_2O)$. By now chemists are aware that the formulas of some carbohydrates, like sucrose, $C_{12}H_{22}O_{11}$, do not quite fit this ratio, but the name has stuck.

Many carbohydrates—glucose and fructose, for example—have the same formulas, and can be distinguished from one another only on the basis of structural formulas. The structural formulas of glucose and fructose in Fig. 5 may look complicated to the uninitiated, but the trained eye of the chemist will notice that each molecule contains six carbon atoms, 12 hydrogen atoms, and six oxygen atoms; that each hydrogen atom is bonded to one other atom, each oxygen atom to two other atoms, and each carbon atom to four other atoms. In other words, the structural formulas agree with the chemical formulas. The bonding potential of each atom is satisfied. The difference between the molecules lies in the arrangement of the atoms within the molecules. These differences in structure give rise to differences in chemical and biological properties.

Simple sugars can combine to form more complex carbohydrates. Sucrose, common table sugar, is made from the two simpler sugars: glucose and fructose (Fig. 5). Simple sugars like glucose are called monosaccharides (single sugars). Sucrose is a disaccharide (double sugar). A long chain of chemically connected single sugar units is called a polysaccharide (literally,

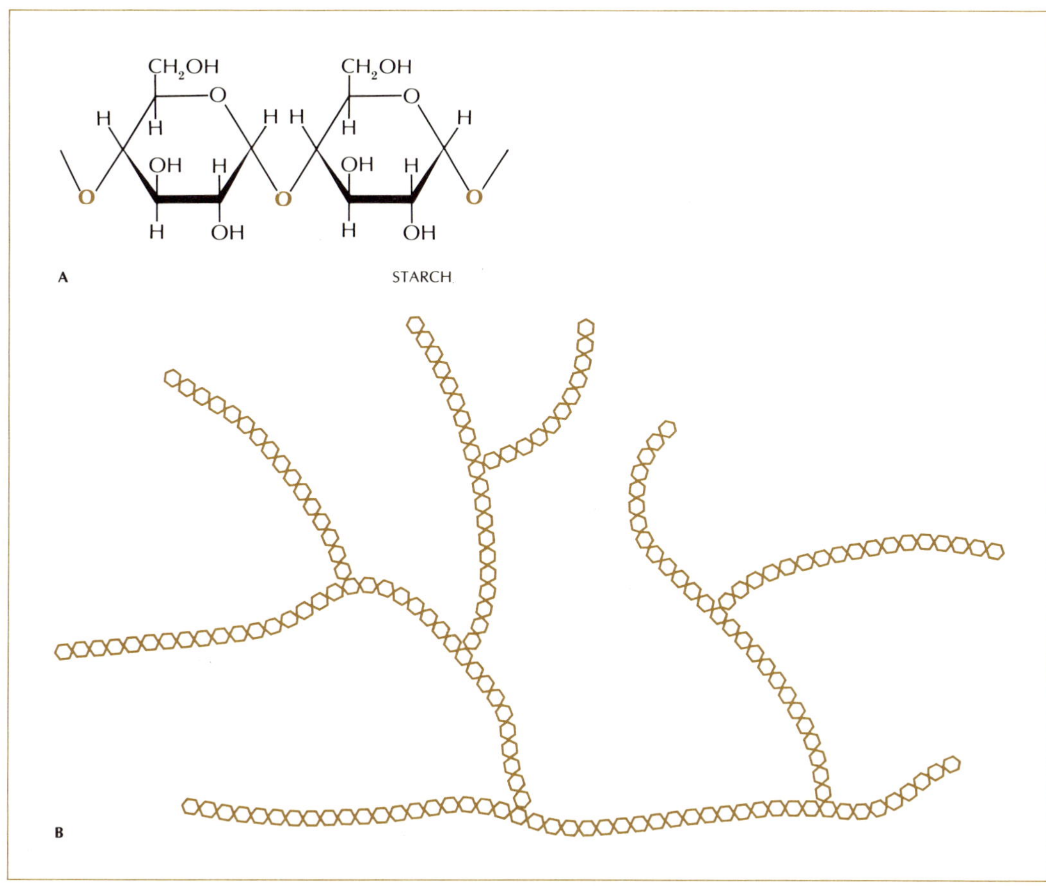

6
Starch molecules (B) are composed of many glucose units linked together (A). Branching points of starch molecule are easily attacked by digestive enzymes.

many sugars). Common examples of polysaccharides are starch, cellulose, and glycogen. The way in which the simple sugars are united in these polysaccharides determines their physical and chemical properties.

Starch consists of long chains of glucose units (Fig. 6). It is relatively insoluble in water and thus can be stored easily within cells. Starch can be rapidly broken down by chemicals in the cell into simpler units which can

7
Glycogen is similar to starch but contains more branching points.

be transported to all parts of the plant where they are used in various ways. In short, starch is an excellent storage form of the useful sugar glucose.

Glycogen, or "animal starch," is the polysaccharide of the animal body. Structurally similar to starch, its chain has many branches (Fig. 7). Its chemical properties are roughly similar to those of plant starches. It plays an important role in sugar metabolism in animals.

Cellulose consists of long straight chains of glucose units linked together somewhat differently from the linkages in starch (Fig. 8). Cellulose is structurally strong and is found as strengthening strands in the walls of most plant cells. Cotton fibers are almost pure cellulose, and wood is about 50 percent cellulose. The slight difference in linkage makes the cellulose molecule difficult for most organisms to break down. The few animals, like cattle and termites, that derive energy from cellulose, do so only because microorganisms in their gut digest it for them (Fig. 9). If the

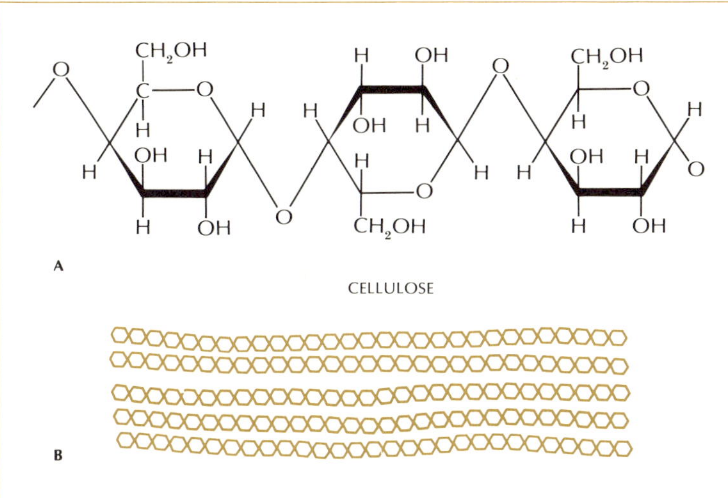

8
Cellulose is also composed of linked glucose units, but the orientation of the units within the chain is different from that in starch and glycogen. This difference explains why the digestive enzymes of most organisms cannot break down cellulose.

9
Large protozoa, *Tryconympha campanulla,* living in the intestine of a termite, digest the cellulose fibers of the wood eaten by the termite. The termite uses the resulting glucose residues as food.

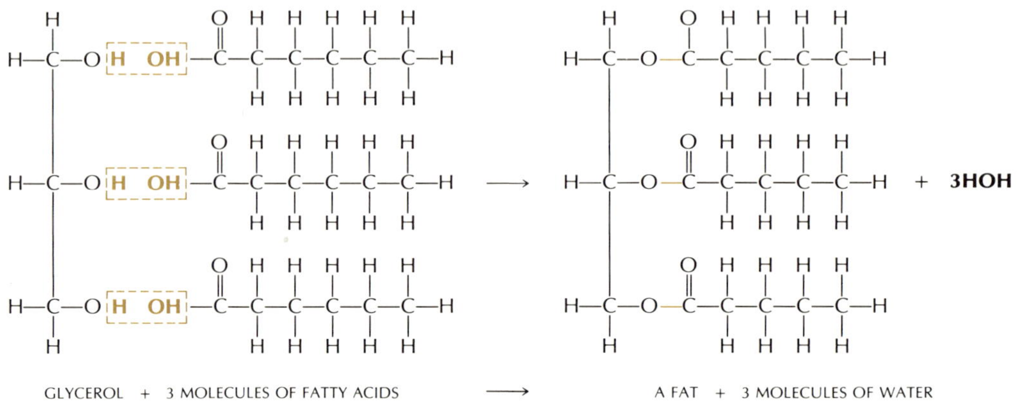

10

Molecule of fat is synthesized when three fatty acid molecules bond to a molecule of glycerol. As in sugars, formation of this linkage involves the splitting out of a molecule of water (color).

microorganisms in the gut of a cow are killed by a dose of antibiotics, the animal may starve to death even though it stuffs its stomach with grass.

LIPIDS

The lipids include such substances as fats and oils, waxes, sterols, and other fatlike substances. Fats are triglycerides, that is, they are formed by the bonding of three molecules of fatty acids to one molecule of glycerol (Fig. 10). Of all foods, fats have the highest energy content per gram, and this has earned them a bad name among weightwatchers. Fats have also been implicated in the disease of atherosclerosis, commonly known as hardening of the arteries. Patients with this disease show a high level of certain compounds in the blood: one of these is cholesterol. In some way, as yet unknown, the synthesis of cholesterol in the body is related to the intake of saturated fats. The food industry has eagerly responded by marketing a variety of cooking products rich in polyunsaturated fats. What is the difference between these two types of fat? It depends on the bonding of the carbon atoms in the fatty acid portions of the fat molecule. Figure 11 depicts the structure of two fatty acids: stearic acid (saturated) and linolenic acid (unsaturated).

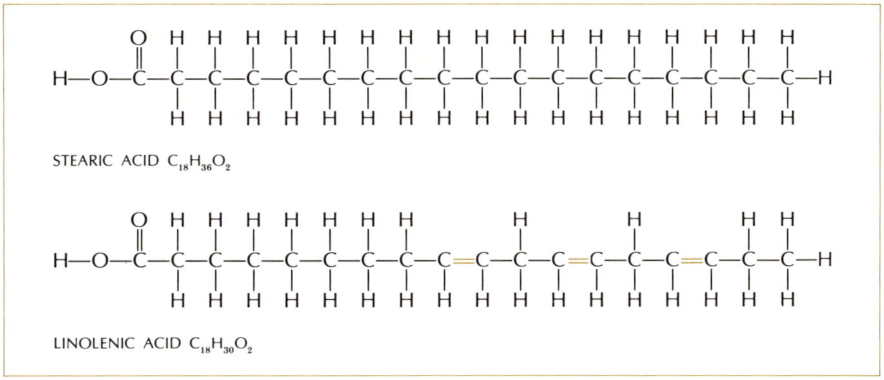

STEARIC ACID $C_{18}H_{36}O_2$

LINOLENIC ACID $C_{18}H_{30}O_2$

11
Fatty acids contain long chains of carbon atoms bonded to hydrogen. Unsaturated fatty acids (such as linolenic acid) contain carbon-to-carbon double bonds (color). Stearic acid is saturated with hydrogen atoms and contains no double bonds.

The main difference between the two structures is that linolenic acid contains double bonds (color) between some of its carbon atoms. The presence of a double bond indicates that the adjacent atoms share four, rather than two, electrons. The structure still satisfies the bonding potentials: one for hydrogen, two for oxygen, and four for carbon. In linolenic acid there are two double bonds between carbon atoms; in the stearic acid molecule there are none. When there are no double bonds, each carbon atom is "saturated" with hydrogen atoms. That is, another hydrogen cannot be added. But if double bonds are present, the carbons are unsaturated; a rearrangement of the carbon-to-carbon bonds would make it possible to add other hydrogens to the molecule. A molecule that contains more than one carbon-to-carbon double bond is said to be polyunsaturated. The presence of double bonds also affects the physical properties of a fat. In general, saturated fats, like beef fat, are solid at room temperature; polyunsaturated fats, like corn oil, are liquid. Medically, however, the exact relationship between the presence of double bonds and hardening of the arteries remains a mystery.

All fatty acids have at one end of the molecule the atomic grouping COOH. Known as the carboxyl group, this part of the molecule has two important features. The first is the double bond between the carbon atom

12
Carboxyl group of a molecule can donate a proton (color) to another molecule. Proton donors are called acids. Molecules containing carboxyl groups are organic acids.

and one of the oxygen atoms (Fig. 12). The second is that the hydrogen atom is held very weakly. The shifting of electrons to form the double bond makes it possible for the molecule to dissociate in such a way that the end hydrogen easily comes off. Remember that a hydrogen atom consists of a proton and a single electron. In this case the proton may be released from the fatty acid molecule while the electron remains behind. This dissociation yields two electrically charged fragments called ions. An ion is a single atom or a group of atoms bearing an electrical charge. The hydrogen atom, having lost an electron, becomes a hydrogen ion carrying a positive charge. Any compound that releases or donates a proton is known as an acid. Linolenic and stearic acids contain carbon-to-carbon linkages (are organic), have the potential of donating protons (are acids), and can combine with glycerol to form a fat (are fatty); thus the name fatty acid.

The properties of any specific type of fat depend upon the kinds of fatty acids linked to the glycerol molecule. All three fatty acids may be identical; all three may be different; two may be the same and one different. Because there are many different kinds of fatty acids, there are many more different kinds of fats. For example, Fig. 13 depicts the structure of one of the lipids in beef fat, and the structure of one of the lipids in linseed oil, a vegetable product.

A more complex group of lipids, known as the phospholipids, are essential to the construction of biological membranes. These molecules contain the element phosphorus. One of the phospholipids, lecithin, is a component of most animal tissues. It is made from two molecules of fatty acid and a nitrogen-containing molecule called choline that is linked to the basic three-carbon alcohol skeleton by a phosphate bridge (Fig. 14). Numerous other kinds of phospholipids are possible, depending upon the

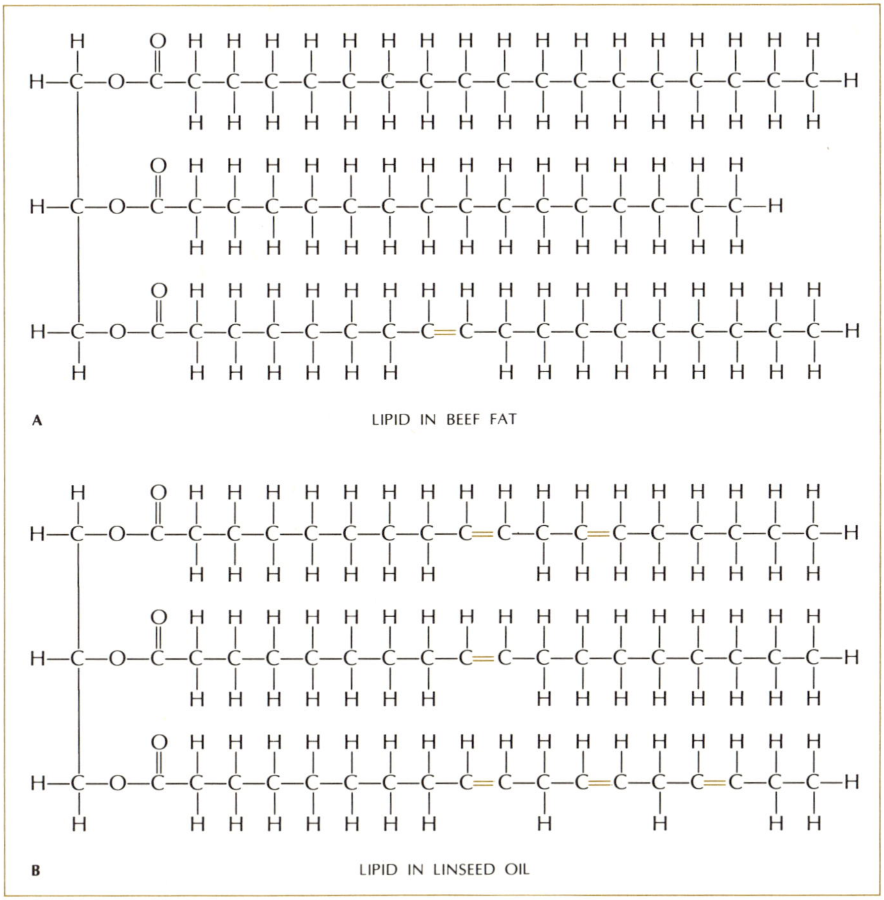

13

Two-lipid molecules. Lipid from beef fat contains mostly saturated hydrocarbon chains. Lipid from linseed oil contains numerous double bonds. Note the characteristic lipid structure: three fatty acids linked to a three-carbon alcohol skeleton.

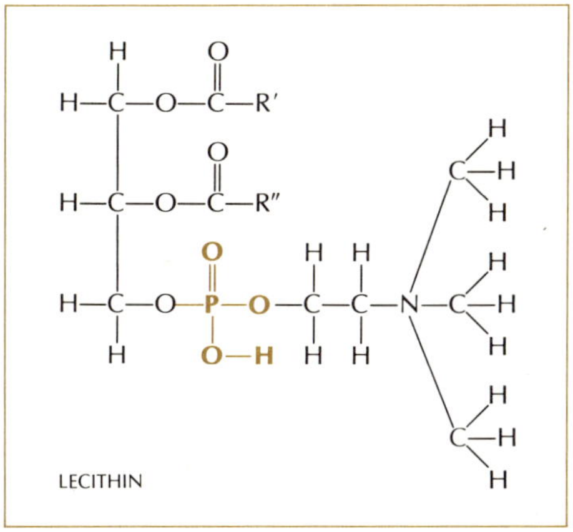

14

Lecithin is a common phospholipid from animal tissue. Like a lipid molecule, it has fatty acid side chains (here abbreviated R' and R") attached to the alcohol skeleton. In lecithin, the third fatty acid has been replaced by a choline molecule (lower right) linked to the alcohol skeleton by a phosphate bridge (color).

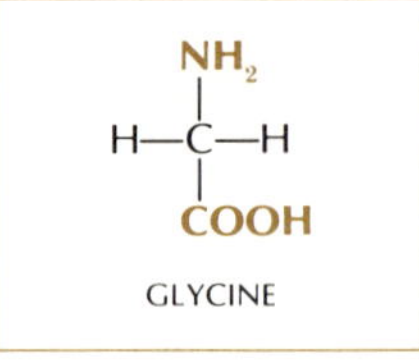

15

Glycine, simplest amino acid, consists of a carbon atom bonded to an amino group and a carboxyl group.

makeup of the section of the molecule attached to the phosphate bridge and upon the number and kinds of fatty acids attached to the alcohol skeleton.

PROTEINS

Proteins are important in the functioning of living organisms in many different ways. They serve as structural components of the cell wall and as a food storage product in many plant cells; as a contractile component of muscle fibers in animals; as a structural part of bone, tendons, cartilage, ligaments, skin, and hair; and as a part of the immunologic disease defence system of higher animals. But their most important contribution to all life is their role as catalysts: compounds that speed up chemical reactions. In the language of the chemist, enzymes are organic catalysts.

Sterile solutions of the sugars sucrose and glucose are quite stable and can be kept for months without apparent change. In reality a few of the sugar molecules are degraded in this time, but the number is so small that the change is extremely difficult to detect. However, if bacteria or fungi are introduced into the solution, the sugars are quickly absorbed by the cells, broken down, and used as an energy source for the growth of these microorganisms. It is the enzymes in these organisms that increase the speed of the chemical breakdown of the sugar molecules to a rate sufficient to maintain the life of the cells. Living systems as we know them simply could not exist without enzymes. Thousands of crucial reactions, from the splitting of cellulose to the clotting of blood, are catalyzed by enzymes. Their diverse functions in biological activity will be discussed frequently in later chapters.

All enzymes are proteins, but all proteins are not enzymes. Most proteins are enormous molecules with molecular weights running into the millions. They are synthesized from smaller units called amino acids. The simplest amino acid, glycine, consists of a carboxyl group and an amino group (NH_2) bonded to a single carbon atom (Fig. 15). In living systems there are about 20 common amino acids. All have at least one carboxyl and one amino group, but differ in the number and kinds of atoms attached to these groups. Table II illustrates the main differences among amino acids.

II

Amino acids all have an amino group
(NH_2) and a carboxyl group (COOH)
bonded to a single carbon (color).
This table presents the name, abbre-
viation, and structure of six of the 20
amino acids found in living organisms.

glycine (gly)

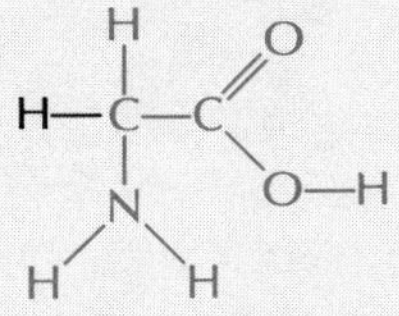

alanine (ala)

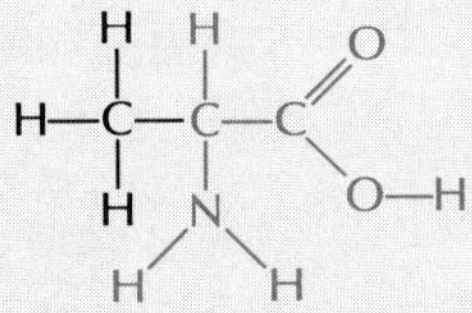

leucine (leu)

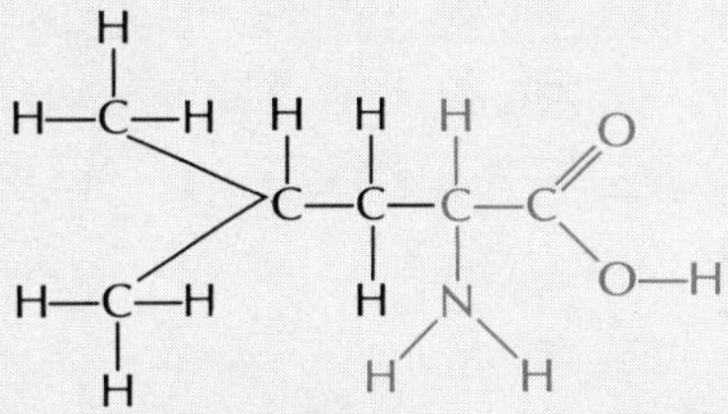

tryptophan (try)

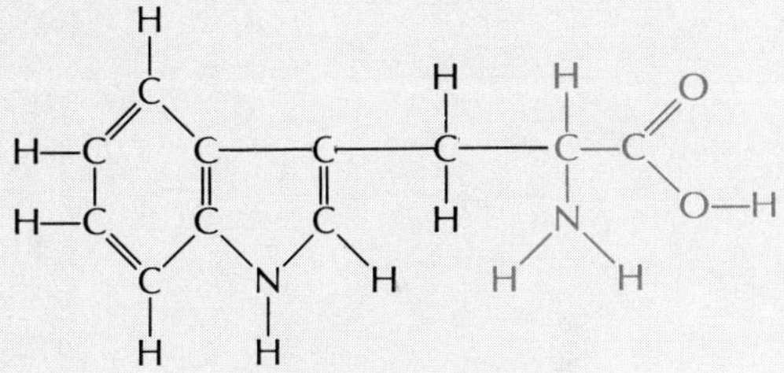

arginine (arg)

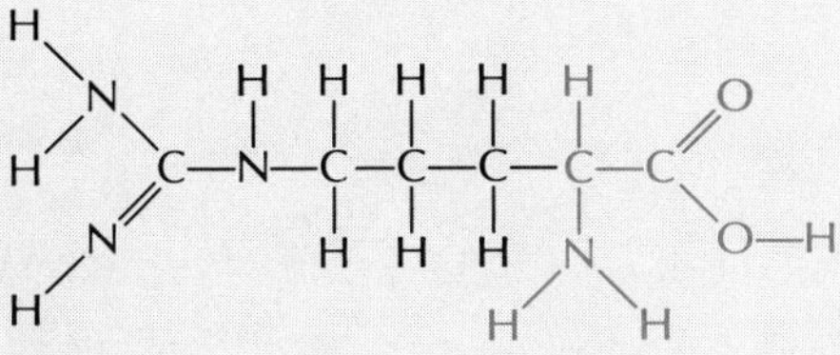

cystine (cys)$_2$

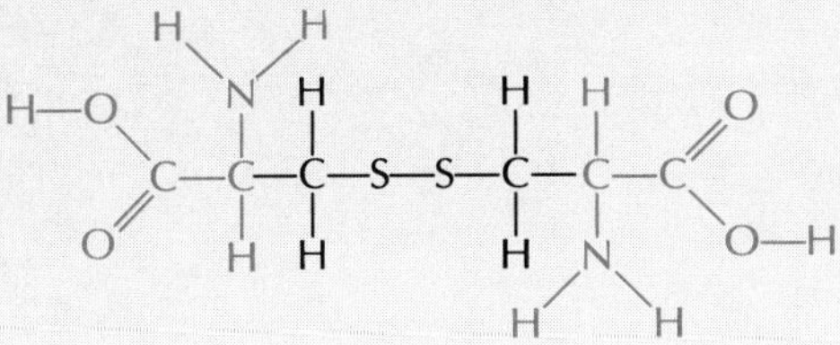

The amino end of one amino acid can be linked to the carboxyl end of another by splitting out a molecule of water (Fig. 16). This linkage is called a peptide bond. Two amino acids joined together in this way form a dipeptide. When more amino acids are linked, the product is a polypeptide. The process of linking many similar units to form a long-chain molecule is called polymerization, and the resulting compound is called a polymer. Proteins are polymers of amino acids, just as starch, cellulose, and glycogen are polymers of glucose.

In the synthesis of proteins it is possible for any of the 20 amino acids to become linked to any other amino acid. Proteins differ from one another in the number and sequence of amino acids in the polypeptide chain. Since each protein contains anywhere from 100 to more than 1000 amino acid units, the possible number of proteins is almost infinite.

NUCLEIC ACIDS

The discovery of the structure of nucleic acids was one of the most exciting advances in biological research in the last three decades. The name nucleic acid resulted from the discovery of these compounds in the nucleus of the cell, and their importance lies in their role in heredity. Nucleic acids are polymers built up from simpler units called nucleotides. Each nucleotide, in turn, is composed of three parts: a sugar, a phosphate group, and an organic "base."

Chemically, a base is the opposite of an acid. An acid is a proton donor; it dissociates to yield a positively charged hydrogen ion. Conversely, a base is a proton acceptor. If the gas ammonia is bubbled through water, the ammonia molecules attract protons from the water molecules (Fig. 17). Two kinds of ions result: the hydroxyl ion (OH^-) and the ammonium ion (NH_4^+). Here the ammonia molecule acts as a base and accepts a proton, while the water molecule acts as an acid and donates a proton.

The chemical structures of the organic bases found in nucleotides are illustrated in Fig. 18. Each of them contains at least one nitrogen atom which, like the nitrogen in ammonia, can accept a proton.

In all nucleotides the phosphate group is always the same. It is attached to a special sugar (Fig. 19). In RNA (ribonucleic acid) the sugar is ribose; in DNA (deoxyribonucleic acid) the sugar is deoxyribose. The gen-

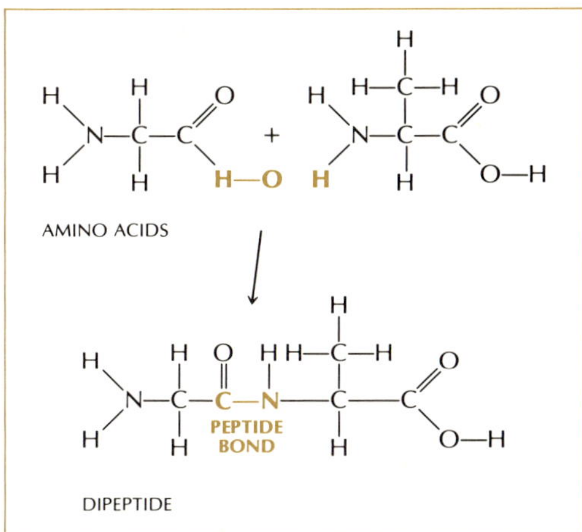

16

Peptide bond links amino acids in protein molecules. Bond is formed between carboxyl group of one amino acid and the amino group of the other, with the splitting out of a molecule of water (color).

17

A base is a proton acceptor. Here ammonia gas and water yield a negatively charged hydroxyl ion and a positively charged ammonium ion.

18

Organic bases are found in nucleotides. Nitrogen atom (color) can accept a proton and act as base. Five common organic bases shown here occur in the genetic molecules of the cell. In the simplified formulas given here, the symbol for the carbon atoms in the ring structures has been omitted.

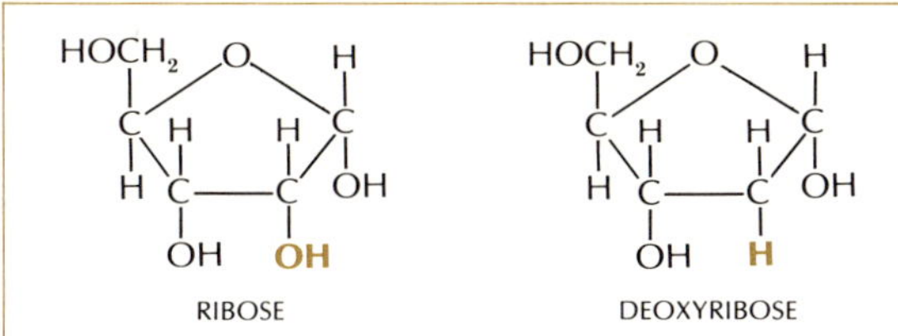

19

Sugars in nucleotides are ribose and deoxyribose. Deoxyribose has one less oxygen atom than ribose. Each has five carbon atoms.

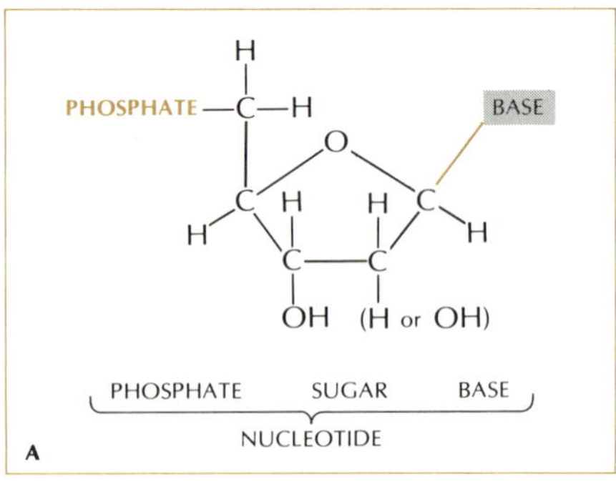

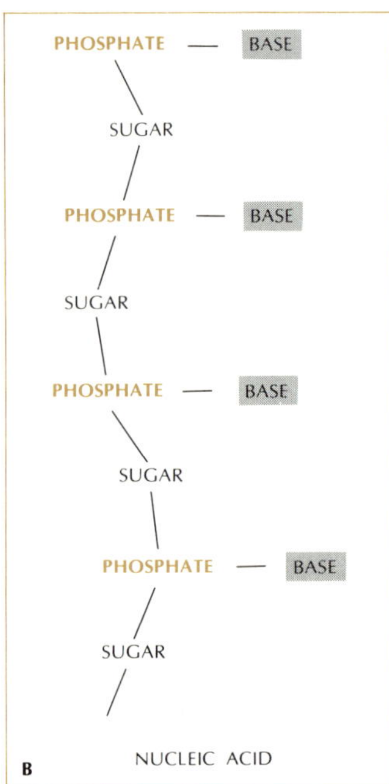

20
Nucleotide (A) is composed of a phosphate group (color), a sugar unit (black) and an organic base. Sugar-phosphate bonds link nucleotides into a nucleic acid strand (B).

eralized structure of all nucleotides can be written as shown in Fig. 20. The same illustration shows how the nucleotide units are linked by the phosphate and sugar portions to form the backbone of the nucleic acid chain. A detailed account of the structure of nucleic acids, their synthesis, and their importance in heredity appears in later chapters.

ADENOSINE PHOSPHATES

All compounds contain chemical energy. According to the Second Law of Thermodynamics, energy is required to link two randomly located atoms into a specific, ordered arrangement. This energy is "trapped" in the chemical bond that holds the atoms together, and is released when the bond is broken.

Adenosine diphosphate (ADP) and adenosine triphosphate (ATP) are called high-energy molecules (Fig. 21). The two terminal phosphate bonds of the ATP molecule behave as if they contain large amounts of energy, although this is not strictly true. The energy is not actually concentrated in the bond itself, but is a result of the positions of the electrons throughout the molecule. Phosphate bonds are called high-energy bonds because much of the energy released when they are broken can be harnessed by living cells to do many different kinds of work. It is biologically useful energy. The energy released by the breaking of some other kinds of chemical bonds is mostly lost as useless heat. It would be more nearly correct to say that the two terminal phosphate bonds of ATP are high, biologically useful energy bonds. Regardless of semantics, ADP and ATP can be regarded as the currency of energy exchange in all living organisms.

21
Adenosine triphosphate (ATP) consists of an adenine unit (right), a ribose unit (center), and three phosphate units (left). ADP, adenosine diphosphate, has one less phosphate group. High-energy phosphate bonds are indicated in color.

READINGS

Baker, J. J. W., and G. E. Allen, *Matter, Energy and Life.* Addison-Wesley, Reading, Mass., 1970 (paper)

> A clearly written introduction to the chemical basis of life processes, including a short introduction to chemistry for the nonchemistry student.

Cheldelin, V. H., and R. W. Newburgh, *The Chemistry of Some Life Processes.* Reinhold, New York, 1964 (paper)

> A rather elementary but concise survey of the basic topics of biochemistry for the biology student.

Dickerson, R. E., and I. Geis, *The Structure and Action of Proteins.* Harper & Row, New York, 1969 (paper)

> An advanced treatment of these remarkably complex giant molecules. The writing is lucid and the illustrations are superb. A challenge, but worth it.

Harper, H. A., *Review of Physiological Chemistry,* 13th ed. Lange Medical Publications, Los Altos, Calif., 1971 (paper)

> The *Whole Earth Catalog* recommends some of the Lange books to do-it-yourself "doctors" in communes. We recommend this one for similar reasons: if you want to look up some additional information about vitamins or hormones or the chemistry of disease and drugs, this paperback presents it briefly, clearly, and with a minimum of technical jargon.

Pauling, L., and R. Hayward, *The Architecture of Molecules,* Freeman, San Francisco, 1964

> One author holds the Nobel Prize in chemistry; the other is an illustrator whose work appears regularly in *Scientific American.* Full color drawings of various inorganic and organic molecules, with short paragraphs of text that cover topics from the periodic table to molecular abnormalities that cause human disease.

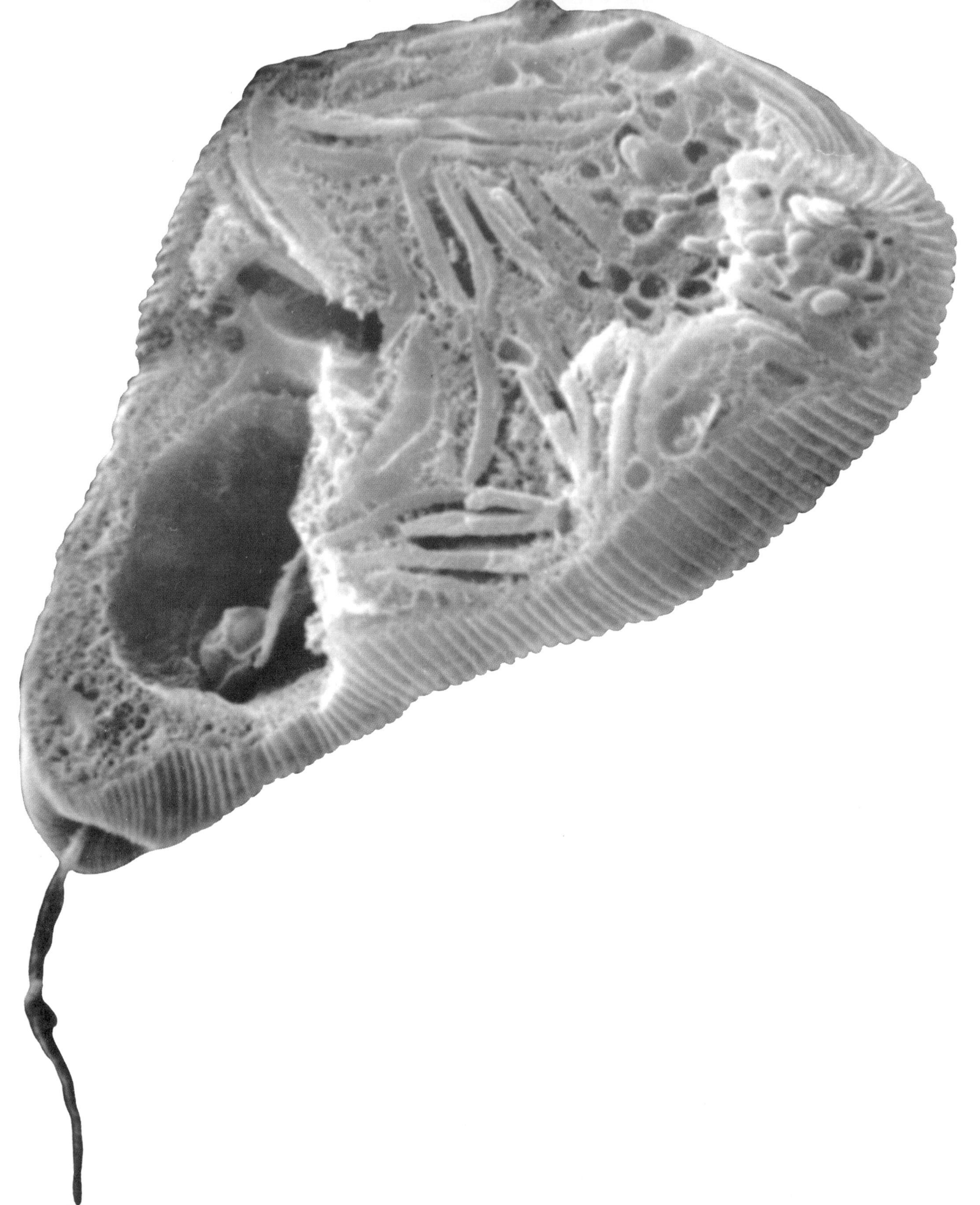

4. THE STRUCTURE OF CELLS

More than a century has passed since Schleiden and Schwann first proposed the cell as the basic unit of life on earth. During those years, research on the cell has thrust in two main directions. Microscopists have explored its structure, and biochemists have traced its functions. This work has confirmed and broadened the early ideas of the cell theory. Today we see the cell as a container and a machine: a container bounded by a membrane that selectively controls the entry and exit of ions, atoms, and molecules; and a machine that carries out all the activities of life. Each cell, to a greater or lesser degree, can live independently. Cells are organisms, and multicellular organisms are communities of cells. All the activities of earthly organisms depend on processes that occur at the cellular level; a few additional properties are derived from the aggregation of cells into multicellular organisms.

The work of the last hundred years has also led to a generalized model of the cell (Fig. 1). Strictly speaking, such a cell appears only in textbooks. Living cells are of two main types: prokaryotic (primitive nucleus) and eukaryotic (true nucleus). Bacteria and blue-green algae are prokaryotes; their genetic material is not segregated within a membrane-bounded nucleus, and they lack some of the internal structures found in eukaryotic cells. Most of the living matter on earth is made up of eukaryotic cells. Most unicellular algae and unicellular animals are eukaryotes, as are all multicellular plants and animals.

◄**Three-dimensional view of a cell** was made with the scanning electron microscope. The ridged outer "skin" of the protozoan *Euglena gracilis* is cut away, revealing the spongy endoplasmic reticulum within. Wormlike objects near the middle of the cell are chloroplasts, and near them are granules of stored food and other substances. Whiplike flagellum that propels *Euglena* through the water extends from an opening at the rear of the organism. Magnification about 5800 ×.

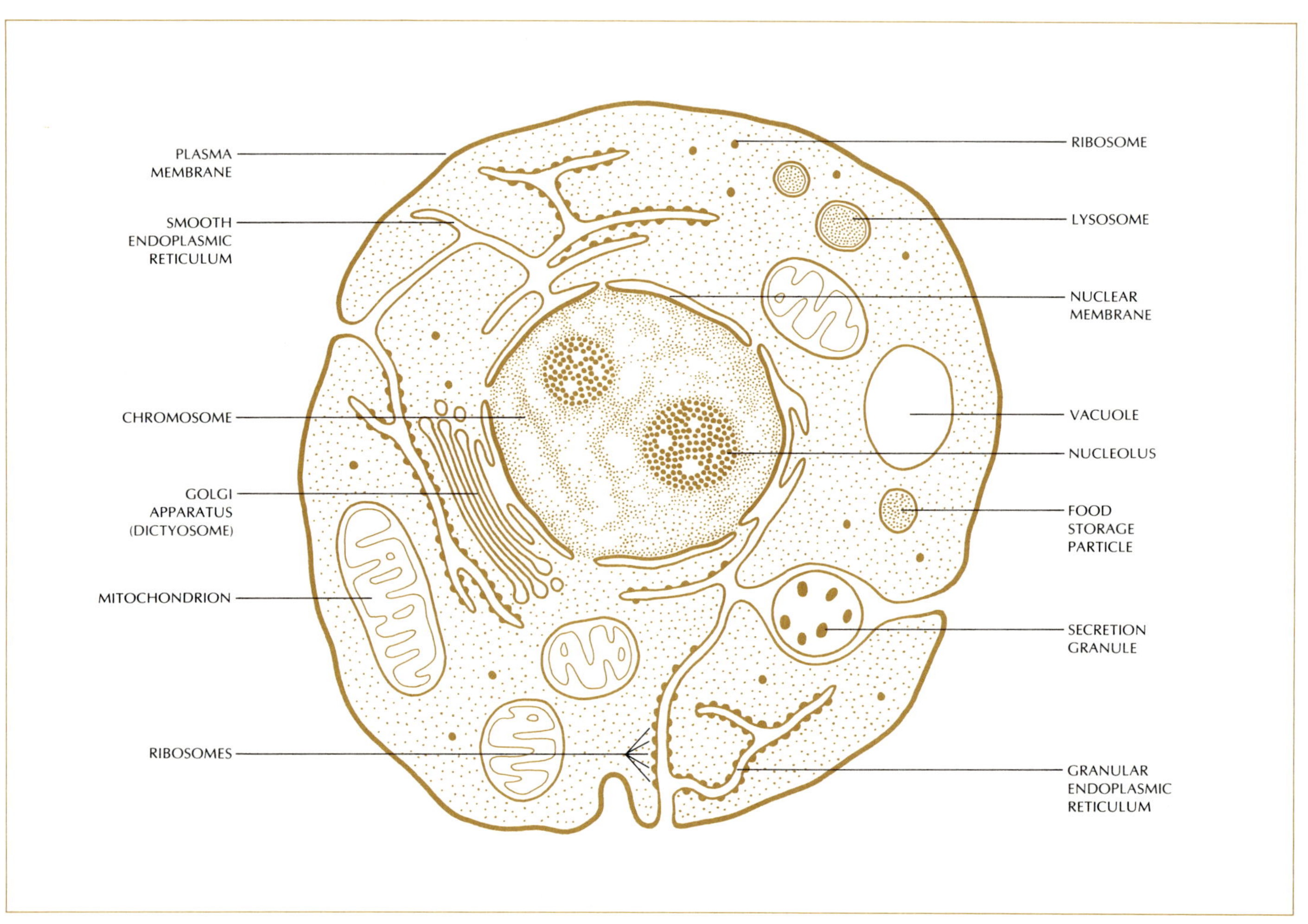

The model cell depicted in Fig. 1 is a eukaryotic cell. In developing the model, biologists have relied heavily on the resolving power of the electron microscope. The light microscope can enlarge a cell about 1000 times, to the apparent diameter of a marble; the electron microscope, more than 100,000 times, to the apparent diameter of a giant beach ball.

1
◀**Anatomy of a cell** is depicted in this schematic
diagram. Labels indicate the principal structures
within the cell.

The cell is a triumph of microminiaturization, an achievement that is
dictated by the requirement that substances must move rapidly from one
site to another within the cell. Cells carry out thousands of chemical reactions per second, and the substances involved in these rapid reactions move
by diffusion. This is a very slow process; a molecule may require hours to
diffuse even one centimeter. Diffusion can match the tempo of cell chemistry only over very short distances, say a few microns. Thus diffusion effectively imposes an upper limit on the size of the cell. This limit applies
even to giant cells, such as a hen's egg or a nerve fiber. The chemically
active part of the egg, the part that will become the future embryo, is a
microscopic disk on the surface of the yolk; the rest of the egg is merely
stored food. The nerve fiber is actually a long tube with a diameter about
that of an average cell; substances that diffuse inward and outward travel
very short distances.

The microminiaturization of the cell derives from the same design
"strategy" that engineers now apply to the design of electronic circuits:
reducing functional components to the molecular level. Into tiny wafers
of semiconducting material, usually silicon, circuit designers can now incorporate up to 30,000 transistors and capacitors per square centimeter
(about the area of a man's fingernail). The components of cells are thousands of times more compact than the components illustrated in Fig. 2.

In terms of the biochemical tasks that it must perform, the cell approaches the theoretical limit of miniaturization. A simple cell needs at
least a few hundred enzymes, many of them giant molecules. Other giant
molecules store genetic information. The remaining biochemical machinery
occupies a certain volume; so do stored food and water. Based on these
requirements, some workers have calculated the minimum volume of a
cell, and their estimates agree closely with the measured sizes of the smallest
known cells.

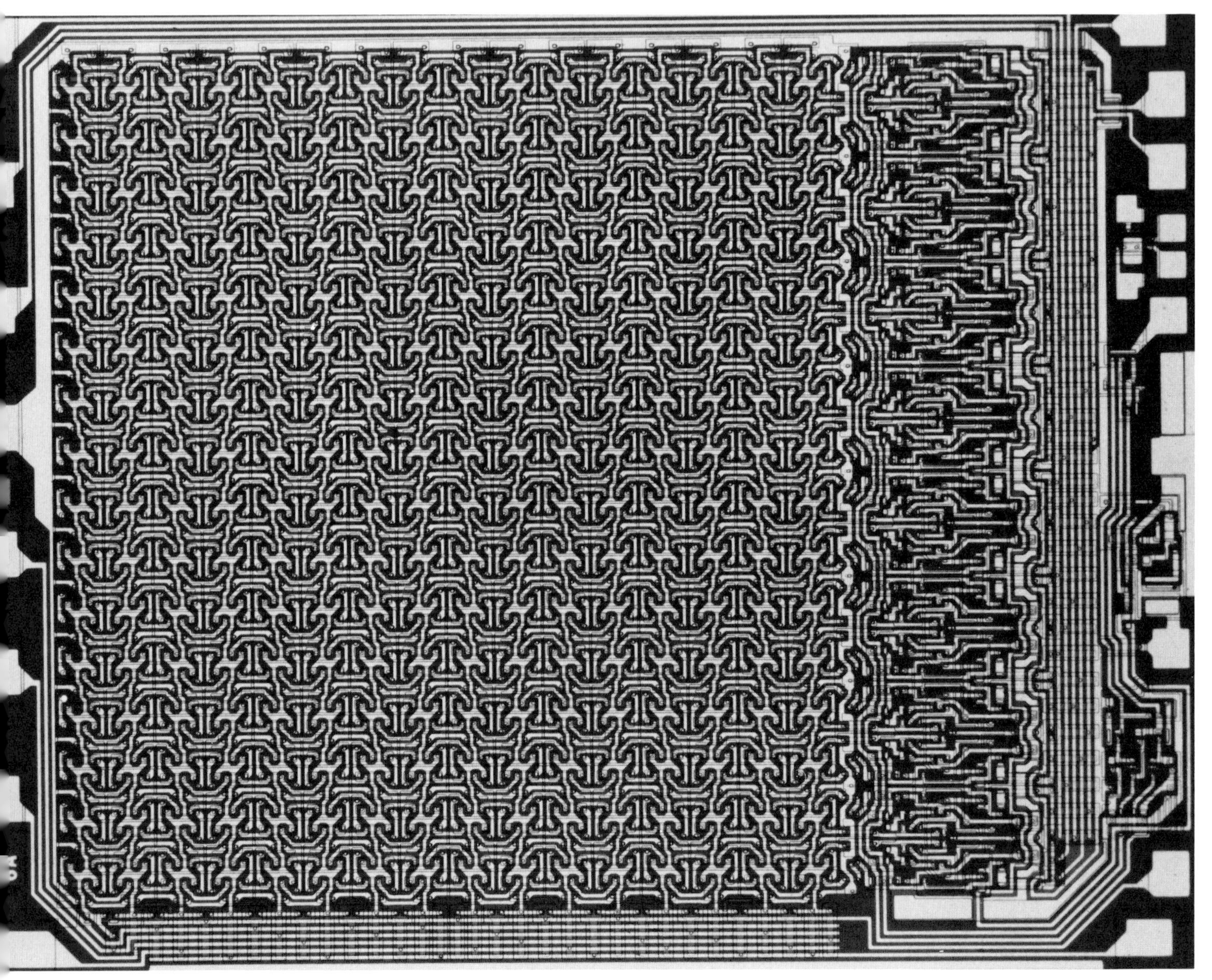

◄**Microminiaturization** of cell can be compared with that of integrated electronic circuits. Computer memory circuit shown here is incorporated into a tiny semiconductor chip that contains almost 1500 transistors, resistors, and diodes. Chip is shown here greatly magnified; its actual size is about that of this black square: ■ Functional units in cell are thousands of times smaller.

THE EUKARYOTIC CELL

As Fig. 1 shows, the eukaryotic cell consists of two main regions, nucleus and cytoplasm. Suspended or dissolved in the gellike sap inside the cell are a variety of chemical substances: ions, compounds, granules, and droplets. One or more sacs (vacuoles) may be found in the cytoplasm. These contain a watery solution of sugars and salts, plus fats, proteins, and other compounds.

At the limit of its resolving power, the light microscope reveals the presence, in both nucleus and cytoplasm, of an assortment of subcellular structures which are collectively named organelles. The electron microscope discloses the so-called "fine structure" of the cell. Perhaps the most striking discovery was that the cytoplasm is not simply an amorphous sap, but consists of an elaborate system of organelles: tubes, channels, plates, bags, and particles. As each was discovered it was named, but only later, after the development of painstaking techniques for separating them from the cell and from each other, did the function of each type of organelle become clear.

THE CELL MEMBRANE

The cell membrane is the gateway to the cell. It is a very thin skin, about one hundred-thousandth of a millimeter thick. If the cell were enlarged to the size of a beach ball, the cell membrane would be about as thick as the skin of a balloon. Through it flows a two-way traffic of ions and molecules entering and leaving the cell. The membrane governs the flow of this traffic. Some substances can traverse the membrane easily, some with varying degrees of difficulty, and some not at all. Some substances can cross the membrane easily in both directions, some pass the membrane more easily in one direction than in the other. In other words, the membrane is differentially permeable to different molecules.

The membrane can play either a passive or an active role in controlling the movement of ions and molecules into and out of the cell. In passive transport, the membrane acts rather like a tight sieve, and particles cross it on their own energy. In active transport, chemical energy from within the cell supplies the energy to "pump" particles across the cell membrane.

DIFFUSION

The process of passive transport involves diffusion. In the language of physics, diffusion is the net movement of molecules by their own kinetic energy from a region where they are in high concentration to one where they are in lower concentration. The key words in this formal definition are net movement, concentration, and kinetic energy.

Kinetic energy is the energy of motion. Molecules of all substances are constantly in random motion. Thus physicists speak of the kinetic energy of individual molecules. This energy is increased by adding heat, or reduced by subtracting heat. At the temperature of absolute zero ($-273°C$) the motion of molecules stops, and their kinetic energy is zero. At any temperature above absolute zero, the thermal movement of the individual molecules leads to diffusion. The rate of diffusion depends on temperature: the higher the temperature, the greater the movement of the molecules and the greater the rate of diffusion. The observed direction of diffusion depends on concentration.

Concentration is not a measure of the number of molecules; it is the measure of the number of molecules in a given volume. If 100 grams of sugar solution contain 10 percent (10 grams) glucose, then the concentration of water is 90 percent. If more water is added, the amount and concentration of water are increased. However, the amount of sugar in the solution remains constant, while its concentration is reduced. Table I illustrates the relationship between amount and concentration. Diffusion depends on concentration, not amount. A saucepan full of water contains

Amount and concentration. Concentration of sugar is reduced as more water is added to solution, although the amount of sugar in solution remains constant.

	Amount	Concen-tration	Amount	Concen-tration	Amount	Concen-tration
sugar	10 grams	10%	10 grams	5%	10 grams	2.5%
water	90 grams	90%	180 grams	95%	360 grams	97.5%

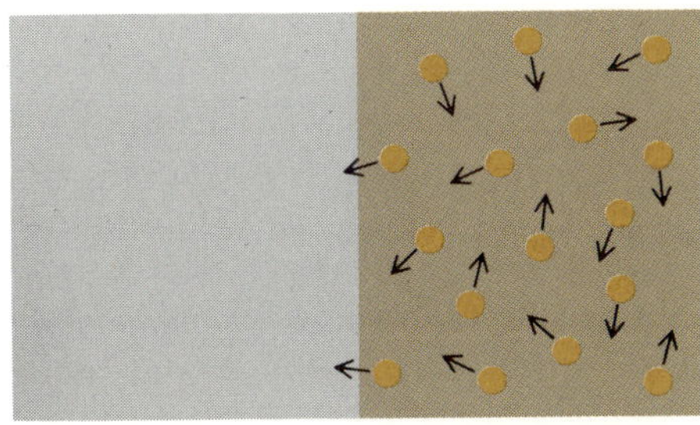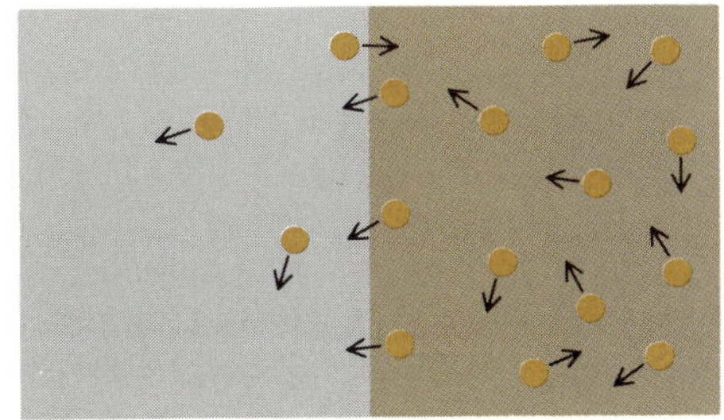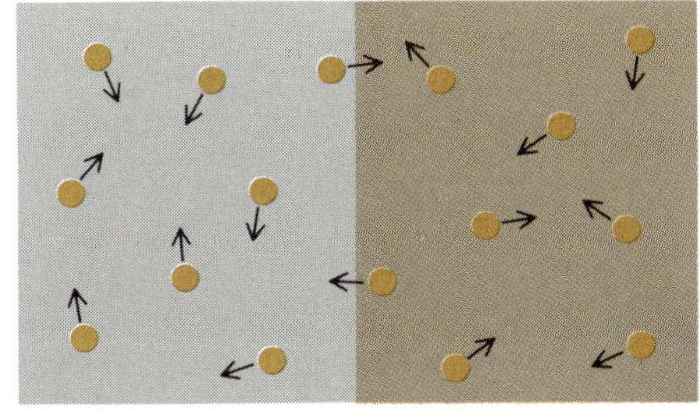

relatively few water molecules compared with the enormous number of water molecules in the air, yet the water diffuses from the pan into the air because the concentration of water molecules in the pan is much higher than the concentration in the air. Molecules always diffuse from a region of higher concentration to a region of lower concentration.

The "net" movement of molecules from one region to another is the total number of molecules moving in one direction minus the number moving in the opposite direction. If the cell membrane acts like a sieve, molecules can move through it in either direction. But if the concentration on one side of the membrane is higher, more molecules will move toward the area of lower concentration than in the reverse direction. When the number of molecules moving in one direction equals the number moving in the opposite direction the net movement is zero. Although molecules are still moving in both directions, the system has reached a dynamic equilibrium in which diffusion no longer occurs. Diffusion is the net movement of molecules. The rate and direction of movement of molecules during diffusion are depicted schematically in Fig. 3.

3
Diffusion drives molecules toward regions of lower concentration. Molecules (colored disks) are continuously moving, as suggested by arrows. Diagrams show two regions separated by a membrane. In diagram at left, concentration of a certain molecule is very high in colored region, zero in grey region. Molecules diffuse through the membrane (center) until concentration in both regions is equal (right).

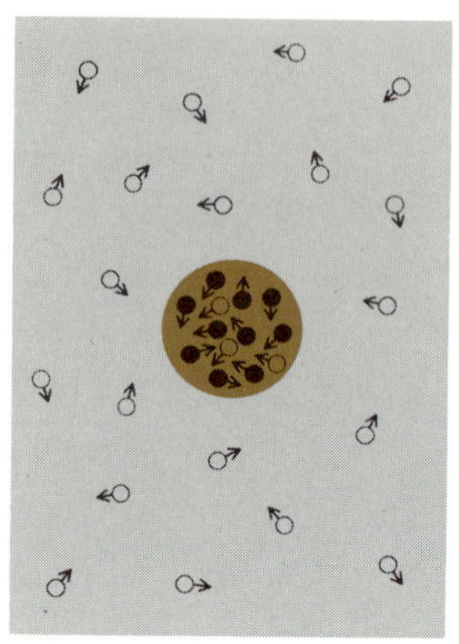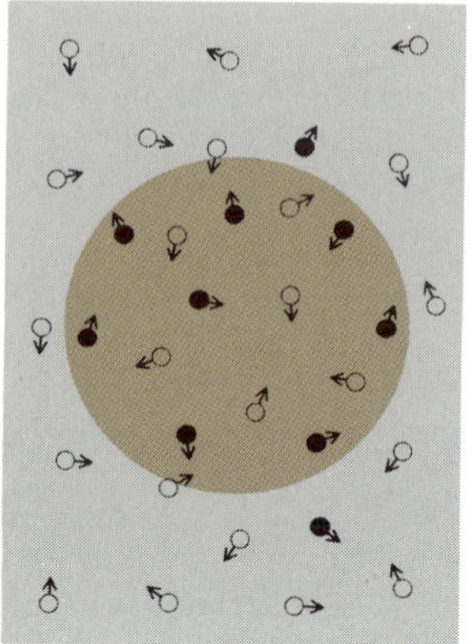

4 Differential permeability. Water molecules (open disks) pass through cellophane more easily than do sugar molecules (black disks). Plastic bag of sugar solution (color) is immersed in a beaker of distilled water (left). Water molecules diffuse into the bag faster than sugar molecules diffuse out. Bag swells (center) and eventually bursts (right).

Each type of molecule diffuses because of its own relative concentration. Thus it is possible for one kind of molecule—water, for example—to diffuse into a cell at the same time that another kind—sugar, for example—diffuses out. This is where the differential permeability of the membrane often intervenes. The effects of differential permeability can be demonstrated by suspending a celluloid bag containing a sugar solution in a beaker containing distilled water. Outside the bag, the concentration of water is virtually 100 percent; inside the bag there is a high concentration of sugar molecules. But the membrane is much more permeable to water than to the sugar molecules. Water diffuses rapidly into the bag, while at the same time sugar molecules diffuse out very slowly. So long as the difference in concentration of water persists, water continues to enter the bag. As long as the difference in concentration of sugar persists, sugar continues to leave the bag. However, since the water diffuses in much more rapidly than the sugar diffuses out, the bag begins to swell, the membrane stretches, and finally the bag bursts under the pressure created (Fig. 4). Living cells may respond in the same manner. If red blood cells, for example, are placed in distilled water, they also rapidly swell and burst.

Diffusion is a consequence of the Second Law of Thermodynamics, which states that the entropy or randomness of a system tends to increase spontaneously. If the concentration of a substance is higher on one side of a membrane, the particles of that substance tend to distribute themselves randomly; that is, enough of them move through the membrane to equalize the concentration on both sides. In terms of entropy, the diffusing particles are moving "downhill." The membrane itself is passive: it supplies no energy to effect the diffusion. In fact, the membrane would have to expend energy to prevent it.

ACTIVE TRANSPORT

This is exactly what happens in living cells. In this process, called active transport, the cell membrane acts as a biochemical pump, forcing certain molecules and ions to move "uphill" against a concentration gradient. That is, the molecules move from a region of low concentration to a region of higher concentration. Active transport makes it possible for a cell to accumulate certain types of molecules and ions even though their concentration outside the cell is extremely low.

Plasma, the liquid component of the blood, contains 20 times more sodium than the interior of a red blood cell; the interior of the cell contains about 20 times more potassium than the plasma. The membrane is passively permeable to both types of ion, but active transport enables the cell to accumulate potassium ions from the plasma, and at the same time to secrete sodium ions into it.

Young marine biologists are sometimes introduced to the wonders of active transport by an encounter with the seaweed *Desmarestia*. The novice may drop *Desmarestia* plants into his collecting bucket along with other prized specimens. To his dismay, a few hours later he finds the bucket full of stinking black sludge. *Desmarestia* cells accumulate large quantities of sulfuric acid. Active transport maintains the high concentration of acid inside the plant. But the energy required to drive the machinery of active transport is regenerated only by living cells. When the plant is collected it begins to die, and its cells cease to produce this energy. Active transport stops, and the acid diffuses out of the cells and effectively digests the specimens in the bucket.

There are numerous kinds of active transport systems. One system facilitates the rapid movement of food molecules, such as glucose, into the cell where they can be used as an energy source. Another system actively pumps amino acids into the cell where they are used in the production of proteins. Still other mechanisms remove waste products from the cell. Active transport depends upon special enzymes in the cell membrane, but the exact nature of the transport mechanism has not yet been discovered. Because of its importance in the life of all cells, it is the subject of intensive research.

Part of the explanation of the complex activity of the cell membrane lies in its structure. Under the light microscope the membrane itself cannot be seen, but the electron microscope reveals its fine structure. In many

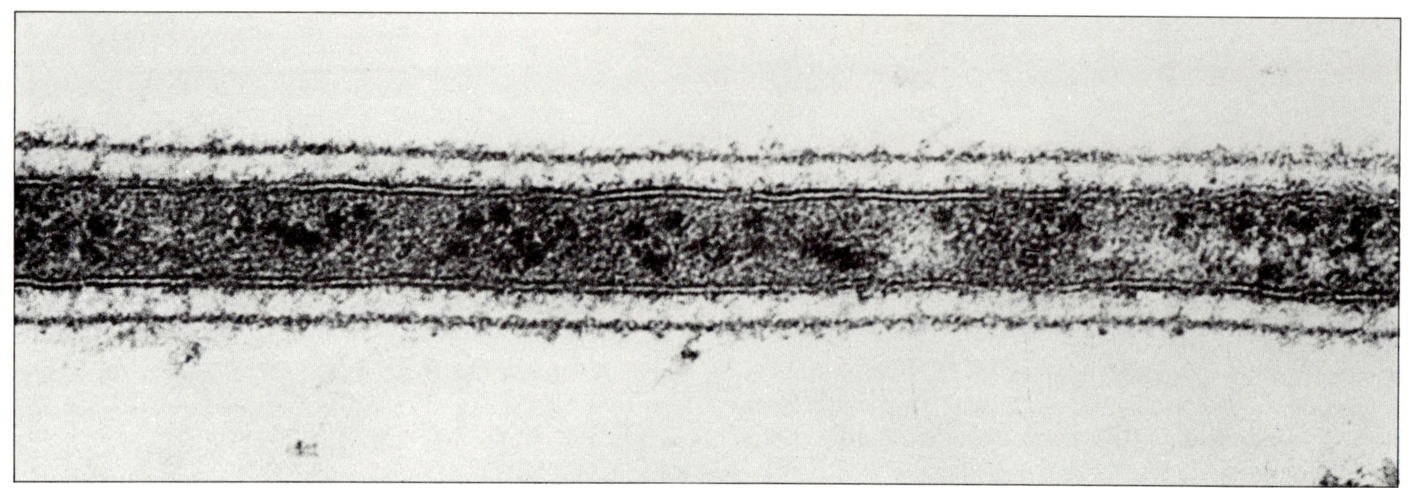

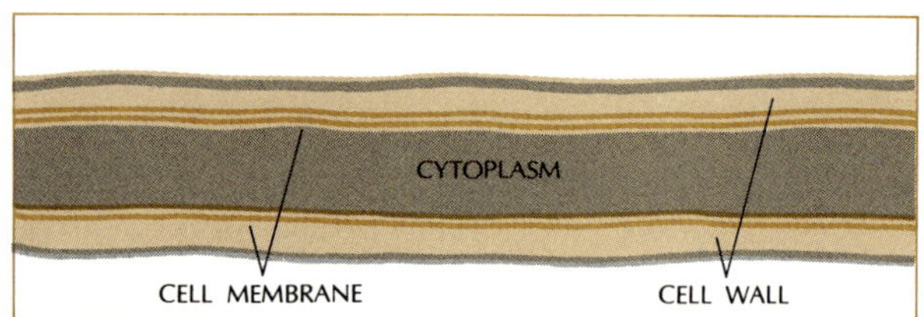

5
Cell membrane appears under the electron microscope as a light grey band sandwiched between two darker bands. This view depicts the cell membrane of a bacterial cell. Light area outside membrane is cell wall. Area between membranes is cytoplasm. Magnification about 160,000 ×.

electron micrographs, the cell membrane shows up as three stripes: a broad light band between two darker ones (Fig. 5). Biochemical analysis has shown that the membrane is made up of lipoprotein: lipid and protein molecules bound together by chemical bonds. Membranes differ from one another in the kinds of lipid and protein that compose them.

At the molecular level, the interpretation of the structure of membranes becomes controversial. One model of the cell membrane is shown in Fig. 6. Two layers of phospholipid molecules (lipids with a phosphate group at one end) are sandwiched between outer layers of protein molecules. According to this model, the two phospholipid layers are attracted to each other because of the fatty acid portions of the molecules. The protein molecules are bonded to the phosphate portions of the phospholipid by electrical forces.

On the basis of recent high power electron micrographs, some workers have inferred that all membranes are composed of a series of repeating units. Experimental evidence suggests that these basic units may consist of clusters of phospholipids and proteins, intimately mixed. The three-dimensional arrangement of these molecules is unknown, but it is thought to be

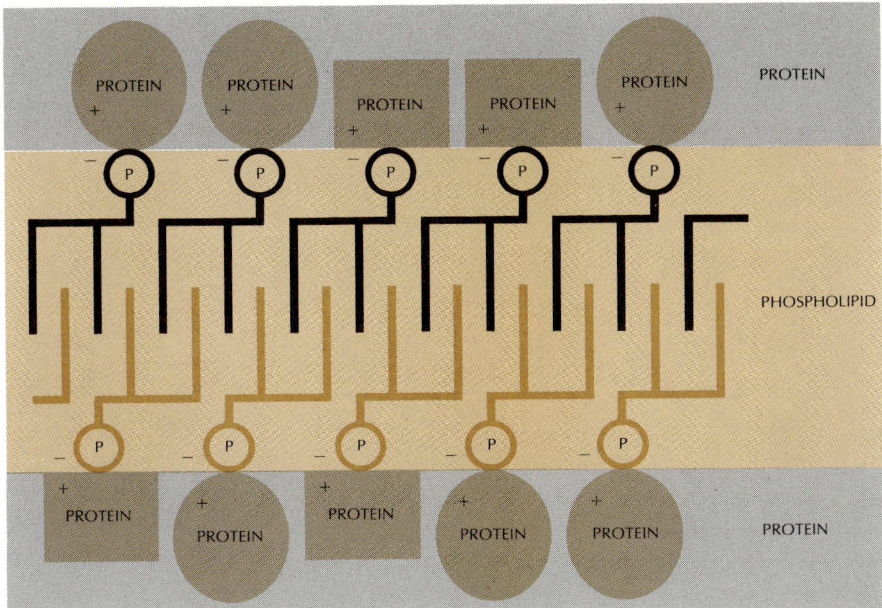

6
Model of cell membrane shows layers of protein molecules (grey)
sandwiching a layer of phospholipid (light color). Negatively charged
heads of phospholipid molecules are bonded to positive sites on
proteins. Hydrocarbon tails of the phospholipid molecules intermesh
(middle).

much more complex than the model shown in Fig. 6. These two conflicting
models are a reminder that in a field as rapidly changing as biology, today's
ideas often become tomorrow's discarded theory. Which of these models
will prevail depends upon further research.

THE NUCLEUS

The nucleus is the control center of the cell. Under the light microscope it
appears as a rather dense, round mass, often lying somewhere toward the
middle of the cell (Fig. 7). Bounded by a membrane, it is filled with a sap
called nucleoplasm. Suspended in the nucleoplasm are long threadlike
organelles called chromosomes (from the Greek words meaning "colored
body," because the chromosomes become deeply stained when the cells
are treated with certain dyes). Made of protein and nucleic acids, the chromosomes direct the life of the cell. The nucleic acids control the synthesis
of proteins, including enzymes. Since almost all biological reactions
depend on one enzyme or another, control of enzyme synthesis gives the

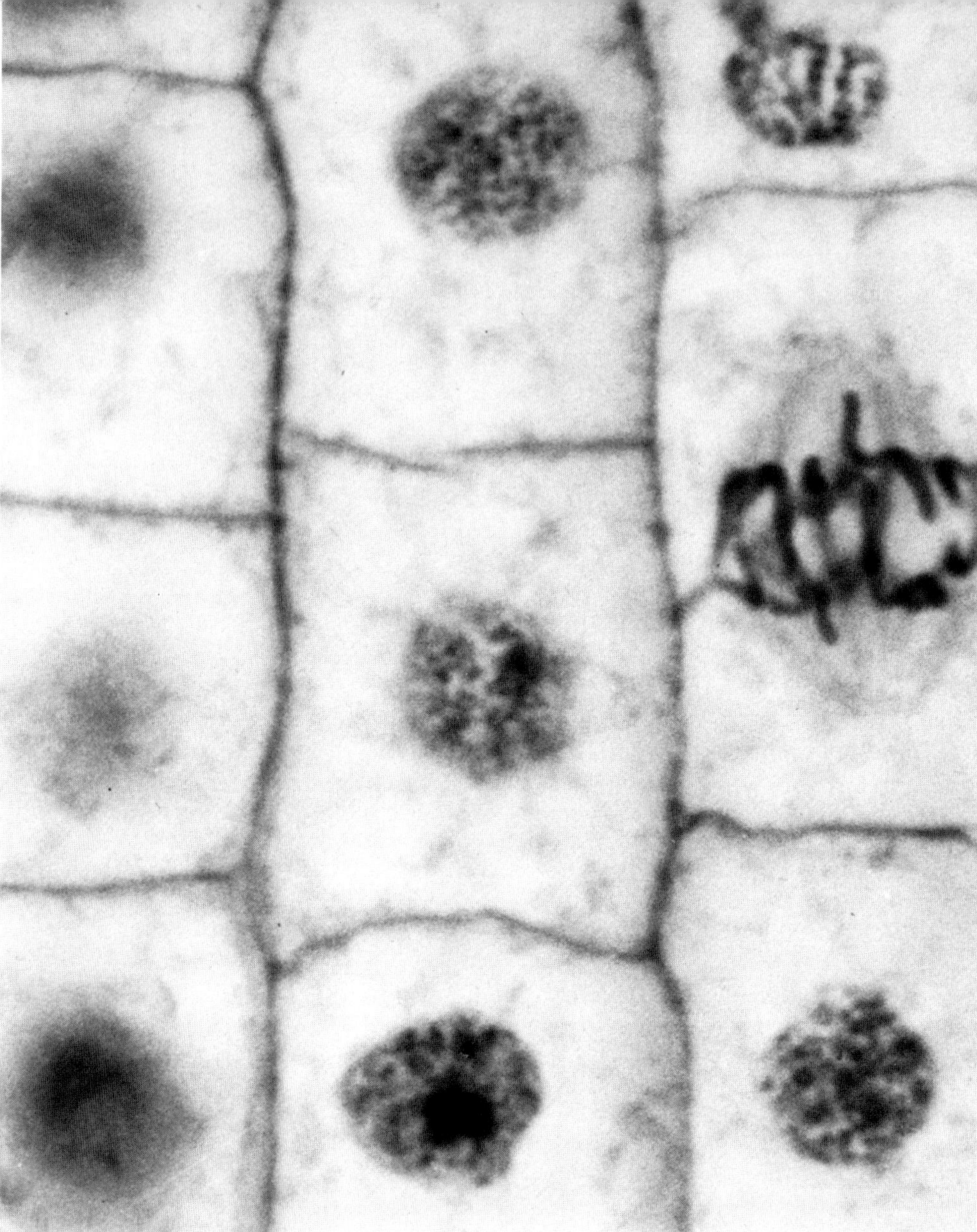

7
Cell nucleus usually appears as a dark round mass in cells stained for viewing under the light microscope. In this photomicrograph of cells from the root of an onion plant, stringy chromosomes are clearly visible in the nucleus of the cell at right. Magnification about 1000 ×.

nucleic acids indirect control of all chemical reactions in the cell. The discovery of the structure and function of nucleic acids is one of the great stories of modern biology, and it is told at some length in later chapters.

The nucleus also contains one or more nucleoli ("little nuclei"). These organelles are tiny globules of protein and nucleic acid. They play a role in the synthesis of small cytoplasmic particles called ribosomes (Fig. 8).

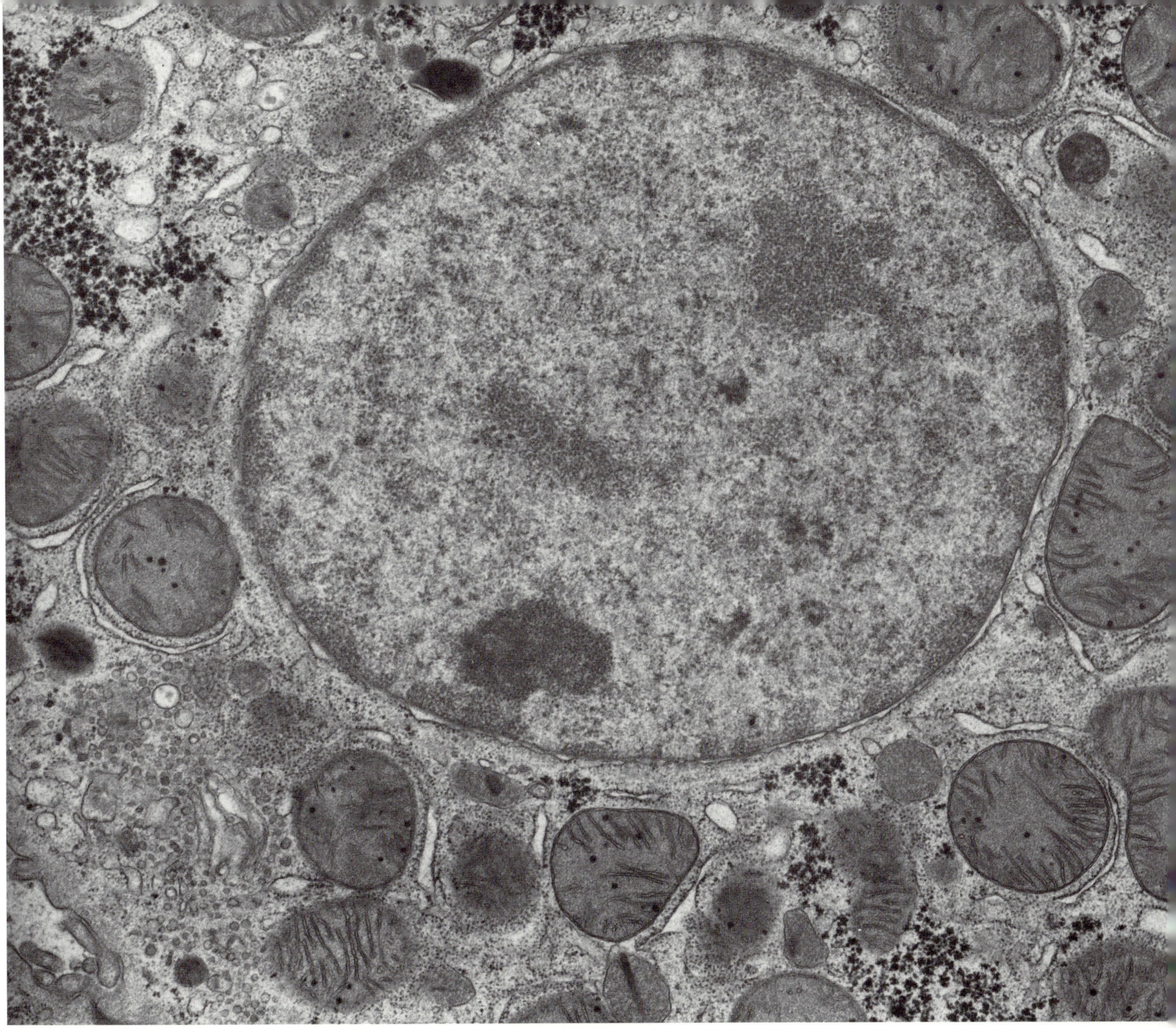

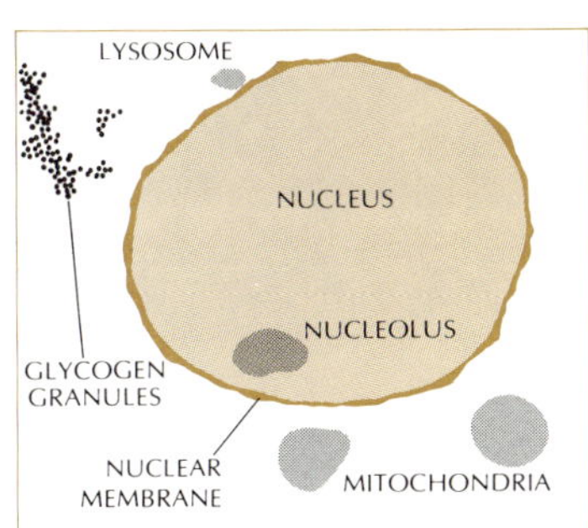

8
Nucleus and nucleolus dominate this electron micrograph of cell from liver of a bat. Note two-layered membrane that surrounds nucleus and the dark glycogen granules at upper left. Magnification about 80,000 ×.

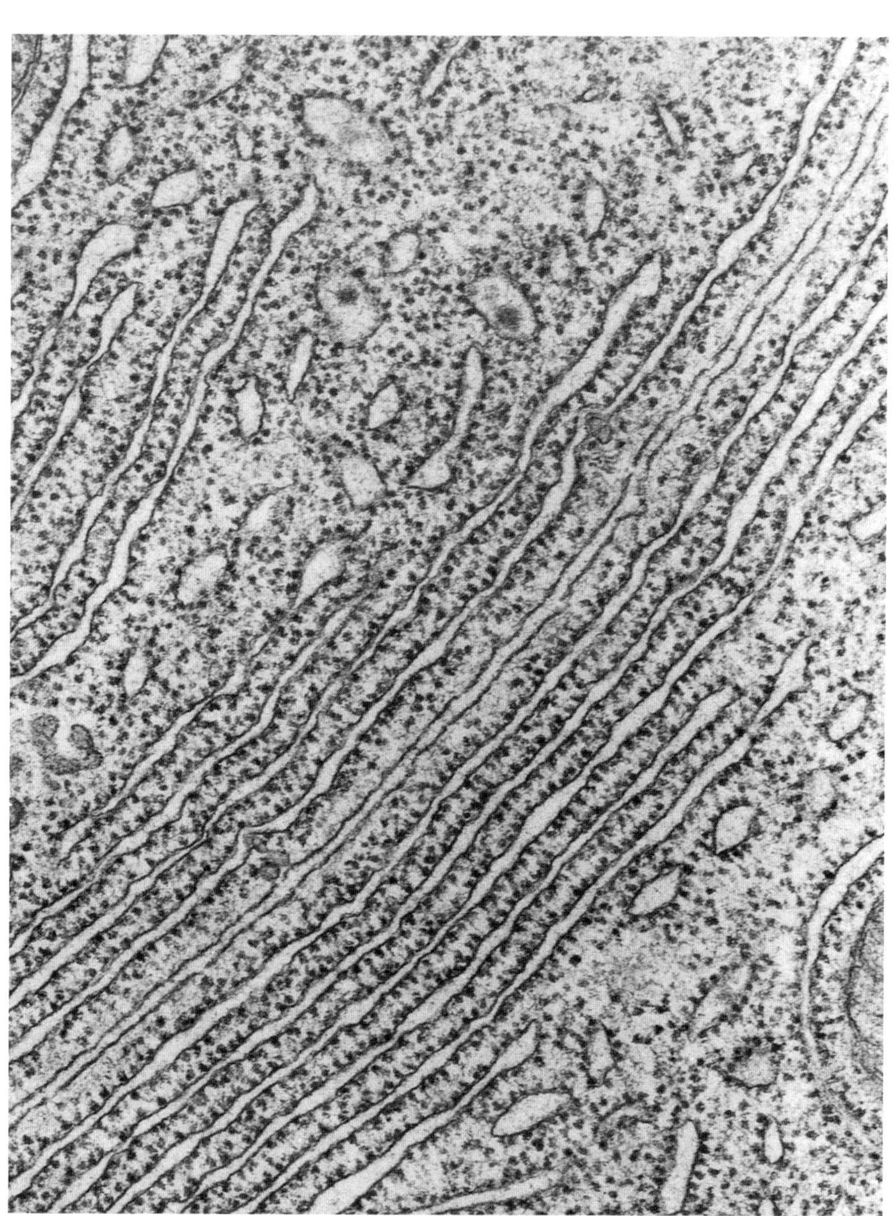

9

Endoplasmic reticulum is network of membrane-lined canals that extends throughout the cell. Dark granules along the membranes are ribosomes. Magnification about 80,000 ×.

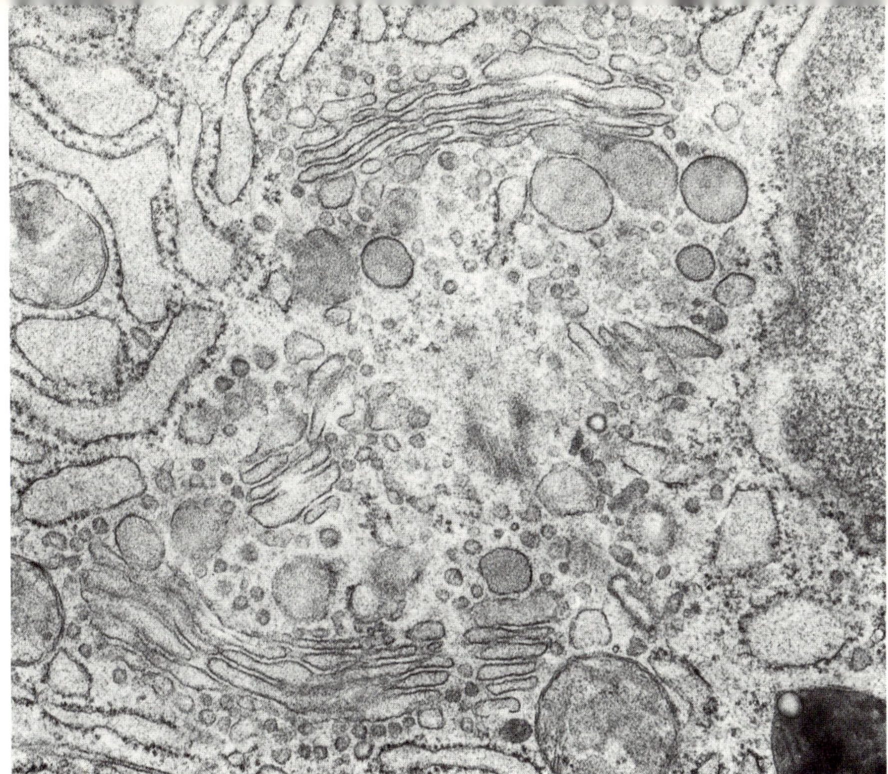

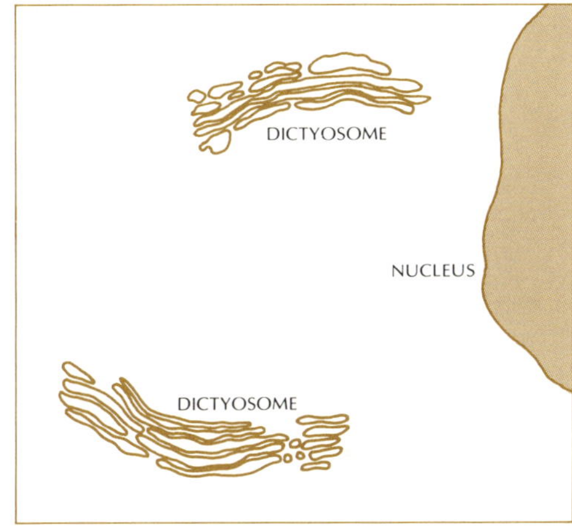

CYTOPLASM

Organelles in the cytoplasm carry out the biochemical directives of the nucleus. But this generalization should not be interpreted too strictly, because the nucleus responds to the feedback of chemical information from the cytoplasm. The two exist as a mutually interdependent partnership.

A labyrinth of membranes branches through the cytoplasm. Known as the endoplasmic reticulum, these membranes are continuous with the cell membrane and the nuclear membrane (Fig. 9). The endoplasmic reticulum serves as a system of canals, enabling substances from outside to penetrate deep into the interior of the cell without first having to cross the rather tight barrier of the cell membrane. The canals also provide a rapid passage to the outside for secretion products, which are synthesized in special membrane structures called dictyosomes or Golgi bodies. Under the electron microscope, dictyosomes often look like a stack of pancakes seen edge-on (Fig. 10). Electron micrographs give the impression that the membranes and the other structures of the cell are rigid and unchanging. In fact, membranes may easily stretch, shrink, rupture, fuse, and grow. The bonds holding the parts of the membrane together are weak.

10
Dictyosomes appear at top center and lower left in electron micrograph of bat plasma cell. Also known as Golgi bodies, dictyosomes are membrane structures that synthesize compounds secreted by cells. Edge of nucleus is visible at far right. Magnification about 80,000 ×.

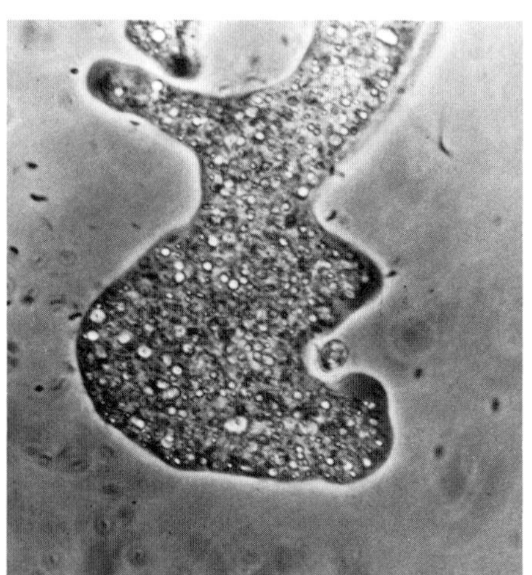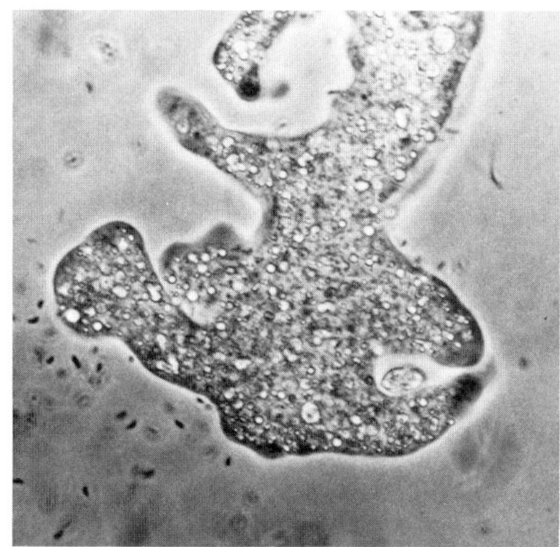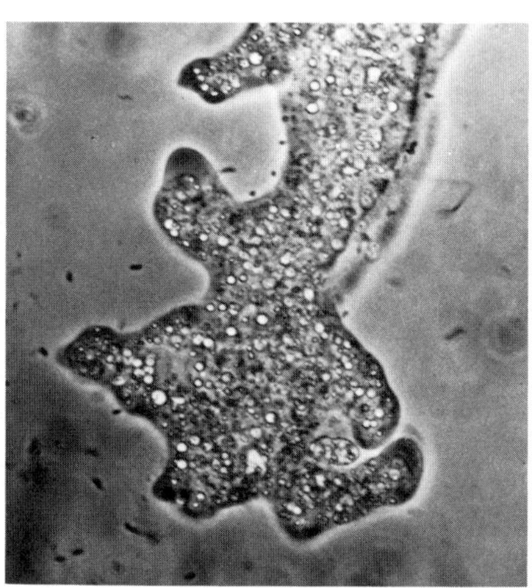

11

Cytoplasmic movement in the protist *Amoeba proteus*. As the *Amoeba* ingests a particle of food, the cytoplasm flows and shifts, rearranging itself until the food is entirely engulfed.

The cell is a dynamic system, constantly in motion. Its structures continually form, divide, disintegrate, and reform. In carrying out its life processes, the cell itself may change shape and its semifluid contents stream from place to place within its boundaries (Fig. 11). But many processes in the cell must take place step by step in a rigidly controlled sequence. This requires a precise arrangement of enzymes within the cell, and the membrane systems act as a scaffolding that maintains this arrangement. Without the order imposed by membrane systems, even the simplest cells could not function.

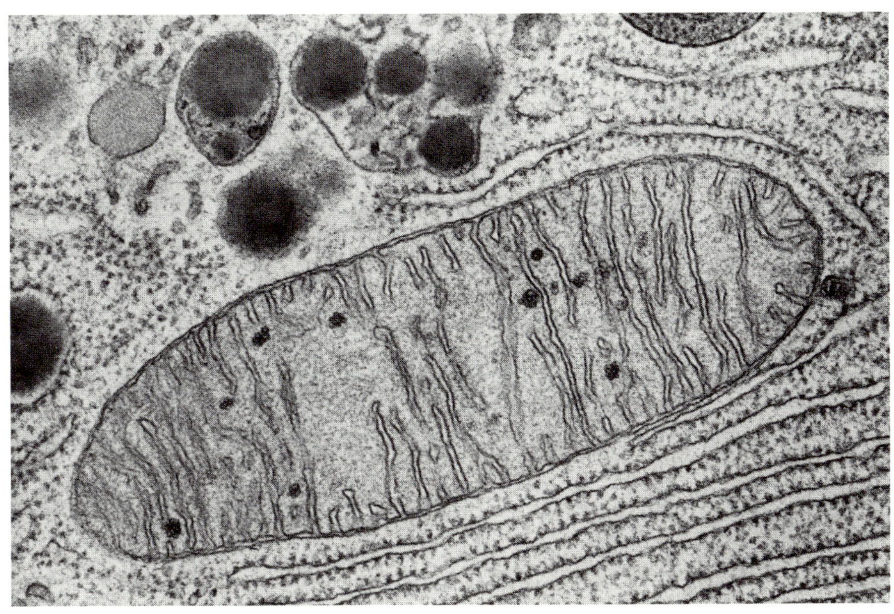
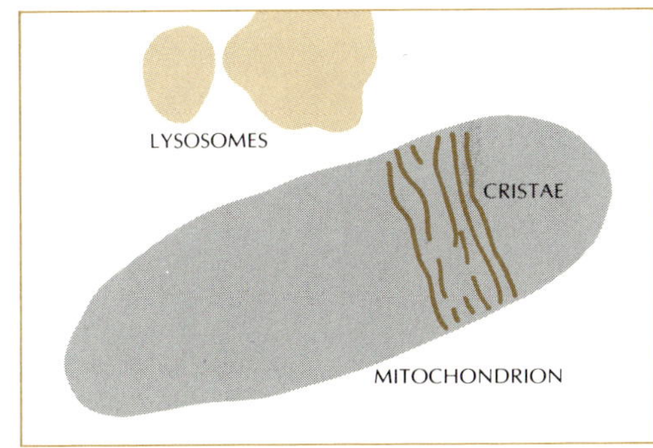

12

Mitochondrion and lysosomes. In electron micrograph of bat pancreas cell, lysosomes appear at top center. Cristae are membrane folds within the mitochondrion.

MITOCHONDRIA

Mitochondria are energy transducers: organelles that convert one form of energy to another. Structurally they consist of two membranes, a continuous outer one surrounding an elaborately folded inner one. These inner folds form structures called cristae that protrude into the interior of the mito-chondrion (Fig. 12). The two membranes differ in the types of enzymes they contain, and thus in the chemical tasks that they perform. The outer membrane contains enzymes that enhance the breakdown of foods into carbon dioxide, with the release of energy. Enzymes of the inner membrane en-hance the conversion of this energy into the energy of ATP.

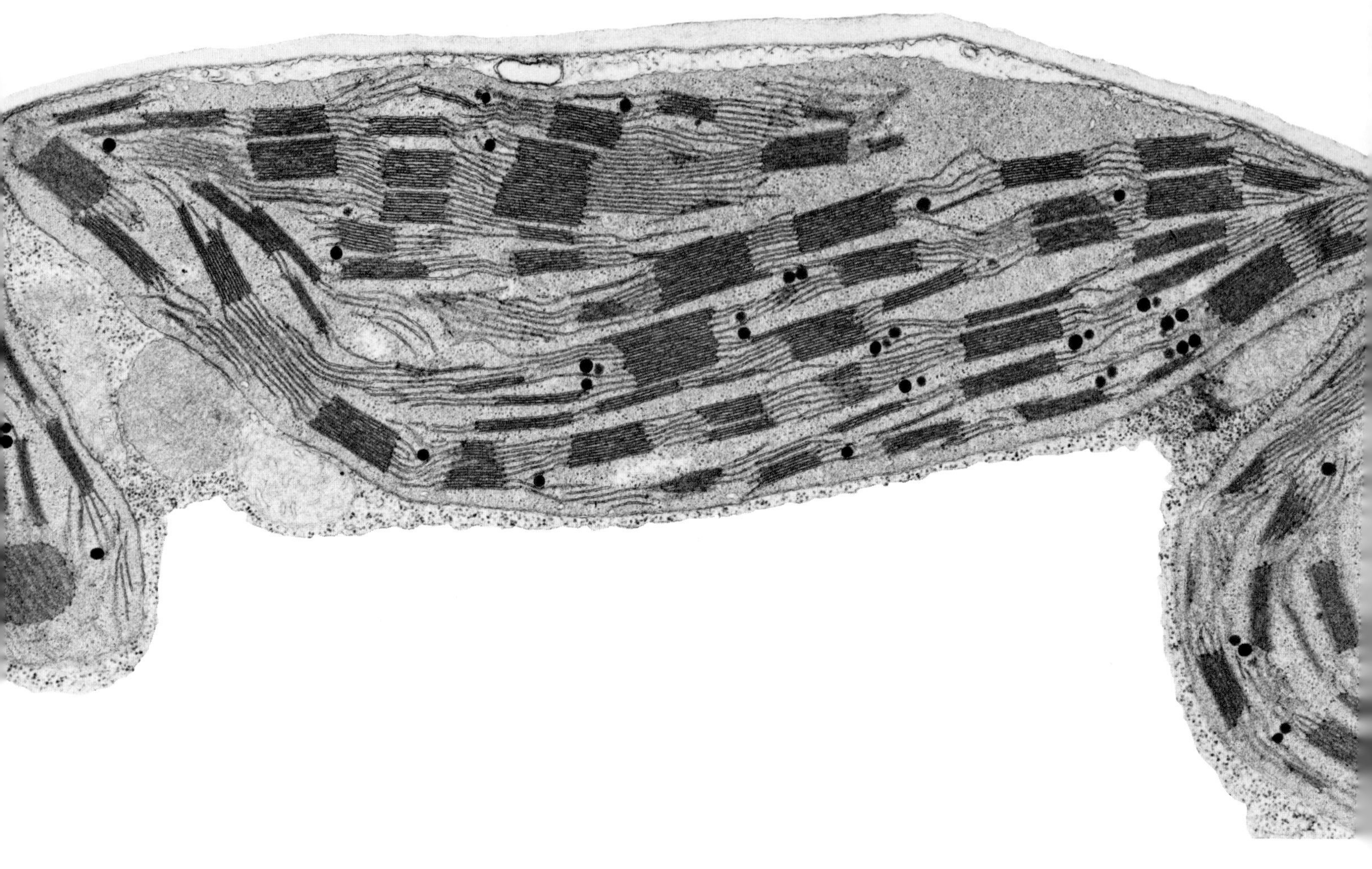

CHLOROPLASTS

In the cells of green plants, chloroplasts act as another kind of energy transducer (Fig. 13). Chloroplasts are the site of photosynthesis. Here light energy is transformed into the chemical energy of organic compounds. This extremely complex process occurs in the internal membranes, where light-trapping molecules such as chlorophyll are intimately associated with enzymatic proteins. The spatial regimentation of specific enzyme molecules on these membranes is essential for the completion of complex reactions that must occur step by step in a rigidly controlled sequence.

13

◄**Chloroplast** converts the energy of sunlight into
chemical energy. This electron micrograph depicts a
mature chloroplast from a blade of timothy grass. The
parallel bands are the membranes which contain the
photosynthetic pigments. Magnification about
15,000 ×.

LYSOSOMES

Many cells, especially animal cells, contain lysosomes, membranous sacs
of digestive enzymes (Fig. 12). When the membranes burst, the enzymes
are released into the cytoplasm. They are important in the chemical breakdown of foods or of foreign substances that have invaded the cell. Experimentally the simultaneous rupture of most of the lysosomes in a cell
leads to the dissolution and death of the cell. The lysosome probably serves
to segregate digestive enzymes and thus to protect the cell from digesting
itself.

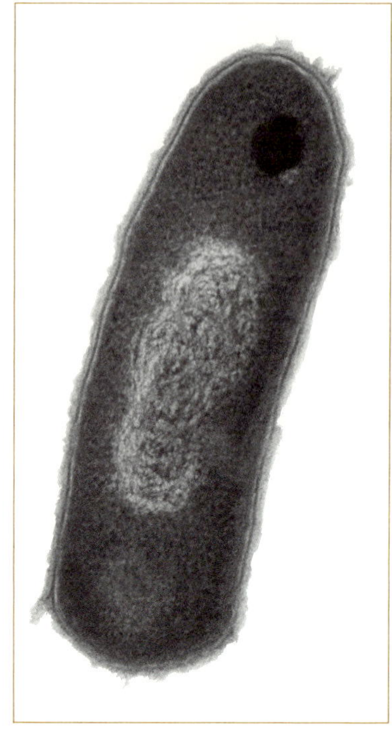 

14
Prokaryotic cells lack nuclei and mitochondria. Left, a bacterial cell. Right, a blue-green alga.

RIBOSOMES

The ribosome is a very small particle composed of protein and nucleic acid (Fig. 9). The name indicates that the nucleic acid contains the simple sugar ribose and therefore is RNA. Large numbers of ribosomes occur free in the cytoplasm. They also cover the surface of some of the membranes of the endoplasmic reticulum. Ribosomes are the site of protein synthesis.

CELL DIVERSITY

The two broad categories of cell—prokaryotic and eukaryotic—encompass a sweeping diversity of cell types. The cells of bacteria and some algae, as we noted earlier, contain no nuclei or mitochondria (Fig. 14). In prokaryotic cells, the functions of these organelles are served by portions of the membrane system.

The internal structure of plant cells follows the pattern of the model cell in Fig. 1, but with certain variations. Plant cells can be distinguished from animal cells by the presence of a thick wall outside the cell membrane (Fig. 15); animal cells have no walls. Moreover, plant cells may contain chloroplasts and starch granules, while animal cells do not.

Functionally, there are important differences between free-living cells and those that form a multicellular organism. Unlike a one-celled animal, for example, a human brain cell must rely on the body to provide it with

15
Plant cell can be distinguished by its thick wall, and
sometimes by the presence of chloroplasts and starch
granules. This picture shows four distinct chloroplasts,
three at left and one at lower right. The lower left
chloroplast contains two starch grains. Nucleus is the
large structure at right center. Magnification about
10,000 ×.

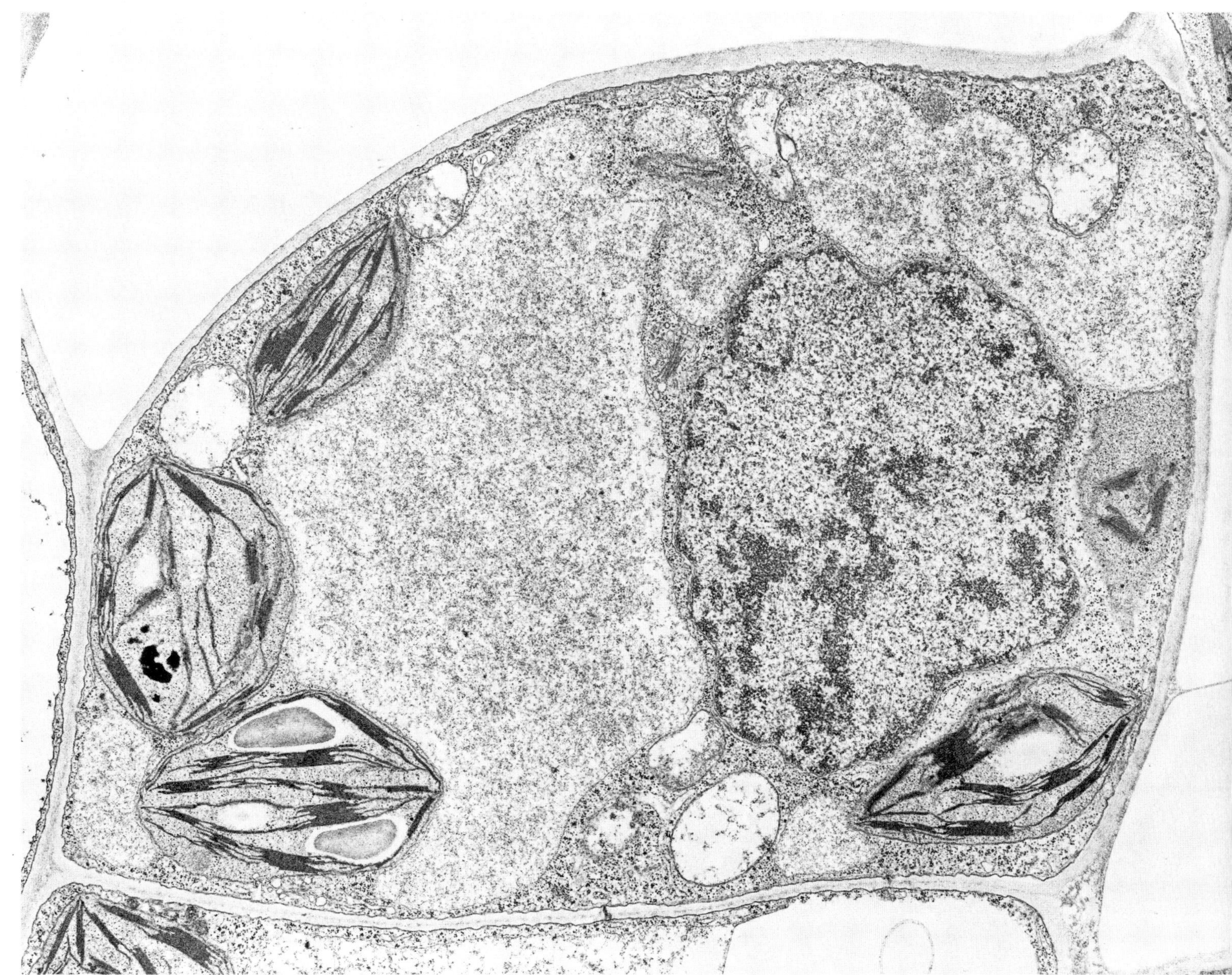

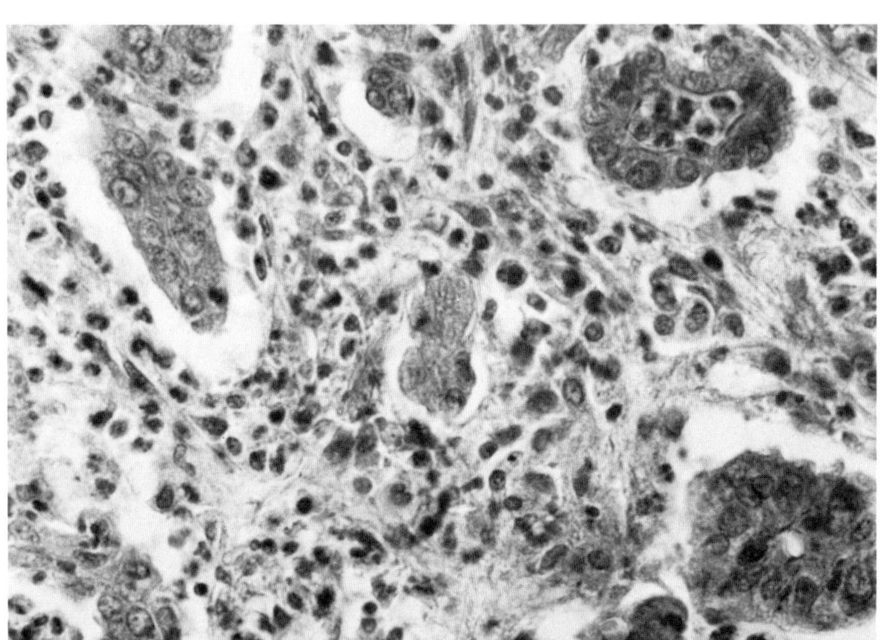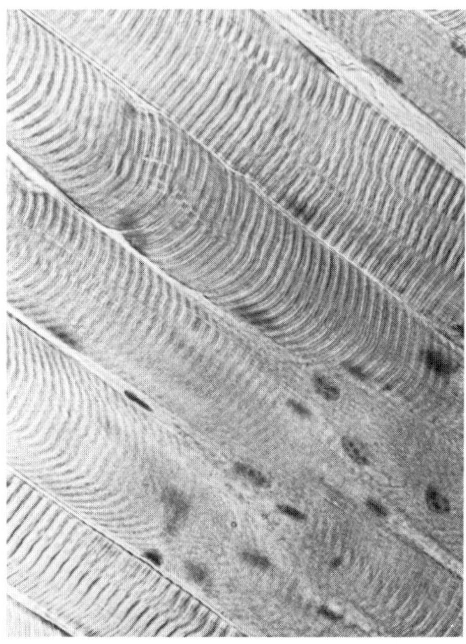

16

Human cells from liver (left) and muscle (right). Liver cells are small and each contains a large nucleus. The skeletal muscle cells are large and elongated with the nuclei lying at the edges of the cells. Magnification about 500 ×.

oxygen, nutrients, and an environment favorable to survival. By itself, the brain cell cannot respond or adjust to changes in the environment; it cannot make all the compounds that it needs; it lacks a complete control system to regulate its chemical machinery. In brief, the brain cell is a specialized part of a specialized organism.

There is considerable diversity in structure among the specialized cells of a single organism; for example, between liver and muscle cells in man, or between the green cells of a leaf and the water-conducting cells of the

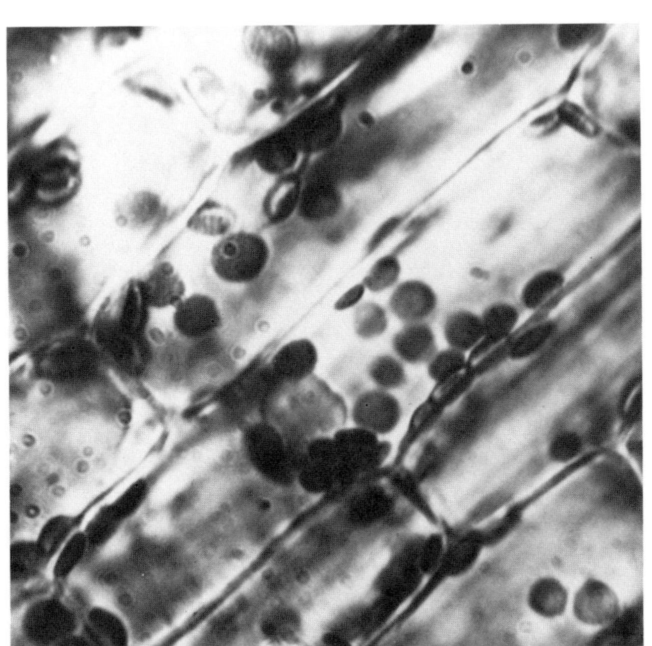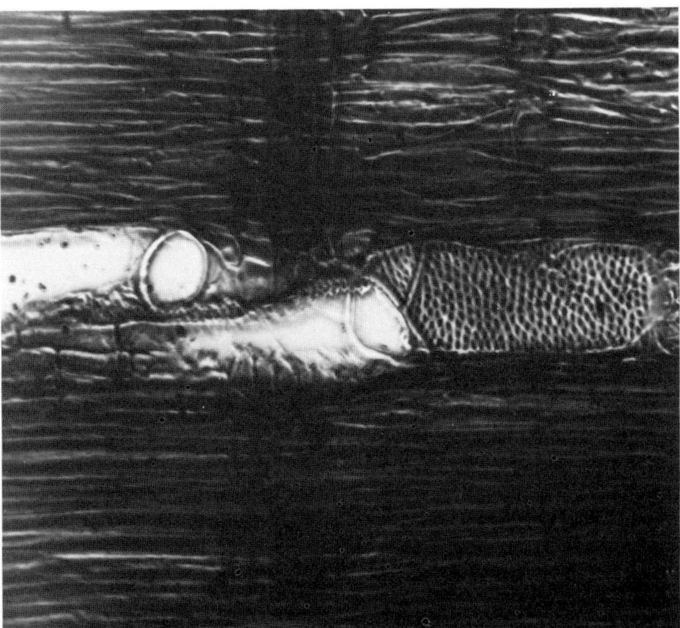

17

Plant cells vary in shape. Leaf cells may contain chloroplasts embedded in the cytoplasm. Water-conducting cells in the stem are hollow tubes which are linked end to end.

trunk. Liver cells are cuboidal and perform a wide variety of synthesis and storage activities; muscle cells are elongated and have the capacity to contract (Fig. 16). Many leaf cells are sausage-shaped, contain chloroplasts, and convert light energy into chemical energy; conducting cells are thick-walled, hollow tubes that conduct water and dissolved nutrients throughout the roots and shoot of the plant (Fig. 17). Other examples are illustrated in Fig. 18; these merely hint at the tremendous range of diversity among cell types.

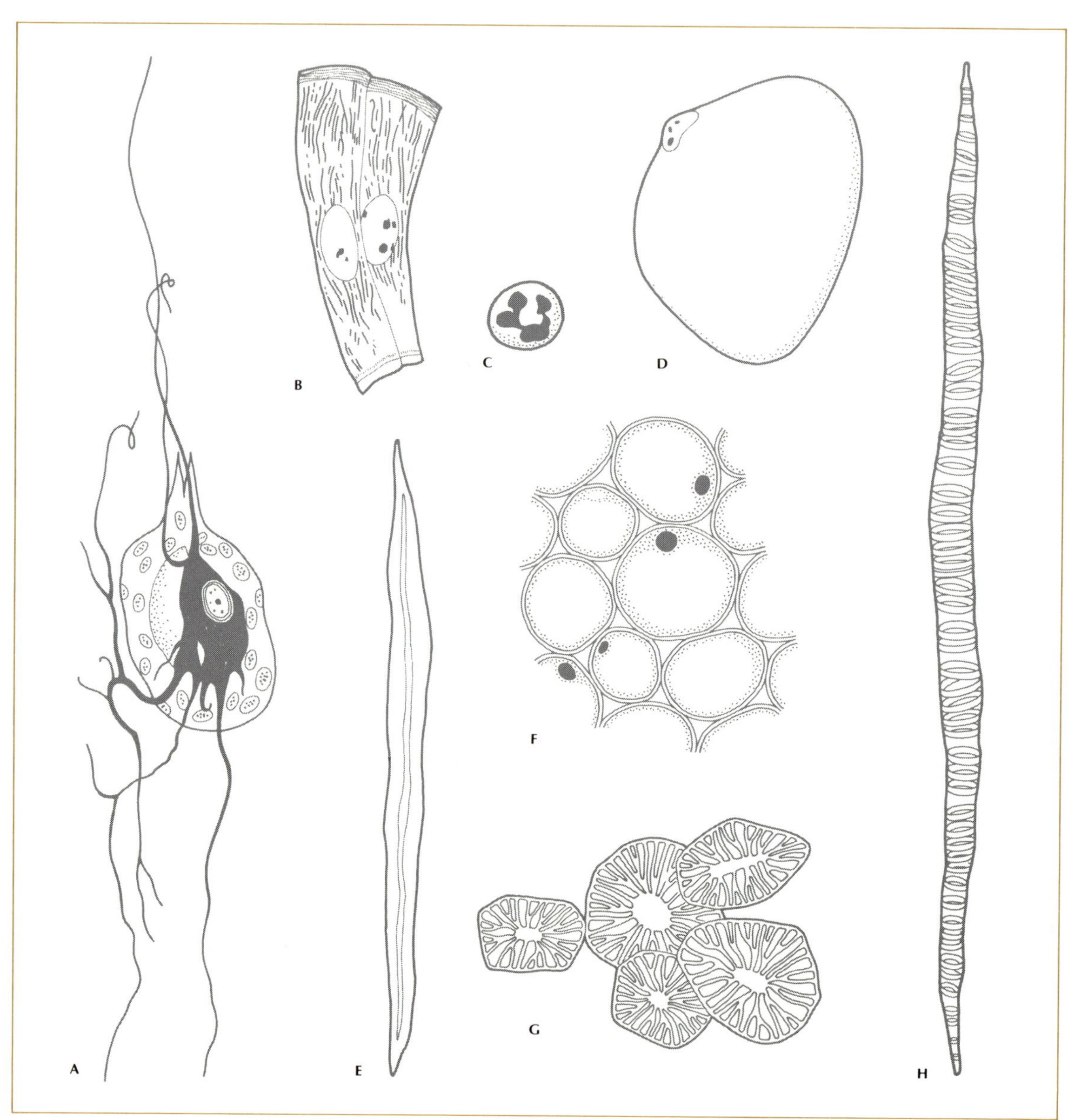

◀ **Cell types.** (A) Human nerve cell. (B) Cells from surface of rat intestine.
(C) Human white blood cell. (D) Human fat storage cell. (E) Plant wood fiber.
(F) Storage cells from pith of plant stem. (G) Stone cells from pear fruit.
(H) Water-conducting cell from trunk of tree.

MULTICELLULARITY

One-celled organisms have successfully colonized virtually every environment on earth. Biochemically they are more versatile than man, because
from a very few nutrients they can synthesize everything they need. And
yet, as we shall see in Chapters 12 and 13, the thrust of evolution is toward
ever more complicated organisms. Given the spectacular biological success
of unicellular life, it is only natural to ask: What are the advantages of multicellularity? The answer must be either that multicellular organisms are
better suited than unicellular forms to survive in certain environments, or
that they are better able to colonize areas completely unsuited for the
growth and reproduction of simpler organisms.

PLANTS

In plants the advantages of multicellularity appear to lie in the ability literally to be in two places at the same time. Single-celled algae survive well in the open ocean and in the open water of lakes because they are continually immersed in water containing appropriate inorganic nutrients, and because currents keep the cells suspended near the surface, where solar energy is continuously available. The terrestrial environment is not well suited to the survival of such forms. The water and nutrients are trapped in the soil, and sunlight cannot penetrate the ground. A single long cell with one end buried in the soil and the other protruding into the atmosphere would seem to be able to exploit the best of both worlds. But diffusion imposes limits on the size of a cell. And in the case of the long thin cell, the transport of nutrients from the buried end to the top, and the transfer of food manufactured at the sunlit end to the growing end underground, would present serious mechanical problems.

Plants were able to colonize the land successfully only by becoming structures composed of many specialized cells. With cells specialized for absorption, for photosynthesis, and for the transport of materials, a plant can simultaneously exploit the nutrients in the soil and the energy of sunlight. The evolutionary invention of a waterproof coating on the cells exposed to air, and the emergence of efficient water absorbing cells in the underground portion, enabled multicellular plants to invade ever more arid environments on land. Further inventions, notably true leaves and roots and woody supporting tissue, allowed the growth of trees to heights measured in tens of meters.

ANIMALS

Among animals multicellularity confers a different sort of adaptive advantage. Unicellular animals are limited to ingesting very small particles of food, often of molecular size. Multicellular animals can take in larger chunks of food, including plants and smaller animals. In the course of evolution, the continued increase in the size of animals and the aggregation of cells into tissues and organ systems made possible efficient digestion of large chunks of food in a specialized digestive tract. Distribution of the digested food through a circulatory system made the food available to cells specialized for movement (muscle), control (nerve), and support (bone).

The invention of lungs and a system for transporting absorbed oxygen to other cells enabled animals to leave the sea and to enter the harsher environments of land and air.

In some species of animal, evolution has carried the process of specialization one step further. In colonies of social insects such as bees, wasps, termites, and ants, certain individuals in each population perform specific tasks. For example, termite societies comprise workers, soldiers, kings, and queens. Biologists once believed that each individual was destined from birth to become a member of one of these castes. But now it is known that a social hormone regulates the development of the young termites. Secreted by soldiers, kings, and queens, it inhibits the transformation of immature individuals into kings, queens, and soldiers but does not inhibit the development of workers. The removal of a soldier, king, or queen decreases the quantity of hormone in the colony, causing an immature individual to develop into an adult of the same caste as the one removed. Thus the population collectively maintains the optimal numbers of specialized individuals necessary for the continuation of the species.

Lastly, the specialization possible in multicellular organisms has led to the emergence of one species of animal endowed with conscious intelligence: man. A highly evolved brain has endowed man with overwhelming biological advantages, including the power to shape his own evolution and to devastate other species. Whether human intelligence represents an evolutionary breakthrough toward some higher form of life, or the instrument that renders the planet unfit for all life, remains to be seen. In the long run, the evolution of human intelligence might prove to have been a biological disaster.

LEVELS OF ORGANIZATION

The living world embodies several levels of organization (Table II). The levels represent degrees of structural complexity: atoms, molecules, organelles, cells, tissues, organs, and organisms. Higher levels comprise a hierarchy of groups of organisms: populations, communities, ecosystems, and biosphere. The higher levels will be discussed in Chapter 15.

Atoms are a basic unit of matter. Aggregations of certain atoms produce molecules with specific chemical and physical properties, and certain of these molecules form organelles, microscopic subcellular structures.

II
Levels of organization of matter and living organisms. Structural complexity increases from top to bottom of table. Life begins at the cellular level.

Level of Organization	Animal Example	Plant Example
atom	hydrogen atom	hydrogen atom
molecule	glucose	glucose
organelle	mitochondrion	mitochondrion
cell	neuron	tracheid
tissue	nerve	xylem
organ	brain	leaf
organism	human	pine tree
population	people of a city	pine trees in a forest
community	humans, pets, pests & other organisms	trees, squirrels, insects & other organisms
ecosystem	New York City	pine forest
biosphere	surface of the world	surface of the world

The cell is composed of various organelles and large numbers of molecules
of organic and inorganic compounds.

The cells of multicellular organisms become specialized as the organism develops from egg to embryo to adult. The cells assume characteristic
shapes and acquire the capacity to carry out one or more functions with
particular efficiency. An association of similar cells comprises a tissue: a
group of cells that cooperate in performing a particular function. In plants,
groups of tracheids (conducting cells) and other cells comprise xylem, a
complex tissue that conducts water and other substances. In animals,
neurons (nerve cells) associate as nerve tissue, which conducts electrochemical signals.

Organs are associations of tissues that cooperate in a particular function. The heart is an organ made up of nerve, muscle, and vascular (conducting) tissues. A leaf is a plant organ comprised of chlorenchyma (green
tissue), parenchyma (storage tissue), and vascular tissue. An organ system

is a cooperative association of organs, such as the nervous system in animals and the flowers of plants. Together the cells, tissues, and organs comprise an organism.

THE NEW BIOLOGY

Classical biology began with the study of organisms, particularly multicellular organisms. Centuries of work yielded a wealth of information about the structure, functioning, and life cycles of these organisms. More important, this work gave rise to the three great unifying ideas of modern biology: cell theory, genetics, and evolution.

In recent decades the emphasis of research has shifted to the cellular and the subcellular level. Powerful new tools and techniques have enabled biologists to take organisms apart carefully, to analyze the structure and function of single cells and organelles at the molecular level. Biology has changed from a descriptive to an analytical discipline, and the shift in perspective has been so dramatic that some authors have called it a revolution. Perhaps this is too strong a term, but there is no doubt that the new approach has yielded a rich harvest of discovery. Biologists have traced the pathways by which organisms obtain energy and expend it to drive their biochemical machinery. The chemicals that carry genetic information have been discovered and analyzed, and the chemical language of the gene has been decoded. Investigators have documented the steps by which cells break down foods and convert them to living matter. Such discoveries have revealed the basic similarity of all living systems on earth, and the principles that underlie the complex phenomena of life.

All these principles can be understood in terms of unicellular organisms, the simplest living systems. A bacterium, for example, contains only about 1/1000th as much genetic material as a human cell. Since researchers rarely work with a complicated system when a simple one will do, many of the discoveries of the new biology result from studies of one-celled organisms. The remaining chapters of this book, which emphasize recent research, will rely heavily on information obtained from studies of unicellular organisms and of isolated tissues and cells from multicellular organisms. With three exceptions, we shall leave the biology of multicellular organisms to later courses. The exceptions are evolution, ecology, and behavior: evolution, because the emergence of multicellular organisms is a crucial part of the story; ecology, because man's careless tampering with nature and his unwillingness to control his own population growth threaten the survival of many species, including his own; and behavior, not only because much of the research in this young field is interesting in itself, but also because the behavior of other species provides a fresh perspective for examining our animal nature.

READINGS

Afzelius, B., *Anatomy of the Cell.* University of Chicago, Chicago, 1966
  The ultrastructure of cells and the functions of the organelles. Written clearly, with a minimum of chemical symbols.

Hurry, S. W., *The Microstructure of Cells.* Houghton Mifflin, Boston, 1964 (paper)
  A blend of excellent electron micrographs and short comments on various cellular organelles.

Jensen, W. A., *The Plant Cell.* Wadsworth, Belmont, Calif., 1964 (paper)
  Good survey of the ultrastructure and function of cells, using plant materials as examples.

Jensen, W. A., and R. B. Park, *Cell Ultrastructure.* Wadsworth, Belmont, Calif., 1967 (paper)
  Excellent electronmicrographs and diagrams of the detailed structure of cells. Short, clear discussions of each subcellular organelle.

Kennedy, D. (ed.), *The Living Cell.* Freeman, San Francisco, 1965 (paper)
  A collection of articles and illustrations from *Scientific American* written at about the level of this book.

Morrison, J. H., *Functional Organelles.* Reinhold, New York, 1966 (paper)
  A discussion of the processes that occur within the structures revealed by the electron microscope.

Pfeiffer, J., *The Cell.* Time Science Library, Time-Life Books, New York, 1964
  The function of cells as components of complex organisms. Good diagrams, color photographs, and a well-written text by a former editor of *Scientific American*.

Porter, K. R., and M. A. Bonneville, *An Introduction to the Fine Structure of Cells and Tissues*. Lea & Febiger, Philadelphia, second edition, 1965
    Brief and informative prose, with photomicrographs by a master of the electron microscope.

Swanson, C. P., *The Cell*. Prentice-Hall, Englewood Cliffs, N.J., 1969 (paper)
    A good survey of the processes of the division, differentiation, and development of cells.

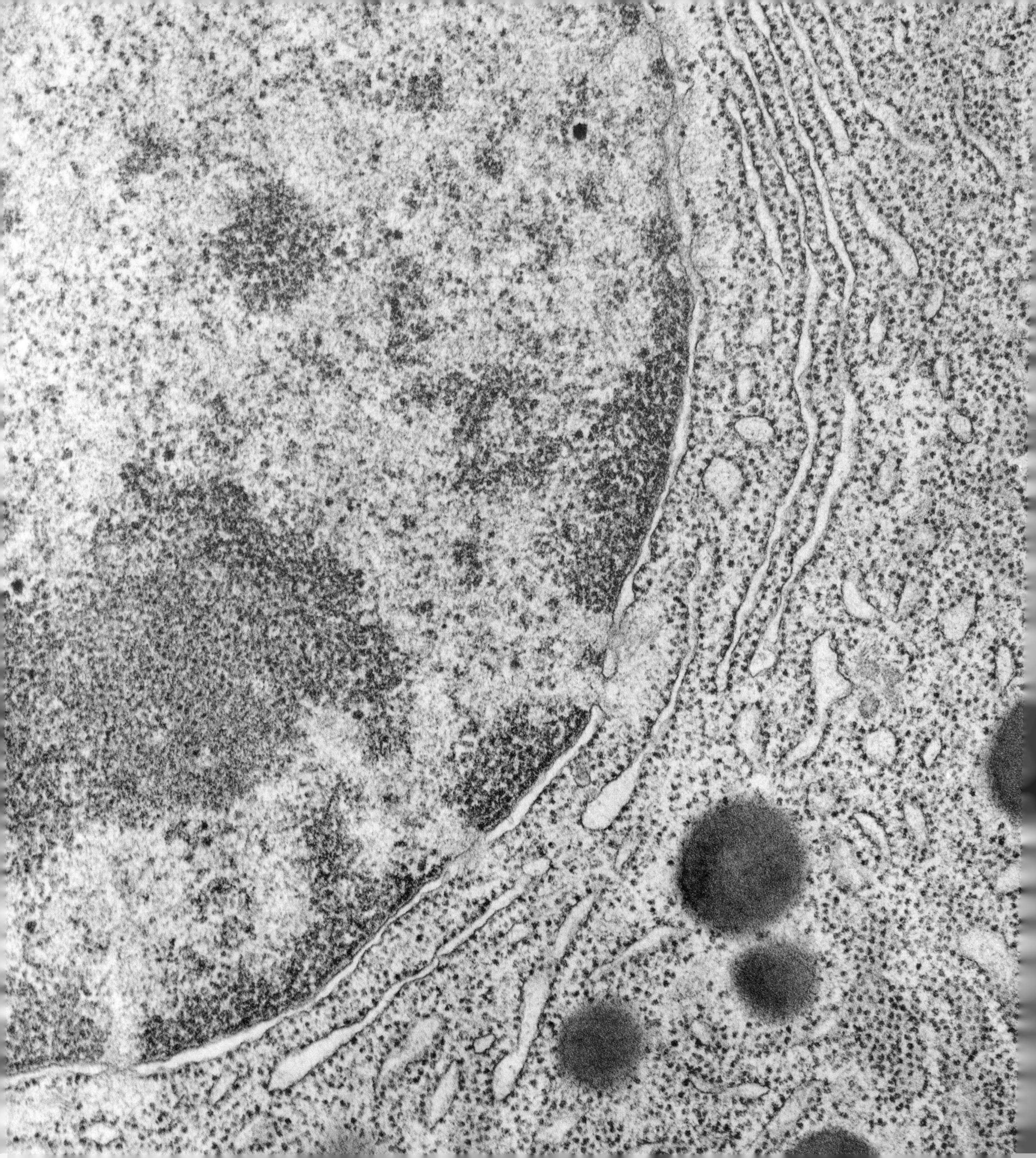

5. THE CELL NUCLEUS

From the viewpoint of thermodynamics, the duplication of a structure as highly organized as a cell appears to be a highly improbable event. A theoretician who did not know it could be done might be tempted to conclude that the task was impossible. Certainly the manufacture of a living cell in the laboratory lies far beyond the most sophisticated efforts of modern chemistry. And yet, a living cell can duplicate itself by cell division in about twenty minutes.

Biologists have long been aware that cells do not reproduce by chance. A random series of chemical accidents could not reproduce the giant molecules of living systems in so short a time, nor with such exquisite accuracy. Obviously the step-by-step assembly of a new organism, from atoms, to molecules, to cell, is a precisely directed process. But directed by what?

Borrowing the language of a branch of mathematics known as information theory, the pioneers of the new biology maintain that the precise duplication of a system as complicated as a living cell requires the transfer of an enormous amount of information, a detailed chemical blueprint. The existence of such a blueprint was strongly suggested by the mechanism of reproduction. A bacterium, or a seed, or a fertilized egg always gives rise to a predetermined type of organism, an organism almost exactly like its parent. Stimulated by the work of the 19th-century Austrian geneticist Gregor Mendel, 20th-century biologists set up countless experiments to answer the questions that his results raised: Can the existence of a blueprint be proved? In what part of the cell does it lie? What is its chemical nature?

◀**Cell nucleus** almost fills this electron micrograph of cell from pancreas of bat. Double membrane surrounds nucleus. Several pores in nuclear envelope can be seen at center and bottom. Magnification about 42,000 ×.

Acetabularia, a marine alga, is one of the largest single-▶ celled organisms. A cluster of them is reproduced here magnified about 10 ×.

The work that answered these questions has been called the greatest scientific breakthrough of the 20th century. It is hard to dispute this evaluation, because the information-carrying molecule is the substance that organizes inanimate matter into a system capable of life.

THE ROLE OF THE NUCLEUS

Thanks to five decades of research, today every biology student knows at least something about the answers to these questions. But this information was not easily won. Important clues about the location of the blueprint within the cell emerged from the experiments of the German biologist J. Hammerling. He used the single-celled marine alga *Acetabularia* (Fig. 1). This plant grows profusely in the warm lagoons of coral islands, where its small umbrella-topped stalks are easily spotted by skin divers. To biologists its importance lies in its ability to grow and regenerate the umbrella if the original cap is cut off. Grafts from one plant can even become integral parts of another.

The experiments by Hammerling involved two species: *Acetabularia mediterranea,* which was first discovered in the Mediterranean Sea, and *Acetabularia crenulata,* which has a crenulated (pleated) cap. Both species contain a large nucleus near the base of the stalk. The problem was to determine whether the nucleus or the stalk (the cytoplasm) contains the information that controls the shape of the cap.

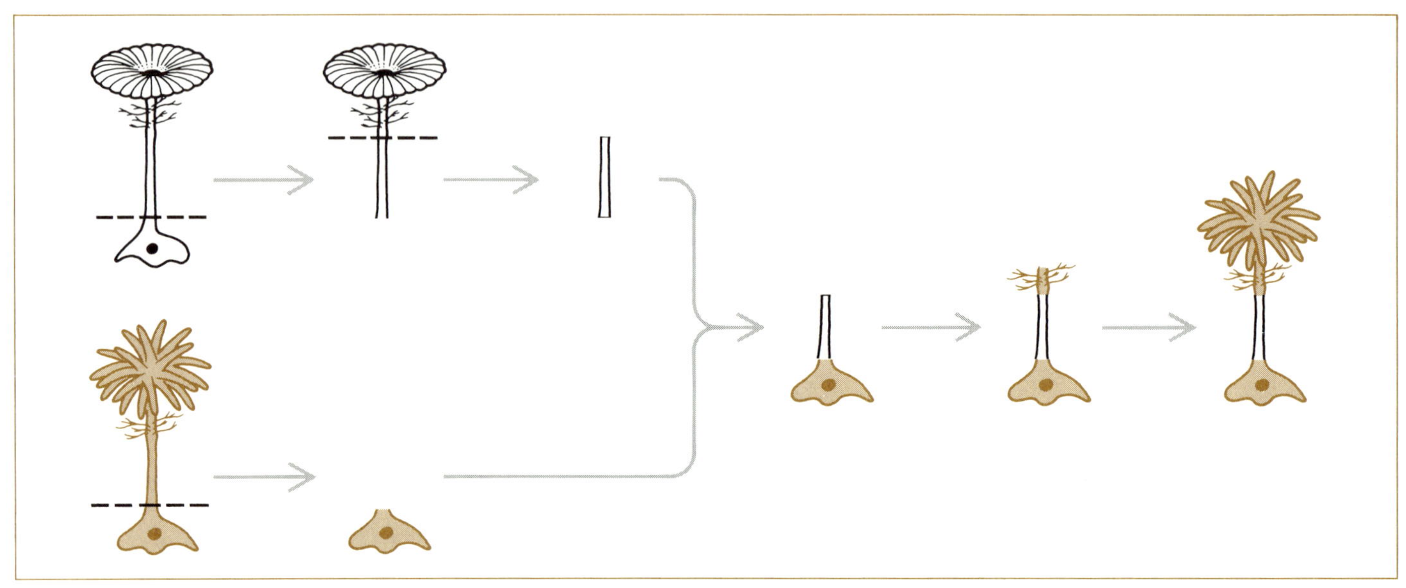

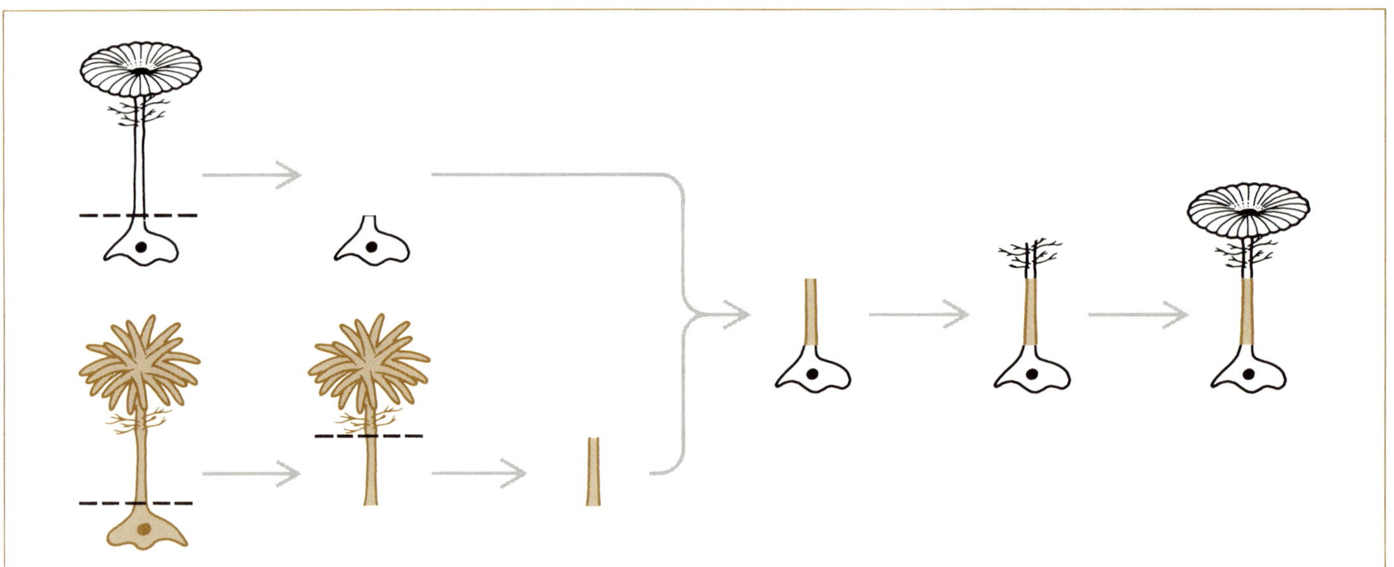

◄**Role of nucleus** in reproduction was demonstrated by grafting stalk from *Acetabularia mediterranea* (black, top left) onto base from *Acetabularia crenulata* (color). Base contains nucleus, shown here as a dark spot. Grafted plant developed crenulated cap (top right). Conversely, *crenulata* stalk grafted onto *mediterranea* base yielded plant with *mediterranea* cap (bottom right).

Plants of the two species were cut into three parts: cap, stalk, and base. The caps were thrown away, and the stalks from one species were grafted onto the bases of the other. Soon new umbrellas began to develop.

Hammerling reasoned that if the blueprint for the cap lay in the cytoplasm, the new umbrella would be the same as that originally carried by the stalk. But if the blueprint lay in the nucleus, the new cap would match the one on the plant from which the base was derived. In one experiment a base from a *mediterranea* plant was grafted onto a stalk from a *crenulata* plant. The umbrella that formed was identical to those found on intact *mediterranea* plants. Information from the *mediterranea* base must somehow have passed through the *crenulata* cytoplasm, causing the regeneration *of a mediterranea* cap. Conversely, another experiment with a base from a *crenulata* plant and a stalk from a *mediterranea* plant resulted in the formation of a *crenulata* cap (Fig. 2).

In repeated experiments, the new caps always corresponded to those of the species from which the base piece was taken, regardless of the source of the grafted middle segment. This result supported the hypothesis that the nucleus contains the blueprint, in short, that the nucleus is the control center of the cell. But the results did not disprove several alternative possibilities: (1) that the control center lies in the cytoplasm near the nucleus, not in the nucleus itself; (2) that the control center is some unobserved structure in the base of the cell; (3) that cutting up the organism destroyed some control mechanism in the stalk, allowing the control mechanism in the base to take over; or (4) that *Acetabularia* is an unusual organism, and consequently results from experiments on it would not apply to other organisms.

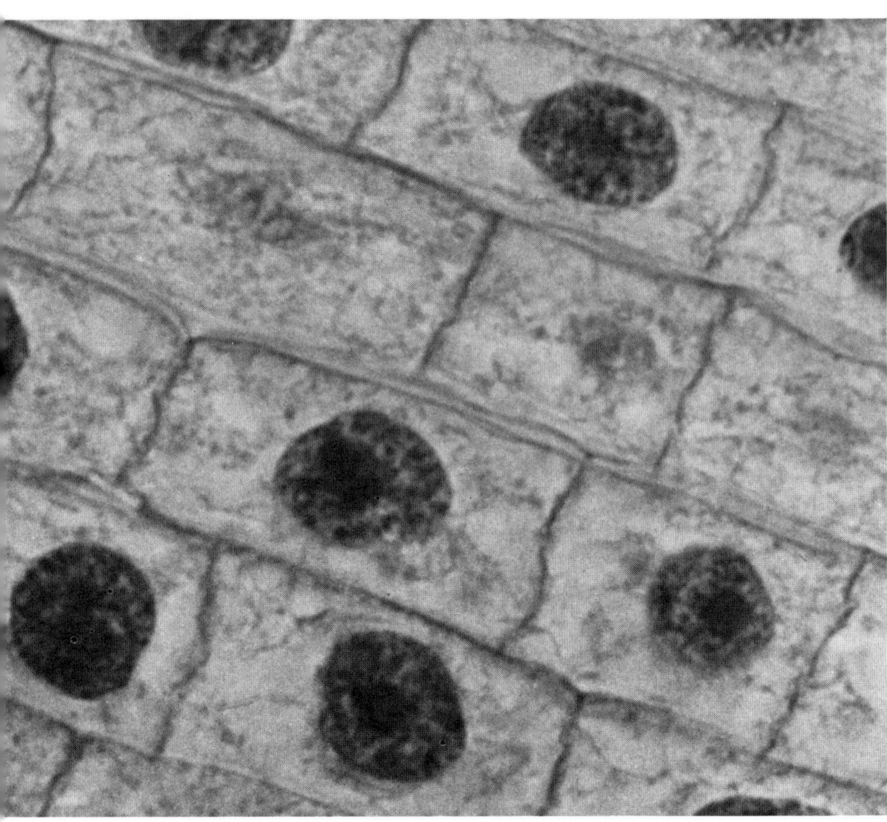

3
No visible chromosomes appear in the nuclei of these cells from an onion plant. Nuclei are dark disks; larger darker spots within disks are nucleoli. Magnification about 1000 ×.

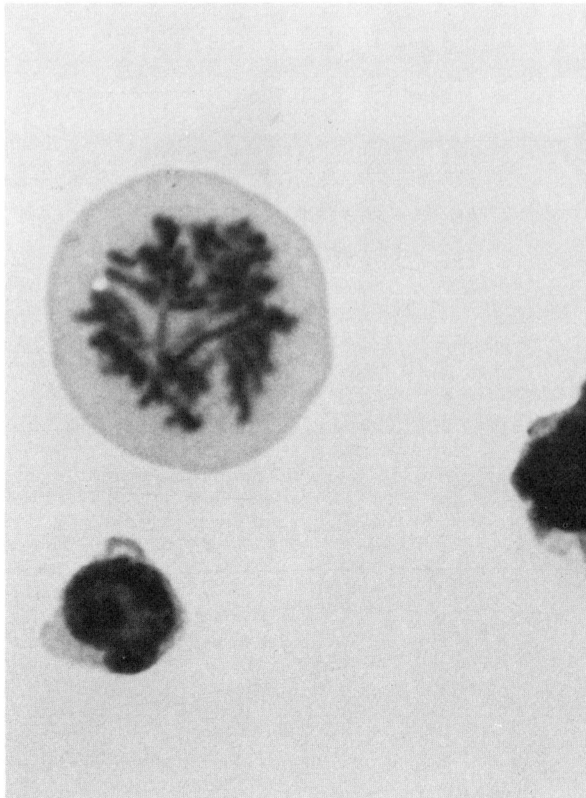

4
Chromosomes are conspicuous in tissues that contain rapidly dividing cells. Animal blood cells shown here are from bone marrow, site where red cells are formed. Cluster of chromosomes appears as a dark mass within the large cell at top.

CHROMOSOMES

Despite such uncertainties, the results were clear enough to encourage further experiments. Assuming the nucleus to be the true control site, the question remained as to what portion of the nucleus contained the blueprint. Two possibilities were the nucleoli and the chromosomes. The nucleoli looked structureless and rather uninteresting, but the chromosomes were observed to undergo striking changes during the process of nuclear division. Efforts were concentrated on studying the chromosomes. However, chromosomes are invisible during most of the life cycle of the cell. A microscopic examination of a slice of tissue taken at random from any mature organism will probably reveal cells whose nuclei contain one or more nucleoli but no visible chromosomes (Fig. 3). Chromosomes are most

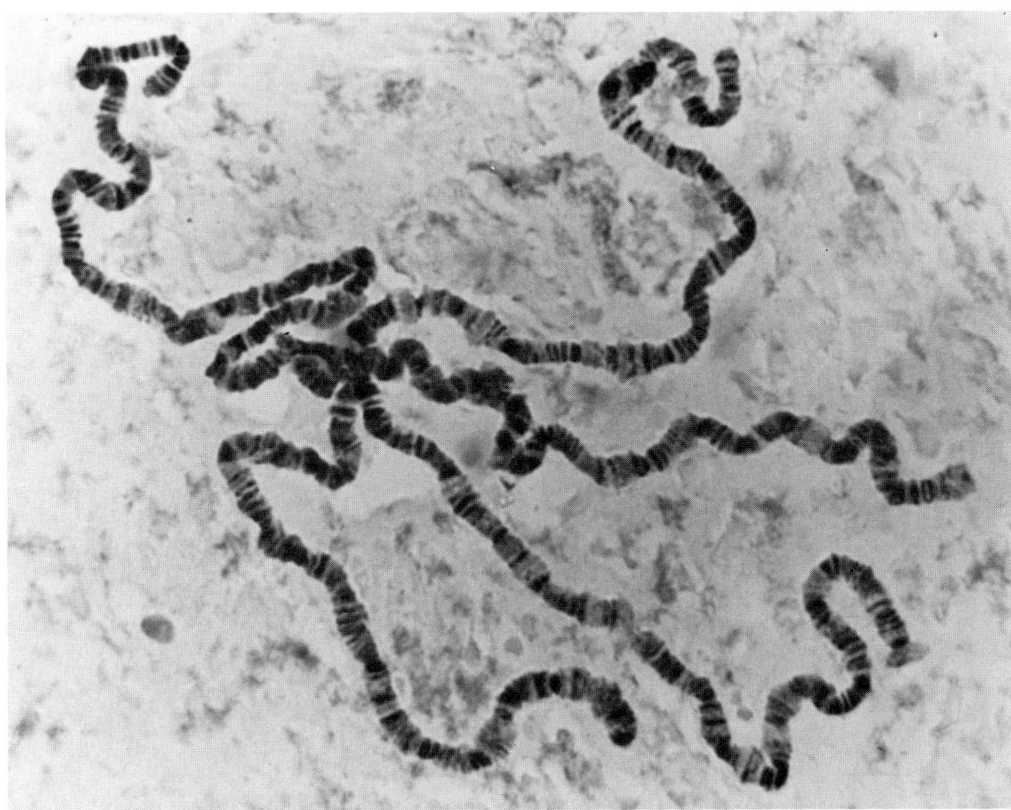

5

Giant chromosomes from salivary gland of fruit fly show as a pattern of light and dark bands. Bands mark location of individual genes.

evident in structures in which cells are rapidly dividing, such as developing pollen grains and actively growing shoot and root tips in plants, or the red marrow of the arm and leg bones of animals (Fig. 4).

Proof of the genetic importance of chromosomes came only after years of observing chromosomes from thousands of different kinds of organisms. Orchids proved notoriously difficult to use because each cell contains a large number of extremely small chromosomes. Conversely, certain species of fly contain giant chromosomes. The salivary glands of the common fruit fly *Drosophila melanogaster* contain chromosomes 100 times larger than those from other parts of the body. The complicated, easily visible patterns of bands on each of these chromosomes (Fig. 5) led early geneticists to

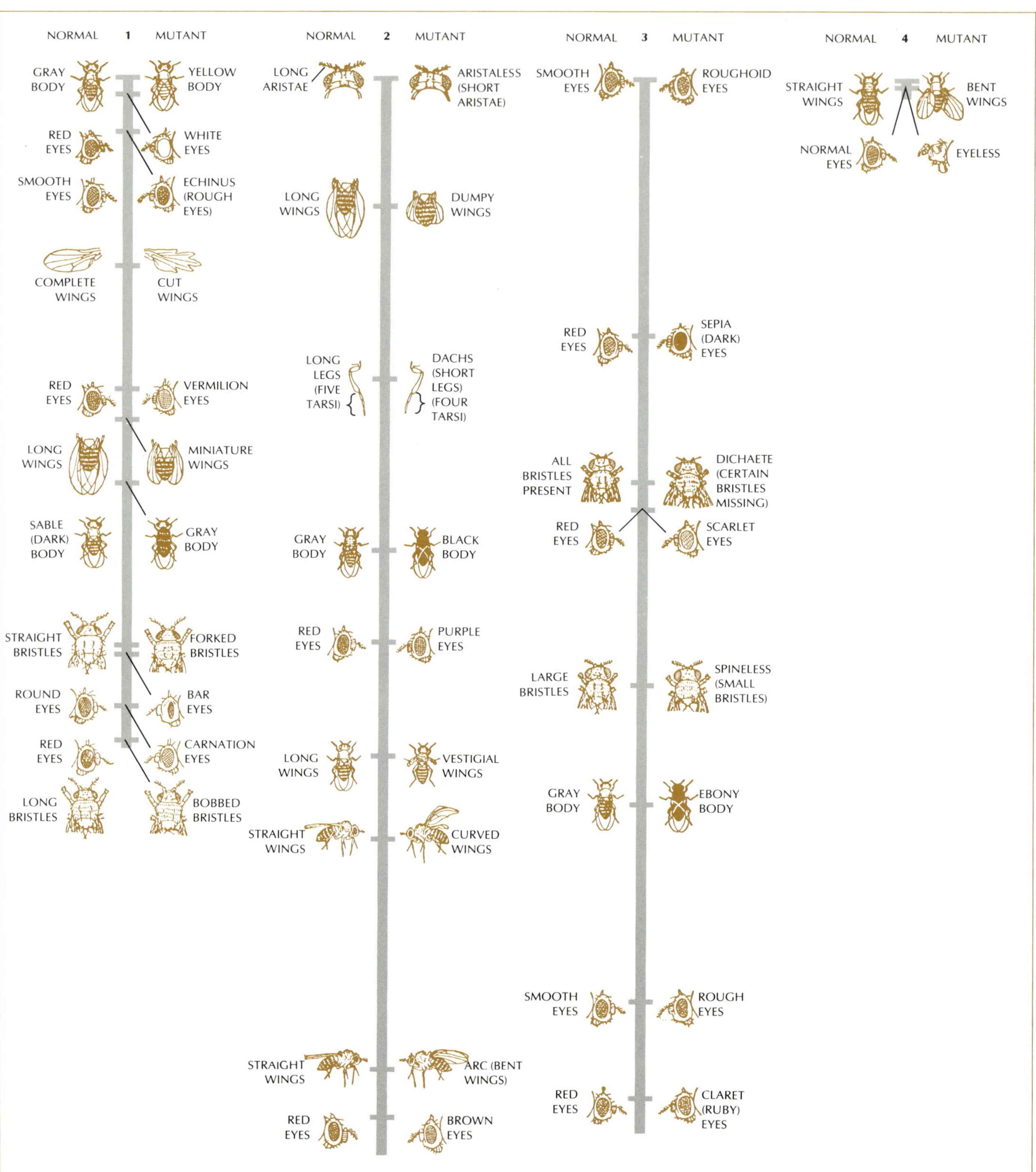

NORMAL   1   MUTANT
GRAY BODY / YELLOW BODY
RED EYES / WHITE EYES
SMOOTH EYES / ECHINUS (ROUGH EYES)
COMPLETE WINGS / CUT WINGS
RED EYES / VERMILION EYES
LONG WINGS / MINIATURE WINGS
SABLE (DARK) BODY / GRAY BODY
STRAIGHT BRISTLES / FORKED BRISTLES
ROUND EYES / BAR EYES
RED EYES / CARNATION EYES
LONG BRISTLES / BOBBED BRISTLES

NORMAL   2   MUTANT
LONG ARISTAE / ARISTALESS (SHORT ARISTAE)
LONG WINGS / DUMPY WINGS
LONG LEGS (FIVE TARSI) / DACHS (SHORT LEGS) (FOUR TARSI)
GRAY BODY / BLACK BODY
RED EYES / PURPLE EYES
LONG WINGS / VESTIGIAL WINGS
STRAIGHT WINGS / CURVED WINGS
STRAIGHT WINGS / ARC (BENT WINGS)
RED EYES / BROWN EYES

NORMAL   3   MUTANT
SMOOTH EYES / ROUGHOID EYES
RED EYES / SEPIA (DARK) EYES
ALL BRISTLES PRESENT / DICHAETE (CERTAIN BRISTLES MISSING)
RED EYES / SCARLET EYES
LARGE BRISTLES / SPINELESS (SMALL BRISTLES)
GRAY BODY / EBONY BODY
SMOOTH EYES / ROUGH EYES
RED EYES / CLARET (RUBY) EYES

NORMAL   4   MUTANT
STRAIGHT WINGS / BENT WINGS
NORMAL EYES / EYELESS

◄**Chromosome map** of fruit fly illustrates positions of genes for several anatomical traits. Map shows normal and mutant genes on four chromosomes.

Centromere is a weakly stained portion of chromosome. Its position divides the chromosome into two arms. When the longitudinal halves of the chromosome separate, it is the centromeres that lead the way.

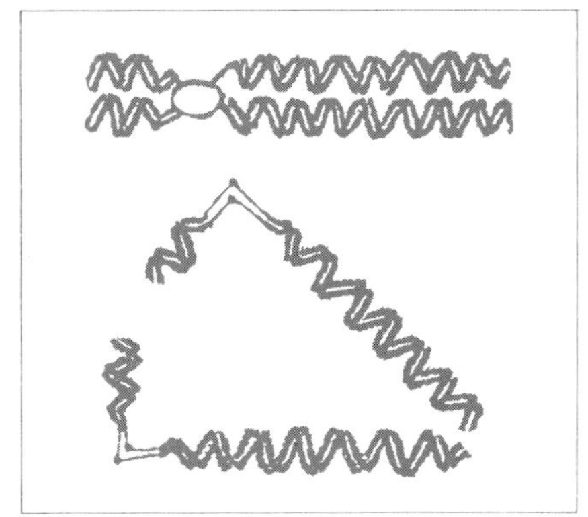

attempt to relate the bands in some way to the appearance of certain features in adult flies. For instance, flies whose chromosomes lacked a certain band had white eyes, while flies with this band had normally pigmented red eyes. Here, unexpectedly, was evidence not only that chromosomes are the structures that control heredity, but also that a specific part of the chromosome controls the appearance of a particular characteristic. By correlating the presence or absence of many characteristics and structures with the corresponding presence or absence of bands on the chromosomes, geneticists developed maps of *Drosophila* chromosomes showing the exact location of scores of individual genes (Fig. 6).

STRUCTURE OF CHROMOSOMES

Except for such unusual examples as those of the fruit fly, most chromosomes are at first glance nothing more than small wormlike threads. But closer examination reveals that they can be distinguished from one another by size and shape. Every chromosome shows at least one constriction somewhere along its length, due to a break in its stainable material. The primary constriction indicates the position of the centromere, a structure that divides the chromosome into two arms (Fig. 7). If the centromere lies in the middle of the chromosome, the arms will be of equal length; if it is nearer to one end, the arms will be unequal. In some chromosomes the centromere may be so near to one end that only one arm is visible. Chromosomes may also have one or more smaller, secondary constrictions. When one lies close to the end of the chromosome, the small piece of chromosome beyond the constriction is called a satellite. The length of the

8

Four pairs of chromosomes from a cell of *Hypochoeris tweedie*. Each chromosome is made up of two identical halves called chromatids. Satellites can be seen on two of the shorter chromosomes.

individual chromosomes, the location of the centromere, the number and placement of the secondary constrictions, and the presence or absence of satellites enables the geneticist to distinguish one chromosome from another.

In the chromosomes from a cell of *Hypochoeris tweedie*, a relative of the sunflower, four distinctly different kinds of chromosomes are recognizable, and there are two of each kind. A closer look reveals that each of these chromosomes is composed of two identical lateral halves, called chromatids, lying next to each other (Fig. 8). The chromatids are strongly attached to each other in the region of the centromere. Weaker forces normally attract the rest of the arms of the chromatids to each other along the rest of their length. At times the chromosome acts as a unit, with the chromatids bound to each other; at other times the chromatids' arms may repel each other in all regions except near the centromere. At still other times the chromatids may separate from each other completely. These characteristic activities of chromatids took on great significance when biologists realized that the "words" of hereditary information, the genes, were arranged in linear order in the chromosomes. By watching the movement of the chromatids, they could actually watch the movement of genes.

The number of chromosomes in a cell is usually characteristic for each species of organism. The cells of the fruit fly have 8 chromosomes; those of man have 46, although until the late 1950's biologists stated flatly that they had 48. Only by judiciously examining many cells from many individuals of the same species can the characteristic chromosome number of that species be established with any confidence. For example, in the common trout *Salmo irideus* there is a variation in chromosome number among cells taken from different tissues. Some cells from early trout embryos contain 59, 61, 62, 63, or 64 chromosomes. Some cells from the spleens of one-month-old trout embryos have 58 chromosomes. Liver cells with 65 chromosomes have been found. But most cells contain 60 chromosomes, and this is now generally accepted as the correct chromosome number for this kind of trout.

Table I lists the chromosome numbers for various plant and animal species. Although the bacteria, some of the simplest organisms known, contain only one chromosome per cell, it is obvious that there is no clear-cut relationship between the chromosome number and the biological complexity of organisms of different species.

MITOSIS AND CELL DIVISION

The process of nuclear division, which includes the division and replication of chromosomes, is called *mitosis* or *karyokinesis*. The division of the cytoplasm is called cell division or *cytokinesis*. Both karyokinesis and cytokinesis are active, continuous processes. The only proper way to grasp the essence of these changes is to examine living cells. The techniques of time-lapse photography make it possible to condense the time scale of the entire process from several minutes of real time to a few seconds of projection time. The sequence of illustrations in Fig. 9 gives only a rough idea of the coiling, the pushing, the pulling, the sudden movements, and the abrupt appearance and disappearance of various structures as they appear on the motion picture screen. Even the study of prepared microscope slides of actual biological material can only suggest the complexity and drama of the process that unfolds in living cells.

Mitosis ends with the formation of two new nuclei with chromosome complements identical to each other and to that of the parent cell. In most cells, karyokinesis triggers cytokinesis. After the chromatids have separated and moved to opposite poles during mitosis, reactions occur in the cytoplasm midway between the two newly forming nuclei. In plant cells a plate

I

Number of chromosomes in the cells of a few plants and animals. Chromosome number is not clearly related to structural complexity of an organism. Both dog and sugar cane contain more chromosomes than man.

Plants	
sugar cane	80
peanut	40
tomato	24
passion flower	18
wood rush	6

Animals	
dog	78
man	46
tree frog	24
housefly	12
mosquito	6

NUCLEAR
MEMBRANE
SISTER
CHROMATIDS
CENTROMERE
1
2
3
4

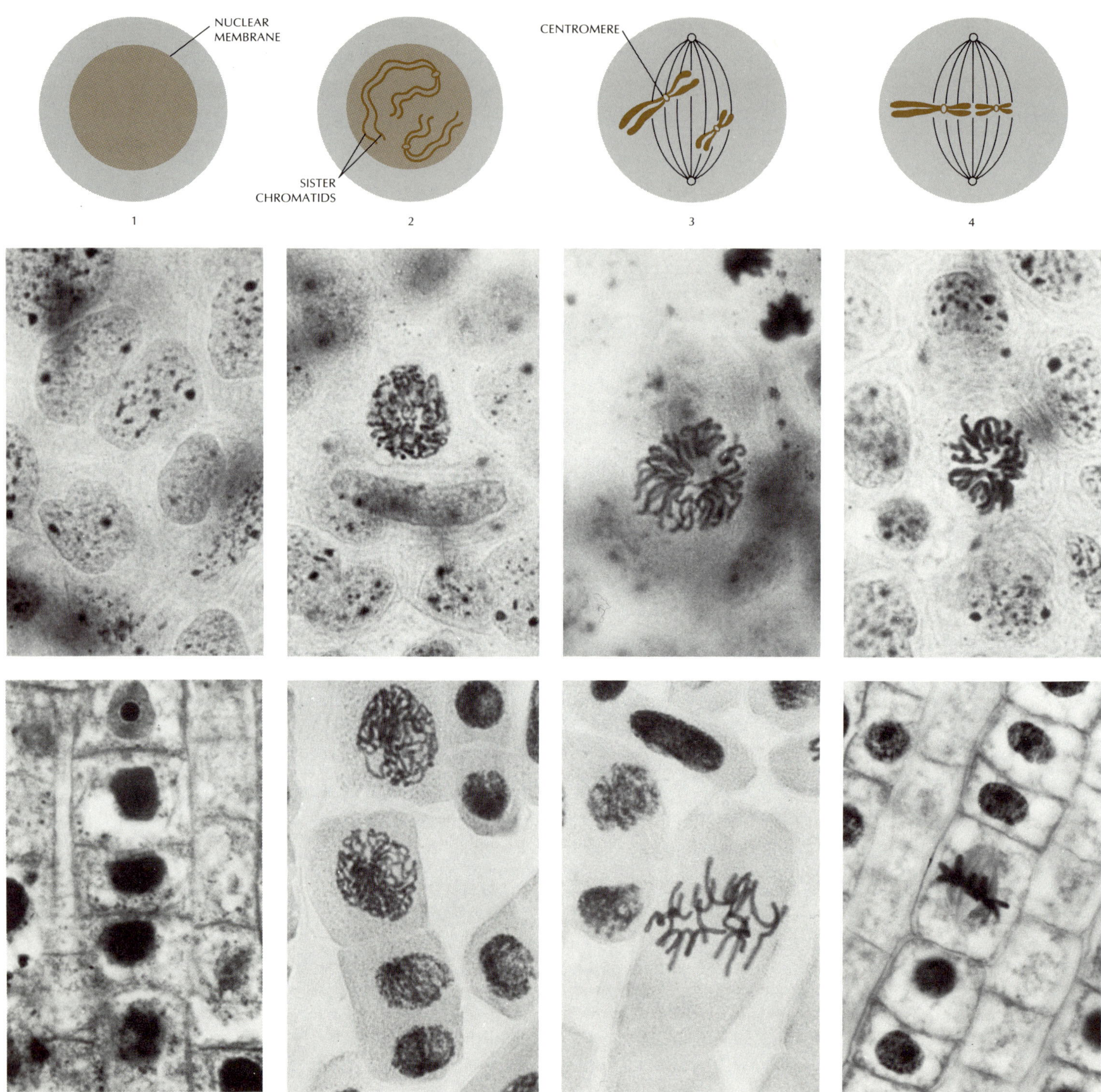

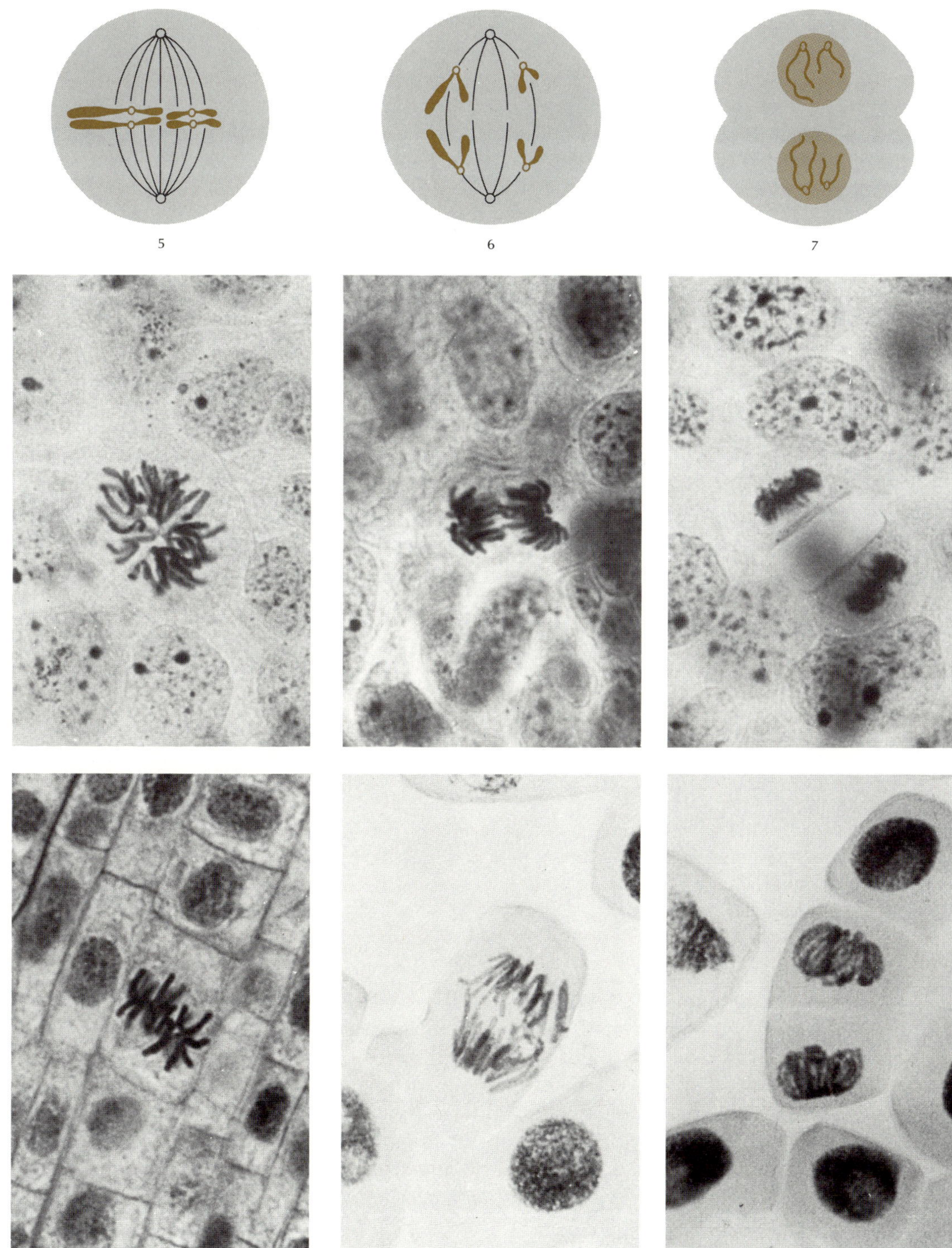

9
Mitosis in animal cells
(top row) and plant cells
(bottom row):
1. As a nucleus prepares to
   divide,
2. the chromosomes appear
3. and thicken.
4. They align in the middle of
   each cell.
5. The chromatids of each
   chromosome separate
6. and migrate to opposite
   ends of the cell,
7. resulting in the creation of
   two nuclei.

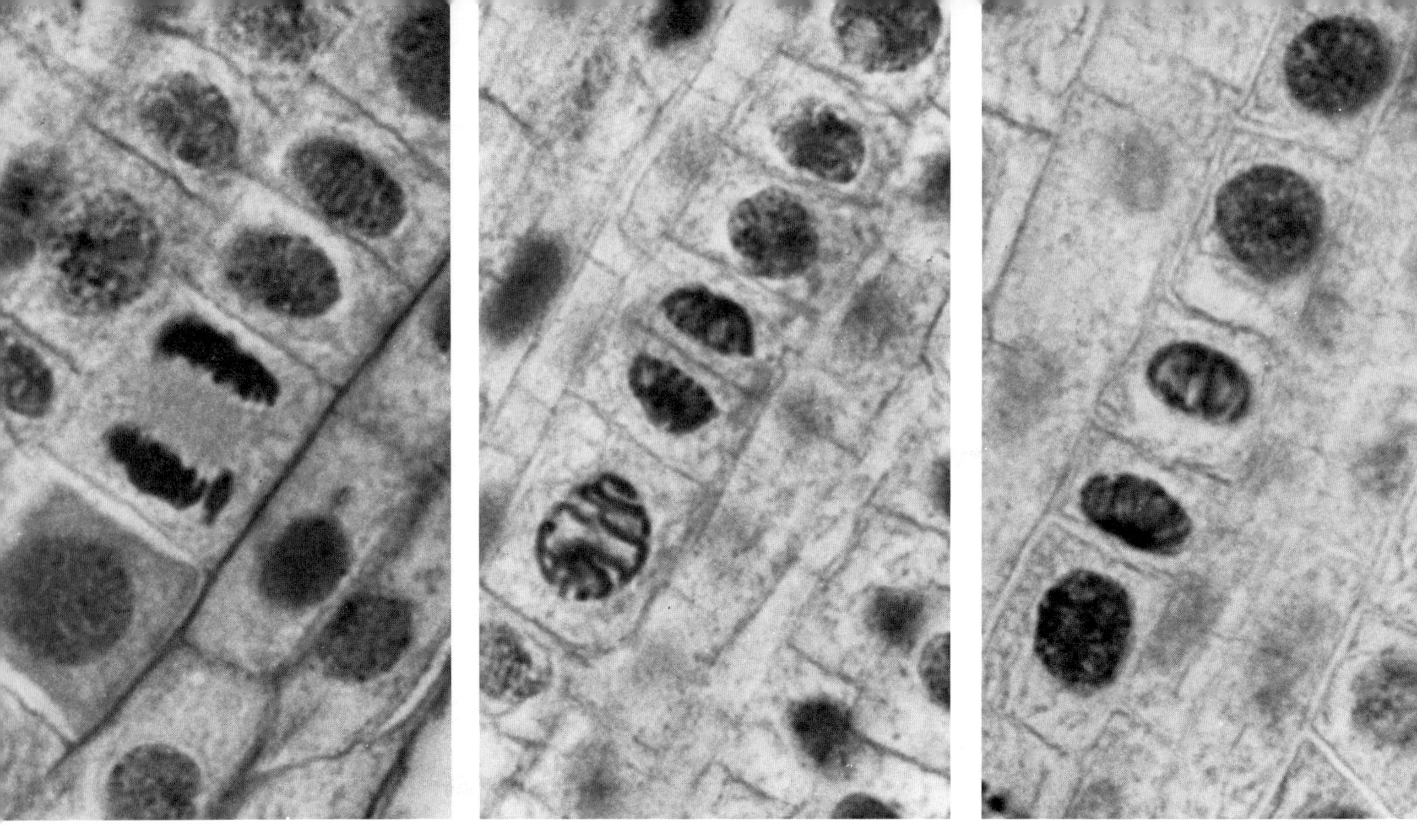

10
Cytokinesis in root-tip cells of garden onion plant. As the two new nuclei are forming during mitosis (left), a cell plate begins to form between them (center), ultimately cutting the original cell into two cells (right).

of cell wall material begins to form in the center of the cell. When the connections are completed, two new cells are formed (Fig. 10).

In animal cells the process of cytokinesis is slightly different. No cell plate forms in the center of the cell; instead, the sides of the original cell pinch inward. Eventually they meet in the center of the cell and fuse, forming two cells (Fig. 11). Most cells divide immediately after mitosis, producing daughter cells, each of which contains a single nucleus. In a few organisms the nucleus divides, but not the cytoplasm; the result is a multinucleate cell (Fig. 12).

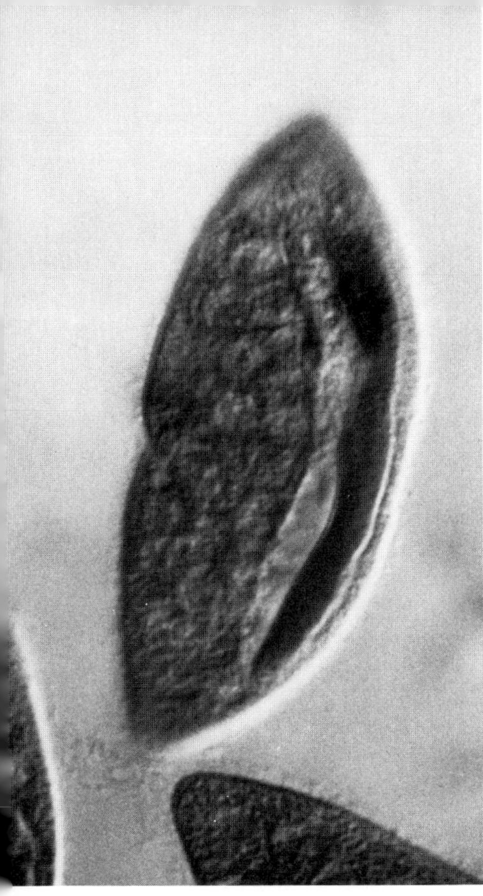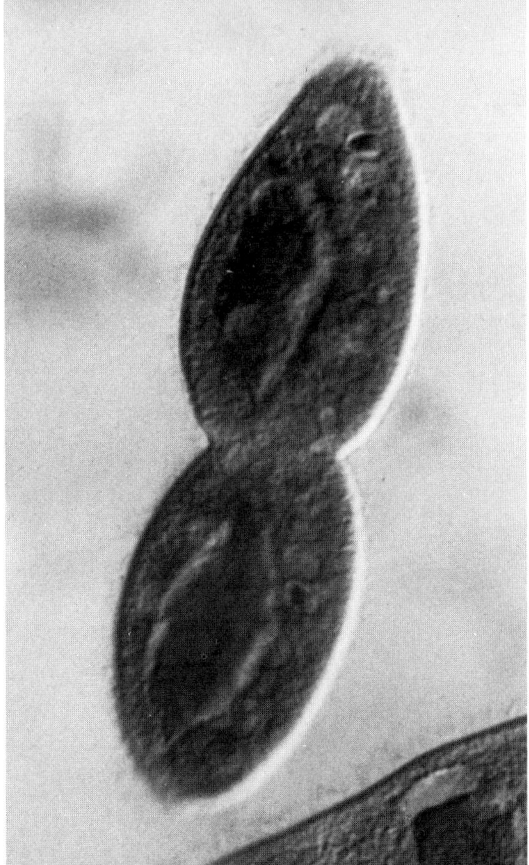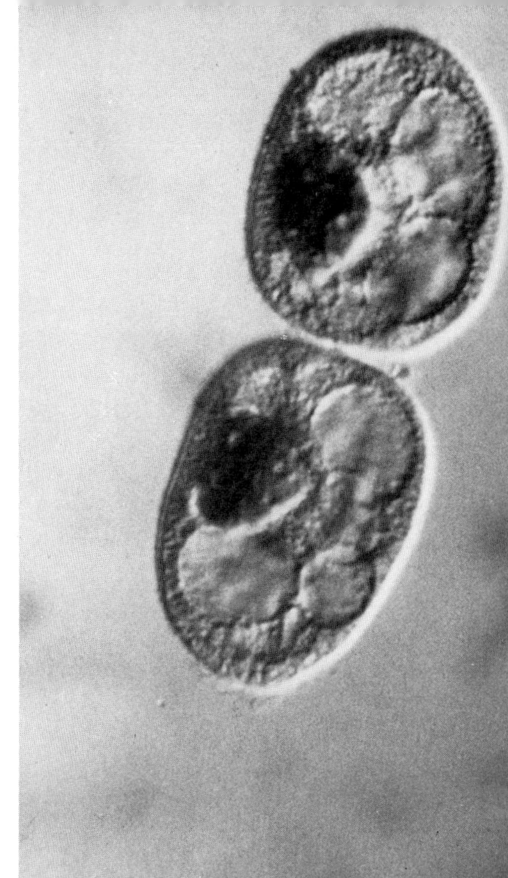

11

Cytokinesis in paramecium is representative of animal pattern of cell division. Following mitosis the cell membrane begins to pinch in around the middle of the cell (left). As the process continues (center), it eventually cuts the cell in two (right).

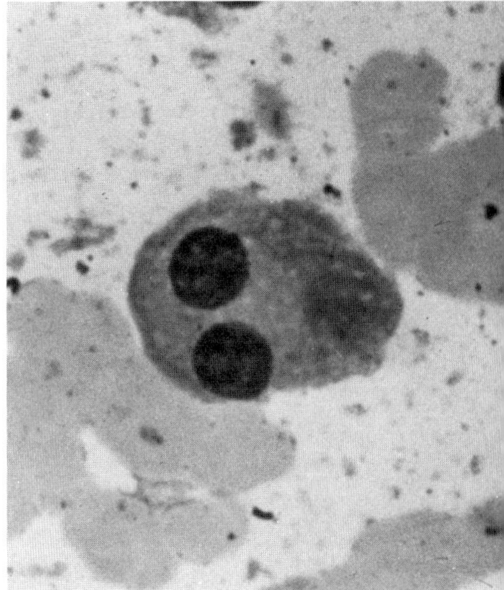

12

Human cell from plasma in bone marrow contains two nuclei.

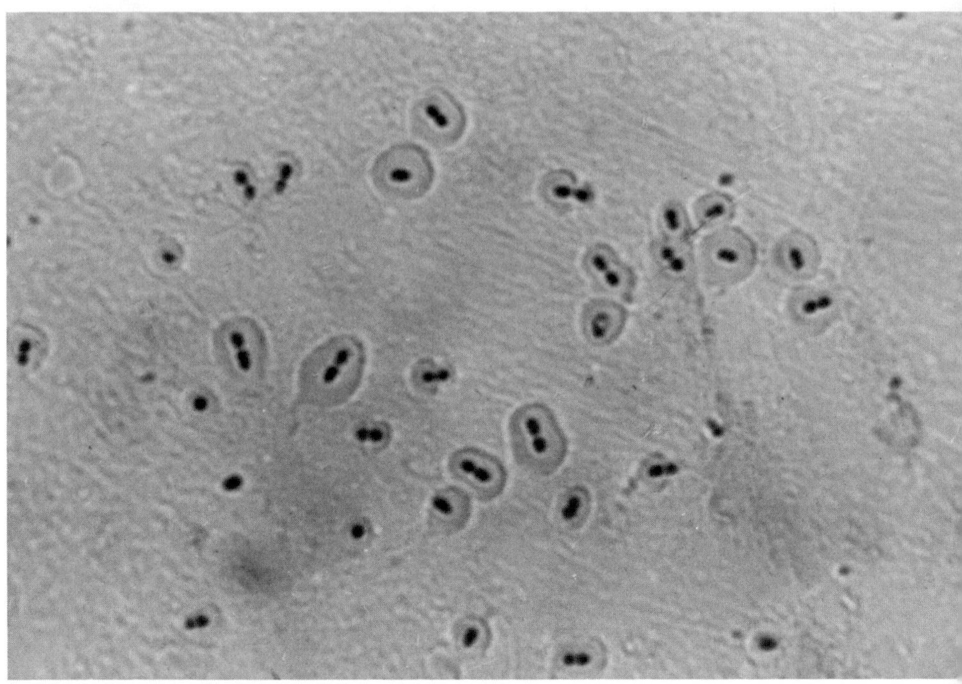

NUCLEIC ACIDS

Once biologists recognized the genetic importance of the chromosomes, the next step was to analyze them chemically. This revealed that they consist primarily of proteins and nucleic acids. Knowing the enormous amount of genetic information that the chromosome must carry, investigators assumed that the protein was the information-carrying component. They reasoned that the 20 different amino acids found in proteins comprised a sort of chemical alphabet, with each amino acid perhaps representing one letter. The order of the amino acids in the protein molecules was thought to be the genetic blueprint. The assumption was logical, but wrong; eventually it became evident that the nucleic acid portion of the chromosome is the genetic material. The protein acts as a regulator that controls the activity of the nucleic acid and also serves as a scaffolding that maintains the structure of the chromosome.

GENETIC TRANSFORMATION

The experiments that led to the realization that nucleic acids and not proteins are of primary genetic importance also shed light on the genetic mech-

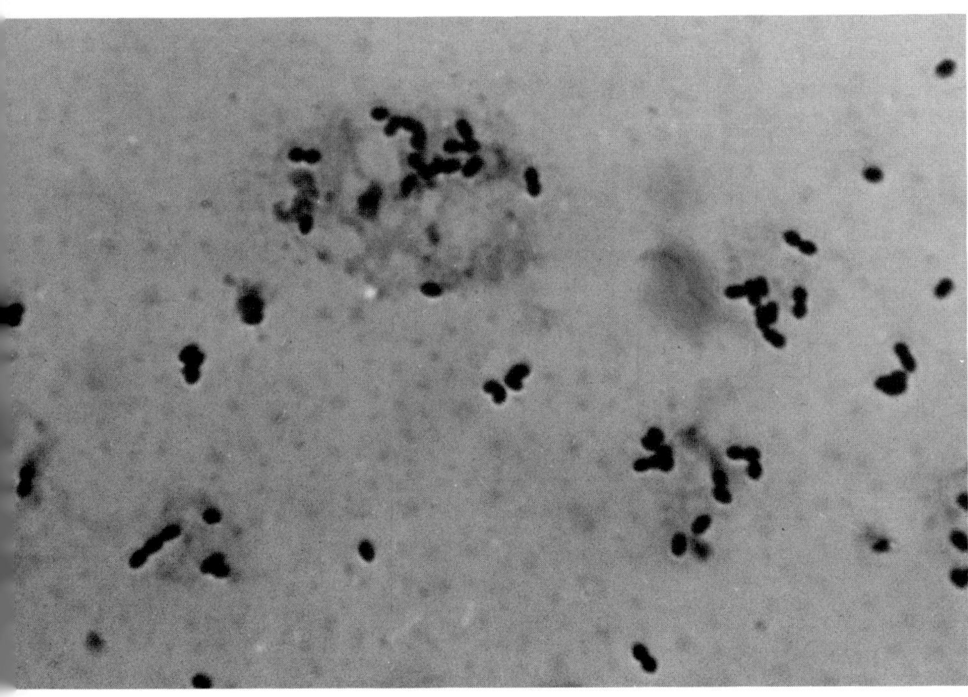

13
Encapsulated bacterium, *Diplococcus pneumoniae* (left), causes pneumonia in man and other animals. Nonencapsulated form (right) is relatively harmless.

anisms of microorganisms. One classic study was carried out in the 1920's by the English biologist Fred Griffith; it concentrated on two strains of the pneumonia-causing bacterium *Diplococcus pneumoniae*. One strain of this bacterium possesses a carbohydrate capsule around each cell. This strain is virulent; that is, as by-products of its metabolism, the cell releases organic compounds that are toxic to man and other animals. The presence of these toxins in the body of the infected animal causes symptoms of pneumonia (Fig. 13). The other strain of *Diplococcus* has no capsule and is much less virulent.

Mice injected with the encapsulated strain soon develop the disease and die; mice injected with the noncapsulated strain remain healthy. If the virulent bacteria are killed by exposure to heat and then injected into a group of mice, these mice also remain healthy. This is not surprising: dead bacteria are unable to produce the toxins that cause the symptoms of pneumonia.

A less predictable result occurred when the experimenter took a culture of living nonvirulent bacteria, mixed it with some dead virulent bacteria, and injected this combination into another group of mice. This procedure

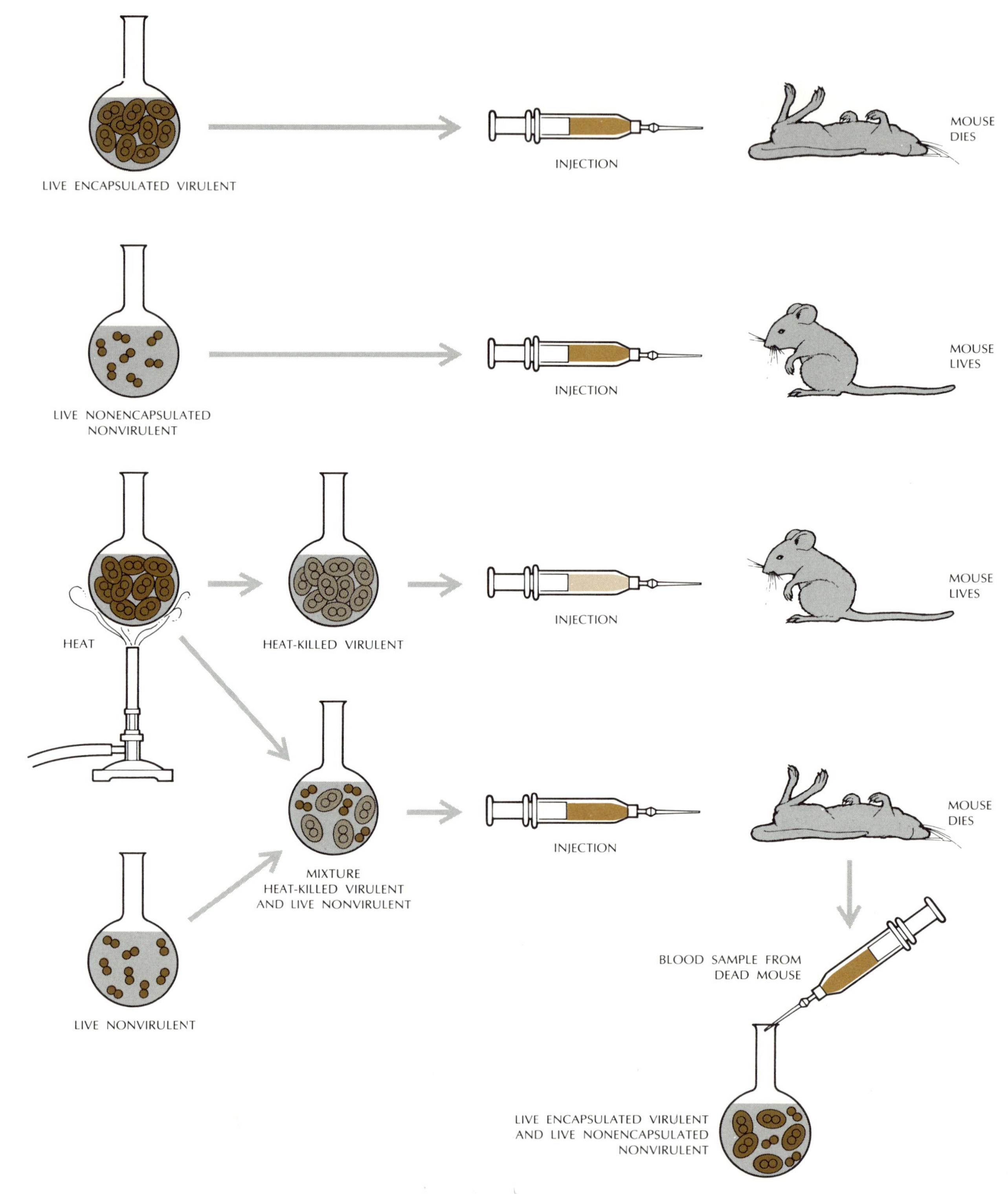

LIVE ENCAPSULATED VIRULENT
INJECTION
MOUSE DIES
LIVE NONENCAPSULATED NONVIRULENT
INJECTION
MOUSE LIVES
HEAT
HEAT-KILLED VIRULENT
INJECTION
MOUSE LIVES
MIXTURE HEAT-KILLED VIRULENT AND LIVE NONVIRULENT
INJECTION
MOUSE DIES
LIVE NONVIRULENT
BLOOD SAMPLE FROM DEAD MOUSE
LIVE ENCAPSULATED VIRULENT AND LIVE NONENCAPSULATED NONVIRULENT

14

◄**Genetic transformation.** Nonencapsulated pneumococcus was transformed into virulent encapsulated strain when mixed with dead virulent strain. Genetic material from dead bacteria entered live ones, transforming some of them into encapsulated strain.

would not be expected to produce an infection. But the mice sickened and died. What happened? Blood samples from the diseased animals revealed the presence of living bacterial cells of both the noncapsulated and the capsulated strains (Fig. 14). No living capsulated cells had been injected, but large numbers of them appeared in the blood of the dying animals. Where had they come from?

Apparently, genetic material from the virulent cells somehow passed from the dead bacteria to the nonvirulent cells, transforming them into disease-producing cells. As the transformed cells multipled they ultimately produced enough toxin to kill the infected mice. This process, which resulted in a change in the genetic potential of the cell, was named *transformation*.

Some biologists interpreted the results differently. They argued that genetic material might not have been transferred at all. Perhaps in the mixed culture the living nonvirulent cells in some way activated the heat-killed cells: unkilled them, so to speak. The reactivated cells then began to multiply and produce toxins. At the time, neither possibility seemed more farfetched than the other.

About five years later J. Lionel Alloway of the Rockefeller Institute for Medical Research set up another series of experiments in an attempt to verify one or the other of these conflicting conclusions. This time the virulent strains were not just heat-killed; they were also ground into submicroscopic pieces. The cell debris was then mixed with a culture of living nonvirulent cells and injected into mice. The mice became diseased, and their blood contained living virulent *Diplococci*. There could be no doubt that some chemical from the dead bacteria had transformed the nonvirulent cells into virulent ones. But what was it?

A decade elapsed before another group of investigators suggested the answer. With a rigorous biochemical technique, Avery, MacLeod, and McCarty of the Rockefeller Institute separated the protein and nucleic acid fractions of the pulverized cells. They mixed the protein with living nonvirulent cells and injected the mixture into mice; the mice remained healthy. But when a mixture of nonvirulent bacteria and the nucleic acid from the virulent strain was injected, the mice developed the symptoms of pneumonia. In their blood were the telltale encapsulated cells. There was little doubt that the nucleic acid from the virulent cells had transformed the nonvirulent cells into cells capable of causing pneumonia.

VIRAL NUCLEIC ACIDS

Supporting evidence for this conclusion emerged after another decade of work on the problem. A team of scientists at the genetics laboratory at Cold Spring Harbor was working with viruses, enigmatic particles that bridge the gap between the nonliving world of chemicals and the living world of active organisms. Unlike bacteria, viruses can be isolated, crystallized, and stored in a bottle for years. When injected into an organism, the viruses become active, invade its cells, and somehow force them to manufacture a new crop of viruses. There are thousands of viruses, ranging from those that cause the common cold or polio in humans, to those that cause plant diseases like mosaic in turnips and bushy stunt in tomatoes. Other types of viruses infect bacteria. They are known as bacteriophages (bacteria eaters) or just "phages."

One type of phage particle consists of only a head and a tail made of protein, and a central core of nucleic acid (Fig. 15). From the examination of numerous electron micrographs of bacteria under attack by phages, the team proposed a theory to explain how these viruses infect a cell and reproduce themselves. They deduced that the tail fibers enable the virus to attach itself to the wall of the bacterium. Then enzymes in the tail digest part of the cell wall, and a contraction of the protein sheath injects the nucleic acid core into the bacterial cell. It takes a phage only about 20

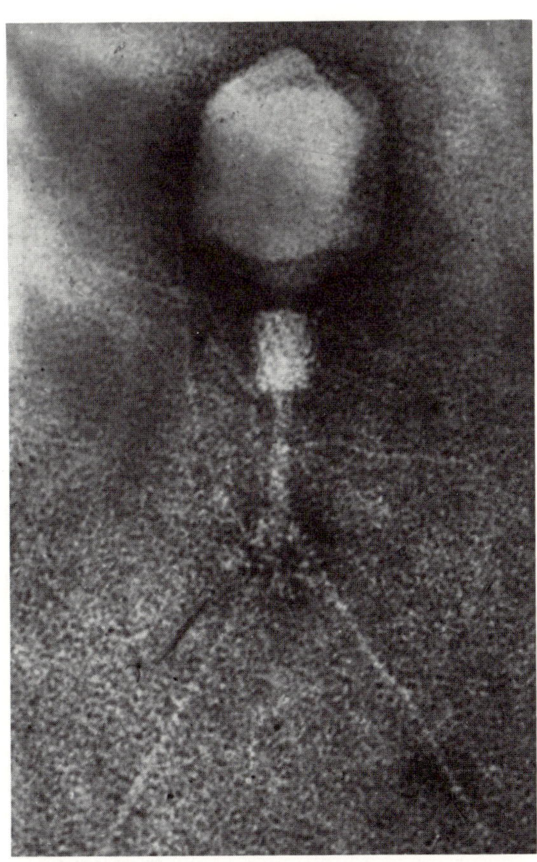

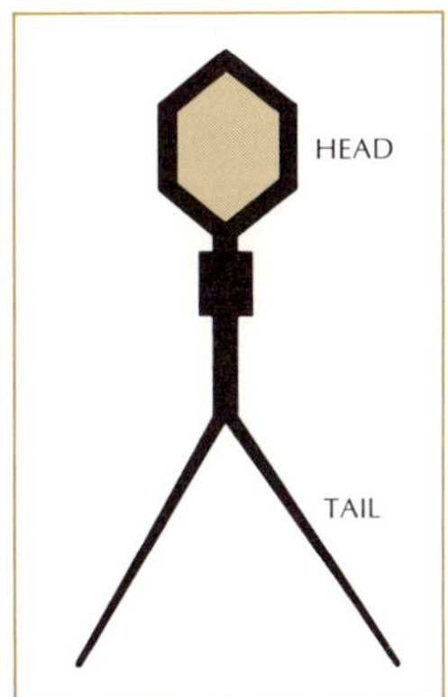

15
Bacteriophage is shown at a magnification of about
350,000 × in this electron micrograph. Diagram
indicates composition of virus. Hexagonal head and
center of shaft contain DNA (color). Outer covering
and tail fibers are protein.

minutes to cause the production of hundreds of fully formed phage particles within the cell. The bacterium then bursts like a seed pod, releasing a swarm of virus particles. Each new virus particle can then infect another bacterial cell. In this process the nucleic acid injected into the bacterial cell seems to overpower the genetic machinery of the cell, suppressing the chemical reactions that produce bacterial protein and nucleic acid in favor of reactions that make new virus protein and virus nucleic acid.

TRACER EXPERIMENTS

Electron micrographs do not show conclusively that only nucleic acid enters the bacterium. Perhaps viral protein enters too. If so, this protein might cause the production of the new viruses. To decide which was the case, the team of A. D. Hershey and Martha Chase at Cold Spring Harbor designed an elegant set of experiments that involved the use of radioactive atoms of phosphorus and sulfur.

The atomic mass number of most atoms of phosphorus is 31; the nucleus contains 15 protons and 16 neutrons. But some phosphorus atoms have an extra neutron, and hence an atomic mass number of 32. Such "sister atoms" are isotopes—atoms that differ from each other only in the number of neutrons in the nucleus. Common sulfur atoms have 16 protons and 16 neutrons, for an atomic mass number of 32. The heavier isotope, sulfur-35, contains 16 protons and 19 neutrons. Extra neutrons often make the heavier isotope unstable. Phosphorus-32 and sulfur-35 (also written as ^{32}P and ^{35}S) spontaneously break down, releasing energy as radiation. Both are beta emitters; that is, they release a high-energy electron which is known as a beta particle.

Biologists use radioisotopes to "label" a particular element or compound. Physiologically there is usually very little or no difference between the metabolism of radioactive and nonradioactive isotopes. Organisms absorb them from nutrients. Then with Geiger counters researchers can trace the passage of the radioactive isotopes through living tissues.

For studies of proteins and nucleic acids, ^{32}P and ^{35}S were logical choices. Proteins contain sulfur atoms but no phosphorus. Nucleic acids contain phosphorus but no sulfur. In the bacteriophage experiments, the nucleic acid core was labeled with ^{32}P; after the labeled viruses infected the bacteria, the cells were agitated in a kitchen blender to shake the protein husks of the viruses off the walls of the bacterial cells to which they were attached. The viral and bacterial particles were separated and analyzed for radioactivity. According to the team's hypothesis, ^{32}P-labeled nucleic acid should have been present inside the bacterial cell. This was exactly the result. There was no doubt that viral nucleic acid had entered the bacterial cell.

The next step was to determine whether the protein of the viral coat might also have entered the bacterium. A batch of viruses were labeled with ^{35}S, which was incorporated into the protein coat. After infection and agitation in the blender, the radioactive sulfur was found outside the bacterial cells. These two results provided powerful evidence that only the viral nucleic acid entered the bacterium, and that it must be the ultimate genetic chemical.

TRANSDUCTION

Other experiments with bacteria and phages, carried out by Norton Zinder and Joshua Lederberg at the University of Wisconsin, revealed another oddity of microbial genetics. Certain phages infect the bacterial cell but do not immediately cause the cell to make new viruses. Instead, the viral nucleic acid apparently attaches itself to the bacterium's own chromosomal nucleic acid and reproduces itself when the bacterial nucleic acid divides. Repeated cell divisions result in a large population of bacteria harboring the viral nucleic acid. At some later time the bacterial cells may suddenly begin to synthesize intact virus particles. The bacterial cells die and rupture, releasing the virus to infect other cells.

Now this is the strange part: during the period when the viral nucleic acid is replicating along with the bacterial nucleic acid, exchanges may occur between the two molecules. Some of the new viruses that eventually emerge may contain a bit of the bacterial chromosome. When these new viruses infect a new bacterial cell, they inject a fragment of bacterial nucleic acid along with their own. This process—the transfer of genes of one bacterium into another by a virus—was named transduction (Fig. 16). It differs from transformation only in that it involves a naturally occurring carrier, in this case a virus. In transformation the nucleic acid is transferred from one bacterium to another without assistance.

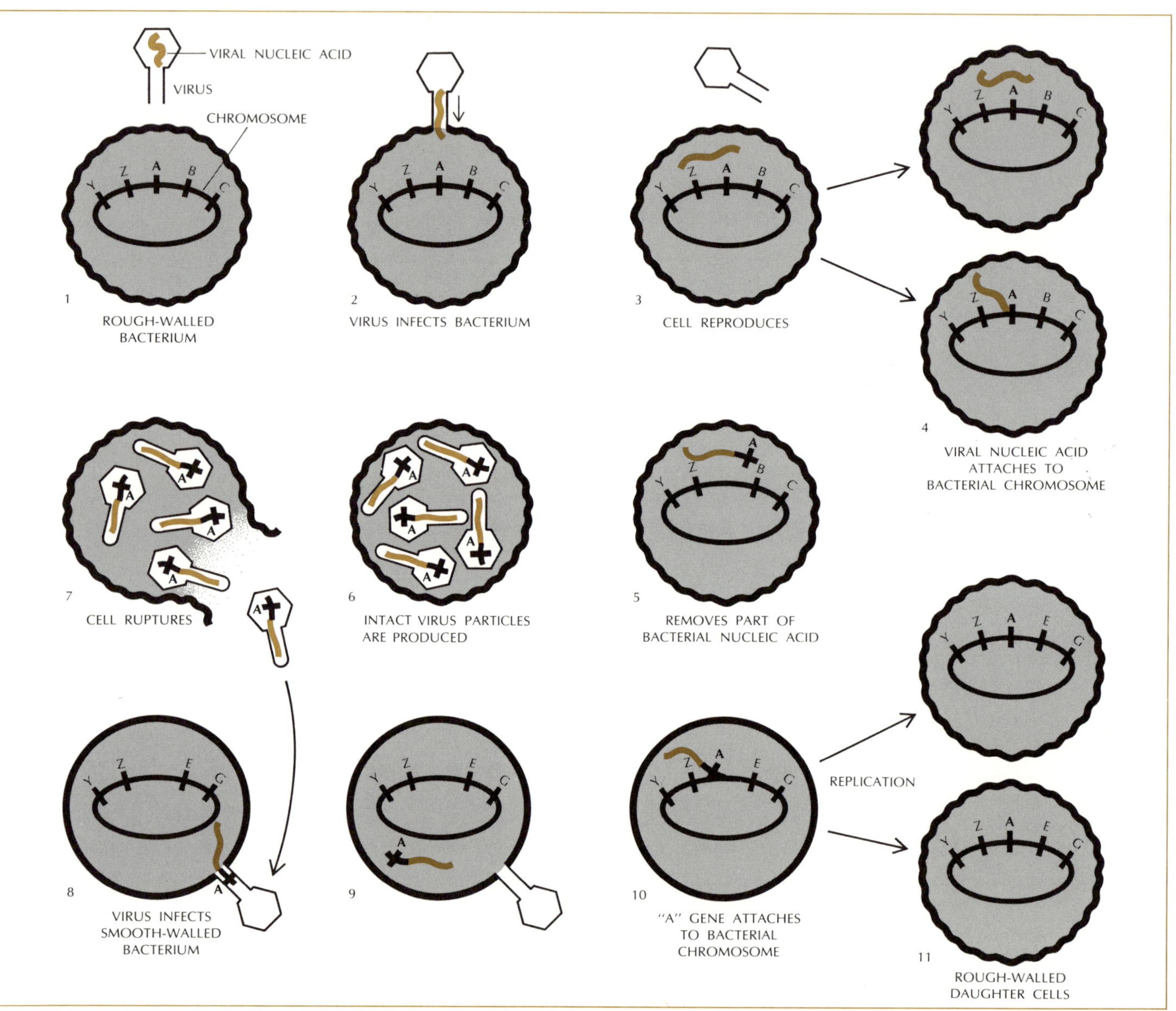

VIRAL NUCLEIC ACID
VIRUS
CHROMOSOME
1
ROUGH-WALLED
BACTERIUM
2
VIRUS INFECTS BACTERIUM
3
CELL REPRODUCES
4
VIRAL NUCLEIC ACID
ATTACHES TO
BACTERIAL CHROMOSOME
5
REMOVES PART OF
BACTERIAL NUCLEIC ACID
6
INTACT VIRUS PARTICLES
ARE PRODUCED
7
CELL RUPTURES
8
VIRUS INFECTS
SMOOTH-WALLED
BACTERIUM
9
10
"A" GENE ATTACHES
TO BACTERIAL
CHROMOSOME
REPLICATION
11
ROUGH-WALLED
DAUGHTER CELLS

16

◄**Genetic transduction** involves the transfer of chromosomal material from one cell to another by viruses.

CYTOPLASMIC INHERITANCE

In the 1960's, the work with nucleic acids and bacteria reached a feverish pace. Investigators around the world are still following up the leads provided by the discoveries of transformation and transduction.

One problem that has disturbed physicians is the striking rate with which bacterial pathogens (disease-causing organisms) acquire resistance to "wonder drugs." Originally, scientists discovered that in a population of bacteria a random genetic change (mutation) might make one cell resistant to a single antibiotic. If this population were exposed to the antibiotic, the other cells would die but the resistant cell would live, reproduce, and create a population of bacteria resistant to the drug. Resistance to a second antibiotic could develop in the same way. But such mutations are extremely rare; they occur only once in about half a billion cell divisions.

Thus researchers in Japan in the 1950's were surprised to discover bacteria that had apparently acquired resistance to four different antibiotics at the same time. This discovery reopened the question of the mechanisms of inheritance. How did bacteria acquire this packaged resistance? A few experiments ruled out transformation and transduction, leaving only one known possibility, that of conjugation.

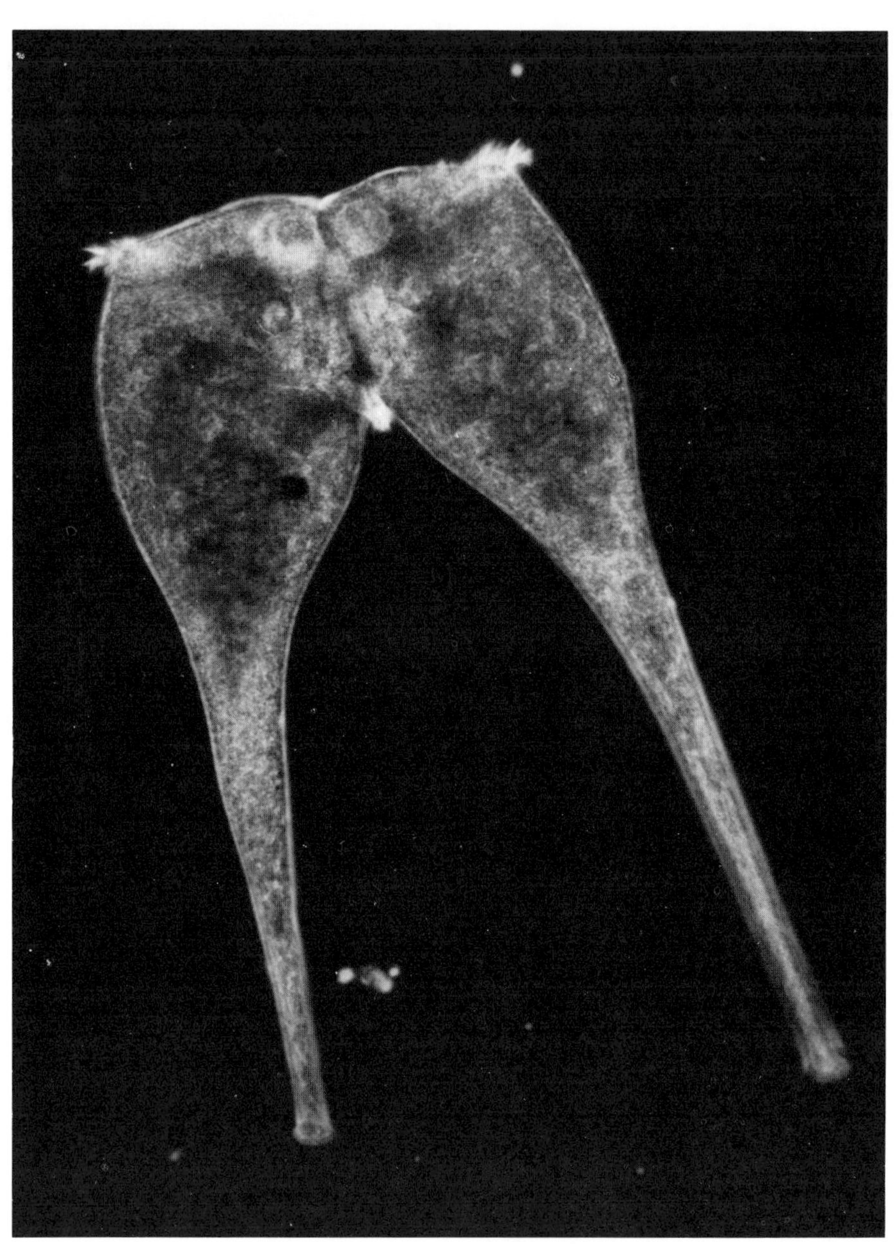

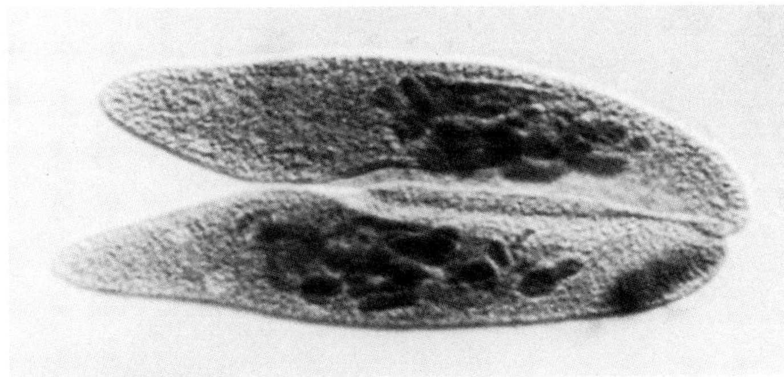

17

Sexual process in protozoa involves the transmission of genetic material from one individual to another. Photos show conjugation between two stentors, trumpet-shaped one-celled protists (left), and between two paramecia, slipper-shaped protists (above).

The reciprocal exchange of genetic material in some protozoa is known as conjugation (Fig. 17). Conjugation occurs between two organisms (like paramecia) in actual physical contact. In bacteria the process is a one-way transfer, not an exchange. A large number of genes, perhaps at times the entire chromosome, may be transferred this way. But investigators found that the chromosomal genes were not involved in the case of drug resistance. Apparently multiple drug resistance is carried by other genes. The genes for resistance to at least five different antibiotics are thought to be on a separate strand of nucleic acid. The R (resistance) factor, as it was named, may be transferred from cell to cell through a conjugation tube. In this way, a bacterial cell can acquire multiple resistance from another bacterium in a single transfer (Fig. 18). Present research is aimed at altering or destroying the R factor, resulting in a much lower prevalence of antibiotic-resistant bacteria. The discovery of cytoplasmic inheritance opened a search for similar genetic mechanisms in other organisms, and many examples have already been discovered.

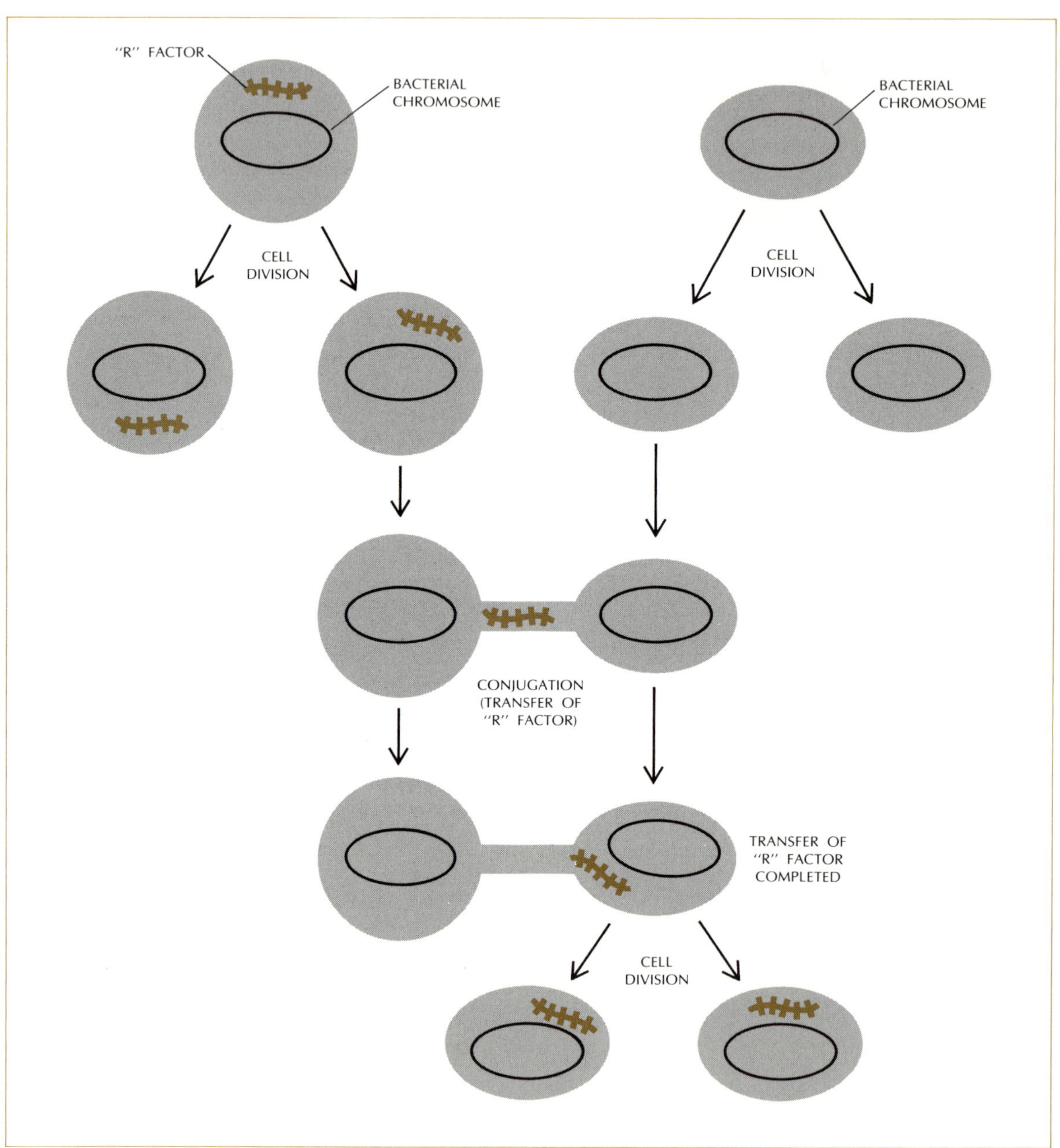

"R" FACTOR
BACTERIAL CHROMOSOME
BACTERIAL CHROMOSOME
CELL DIVISION
CELL DIVISION
CONJUGATION (TRANSFER OF "R" FACTOR)
TRANSFER OF "R" FACTOR COMPLETED
CELL DIVISION

18

◄**Bacteria transmit resistance** to several antibiotics by transferring a cytoplasmic
genetic R factor (color) to another bacterium via conjugation. R factor is not
carried by bacterial chromosome.

READINGS

Beadle, G. W., and M. Beadle, *The Language of Life*. Doubleday, Garden City,
N.Y., 1966 (paper)
    Genetics for the layman, interestingly written by a Nobel laureate and his wife.

Bonner, D. M., and S. E. Mills, *Heredity*. Prentice-Hall, Englewood Cliffs, N.J.,
1964 (paper)
    The cellular and biochemical approach to heredity, based mostly on experi-
    ments with microorganisms; but two chapters concern man.

Fraser, D., *Viruses and Molecular Biology*. Macmillan, New York, 1967 (paper)
    Good descriptions of major experiments with viruses, and the implications of
    the results for the field of biology in general.

Wallace, B., *Chromosomes, Giant Molecules and Evolution*. Norton, New York,
1966 (paper)
    A short book with many well-explained figures. Written with a minimum of
    confusing terminology, but full of information.

Wolf, G., *Isotopes in Biology*. Academic Press, New York, 1964 (paper)
    Good general introduction to a technique that has made it possible to unravel
    the intricate chemical processes of living systems.

6. THE DOUBLE HELIX

The results of the transformation experiments set off an avalanche of research. Were nucleic acids the hereditary materials in other organisms? Studies of cells from microorganisms, plants, and animals soon established beyond a doubt that nucleic acids control the heredity of all known cells.

Several researchers were quick to guess that the explanation of the hereditary activity of nucleic acids lay in their structure. In the early 1950's the architecture of these giant molecules was still a mystery. What was their overall shape? A strand? A coil? A sphere? How were the subunits —the sugars, the phosphates, and organic bases—arranged? And most important, how could the structure of these molecules encode the information of heredity?

The story of the discovery of the structure of nucleic acids has become a modern classic in the history of science. At a single stroke it answered the basic questions about the nucleic acid molecules, suggested the way that the molecule could duplicate itself, and led to the breaking of the genetic code.

THE WATSON-CRICK MODEL

This profound discovery was the work of a 25-year-old Ph.D. from Indiana and a Cambridge University biophysicist who, at the age of 35, was still working toward his doctorate. Their names were James D. Watson and Francis H. Crick. They published their landmark paper (in 1953) in the

◄**Molecule of DNA** from a bacteriophage looks like a tangled string in this electron micrograph. Protein husk of virus is at center. This classic picture gives some idea of the enormous size of a single DNA molecule. Magnification 100,000 ×.

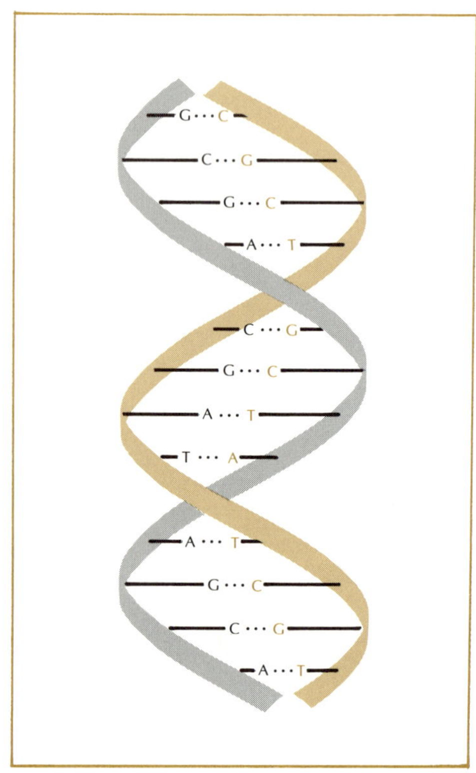

1

Watson-Crick model depicts DNA molecule as a double helix: a structure rather like a twisted ladder. Hydrogen bonds—the rungs of the ladder—link nucleotide bases of the two polynucleotide strands. Adenine, guanine, thymine, and cytosine are designated by the letters A, G, T, and C. This schematic diagram depicts only a fragment of the enormous DNA molecule.

British scientific journal *Nature*. Entitled "Molecular Structure of Nucleic Acids: A Structure for Deoxyribose Nucleic Acid," the paper was a triumph of analysis and intuition. Watson and Crick relied heavily on the experimental results of others: particularly upon the hypothesis that the nucleotide units were linked by bonds between the sugar of one nucleotide and the phosphate groups of the next; upon data from X-ray diffraction experiments that established limits for the width of the molecule; and upon the finding that in deoxyribonucleic acid (DNA) the number of molecules of one of the organic bases, adenine, equals the number of molecules of another, thymine, and the number of guanines equals the number of cytosines. Then came the task of assembling this disconnected information into a meaningful model. Others had tried and failed. Using sheet metal, rods, and wires, Watson and Crick constructed in the laboratory a physical model of the DNA molecule.

The model represented DNA as two strands of polynucleotides arranged in a double helix (double corkscrew). The helices are held together by weak bonds between the bases of the nucleotides opposite them (Fig. 1). According to the model, adenine must bind to thymine, guanine to cytosine.

The nucleotide sequence of one strand determines automatically the nucleotide sequence of the other; the strands are completely complementary. In an understated postscript Watson and Crick pointed out the direction of subsequent research: "It has not escaped our notice that the specific pairing we have postulated immediately suggests a possible copying mechanism for the genetic material."

REPLICATION OF DNA

Two copying mechanisms seemed likely: either the DNA helix reproduces itself as a whole, or it splits into two strands and each reproduces a new half. Matthew Meselson and Franklin Stahl of the California Institute of Technology designed an ingenious experiment to decide between these two alternatives. Their experiment posed some formidable problems of technology. The first was to "label" the strands of the helix. Only by identifying the position of the original two strands after replication could the investigators reconstruct the mechanism of reproduction. Meselson and Stahl noted that the bases along the backbone of the DNA molecule contain many nitrogen atoms, and they decided to label the strands with "heavy" nitrogen.

2
Ultracentrifuge can reach speeds of over 65,000 revolutions per minute and centrifugal force as much as 420,000 times that of gravity.

Most nitrogen atoms contain 7 protons, 7 electrons, and 7 neutrons; their atomic number is 14. Heavy nitrogen has an atomic number of 15 because its nucleus contains an extra neutron. Unlike the radioisotopes phosphorus-32 and sulfur-35 used in the bacterial transduction experiments (Chapter 5), the isotopes of nitrogen are both stable: neither emits radiation. Instead of tracing the labeled isotopes with a Geiger counter, the investigators had to locate them by weight. With a research tool known as an ultracentrifuge it is possible to separate molecules containing a "heavy" isotope from similar molecules containing a lighter one (Fig. 2).

The nitrogen-containing compounds are added to ultracentrifuge tubes containing a special solution. The tubes are spun at very high speed: roughly 60,000 revolutions per minute. Heavier compounds move toward

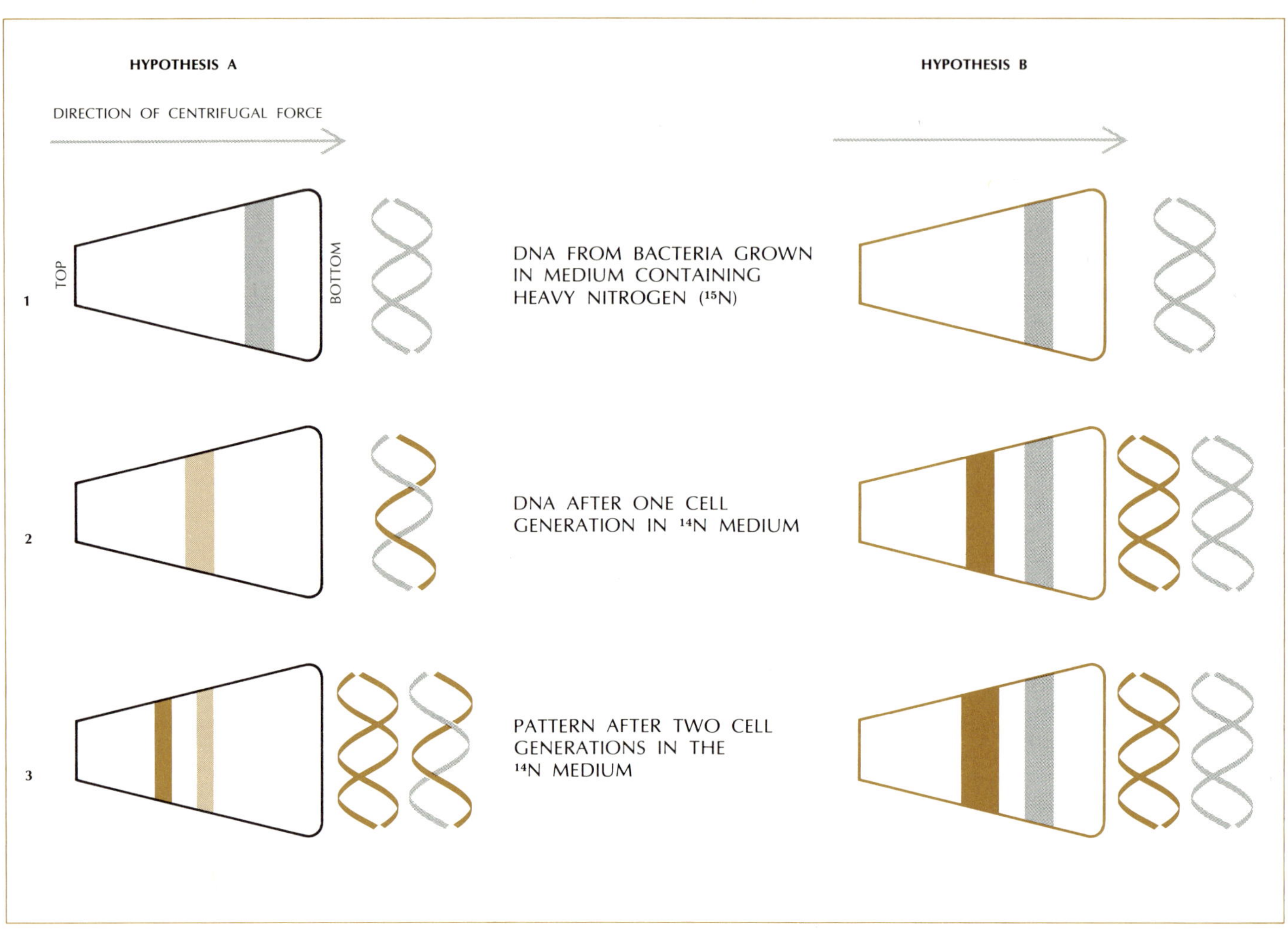

the bottom of the tube. DNA molecules containing heavy nitrogen concentrate in a layer near the bottom of the tube while those containing regular nitrogen accumulate in another layer near the top. If each DNA molecule contained both isotopes, a single layer would form in an intermediate position.

When bacteria grow in a culture medium containing heavy nitrogen, the DNA that they synthesize contains heavy nitrogen; if they are grown in a medium containing nitrogen-14, their DNA will contain nitrogen-14.

3

◄ **Meselson-Stahl experiment** decided between alternative hypotheses about DNA replication. At right are patterns in centrifuge tubes predicted if DNA helix reproduces itself in one intact piece (Hypothesis B). At left are patterns predicted if DNA helix splits and reproduces separate strands (Hypothesis A). Colored bands indicate layers of DNA. Double helices indicate DNA molecule types. Black tint indicates labeling with ^{15}N; solid color labeling with ^{14}N; color tint labeling with a mixture of the two isotopes. Hypothesis A proved correct.

Meselson and Stahl began by growing cells of the bacterium *Escherichia coli* in nutrient media containing heavy nitrogen; eventually all the cells were uniformly labeled with atoms of nitrogen-15. Then the bacteria were abruptly transferred to a culture medium containing only nitrogen-14 atoms. At specific time intervals the growing population was sampled and the relative amounts of DNA containing nitrogen-14 and nitrogen-15 were analyzed. At first all the DNA was labeled with nitrogen-15, and it showed up as a single dense layer in the ultracentrifuge tube. After the population had doubled in the nitrogen-14 medium (one generation later), again a single layer of DNA appeared, but it was farther from the bottom of the tube, and therefore contained lighter DNA molecules. After the growing cells had doubled again in the nitrogen-14 medium, two distinct layers appeared in the tube, one at the middle position and one near the top. Which of the two hypotheses about the helix best fits this experimental evidence?

If the entire double helix were replicated as a single unit, we should expect the results shown in Fig. 3b. But these are not the results actually obtained; thus the single-unit hypothesis is wrong. If the two strands of the helix separate during replication, and each half directs the synthesis of another half, we should expect the results shown in Fig. 3a. These are the results observed; the two-strand hypothesis stands accepted.

It is now assumed that DNA replication begins with the untwisting of the helix and the breaking of the bonds between base pairs. This exposes the unbonded bases on the separated strands. The exposed nucleotides bond to complementary ones from the surrounding cellular "soup." The

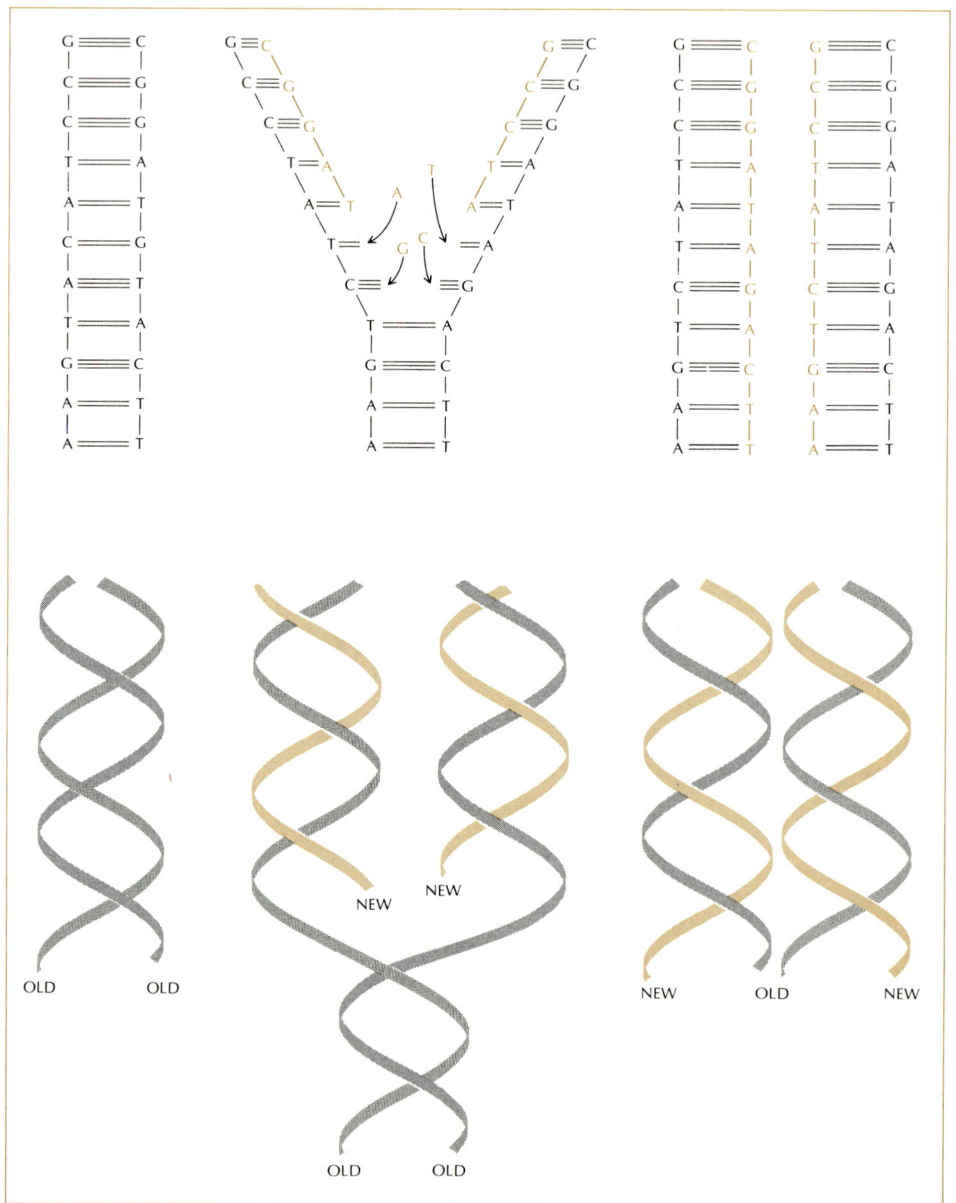NEW
NEW
OLD
OLD
OLD
OLD
NEW
OLD
NEW

4

◄**DNA replication** begins with untwisting of helix (center). Helix acts as template, and exposed nucleotides bond to complementary nucleotides in the cellular "soup," forming new strands. Letters in top diagrams indicate sequence of nucleotide bases. Diagrams below show untwisting and twisting of helices during replication. Original nucleotide strands are shown in grey, new strands in color.

added nucleotides link together to produce complete new DNA molecules identical to the original one. Figure 4 shows the process of DNA replication step by step.

Such an exquisitely simple and subtle mechanism is foolproof enough to guarantee the constancy of the DNA molecule, generation after generation, millenium after millenium. The genetic molecule is a "bank" of information; the survival of the species depends on its stability over eons. However, absolute stability is not necessary, and probably not desirable. Changes do occur in the DNA molecule.

MUTANTS

In a petri dish of bacterial colonies descended from a single cell, once in a while a new kind of bacterium appears. Something in its heredity is different from all the others. Something must have gone wrong with the DNA replication process. A mistake or, in the language of the geneticist, a mutation has occurred. We call the new type a mutant and either wash it down the sink or select it for further study.

Some of the physical and chemical agents known to cause mutations include X-rays, cold shock, and mustard gas. In one way or another, all cause the DNA transcription mechanism to falter, substituting a "wrong" base pair in the new DNA molecule. The change of only a few atoms in the DNA molecule can mean the difference between a green and an albino corn plant, or between a normal child and a mentally retarded child. To understand why such a small change can cause such drastic differences, we must trace the pathway that translates the information stored in the DNA molecule into the anatomy and physiology of a whole organism.

PROTEIN SYNTHESIS

The synthesis of proteins in the cell is controlled by the information encoded in DNA. Most of the DNA is in the nucleus; most protein synthesis occurs on the ribosomes in the cytoplasm. How does the cell transmit information from nucleus to cytoplasm? This involves another nucleic acid called ribonucleic acid, or more commonly RNA (Fig. 5). It differs from DNA in three main ways: it contains the sugar ribose instead of deoxyribose, it contains the base uracil instead of thymine, and it is a single chain instead of a combination of two chains.

To reach the ribosomes, the information in DNA must first be imprinted on an intermediary molecule called messenger RNA (m-RNA). Imprinting proceeds in a rather formal sequence. First the DNA molecule untwists, as in replication. But instead of linking to deoxyribose nucleotides, the exposed bases bond to ribose nucleotides. As the DNA chains separate, the

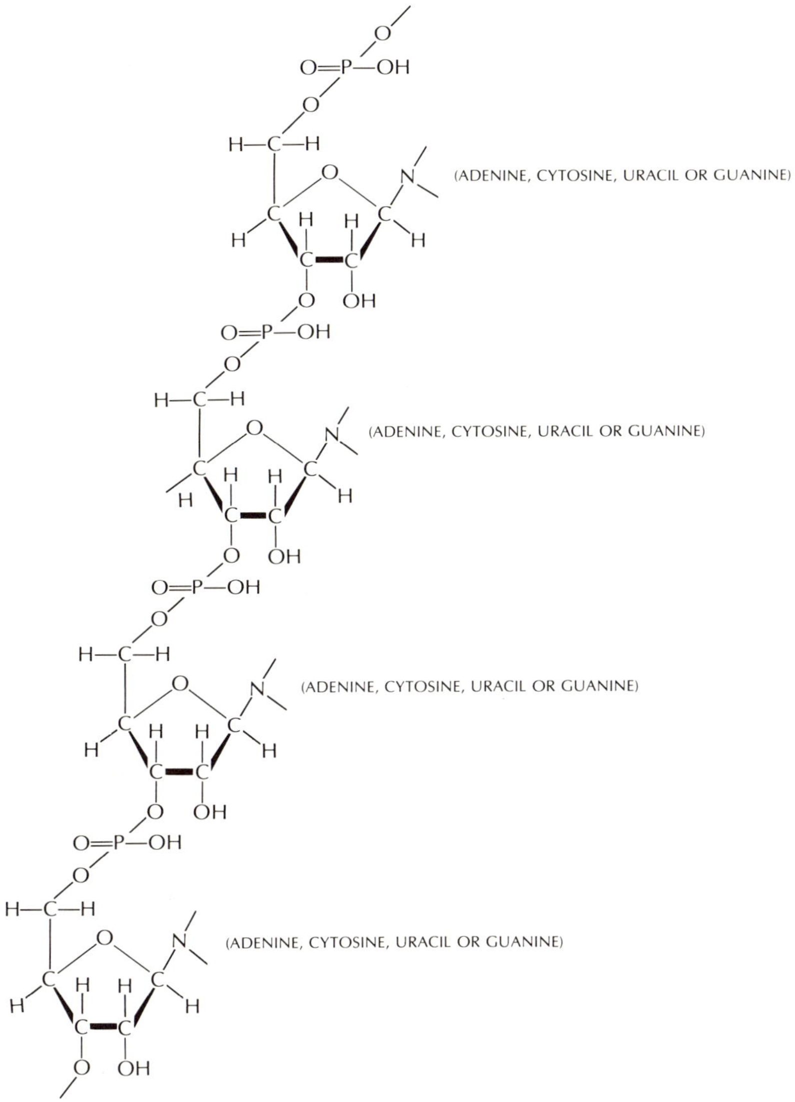

5
Ribonucleic acid (RNA) comprises a single strand of ribose units linked by phosphate groups. Attached to each ribose is one of four nucleotide bases. Three of them (A, G, and C) also occur in DNA. The fourth, uracil (U) occurs only in RNA. Components of RNA molecule are shown on the opposite page, and a fragment of the RNA chain is shown on this page.

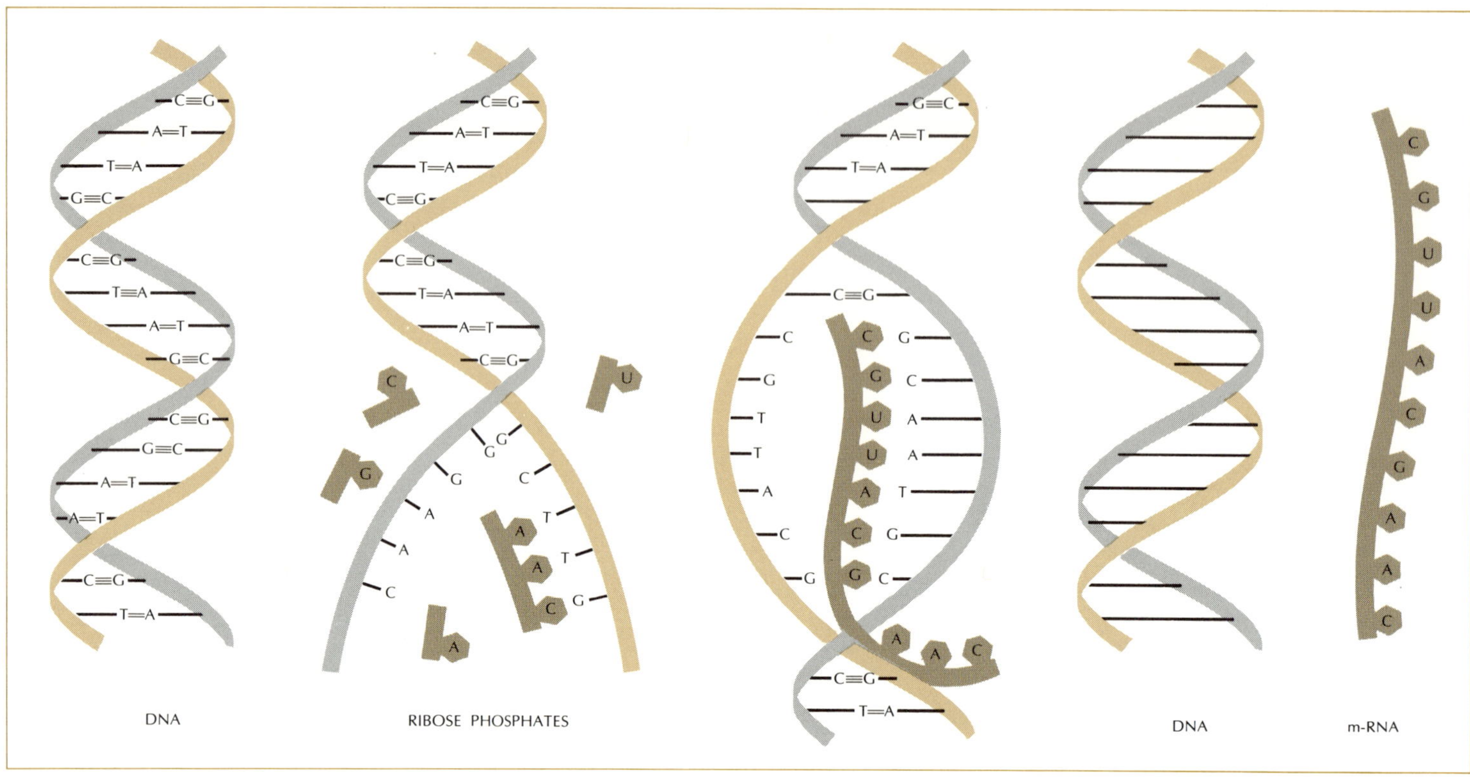

ribose nucleotides attach one at a time to the exposed bases of one of the DNA strands. The bonding rules laid down by Watson and Crick still hold, except that the ribose nucleotide containing uracil (not thymine) bonds to the adenine base of the DNA strand.

Next, the backbone portions of the ribose nucleotides bond together into a polynucleotide chain, a single growing thread of RNA. When the RNA molecule separates from the DNA strand upon which it was made, the two strands of the DNA molecule reunite (Fig. 6). The sequence of bases in the DNA molecule determines the sequence of bases in the new RNA molecule. The information is transferred from DNA to RNA but the structures of the two molecules are not identical. The base sequence of the RNA is complementary to that of the DNA which governs its synthesis. Once the m-RNA has been fashioned and released, the DNA molecule again becomes available to direct the synthesis of additional m-RNA molecules.

◄**Messenger RNA** copies the information encoded in DNA. DNA helix untwists (left center) and ribose phosphates in the cellular soup attach themselves to the appropriate bases, forming an RNA chain (center). DNA molecule retwists into helix (right center), and chain of m-RNA (right) carries the genetic information to the ribosomes.

The m-RNA migrates from the nucleus into the cytoplasm and becomes attached to a group of ribosomes. The ribosomes contain yet another kind of RNA, appropriately named ribosomal RNA (r-RNA). By some still unknown mechanism, the ribosomal RNA bonds to the messenger RNA. The actual synthesis of a protein involves still a third kind of RNA. This RNA is called transfer RNA (t-RNA) because it apparently carries amino acid molecules from the cell soup to the m-RNA–ribosome complex. To create a complicated protein molecule from an assortment of simple amino acids requires the expenditure of energy. It is the molecule ATP that supplies this energy. Specifically, ATP drives the reaction that combines each amino acid with its own specific t-RNA molecule. The result is a battery of activated molecules that attach to the messenger RNA.

How does each t-RNA molecule carry its amino acid to the right position on the strand of m-RNA? Evidently one end of the t-RNA molecule (the end opposite the one that carries the amino acid) has a reactive site

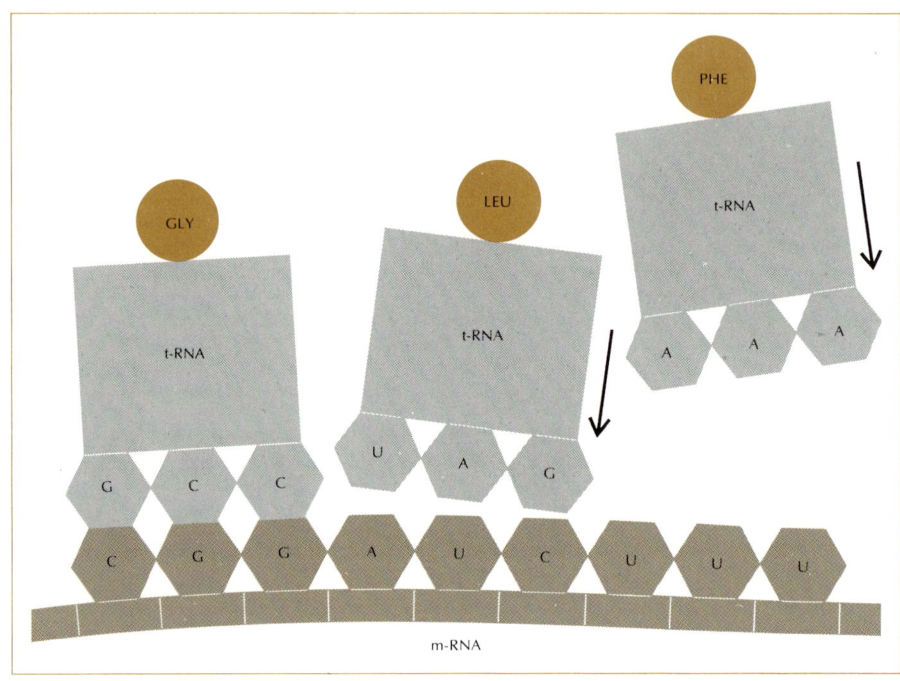

7

Transfer RNA carries amino acids from the cell soup to the m-RNA linked to the ribosome. Complementary bases on the two RNAs linked together, aligning the amino acids (colored disks) in the proper sequence for synthesis of polypeptide chain.

of exposed nucleotides. These nucleotides bond to the m-RNA only at a position marked by complementary nucleotides (Fig. 7). When two such activated molecules lie next to each other on the m-RNA chain, the amino acid portions unite by the formation of a peptide bond. As the amino acids link, the various transfer RNA molecules are released and can pick up other activated amino acid molecules, bringing them in turn to appropriate locations on the m-RNA molecule. As the number of peptide bonds increases, the developing protein chain grows in length. The steps of protein synthesis are depicted in Fig. 8.

170   Chapter 6. The double helix

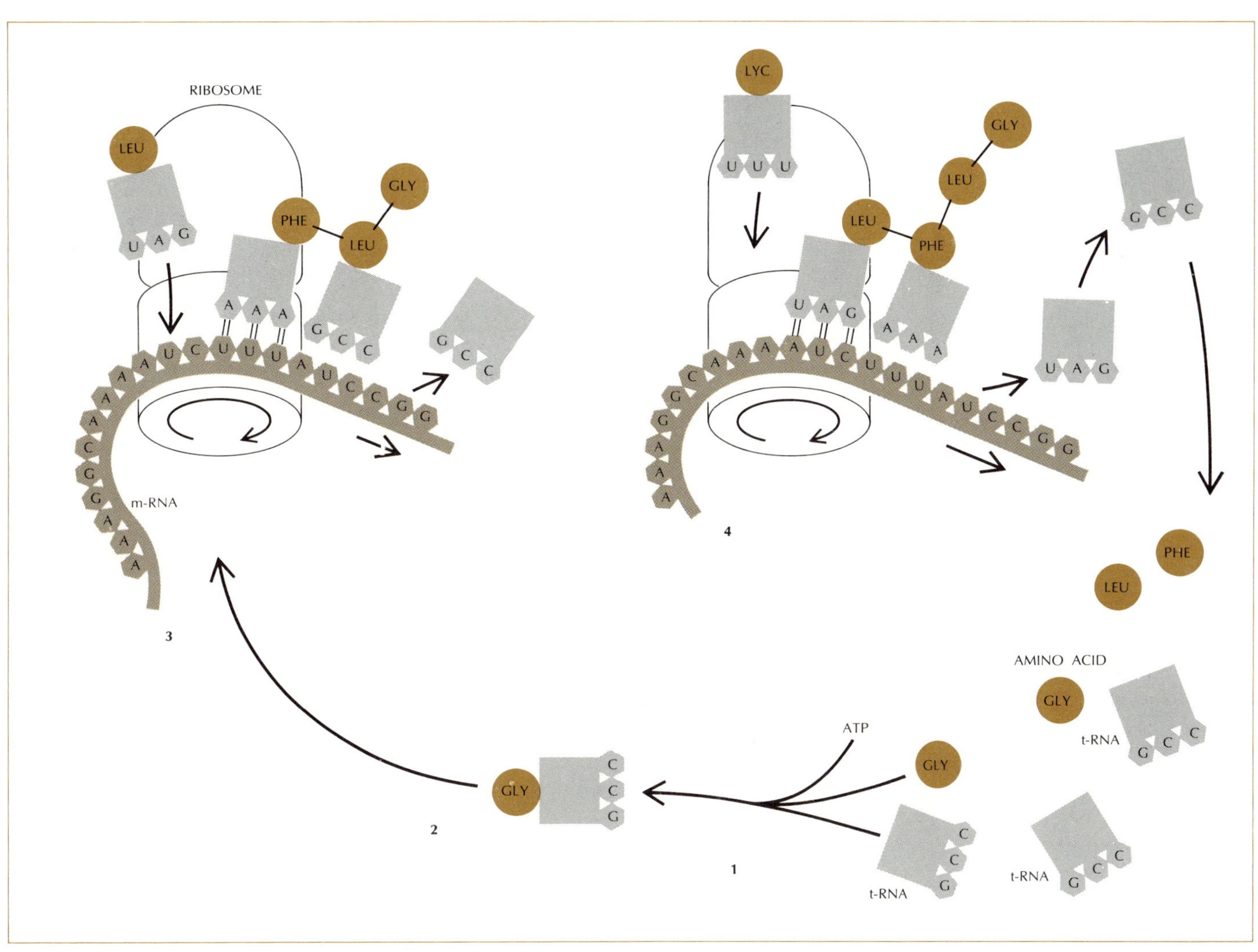

8
Protein synthesis is a cycle that begins with amino acids and ATP (1) being carried to the ribosomes by t-RNA (2). The amino acids are arranged according to information in m-RNA (3). They link together into a polypeptide chain which separates from the ribosome (4). The t-RNA molecules are recycled.

<table>
<tr><td rowspan="2">FIRST BASE OF
m-RNA
TRIPLET</td><td colspan="4">SECOND BASE OF m-RNA TRIPLET</td><td rowspan="2">THIRD BASE OF
m-RNA
TRIPLET</td></tr>
<tr><td>U</td><td>C</td><td>A</td><td>G</td></tr>
<tr><td rowspan="4">U</td><td>PHENYLALANINE</td><td>SERINE</td><td>TYROSINE</td><td>CYSTEINE</td><td>U</td></tr>
<tr><td>PHENYLALANINE</td><td>SERINE</td><td>TYROSINE</td><td>CYSTEINE</td><td>C</td></tr>
<tr><td>LEUCINE</td><td>SERINE</td><td>(END CHAIN)</td><td>CYSTEINE</td><td>A</td></tr>
<tr><td>(OR START CHAIN)</td><td>SERINE</td><td>(END CHAIN)</td><td>TRYPTOPHANE</td><td>G</td></tr>
<tr><td rowspan="4">C</td><td>LEUCINE</td><td>PROLINE</td><td>HISTIDINE</td><td>ARGININE</td><td>U</td></tr>
<tr><td>LEUCINE</td><td>PROLINE</td><td>HISTIDINE</td><td>ARGININE</td><td>C</td></tr>
<tr><td>LEUCINE</td><td>PROLINE</td><td>GLUTAMINE</td><td>ARGININE</td><td>A</td></tr>
<tr><td>LEUCINE</td><td>PROLINE</td><td>GLUTAMINE</td><td>ARGININE</td><td>G</td></tr>
<tr><td rowspan="4">A</td><td>ISOLEUCINE</td><td>THREONINE</td><td>ASPARAGINE</td><td>SERINE</td><td>U</td></tr>
<tr><td>ISOLEUCINE</td><td>THREONINE</td><td>ASPARAGINE</td><td>SERINE</td><td>C</td></tr>
<tr><td>ISOLEUCINE</td><td>THREONINE</td><td>LYSINE</td><td>ARGININE</td><td>A</td></tr>
<tr><td>METHIONINE</td><td>THREONINE</td><td>LYSINE</td><td>ARGININE</td><td>G</td></tr>
<tr><td rowspan="4">G</td><td>VALINE</td><td>ALANINE</td><td>ASPARTIC ACID</td><td>GLYCINE</td><td>U</td></tr>
<tr><td>VALINE</td><td>ALANINE</td><td>ASPARTIC ACID</td><td>GLYCINE</td><td>C</td></tr>
<tr><td>VALINE</td><td>ALANINE</td><td>GLUTAMIC ACID</td><td>GLYCINE</td><td>A</td></tr>
<tr><td>VALINE</td><td>ALANINE</td><td>GLUTAMIC ACID</td><td>GLYCINE</td><td>G</td></tr>
</table>

◄**The genetic code** is composed of three-letter words called triplets. The letters represent the nucleotide bases uracil, cytosine, adenine, guanine. Each triplet specifies a single amino acid or a single instruction (such as "End chain"). To decode a triplet, read this grid across (left to right), then down, then across (right to left). Thus the triplet UGG designates the amino acid tryptophane (solid-colored block). Code is the same in all organisms, from bacteria to man.

THE GENETIC CODE

The sequence of nucleotides in the DNA molecule is a kind of molecular alphabet that spells out the formula for manufacturing a specific protein. Although proteins are synthesized from about 20 different kinds of amino acids, the DNA molecule contains only four different kinds of nucleotides. These are the "letters" of the genetic alphabet. At first it might seem impossible that a four-letter alphabet could govern the protein code that uses 20 different "letters." But the Morse Code uses only two symbols, a dot and a dash, to represent our 26-letter alphabet, and a binary computer uses only a zero and a one to perform its enormous repertory of calculations.

Several groups of investigators tackled the problem of translating the DNA code into the amino acid sequence of the protein molecule being synthesized. They found that each "word" of the DNA code is a sequence of three nucleotides. Each word specifies a single amino acid or a single instruction (Fig. 9). For instance, if the DNA code is

AACCCGGATAAAATC,

the complementary code of the m-RNA would be

UUGGGCCUAUUUUAG.

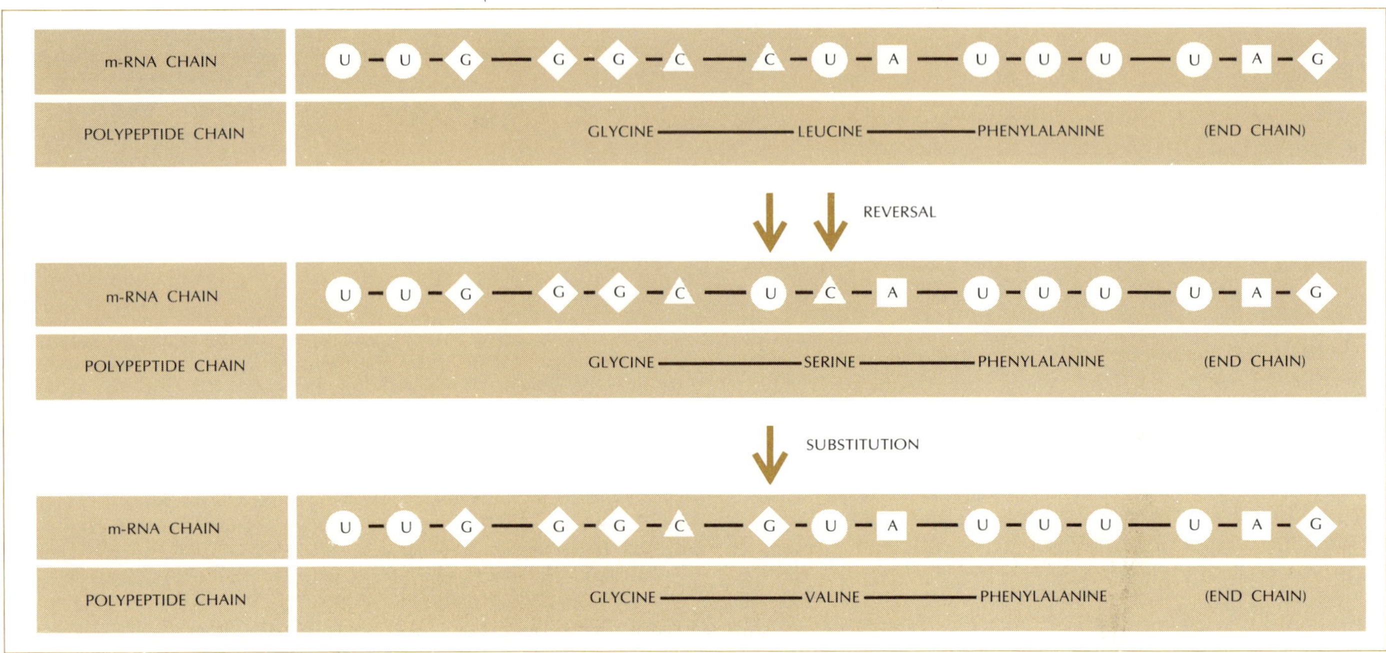

10

Mutations can result from errors in code, such as reversal of two letters or substitution of a
different letter. In this schematic diagram, the "correct" code is at top. Reversal of uracil and
cytosine (middle) causes insertion of serine instead of leucine in polypeptide chain. Substitution
of guanine for cytosine leads to insertion of valine instead of leucine (bottom). Such errors can
impair the effectiveness of resulting enzymatic protein.

The synthesizing mechanism on the ribosomes would read this code three
nucleotides at a time, interpreting the RNA message as follows:

UUG (start chain), GGC (insert glycine), CUA (add leucine),
UUU (add phenylalanine), UAG (stop chain).

The result would be a polypeptide chain of three amino acid units, arranged
in a specific order. But a small change in the sequence of nucleotides,
either in the DNA or in the m-RNA, could result in the formation of a very
different peptide chain. For example, if only two bases in the m-RNA are
reversed in sequence, the resulting chain would be glycine–serine–phenylalanine; or if guanine were somehow substituted for cytosine in the chain,

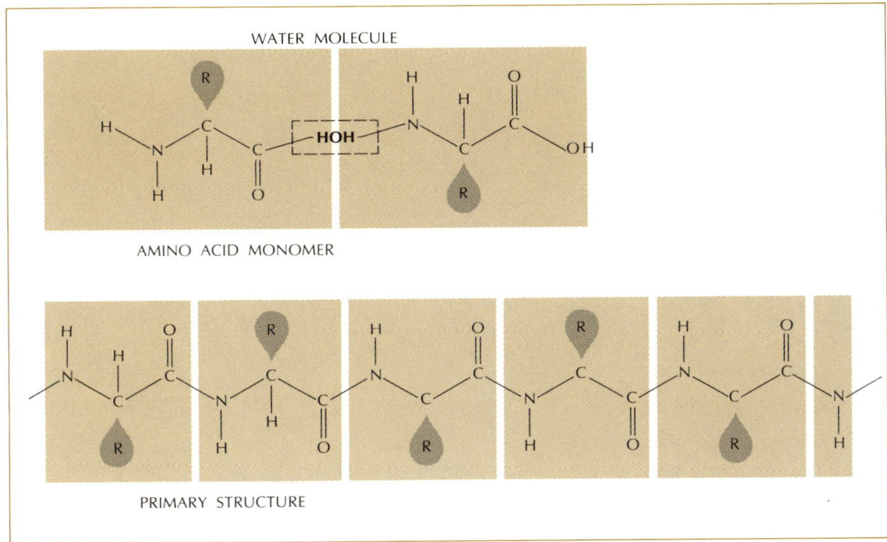

11

Primary structure of a protein is the sequence of amino acid residues that comprises its polypeptide chains. Residues are linked by peptide bond, which is formed by the splitting out of a water molecule from two amino acid monomers (top). Polypeptide chain (bottom) is a polymer of amino acid units. Letter R designates different side chains of amino acids.

the resulting sequence in the protein would be glycine–valine–phenylalanine (Fig. 10). How could such a minor change affect an entire organism? The answer lies in the interrelationship between structure and function in enzymatic proteins.

LEVELS OF PROTEIN STRUCTURE

Protein molecules are not quite so simple as we have implied. Their structure involves at least three levels of complexity. We have discussed only their primary structure: the sequence of amino acids in the polypeptide chain (Fig. 11). The secondary structure of a protein refers to the configura-

12

Secondary structure of proteins is the coiling of the polypeptide chain, usually into a helix, a ring, or a ribbon. Shape of the helix fragment shown here is maintained by hydrogen bonds (dotted lines). View at left shows the amino acid residues within polypeptide chain. At right the same helix is shown more schematically.

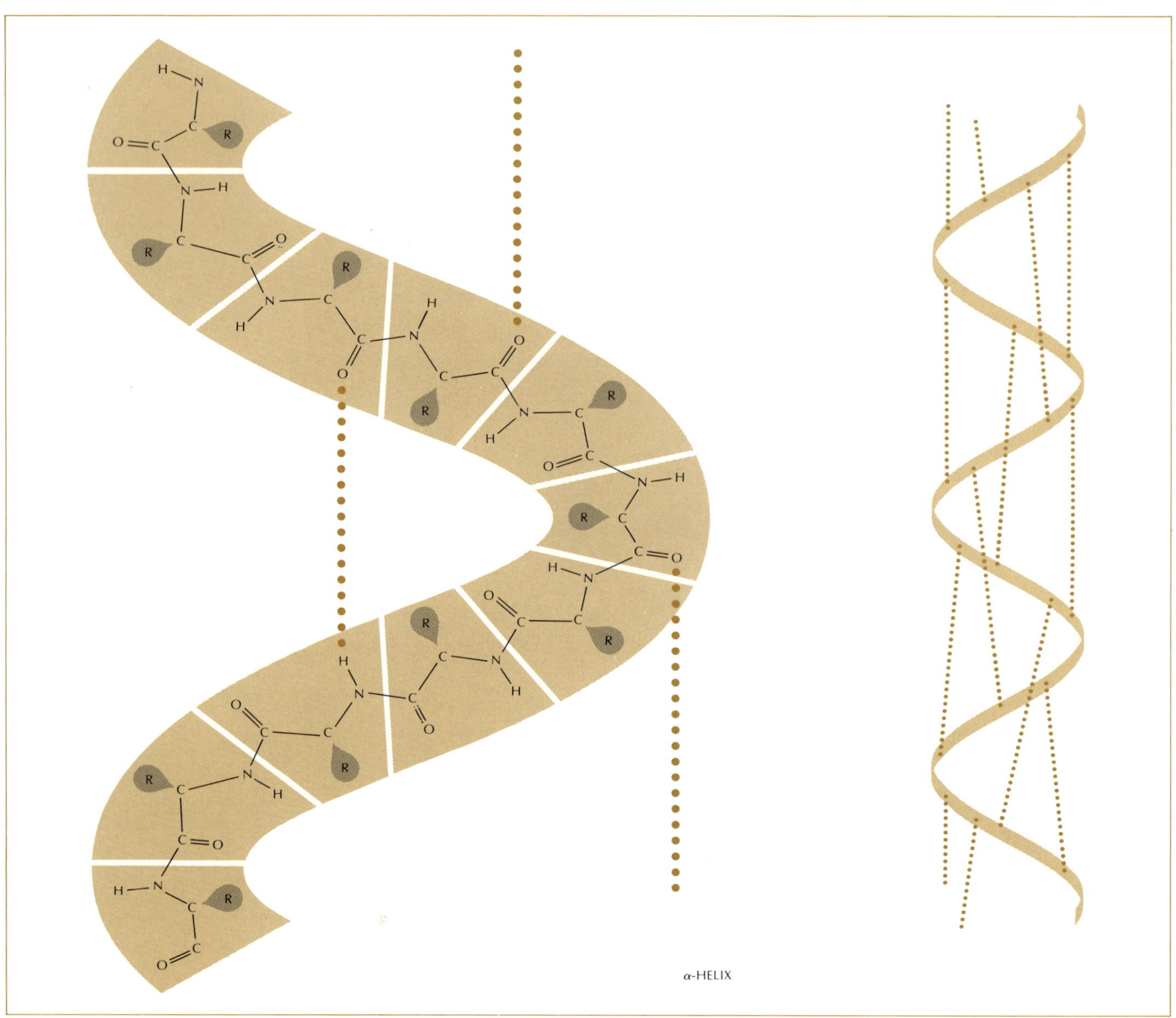

13

Tertiary structure of protein refers to the three-dimensional orientation of polypeptide chains. Some proteins found in skin and tendon are fibrils made up of different helices (upper left). Feathers and silk contain pleated-sheet proteins. Most enzymes are globular proteins.

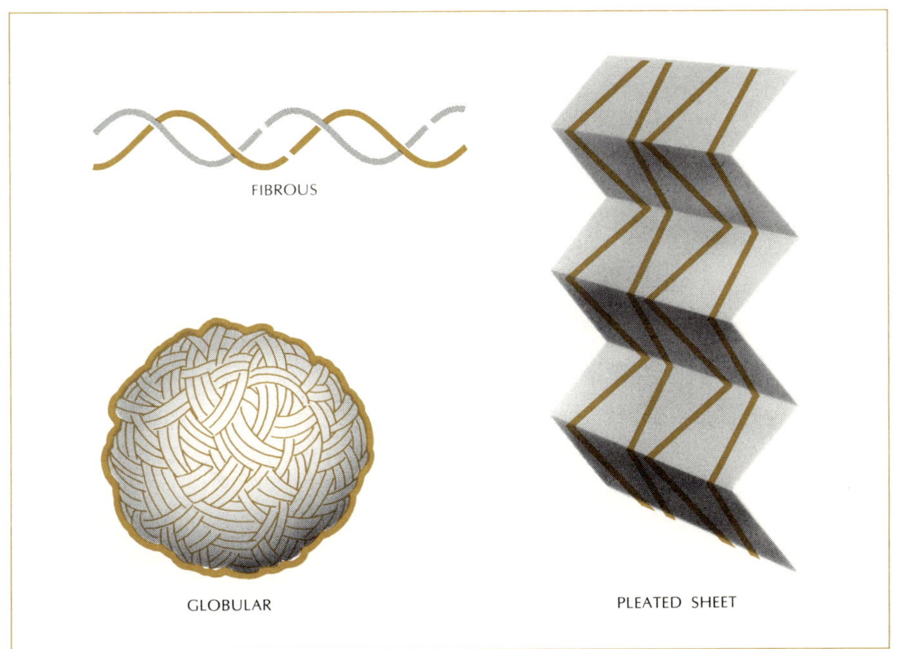

tion resulting from weak cross-bonding between portions of the chain. One example is the corkscrew (helical) shape of some polypeptide chains (Fig. 12). The arrangement of the chains in space is the tertiary structure (Fig. 13), which results from cross-links between two or more separate polypeptide chains. In some proteins the chains are arranged in pleated sheets or in spheres.

From all these theories, facts, and interrelated bits of evidence, one can fit together an explanation of the differences between normal and abnormal organisms. A simple example involves two closely related types of pea plant. One type grows much taller than the other, even in identical environments. Botanists have discovered that a certain chemical, first isolated from a fungus that infects rice plants, can cause the dwarf plants to grow as

14
Effect of gibberellic acid on pea plants. All plants in this photo are dwarf species. Alternate rows have been treated with gibberellic acid and have grown as tall as nondwarf species.

GIBBERELLIC ACID

15
Gibberellic acid is a growth hormone in many plants. It is normally present in tall peas but not in dwarf pea plants.

tall as the others (Fig. 14). It is now known that this chemical, gibberellic acid, is produced in the tall plants, but not in the dwarf plants (Fig. 15).

Such complex chemicals are produced in cells by a complicated series of reactions. Each reaction will "go" only in the presence of a specific enzyme. Most enzymes are globular proteins. The catalytic activity of the enzyme depends on the structure of a very small part of the molecule called the active site. The site consists of a precise arrangement of several amino acid residues. Drastic disruption of the primary, secondary, or tertiary structure that affects the arrangement of residues comprising the active site will destroy the enzymatic properties of the entire molecule.

The dwarf pea plants are unable to synthesize one or more of the enzymes that speed the synthesis of gibberellic acid. There is something wrong with the enzymatic protein-synthesizing machinery of the dwarf plants. This might indicate an abnormality in the DNA of the nucleus. An abnormal nucleotide sequence in the DNA of the dwarf plants would be transcribed by messenger RNA. This message would be carried to the ribosomes where an enzymatic protein with an error in its amino acid sequence would be

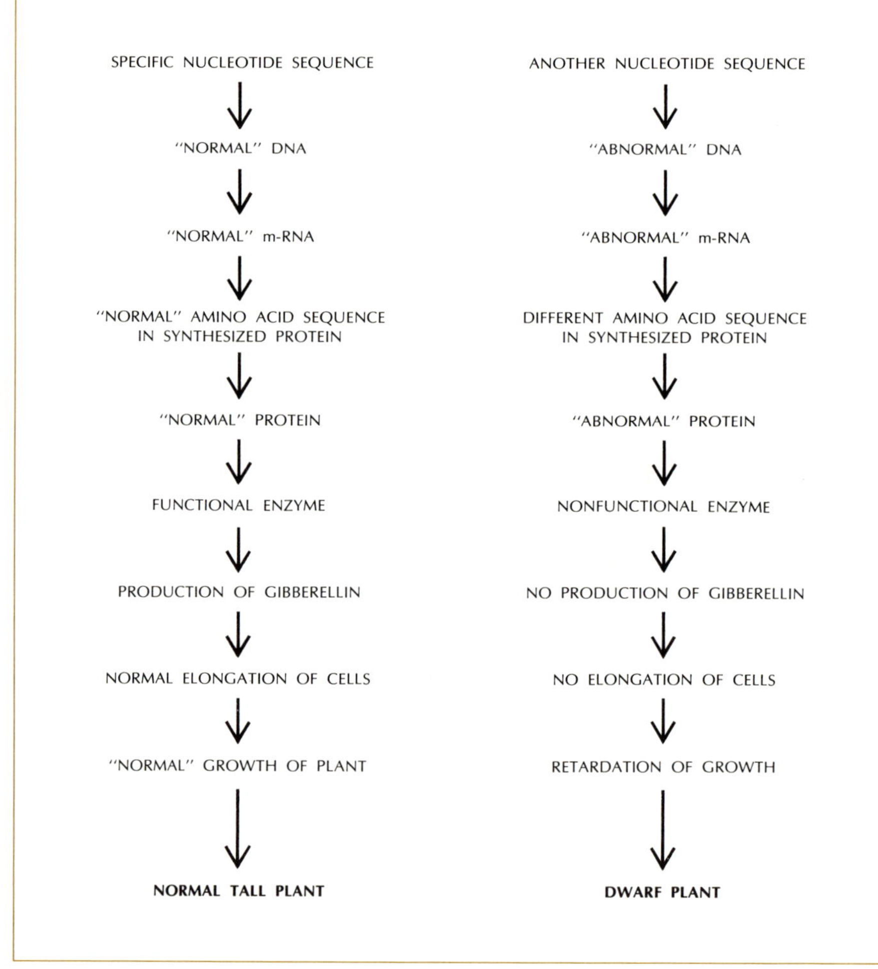

SPECIFIC NUCLEOTIDE SEQUENCE

"NORMAL" DNA

"NORMAL" m-RNA

"NORMAL" AMINO ACID SEQUENCE
IN SYNTHESIZED PROTEIN

"NORMAL" PROTEIN

FUNCTIONAL ENZYME

PRODUCTION OF GIBBERELLIN

NORMAL ELONGATION OF CELLS

"NORMAL" GROWTH OF PLANT

NORMAL TALL PLANT

ANOTHER NUCLEOTIDE SEQUENCE

"ABNORMAL" DNA

"ABNORMAL" m-RNA

DIFFERENT AMINO ACID SEQUENCE
IN SYNTHESIZED PROTEIN

"ABNORMAL" PROTEIN

NONFUNCTIONAL ENZYME

NO PRODUCTION OF GIBBERELLIN

NO ELONGATION OF CELLS

RETARDATION OF GROWTH

DWARF PLANT

16

◀ **Sequence of reactions** responsible for tall and dwarf pea plants is summarized schematically. DNA of dwarf plants may lack information for correct synthesis of one of the enzymatic proteins that enhances the production of gibberellic acid. Without the hormone, the cells of the pea plant do not elongate, and plant growth is retarded.

synthesized. Such an error could disrupt the spatial arrangement of the residues comprising the active site of the enzyme, hence inactivating it. Without the enzyme, the growth-promoting chemical gibberellin would not be made. And without the gibberellin the growth rate is reduced, and the plant remains a dwarf. A step-by-step comparison of the reactions in tall and dwarf plants appears in Fig. 16.

METHEMOGLOBINEMIA

A classic example of a mutation affecting a structural protein occurs in the disease known as methemoglobinemia (met·hemo·globin·emia). The blood of most humans performs well the function of picking up oxygen in the lungs, transporting it to all the cells of the body, and there exchanging it for the waste product carbon dioxide. The oxygen carrier is hemoglobin, which is found in the red blood cells. Hemoglobin consists of an iron-containing group (heme) and a structural protein (globin). The ability of the hemoglobin molecule to combine with oxygen depends on the three-dimensional structure of the intact molecule; the iron-containing group is the reactive site. To carry oxygen, the iron atom must be free to gain and lose electrons, that is, to change easily from the ferrous (Fe^{+++}) to the ferric (Fe^{++}) state and back again. In normal hemoglobin (HbA) nothing interferes with this process. In recent years many kinds of abnormal hemoglobins have been discovered; one of them is methemoglobin (HbM). In this molecule the iron atom is fixed in the ferric state and hence is inhibited from combining with oxygen molecules in the lungs. As a result humans whose blood contains HbM molecules look anemic. They have a slate-gray complexion, tire easily, and become short of breath after only mild exercise. In typical anemia, these symptoms are due to a shortage of hemoglobin in the blood; in this case, however, the problem is the presence of the wrong kind of hemoglobin.

By painstaking analysis it is possible to pull apart a molecule as complex as hemoglobin and to discover its structure. Investigators found that the structure of HbM differs from HbA in only one way: one histidine unit located near the iron atom in the polypeptide chain of HbA is replaced by a tyrosine unit in HbM. Because of its electrical properties, the single tyrosine unit in that strategic location can prevent the iron molecule from changing to the ferrous state and combining with oxygen. Among the hundreds of amino acid residues that comprise the hemoglobin molecule, only one is out of place, but it is enough to make the difference between a normal individual and a sickly one. The biochemical basis of methemoglobinemia adds support to the hypothesis that a mistake in only one nucleotide of DNA, resulting in the substitution of one amino acid for another in a protein molecule, may have drastic consequences for the health and well-being of an organism.

The list of diseases known to be caused by errors in the chemistry of heredity grows at an increasing rate. Sickle cell anemia, a disease marked by blockage of the blood vessels, swollen joints, extreme pain, and ultimate death, is caused by a similar mistake in the architecture of the hemoglobin molecule. Human albinism is due to the absence of the enzyme that catalyzes the production of melanin, the normal coloring pigment in the skin. Phenylketonuria, which leads to mental retardation, is the result of faulty enzymes in the reaction pathway that breaks down the amino acid phenylalanine. The causes of all these diseases are traceable to chemical mistakes in the DNA molecules: mistakes that can be inherited by the offspring.

Medicine has made great progress in the early detection of some of
these diseases. For instance, a simple blood test can detect the abnormal
chemicals present in the blood of a newborn infant suffering from phenylketonuria. In the diseased infant the cells convert phenylalanine to toxic
by-products that may permanently damage the brain. Preventative treatment consists of strict regulation of the baby's diet; phenylalanine is removed from all its food: no phenylalanine—no toxic degradation products.
The child develops a normal mentality. Of course, this treatment does not
cure the disease; it merely averts its worst symptoms.

The recognition of specific errors in the enzyme molecules related to
each malady is only one step toward the conquest of these diseases. Only
by correcting inborn errors in the genetic material itself can we hope to
eliminate these diseases from future generations. When this objective is
finally realized, we shall control our genetic future. We shall no longer be
passive products of our genetic heritage. We shall be able to add, delete,
change, and rearrange our heredity for the benefit of all mankind, a possibility that is already near enough to have aroused sharp controversy (see
Chapter 13).

READINGS

Barry, J. M., *Molecular Biology: Genes and the Chemical Control of Living Cells.*
Prentice-Hall, Englewood Cliffs, N.J., 1964 (paper)

> A clear presentation of the chemical makeup of DNA and its relationship to the
> production of proteins and the functioning of cells.

Sinsheimer, R. L., *The Book of Life.* Addison-Wesley, Reading, Mass., 1967 (pam-
phlet)

> Imaginatively illustrated and clearly written discussion of the way in which
> DNA governs the heredity of organisms.

Watson, J. D., *The Double Helix.* Signet Books, New York, 1969 (paper)

> The story of the discovery of the structure of DNA by one of the men who did it.
> A good picture of the sometimes all-too-human strengths and weaknesses of
> men involved in scientific investigation.

Watson, J. D., *Molecular Biology of the Gene,* 2nd ed. Benjamin, New York, 1970
(paper)

> Not intended for the beginner, this is a detailed picture of gene function.
> Clearly written, heavily illustrated, no mathematics.

7. REPRODUCTION

An organism that is here today and gone tomorrow without leaving off-spring is an evolutionary failure. Every successful generation leaves a new generation to keep the thread of life intact.

In the course of evolution, there has arisen an almost incredible diversity of reproductive mechanisms. Biologists classify these into two basic types: sexual and asexual. The most important difference between the two is that the sexual process involves the transfer or exchange of genetic material while the asexual process does not. Although both types of reproduction result in the creation of new individuals, sexuality offers a considerable evolutionary advantage: it creates new combinations of genetic information, in addition to the changes caused by mutation. As millenia pass and the environment changes, the new combinations which are best suited for survival persist; the others do not. If all offspring were virtually genetically identical to the parents, as they usually are in asexual reproduction, the evolutionary flexibility of the species would be considerably reduced. Thus it is not surprising that most species now alive have some method of transferring or exchanging genetic information.

It would be presumptuous to conclude that any one of the many reproductive schemes used by modern organisms is "better" than the others. All have survival value in a particular environment. Some plants which reproduce sexually in temperate regions reproduce asexually in the arctic. Many plants and animals reproduce asexually for most of the year, producing eggs and sperm only in response to environmental stress. Each pattern

◀**Human egg** was fertilized about 12 hours before this photomicrograph was taken. Two small spheres at left are polar bodies, nuclear material discarded by the maturing egg during meiosis. Egg is about the size of the period at the end of this sentence.

1

Virus attacks bacterium by attaching itself to the cell wall. A single virus ▶
(lower left) has attached itself to an *E. coli* bacterium, a familiar and harmless
inhabitant of the human intestine. After about 20 minutes, the infected
bacterium disintegrates, releasing a swarm of viruses (right).

of reproductive behavior, improved over a long period of time, has enabled
a species to endure for thousands of generations despite the deaths of individual organisms.

Undoubtedly, reproductive mechanisms are still under development.
The slow, complicated pattern of human pregnancy and birth certainly
seems to leave room for improvement. So far, it is merely the most successful result of evolutionary trial and error. Subtle changes are surely taking
place in the process, but the existence and significance of these changes are
difficult to detect because they are so small and, by our standards of time,
occur so rarely. By comparing the mechanisms of reproduction employed
by some representative living systems now on earth we can reconstruct the
major changes that must have taken place in the past, and we can guess at
the types of change that may occur in future eons. In the case of flowering
plants at least, the trend of future change has already become clear.

ASEXUAL REPRODUCTION

Viruses exhibit one of the simplest kinds of asexual reproduction. Apparently an inert chemical system under most conditions, a virus suddenly
springs into "life" when it invades an appropriate host cell. Seizing control
of the cell's biochemical machinery, it efficiently directs the synthesis of
replicas of its own structural components and the enzymes necessary to
assemble the components into viruses. The sequence involves the reproduction, first of nucleic acids, then of proteins, and ultimately of hundreds
of intact viruses (Fig. 1). Although the entire process may take little more

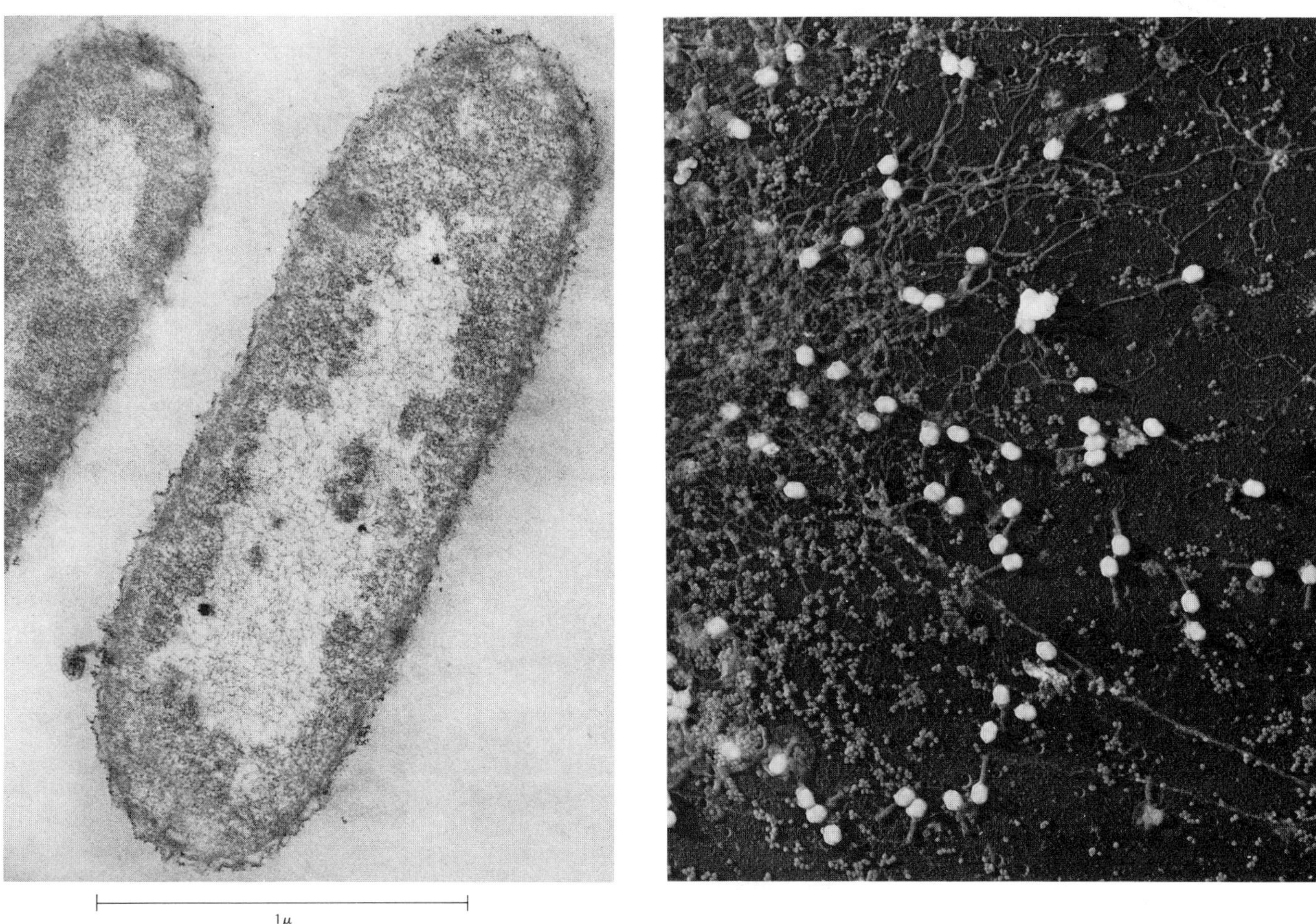

than half an hour, it involves the precise linking of many complex chemical reactions. Only recently has this process become the subject of experimental programs. Its exact nature is not yet understood.

BACTERIA AND BLUE-GREEN ALGAE

Reproduction in the prokaryotes, the simplest unicellular organisms, is a trifle more complicated. The reproduction of a single simple virus particle may yield hundreds of new viruses; the reproduction of a single bacterium yields only two cells. Like virus replication, the process appears deceptively simple. In less than 30 minutes a bacterial cell may divide into two new cells that are structurally, physiologically, and genetically identical to each other and to the parent cell. First the central nuclear region pinches in two. Then a plasma membrane forms, separating the cell into two halves. Fi-

Asexual propagation of plants is a common horticultural practice. Small pieces of stem can produce new roots from the lower cut surface (left) and a new shoot system from the upper portion (right). In this way numerous new plants can quickly be produced from a single parent plant.

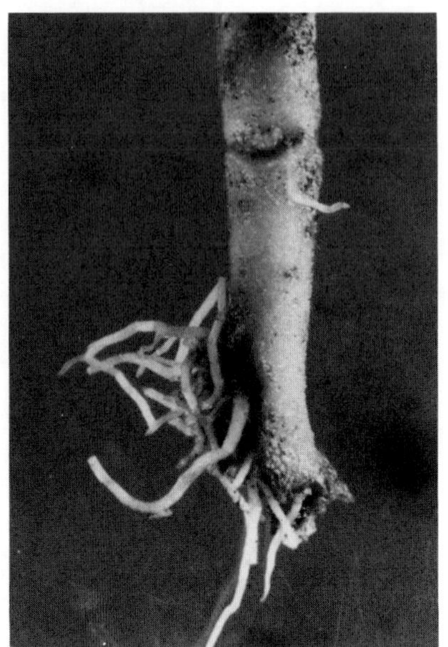

nally a cell wall appears, splitting the membrane lengthwise. The two daughter cells each have a nuclear and a cytoplasmic region, plasma membrane, and cell wall. The process resembles mitosis, but the term is not applied here because, as we recall, bacteria and blue-green algae have no nuclear membrane and no large chromosomes like those of higher plants and animals. Some features of mitosis must occur. Each of the two daughter cells not only acquire half of the nucleic acid of the parent, but exactly the correct half: a full set of genetic information.

THE HIGHER PLANTS

Mitosis is the basis for the asexual reproduction of all eukaryotes. The home gardener is familiar with the practice of increasing the number of such house plants as geraniums, African violets, ivy, and philodendrons by cutting up one plant and then planting the pieces. The cells of each piece of the parent plant undergo numerous mitotic divisions, and differentiate to form the organs of a complete new plant. A stem from a geranium plant, for example, develops leaves on the aerial portion and roots on the base (Fig. 2).

In nature some plants reproduce by a similar mechanism. Strawberry plants send out runners over the surface of the ground that form complete new plants at their tips; the runner may then die, leaving the plants as sep-

Underground stems called rhizomes are one method of reproduction in grasses and certain other plants. Eventually the rhizomes may die (left), separating the parent from the newer plants.

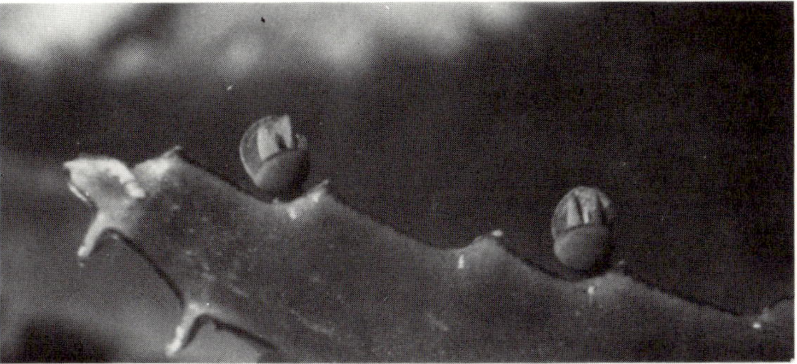

4

Plantlets on bryophyllum leaf fall to the ground and develop into separate plants. Since these plantlets can be produced on a leaf that is pinned to a curtain, the names "air plant" and "live-forever" have been given to this plant.

arate organisms. Plants like ferns, grasses, and iris send out underground stems called rhizomes, which grow horizontally through the soil, producing upright stems at various intervals. The rhizomes between these upright sections may die, leaving separate plants (Fig. 3). A few types of plant, including some ferns and flowering plants, produce minute plantlets along the margins of the leaves. The plantlets eventually drop off and grow into separate plants (Fig. 4).

One of the most efficient methods of asexual reproduction occurs in the fungi, a group that includes mushrooms and puffballs. During one portion of their life cycle, these plants produce by mitosis a vast number of small, easily dispersed structures called spores. Each spore has the poten-

5

◄**Giant puffball** is largest edible fungus. Outer tissue encloses spore-bearing region. This specimen, smaller than the one described in text, grew in Connecticut.

tial to develop into a new individual by further mitotic divisions. It has been estimated that one giant puffball collected many years ago in New York, four feet across and one foot high, contained about 160 trillion spores, each capable of growing into a new puffball (Fig. 5).

MULTICELLULAR ANIMALS

There are numerous cases of asexual reproduction in the animal kingdom. Sponges, flatworms, hydras, sea anemones, hairworms, and their relatives can all produce new individuals from pieces of the parent organism. This reproductive potential had annoying economic repercussions for the uninformed shellfish harvesters of the past. They knew that starfish feed voraciously on clams and oysters, and could sharply reduce the harvest from the shellfish beds. Their solution was to kill all the starfish they could catch. Their method was to cut off the arms of the starfish and throw the pieces back into the sea. But each arm can develop into a new starfish if a portion of the central disk is attached. This practice had a devastating effect on the oyster and clam populations. In their effort to reduce the starfish population, the fishermen were actually increasing it.

In more complex animals the ability to produce entire organisms by asexual means from a portion of the parent body has disappeared. All retain, to a degree, the ability to regenerate parts. Lobsters can regenerate a lost claw; the sea cucumber can regenerate its whole gut; salamanders can regenerate parts of the central nervous system, like the optic nerve. Even humans can regenerate most tissues. The embryos of higher animals show much greater regenerative powers than the adults. If the limb bud of a chick embryo is amputated, the embryo will grow a normal wing from a new bud. The removal of a fully developed wing produces a one-winged chicken. The progressive loss of regenerative ability as an embryo matures into an adult roughly parallels the loss of regenerative ability found among adult organisms of increasing evolutionary complexity. The reason for this loss remains an enigma; to embryologists it is one of the central mysteries of development. To physicians, there would be obvious advantages to discovering the mechanism that could temporarily reestablish in the adult the embryonic potential to regenerate organs and limbs.

SEXUAL REPRODUCTION

Although some organisms have lost the ability to reproduce asexually, almost all species possess a sexual mechanism that involves the transfer of genetic information and the production of individuals with new genetic combinations. The sexual process may or may not directly result in reproduction (the production of new individuals) but it does involve the mixing of genetic material from two different individuals.

CONJUGATION

Bacteria have only a single chromosome. The sexual process in bacteria is called conjugation and involves:

1. The coupling of a male and a female cell.
2. The transfer of a part of the chromosome from the male cell into the female cell.
3. Some type of recombination of the two chromosomes.
4. The production of a cell with a new genetic composition.

Strictly speaking, bacterial conjugation does not involve the exchange of genetic material; there is only a one-way transfer of a single chromosome, or a part of the chromosome, from the male to the female (Fig. 6).

In protozoa, conjugation involves an actual exchange of chromosomal material between two cells. Two cells initiate the process, and after the exchange there are still only two cells. The cells are now both different in their genetic complement, but by no stretch of the imagination can it be said that the cells have reproduced. Only later does each of the cells reproduce by the simple process of mitosis. Note the distinction between sexual

**6
Conjugation in bacteria.** Genetic material from the larger male cell on the top
passes through a narrow conjugation tube into the small female cell on the bottom.

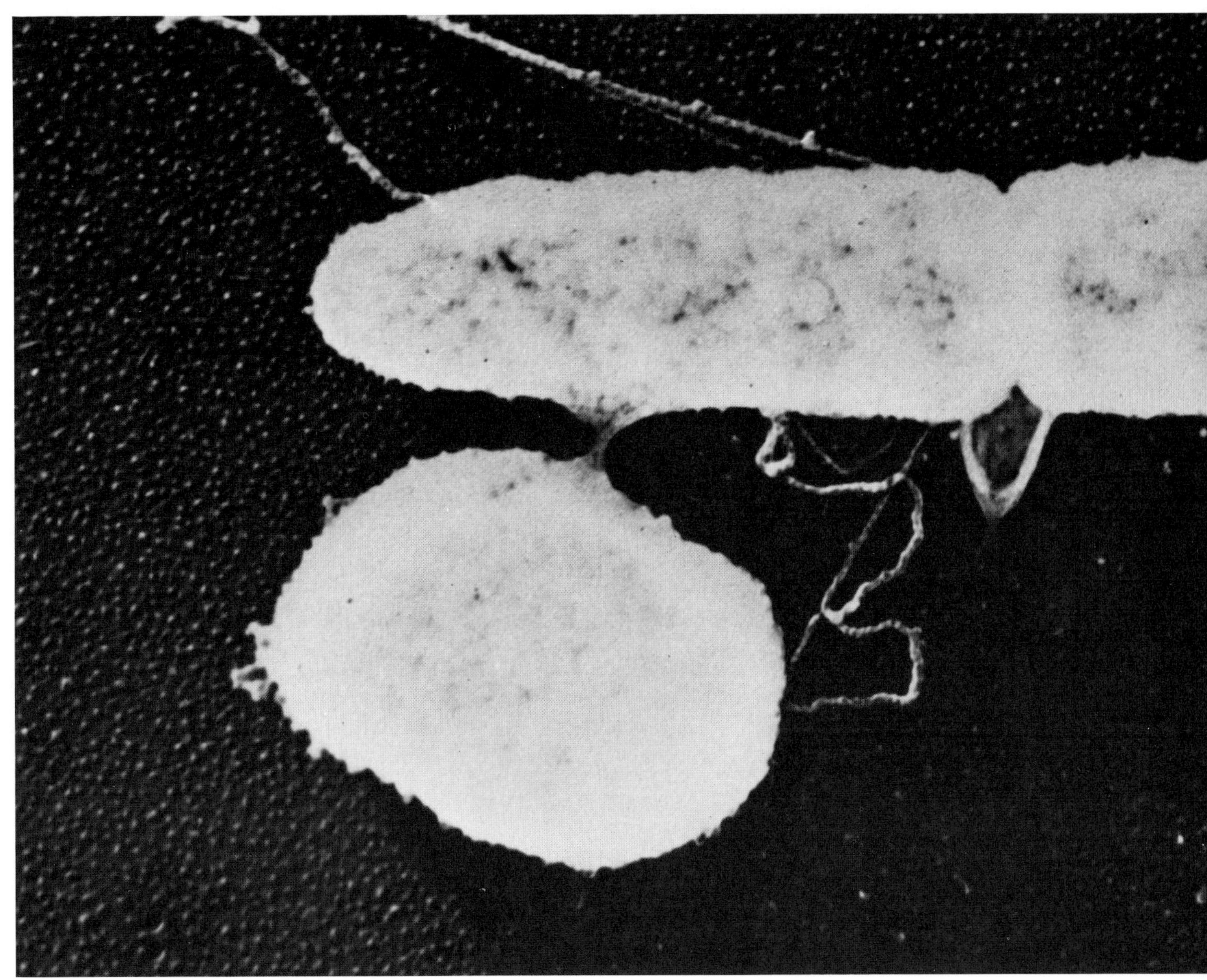

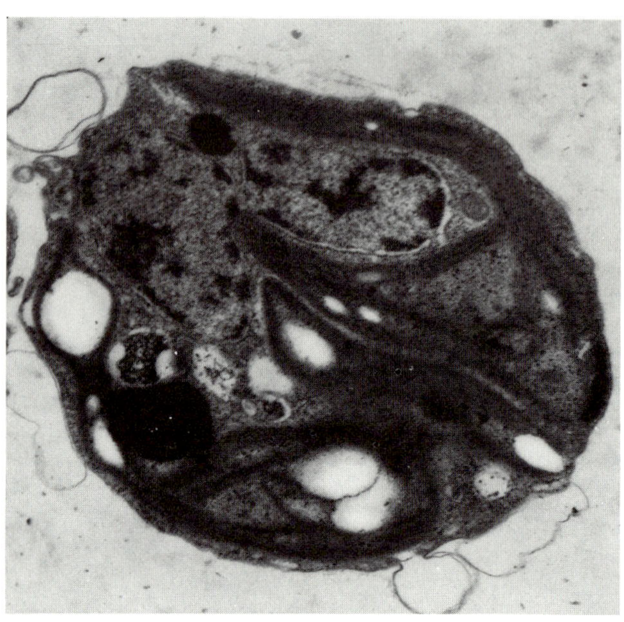

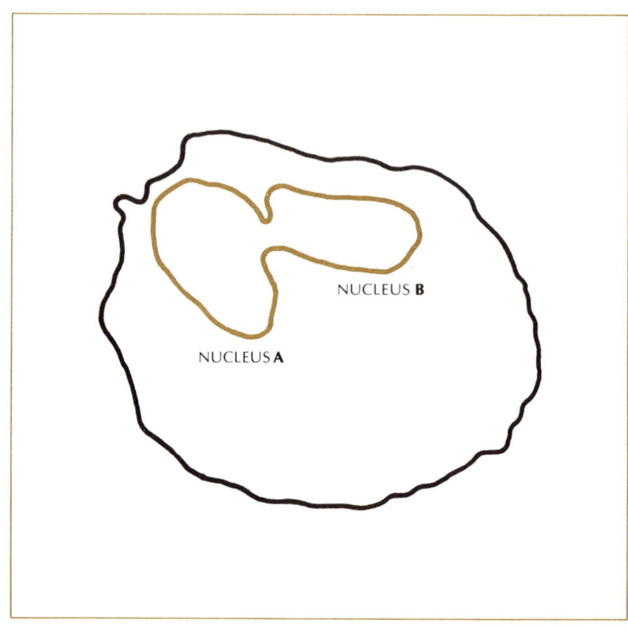

7
Fusion of nuclei in *Chlamydomonas*. After two gamete cells fuse, the nuclei begin to unite (above). Fusion continues (top right) until a single large nucleus is formed containing the contents of the original two nuclei (bottom right). In the case of this *Chlamydomonas* species, each gamete nucleus contributes one nucleolus to the final zygote nucleus.

genetic recombination and reproduction: sex is a process that yields new genetic combinations; reproduction is a process that yields new individuals.

FERTILIZATION

Many organisms exhibit a slightly different sexual pattern. The entire process can be easily observed in the laboratory by using cultures of the unicellular green alga *Chlamydomonas*. Characteristically these organisms multiply by simple mitotic divisions. When the nitrogen content of the water falls below a critical level, the vegetative cells of some species become gametes (sexual reproductive cells): they acquire the potential to fuse in pairs. The first indication of this process is the clumping of many cells into a writhing mass. Soon after this, the mass separates into swimming pairs of cells united by the tips of their flagellae. This event is called fertilization. Next, their nuclei fuse, producing a single nucleus containing the genetic material from both gametes (Fig. 7).

To most people the term fertilization implies the fusion of a large, nonmotile egg and a small, free-swimming sperm cell. Actually, fertilization is the fusion of two gametes regardless of size or shape. In the case of *Chlamy-*

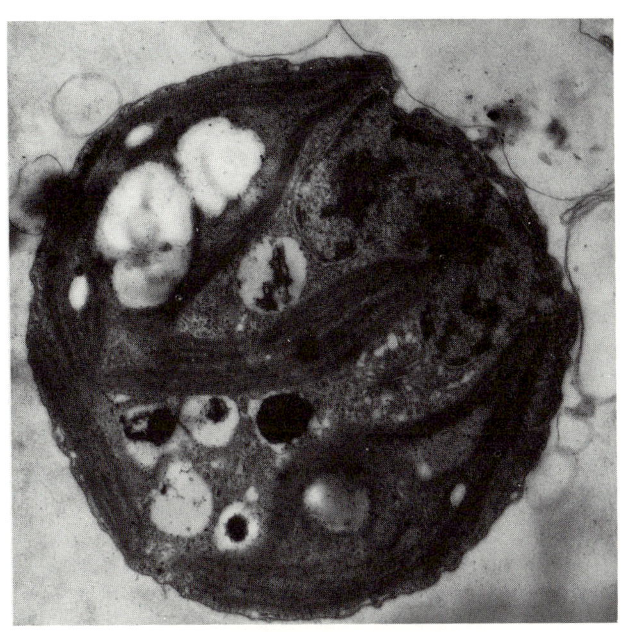
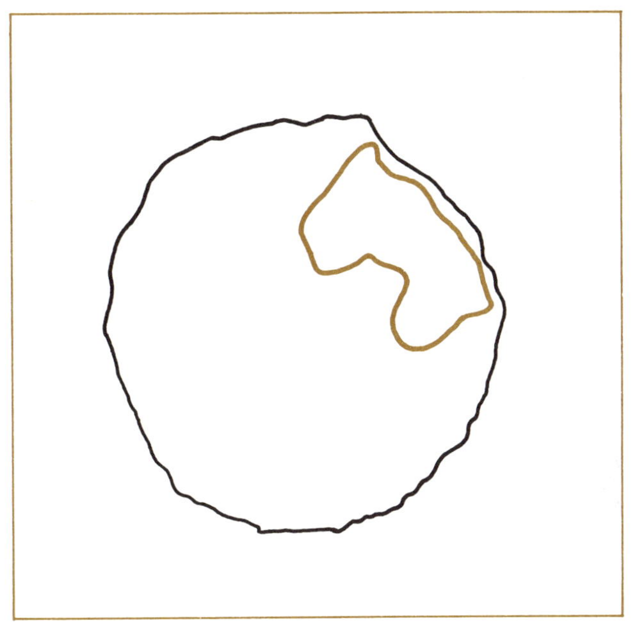
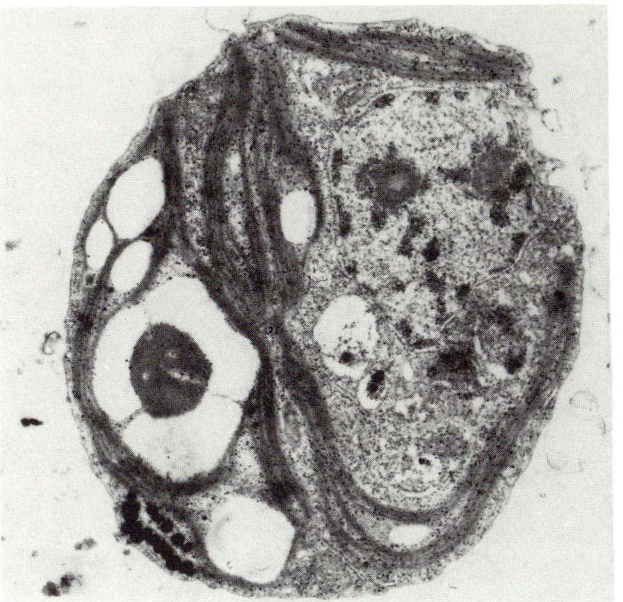
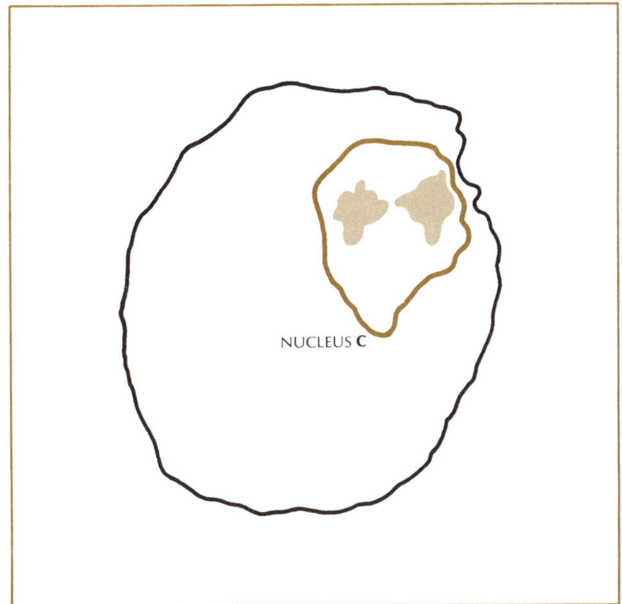
NUCLEUS C

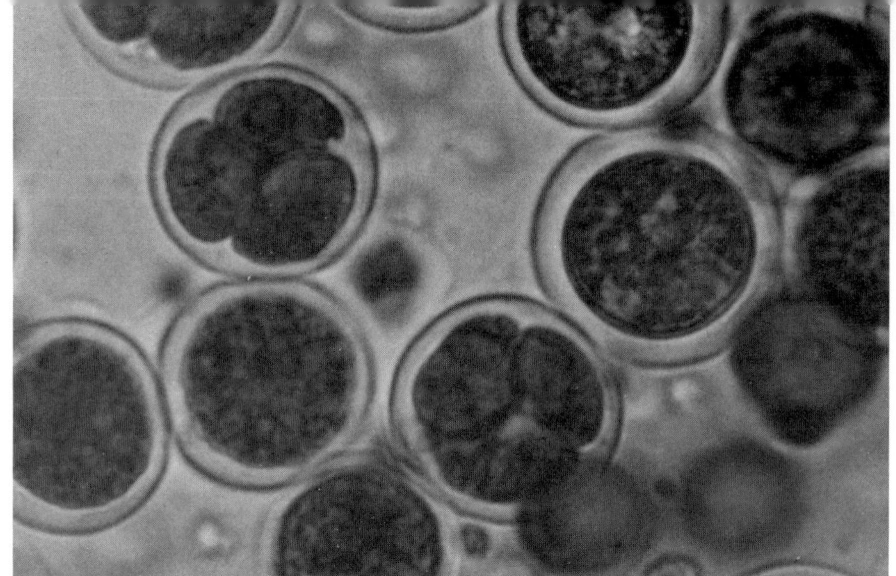

domonas, the terms egg and sperm do not apply. Instead, two motile cells of equal, or nearly equal size fuse. The symbols + and − are used to designate the opposite sex, or mating types. Where there is a definite structural differentiation between the gametes, the symbol ♂ (male) and ♀ (female) are used. In all species, the cell that results from fertilization is called a zygote.

The next step in the life cycle of *Chlamydomonas* is the growth of a thick resistant coat around the zygote (Fig. 8). This enables the zygote to endure for a time in a harsh environment. Certain unknown changes in the environment trigger the division of the protoplast of the zygote into four cells. The wall around the zygote then bursts, releasing the new cells to begin their active motile lives as vegetative cells. These cells divide by mitosis, increasing the population. The sexual phase recurs when the nitrogen concentration drops again. In nature vegetative divisions of *Chlamydomonas* may occur once a day, and the sexual process probably only once a year. Asexual reproduction serves chiefly to increase the size of the population; the sexual process provides a way to survive adverse conditions and, more important, provides the raw material for evolutionary adaptation by producing organisms with new combinations of genetic material.

SEX AND CHROMOSOME NUMBER

Fertilization produces a cell with double the number of chromosomes of each of the gametes. If a gamete with eight chromosomes fuses with another, the result is a zygote with 16 chromosomes. If this zygote were to divide by mitosis its daughter cells would each have 16 chromosomes. If these cells in turn became gametes and fused in pairs, the result would be

8

a zygote with 32 chromosomes, and so on. In other words, the chromosome
number would double each time the sexual process occurred. Actually this
does not happen. We have already seen that, within limits, each species
of organism retains a specific chromosome number, generation after generation. Sexual reproduction must involve some process that can halve the
chromosome number. This process is called meiosis.

MEIOSIS

Meiosis is crucial to the sexual reproduction of most species, including
man. Like mitosis, meiosis is a process in which the nucleus divides and
reproduces itself. Unlike mitosis, the new nuclei contain only one half the
chromosome complement of the original nucleus.

Meiosis begins with a series of invisible changes in the cell. DNA is
replicated; proteins are synthesized; enzymes are activated. The chemical
trigger that determines whether a cell will undergo mitosis or meiosis is
completely unknown. It is known that the preparatory phase of meiosis
is substantially longer than that of mitosis. This indicates that meiocytes
(cells about to undergo meiosis) are really differentiated cells. An undifferentiated cell can divide and continue to divide by mitosis. If a cell
matures into a specialized structure like a neuron, a blood cell, a tracheid,
or a root hair, we say that it has differentiated. It loses the power to divide.
In short, once a cell begins to develop and differentiate it is committed; it
cannot dedifferentiate. Similarly, once a cell enters the first stages of meiosis
it also is committed; it never again reproduces by mitosis. A meiocyte has
only one path open to it: it will produce four germ cells, each with half the
chromosome number of the meiocyte.

Although we do not know the chemical reactions involved in the process, we can describe the visible changes observed in cells undergoing meiosis. First the cell appears to be identical to all other undifferentiated cells. The nucleus is large and apparently structureless under the light microscope (Fig. 9a). The first visible sign of change is the appearance of long threads in the nuclear region (Fig. 9b). As these become more evident, the threads begin to pair (Fig. 9c). This is the first event visually different from those of mitosis. Because of its uniqueness and because of its importance to the whole course of the meiotic process, let us take a closer look at the pairing of chromosomes. The number of chromosomes in a gamete may be represented by the letter N. In *Chlamydomonas* $N = 8$; in humans, $N = 23$. The number of chromosomes in the zygote is the sum of the chromosome numbers of the two gametes that fused to produce it: $N + N = 2N$. In short, fertilization doubles the chromosome number. Meiosis reduces the 2N chromosome number, called the diploid number, to the 1N number, called the haploid number.

One of the two sets of chromosomes in the meiocyte comes from each parent. Close examination of these sets of chromosomes (Fig. 9d) reveals that each chromosome derived from one parent has a look-alike twin in the set derived from the other parent. Not only are these chromosomes alike in structure, but they are also similar genetically. Both chromosomes of the pair contain genes for the same characteristics. Such structurally and genetically similar chromosomes are called homologues (from the Greek prefix homo, meaning alike). In the meiocytes of humans there are 23 pairs of homologous chromosomes, 46 chromosomes in all. In the worm *Ascaris* there are two pairs of homologues, or 4 chromosomes.

Meiosis in lily. Cells in the male flower parts (A) begin to differentiate. Chromosomes appear as thin strands (B). Synapsis of homologous chromosomes (C) is followed by extreme thickening and shortening of the paired chromosomes (D) (continued on next page).

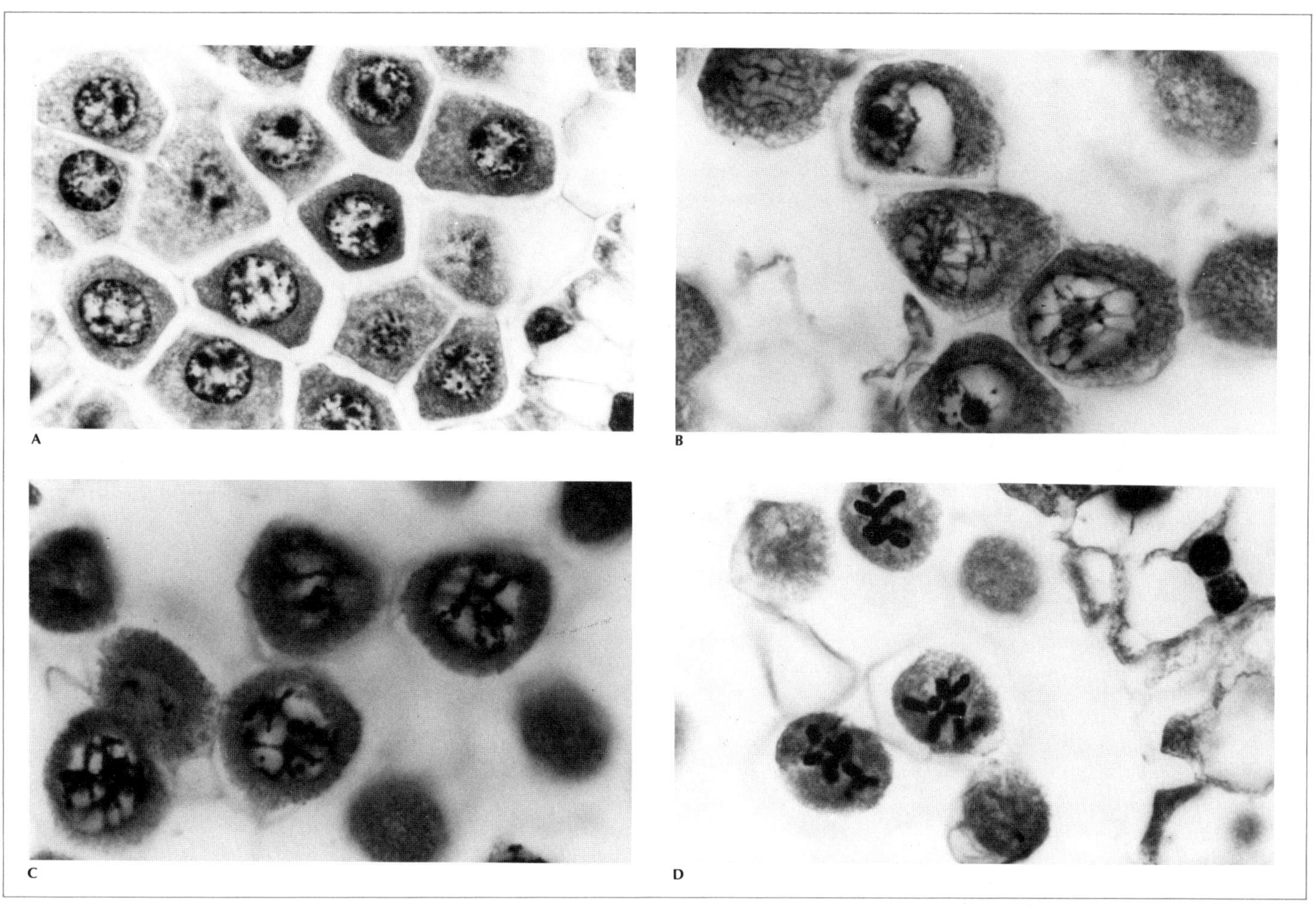

E

Meiosis in lily (cont.) Crossing over results in exchanges of pieces of one chromatid with a chromatid of its homologue (E). The paired chromosomes align themselves at the center of the cell. Some have only one or two cross-over points, others have many (F). The homologous chromosomes separate, bearing the exchanged chromatid portions (G). Cytokinesis produces two cells, each with half the chromosome number of the parent cell (H). The second division of meiosis involves the separation of the chromatids of each chromosome (I). A second cytokinesis produces four cells (J). Due to the exchanged portions of chromatids, no nucleus contains a complement of chromosomes exactly like that of any other.

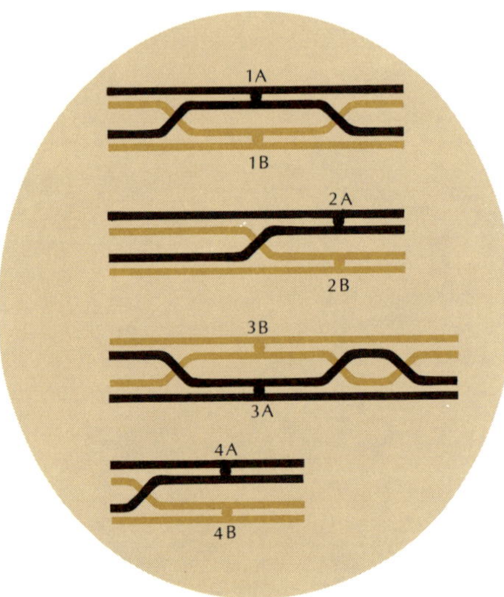

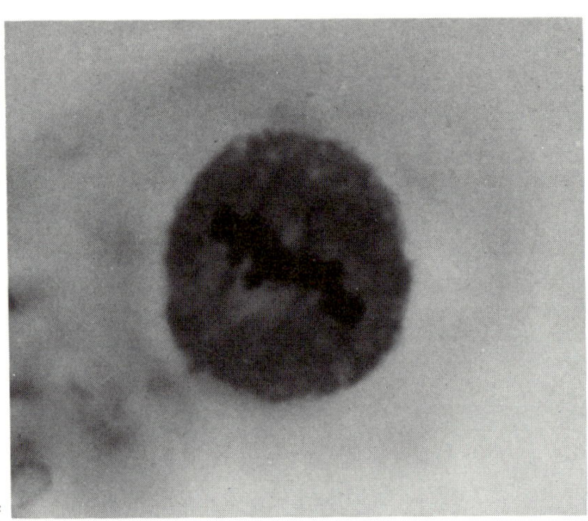

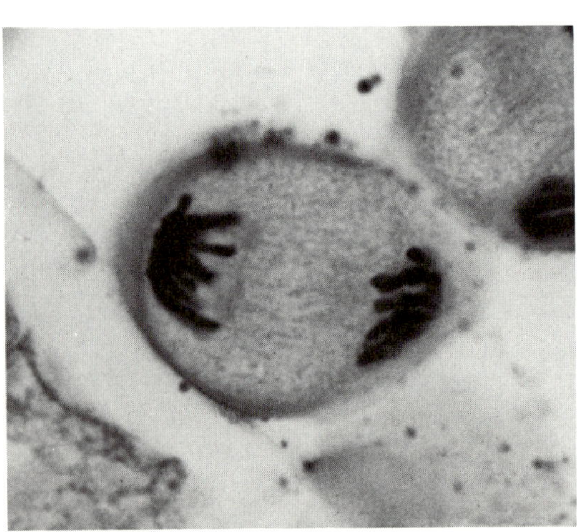

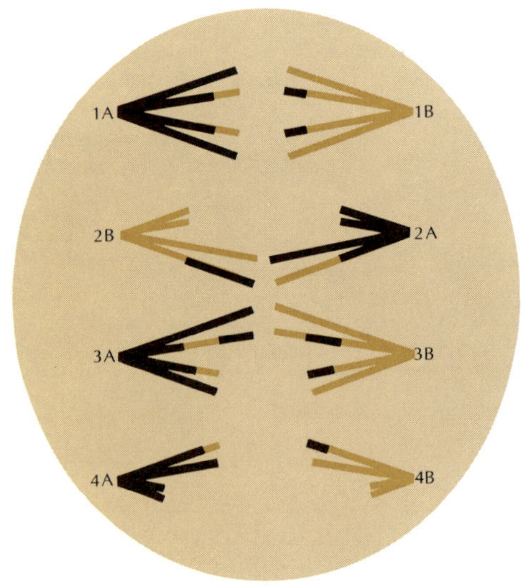

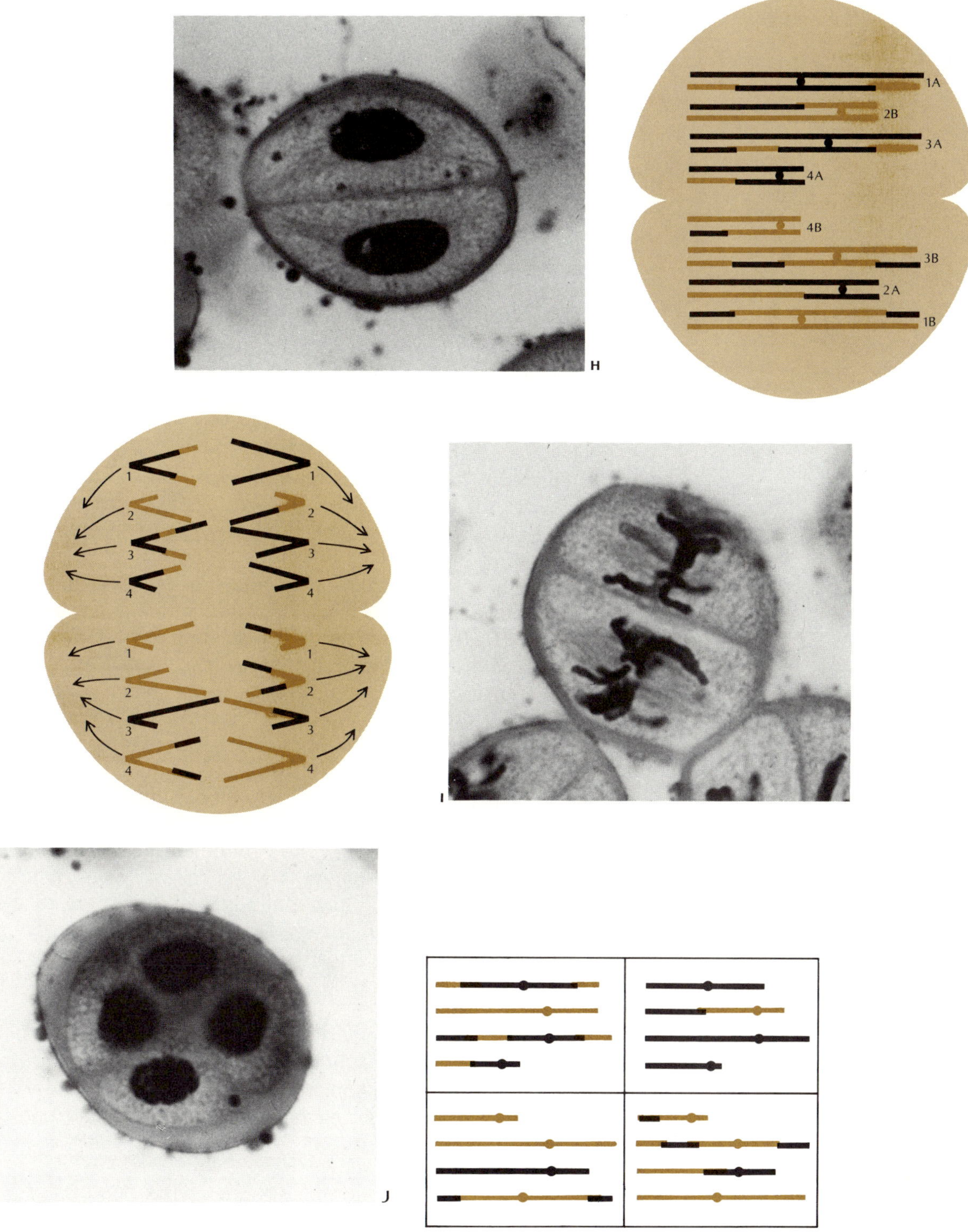
H
1A
2B
3A
4A
4B
3B
2A
1B
I
J
1
2
3
4
1
2
3
4
1
2
3
4
1
2
3
4

The pairing of homologous chromosomes in the first stages of meiosis is called synapsis. It is an intimate association of the two homologues point for point, or more specifically gene for gene. A gene that controls eye color on one chromosome pairs with the gene that governs eye color on its homologue. When synapsis is complete the genes of each chromosome are paired with the corresponding genes of its homologue.

Since each chromosome is composed of two identical chromatids, each pair of synapsed chromosomes is really a closely bonded group of four chromatids. After synapsis, the chromatids break. When breaks occur simultaneously at the same region of two chromatids, the broken ends fuse in such a way that part of one chromatid becomes attached to part of another (Fig. 9d). In short, the chromosomes have exchanged parts. This process, which is called *crossing over*, results in the formation of chromatids with an arrangement of genes different from that of the original chromatids.

During the process of synapsis and crossing over the chromosomes gradually condense and thicken. Then the homologous chromosomes begin to separate. But they cannot achieve complete separation immediately because they are still held together at the cross-over sites. At this stage each cross-over point causes a X in the visual appearance of the chromosome pair (Fig. 9e). This is called a chiasma (from the Greek letter chi, X). There may be one or more chiasmata in a pair of synapsed homologous chromosomes.

Next the paired chromosomes move about until they become arranged on the equatorial plate (Fig. 9f). Spindle fibers become evident and soon the paired chromosomes separate. The centromere of one of the paired chromosomes leads the way to one pole of the cell while the centromere of the other chromosome of each pair pulls that chromosome to the opposite pole (Fig. 9g). It must be emphasized that the chromosomes are now different from what they were originally. They have exchanged parts with their homologues and so have a genetic makeup different from that of the original chromosomes.

When the chromosomes reach the poles, cytokinesis usually occurs. Two cells are formed (Fig. 9h). The chromosomes in each of these cells line up on their respective equatorial plates. The chromatids of each chromosome separate, one from each chromosome going to one pole of the new cell and the other going to the opposite pole (Fig. 9i). Cytokinesis occurs again (Fig. 9j). Distinct nuclei are formed, and meiosis is complete.

There are now four cells, each containing the haploid number of chromosomes. Moreover, each of the four cells is genetically different from each of the others. The subsequent fate of these cells depends upon the life cycle of the species. A comparison of the life cycles of four representative organisms—a green alga, a fern, a flowering plant, and a human—illustrates the range of diversity in patterns of sexual reproduction.

GREEN ALGAE

The life cycle of *Chlamydomonas* was described earlier in this chapter. Here we need only add that each of the free-swimming vegetative forms is a haploid (1N) cell that can divide mitotically to produce two new haploid individuals. In a nitrogen-poor environment the cells become gametic, and can fuse to produce a diploid (2N) zygote. Meiosis in the zygote yields four haploid vegetative cells. These can divide by mitosis, increasing the population and completing the cycle. Two aspects of this cycle differ from the pattern found in most other organisms: (1) the zygote is the only diploid structure; the vegetative cells are haploid; and (2) meiosis occurs in the zygote rather than in the formation of gametes.

FERNS

The fern plant seen growing in the woods or in the home characteristically possesses dissected (segmented) leaves called fronds. In contrast to *Chlamydomonas,* in which the only diploid structure is the zygote, all cells of the fronds and the other parts of the fern plant contain the 2N diploid chromosome number. At maturity the fronds may develop structures called sporangia (spore sacs). Inside these sacs certain cells undergo meiosis, producing spores with the N haploid chromosome number. When the sporangium breaks open the spores can blow away (Fig. 10). Those that land in moist soil germinate (sprout), and after many mitotic divisions become small heart-shaped plants called gametophytes (gamete-producing plants). Each cell of these plants has the haploid chromosome number. Some of these cells produce sperm, others eggs. When an egg is fertilized by a sperm the result

10
Sporangia are sacs containing spores.
Eventually the sacs break open and the
spores blow away. Sporangia of ferns often
occur in clusters on the underside of the
fronds.

is a diploid zygote. This germinates, and after many mitotic divisions a familiar diploid fern plant is produced, similar to the plant with which the cycle began (Fig. 11). Since these plants later produce spores, they are called sporophytes (spore-producing plants).

Note that the life history of ferns includes two kinds of plants: a diploid sporophyte, the typical fern plant, and a haploid gametophyte, an insig-

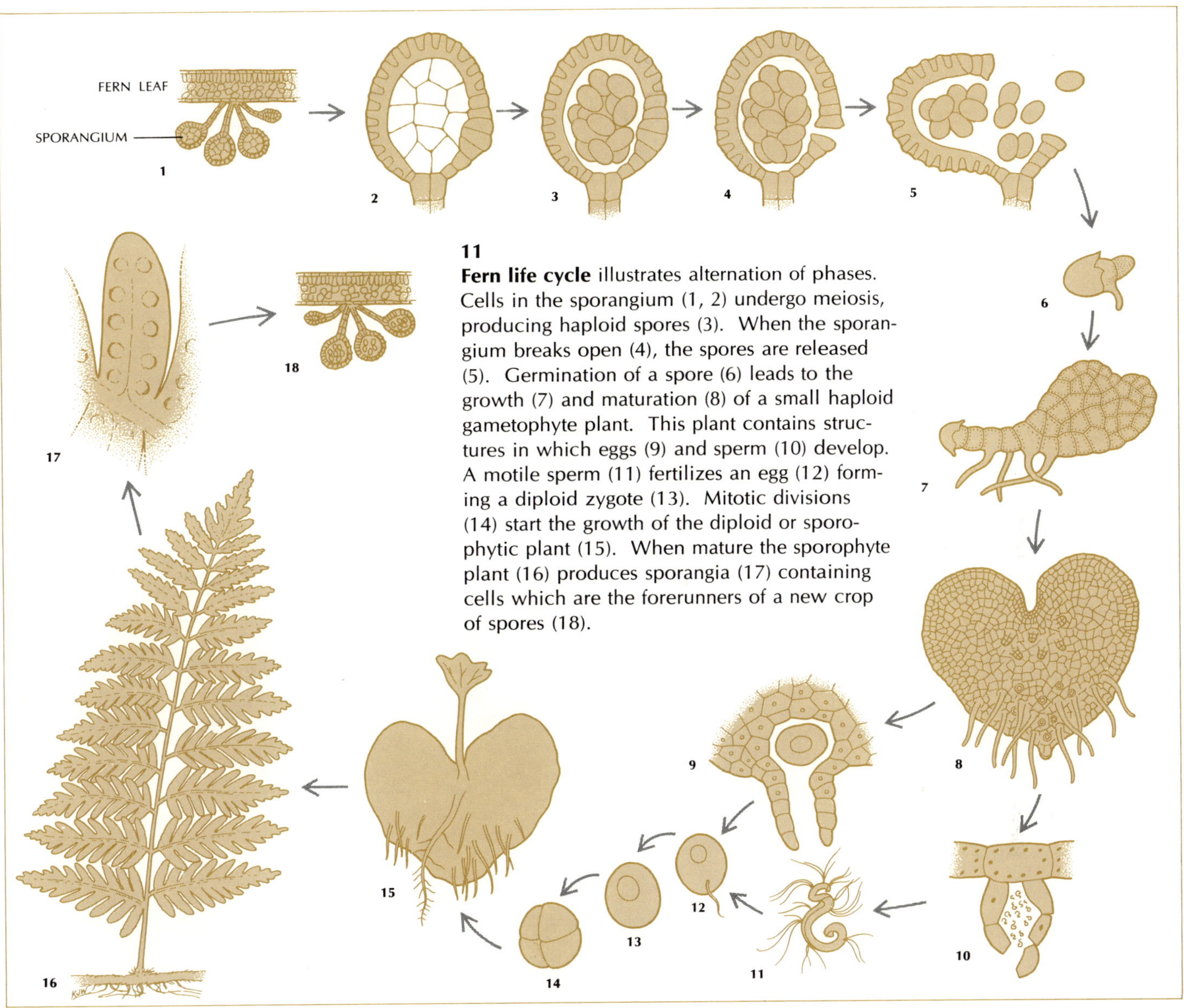

11

Fern life cycle illustrates alternation of phases. Cells in the sporangium (1, 2) undergo meiosis, producing haploid spores (3). When the sporangium breaks open (4), the spores are released (5). Germination of a spore (6) leads to the growth (7) and maturation (8) of a small haploid gametophyte plant. This plant contains structures in which eggs (9) and sperm (10) develop. A motile sperm (11) fertilizes an egg (12) forming a diploid zygote (13). Mitotic divisions (14) start the growth of the diploid or sporophytic plant (15). When mature the sporophyte plant (16) produces sporangia (17) containing cells which are the forerunners of a new crop of spores (18).

nificant heart-shaped plant. This reproduction pattern is known as alternation of phases, and it is common to most plants. It confers on them some of the advantages of both asexual and sexual reproduction: asexual, which, via spores, can rapidly increase the population and disperse it widely; and sexual, which leads to new combinations of genes and hence increases the chances for evolutionary adaptation.

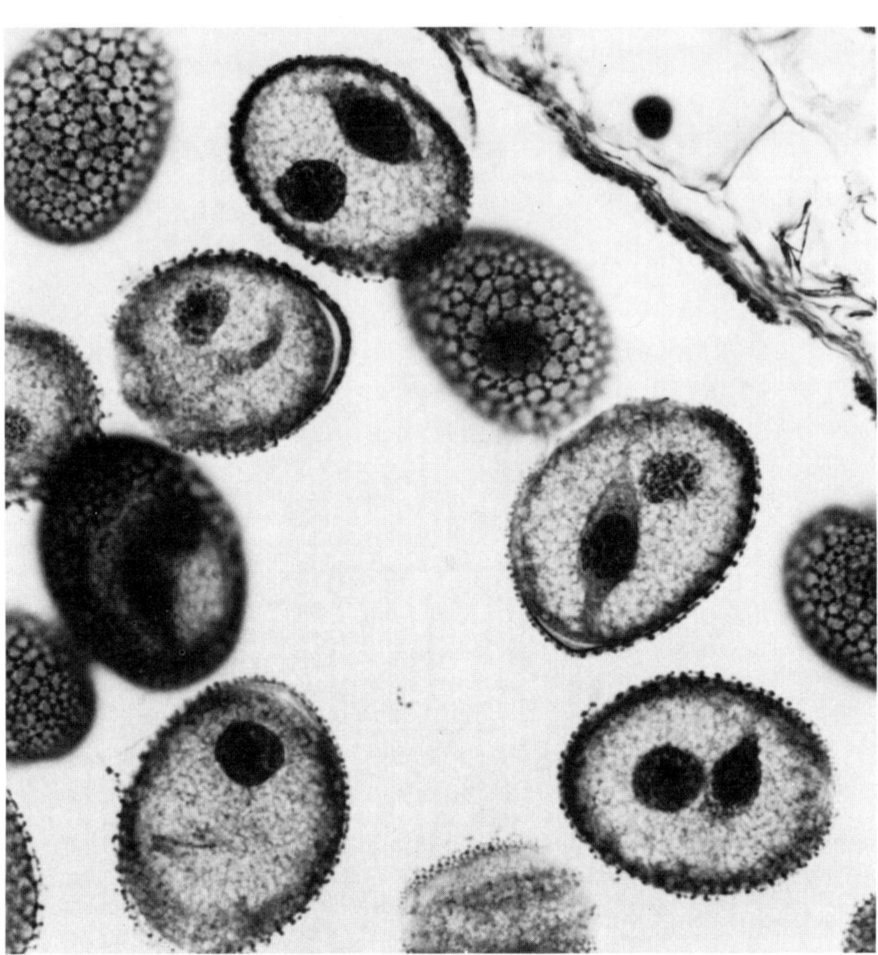

12
Pollen grains from a lily. These are the male
gametophyte plants. Each contains two
nuclei. The black network is the reticulated
wall of the pollen grains as seen in surface
view. Magnification about 300 ×.

FLOWERING PLANTS

The life cycles of flowering plants also involve an alternation of phases.
In flowering plants the diploid phase is even more dominant. The haploid
gametophyte phase is reduced to an even greater degree than in ferns
(Fig. 12). Palm and oak trees, rose and corn plants are all sporophytic plants.
Each of their cells, including the structural cells of the flowers, contain the
2N chromosome number. Special cells in the male parts of the flower (the
stamens) and in the female parts (the pistils) divide by meiosis. In the stamen
each male meiocyte produces four microspores (small spores), each with
the haploid chromosome number. The microspores then divide mitotically,

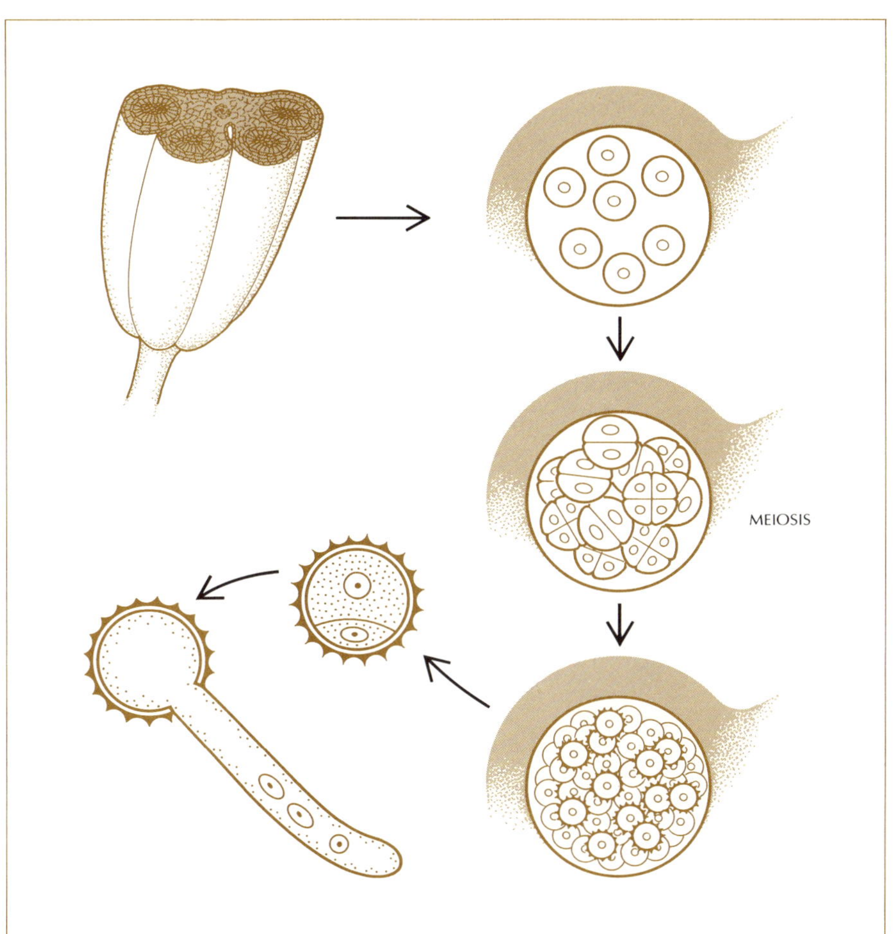

13
Stamens contain cells which undergo meiosis, producing microspores. The surface of the microspore becomes sculptured, and the contents divide mitotically, producing two cells. These pollen grains, as they are now called, are released from the stamen. Upon germination, one of the nuclei divides, producing two sperm nuclei.

ultimately producing structures called pollen grains. These are really male gametophytes, each composed of only two cells. When the pollen grain germinates, one of these cells divides mitotically, producing sperm cells (Fig. 13).

In the female part of the flower, special cells undergo meiosis, producing megaspores (giant spores). Three of the four megaspores produced by meiosis usually disintegrate; the remaining megaspore divides mitotically, ultimately producing the female gametophyte plant or, more specifically, a structure called the embryo sac. It will later envelop the developing plant

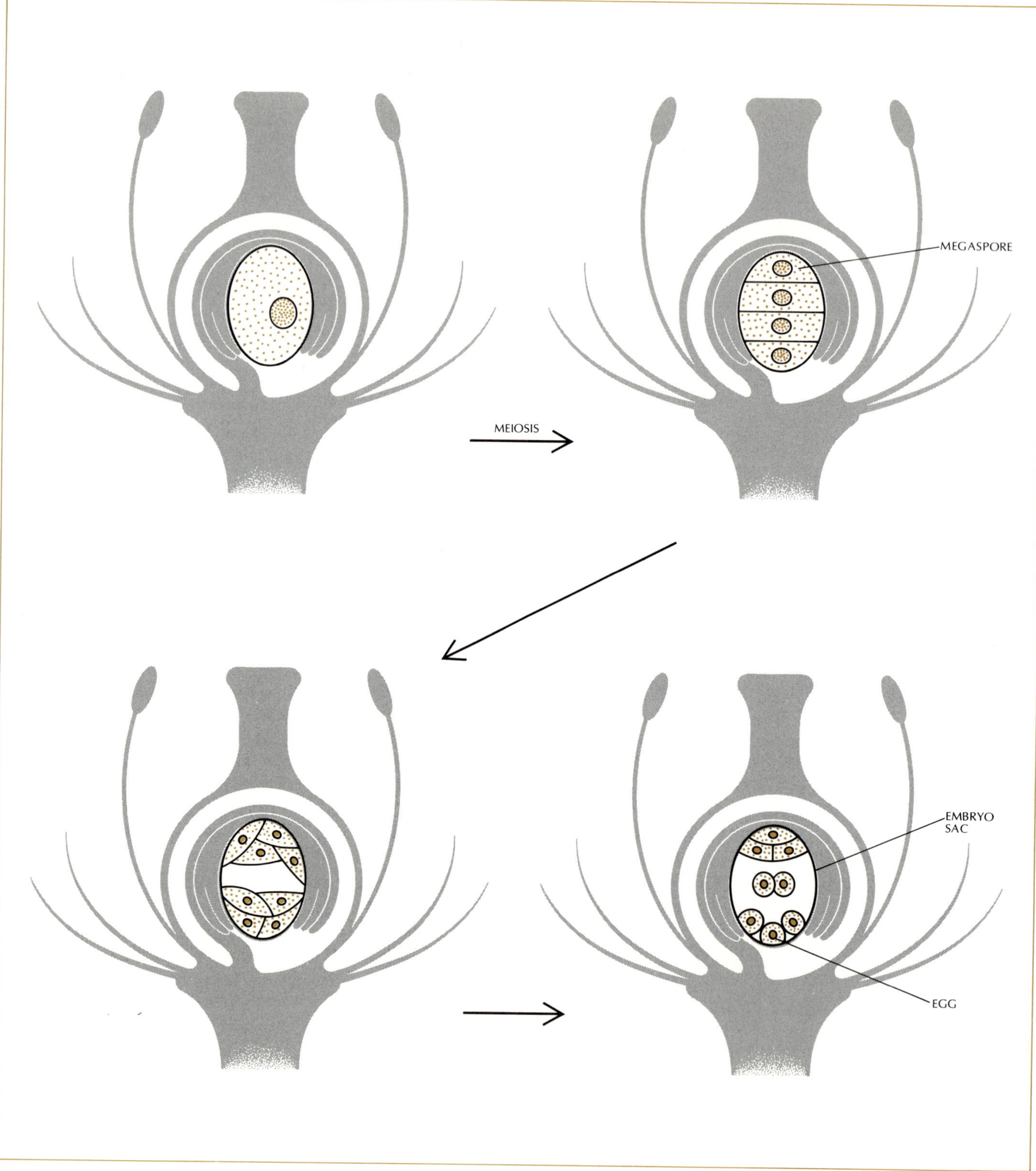

MEGASPORE
MEIOSIS
EMBRYO
SAC
EGG

14

◄ **Pistil** of flower contains cells which undergo meiosis, producing
megaspores. Mitotic divisions of one of the megaspores give rise
to numerous haploid cells, one of which is the female gamete or egg.

embryo. Early in its life it consists of only a few cells, one of which is the
egg (Fig. 14).

Many processes lead to the production of the new embryo. The first is
pollination. In this process the male gametophyte plant (the pollen grain) is
transferred from the stamen, where it was produced, to the tip of the pistil.
There it sends out a tube that grows down through the pistil until it en-
counters and penetrates the embryo sac. The two sperm in the pollen tube
enter the sac, and one fuses with the egg. Fertilization results in the production of a diploid zygote. The zygote remains inside the embryo sac and
divides mitotically. As the divisions continue, a small but recognizable
embryo develops (Fig. 15). This is the first stage of the new sporophyte
plant. The embryo, the embryo sac, and the surrounding tissues mature
into a seed. This, at some future time, germinates, and the embryo grows
into a mature sporophyte plant (Fig. 16).

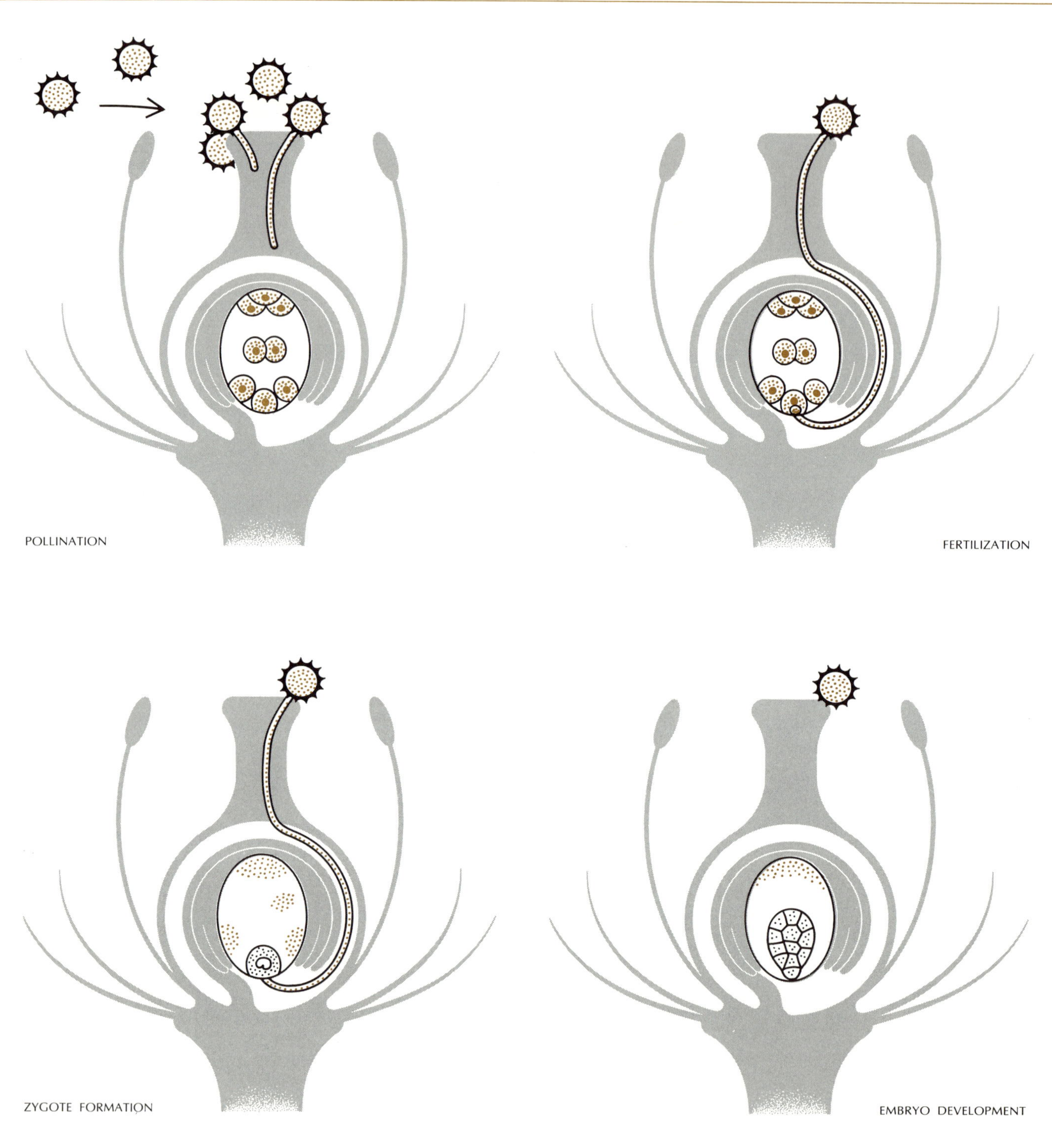
POLLINATION
FERTILIZATION
ZYGOTE FORMATION
EMBRYO DEVELOPMENT

15

◀**Pollination** is the transfer of pollen from the stamen to the pistil. A pollen tube grows down through the tissue of the pistil into the embryo sac. The fusion of a sperm from the pollen grain with the egg in the embryo sac is called fertilization. This results in the formation of a diploid zygote. By mitotic division, the zygote develops into a diploid embryo.

16

Mature embryo of a lima bean consists of a small root, a shoot with small veined leaves, and two large, fleshy, seed leaves.

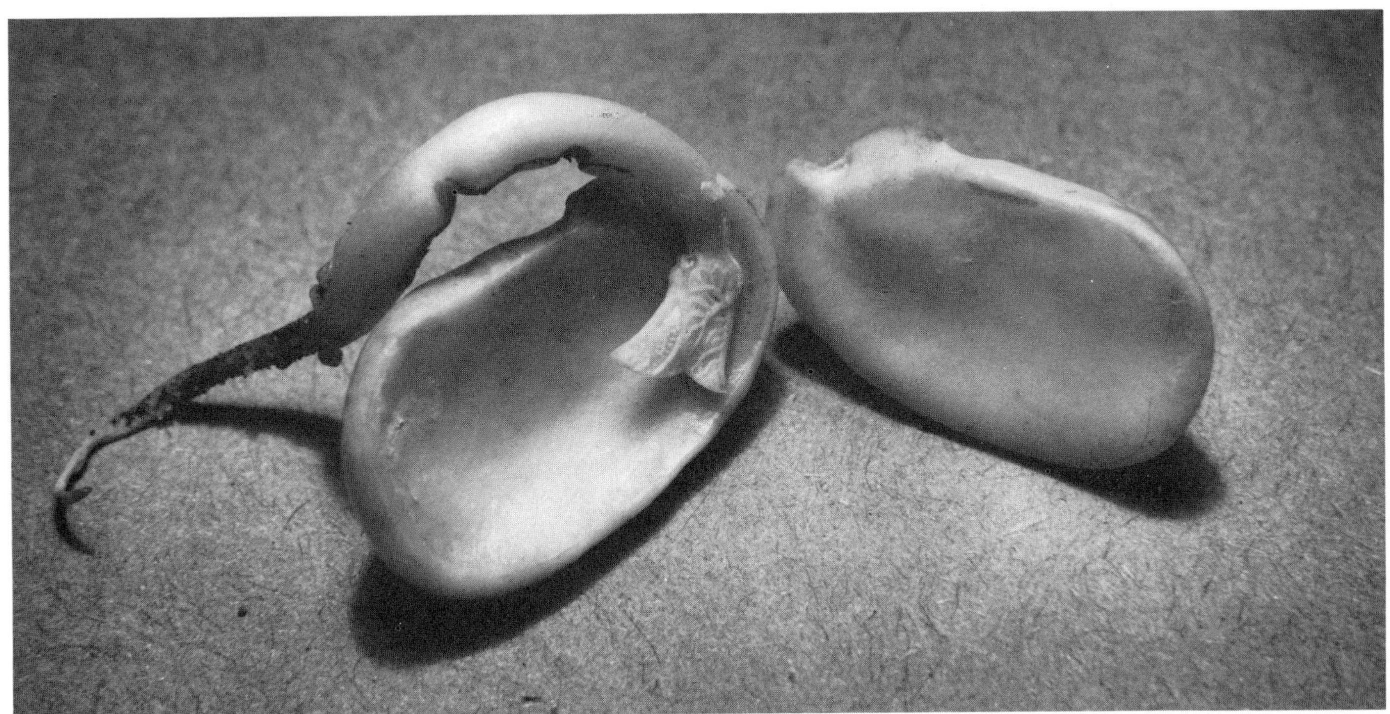

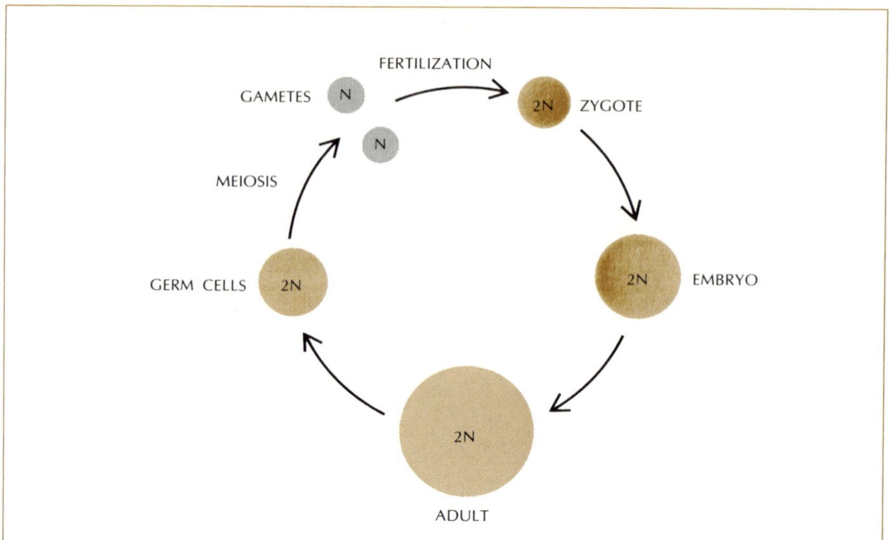

17

Life cycle of animals is characterized by a reduced haploid phase. Gametes are the only haploid cells.

ANIMALS

In the life cycles of animals only the gametes themselves are haploid (Fig. 17). The cells of the mature organism are diploid. Within the sex organs, special cells undergo meiosis, producing either eggs or sperm. The union of egg and sperm produces a zygote that divides mitotically and develops into an embryo. Each cell of the embryo is diploid because all were derived by mitotic divisions from the diploid zygote. The embryo matures into an adult organism capable of producing gametes, and the cycle is repeated. In the animal life cycle there is really no haploid phase, only haploid gametes. There is no true alternation of phases in the animal reproductive pattern because the gametes fuse without first dividing.

The reproductive pattern of man is basically similar to that of all other sexually reproducing animals: a haploid sperm and a haploid egg unite, producing a diploid zygote. This cell divides mitotically, initiating the developmental sequence that leads to the production of a new organism. Since the growth of a human being is rather slow, the ability to produce functional gametes is not reached until about one-sixth of his expected life span has been completed. Puberty is the name of this period in which the growing child attains the ability to commence reproductive activity. In man, puberty is not reached until about 13 to 15 years after birth, much later than any other animal. Peak sexual maturity is reached somewhere between 16 and 22 years of age.

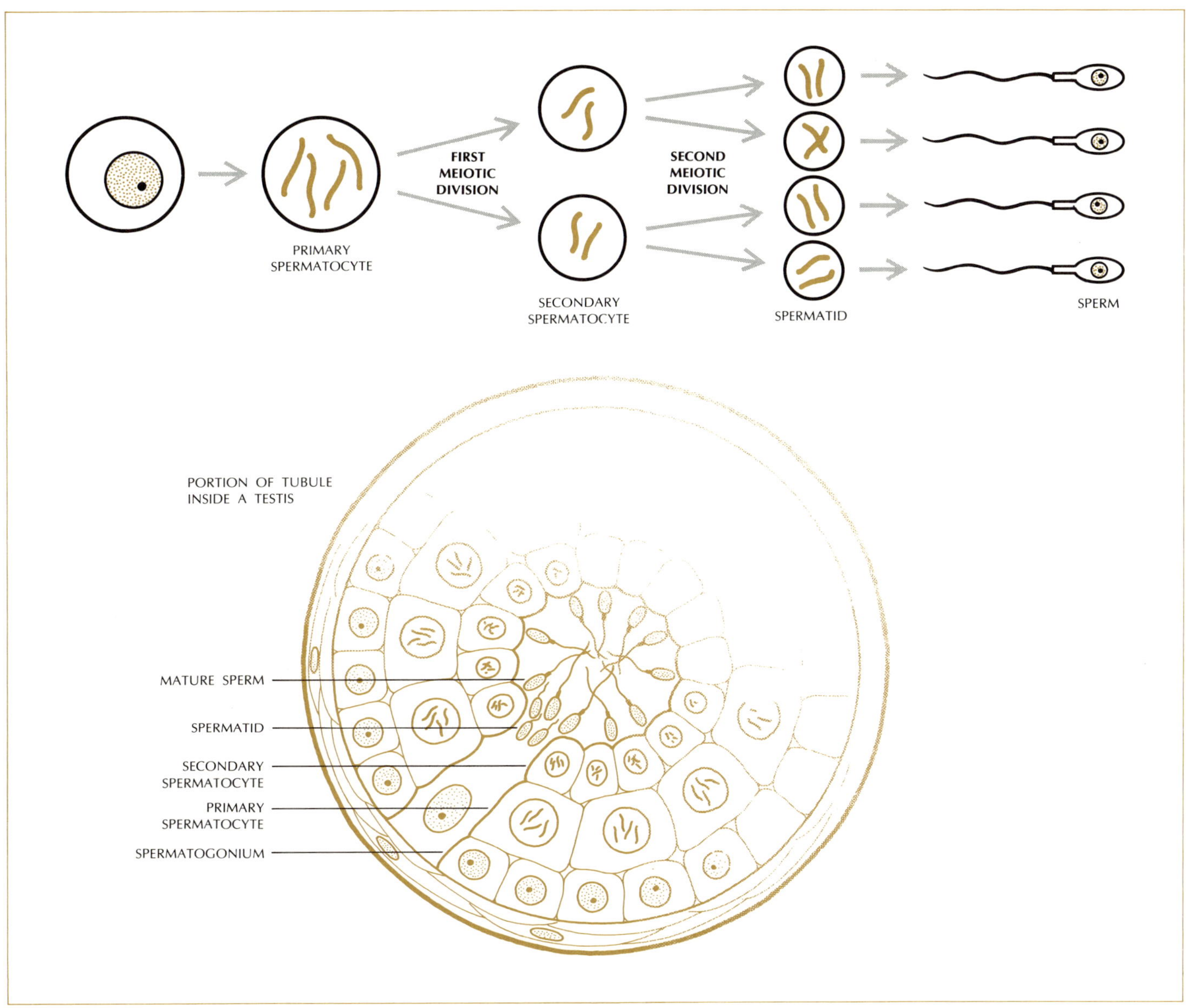

From the time of puberty until old age, human males continually produce functional spermatozoa in specialized tubules of the male sex organs, the testes. In these tubes certain cells, called spermatogonia, undergo meiotic divisions, producing new cells which mature into fully formed sperm (Fig. 18). In human females, the story is quite different. At the time

18
Formation of human sperm occurs in tubules of the testis. Diploid cells (primary spermatocytes) undergo meiosis, producing haploid spermatids which mature into sperm.

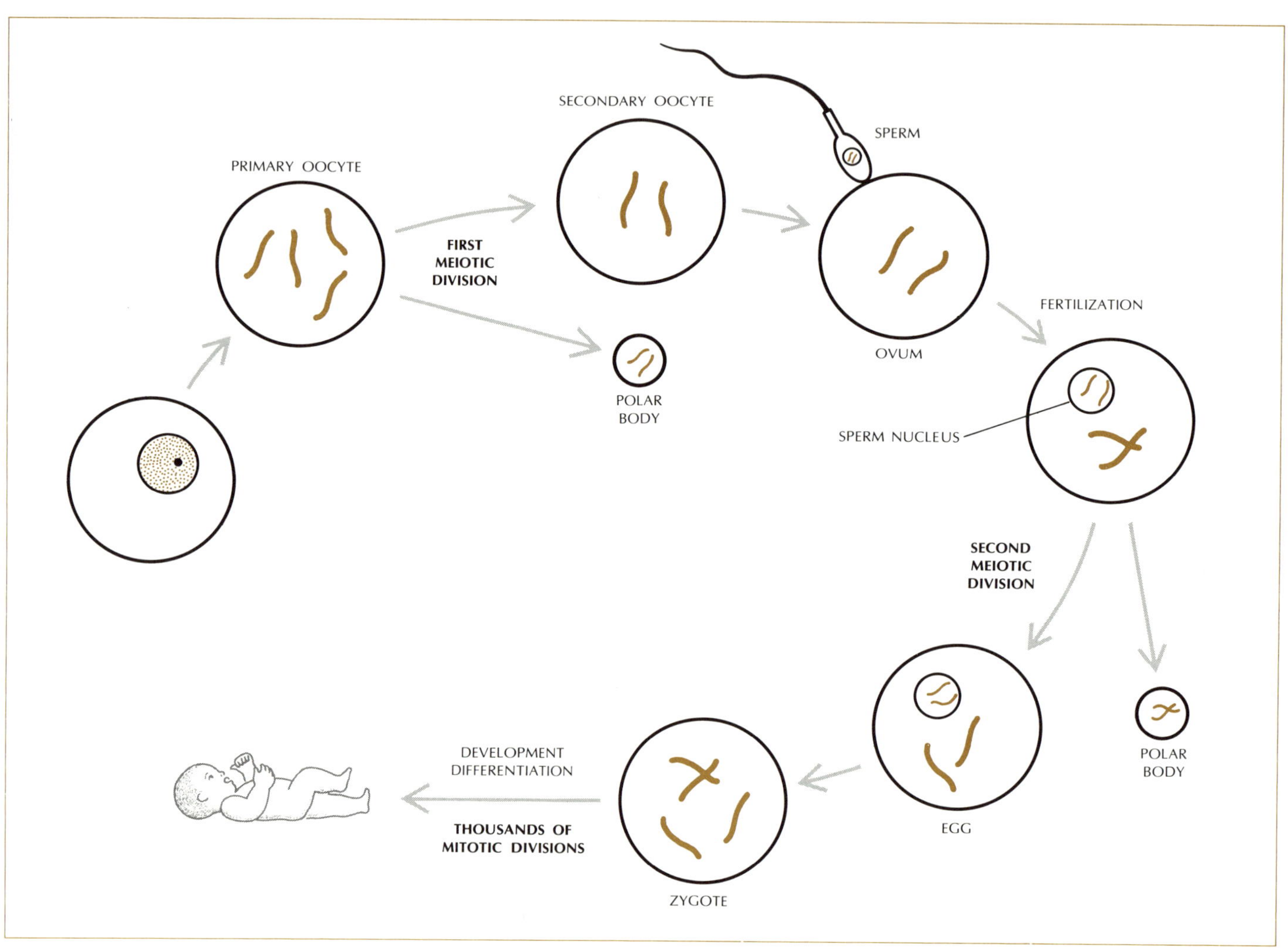

19

Formation of human eggs occurs in the ovary. A diploid primary oocyte divides meiotically, producing a polar body and a secondary oocyte. A sperm fertilizes the secondary oocyte, or ovum as it is now called. The second division of meiosis results in the production of another polar body and the egg. The sperm nucleus fuses with the nucleus of the egg, producing a diploid zygote.

of birth, the sex organs of the female, the ovaries, already contain all the primary oocytes (the forerunners of the mature egg) that she will ever produce. The number is large, perhaps as high as 700,000. During the first 10 years of her life, the number of these specialized cells rapidly declines. Many more are destroyed during the sexually active years. When the woman is from 40 to 50 years old, during the period known as the "change of life" or menopause, all the remaining oocytes disintegrate. Perhaps only 1 out of every 2000 oocytes produced in the ovaries is capable of being fertilized.

Prior to the release of the oocyte from the ovary, the first meiotic division takes place. This division produces a small nonfunctional cell called a polar body and a large cell called the secondary oocyte or ovum. The release of one of these secondary oocytes from the ovary is called ovulation. Once an ovum is released from the ovary it lives for about 24 hours. It passes from the ovary into the Fallopian tube where it may meet one of the 400 million sperm introduced into the female by the male during the act of copulation. Only after the sperm has penetrated the ovum does the second meiotic division finally take place. In this process the ovum divides unequally, producing another small polar body and a large cell, the egg. When the nucleus of the sperm fuses with the nucleus of the egg cell, fertilization is complete. Soon thereafter, the zygote begins to divide mitotically, beginning the developmental process that, in about nine months, will result in the birth of a new, intact functioning human organism (Fig. 19).

HAPLOIDY VERSUS DIPLOIDY

Only the simplest organisms are typically haploid. (We often speak of an organism being haploid when we really mean that each cell of the organism is haploid.) Bacteria, many algae, yeasts and other fungi, green moss plants and their relatives, are all haploid organisms. The more complex organisms are typically diploid. Why? What benefits does diploidy confer on an organism?

To answer this question it is meaningless to compare a haploid moss plant with a diploid cow. It is more appropriate to compare haploid and diploid organisms of the same species. Although no one has ever seen a haploid cow, a few higher organisms regularly do produce haploid offspring. In some cases laboratory techniques can produce haploid individuals from diploid parents. Conversely, normally haploid organisms can at times be made diploid by the use of chemicals, such as colchicine that permit mitosis but prevent cytokinesis. A combination of these approaches can lead to a better understanding of the significance of the chromosome number in the plant and animal kingdoms.

Colchicine is a compound extracted from the meadow saffron plant. When applied to cells about to enter mitosis it inhibits the function of the spindle fibers. The result is the doubling of the chromosome number without the subsequent division of this material into two separate nuclei. A single cell results, with twice the chromosome number of the original cell.

If haploid *Chlamydomonas* cells are placed in a solution of colchicine it is possible to produce a population of diploid organisms. A comparison of these abnormal cells with the normal haploid ones reveals that the diploid cells are bigger, contain two starch-producing structures instead of the usual one, and are no longer motile. These abnormal cells can grow and divide asexually, producing more diploid cells. But in almost all cases, after a few generations the cells revert back to the normal haploid condition. Spontaneous diploids are occasionally found in cultures of *Chlamydomonas* growing under strong light, but when the light is reduced, diploid cells disappear from the population. The change from haploidy to diploidy in the organism is not stable. Perhaps diploidy confers some slight benefit on cells growing under unusual conditions, but undoubtedly the haploid state is more beneficial and stable to *Chlamydomonas* organisms growing under the normal range of environmental conditions.

The green moss plants found in nature are usually haploid, but at times diploid ones may develop. The diploid plants are usually larger and have larger cells; there may be other superficial differences in structure or growth. These plants grow well and are able to reproduce as easily as the haploid plants. The change from haploidy to diploidy is stable, generation after generation. The evolutionary and ecological implications of this change are unknown.

In higher plants, when colchicine is applied to normal diploid cells just as they are about to divide, some of the resulting cells have four times the haploid number of chromosomes. Cells with more than two sets of chromosomes are called polyploids, as are plants derived from such cells. Characteristically these plants are larger, more vigorous, have bigger leaves and

20
Polyploid lily is taller than the diploid form
at right. It also has larger flowers and leaves.

showier flowers and fruit than the normal diploid plants (Fig. 20). Such differences are stable, generation after generation, and are striking enough that
sometimes we can actually call the polyploid plant a new species. By carefully selecting naturally occurring polyploids and by inducing polyploidy
artificially in domesticated plants, geneticists and plant breeders are continually producing new varieties of plants useful to man. At present nearly
half of the important cultivated plants are polyploids of one sort or another.

POLYPLOID ANIMALS

The cells of most animals are normally diploid, although polyploids are not unknown. However, the doubling of the chromosome number beyond the usual 2N condition in animals seems to upset the delicate balance of the genes that govern sex; the result is reproductive sterility. It appears that only animals not dependent upon fertilization for the continuation of the species can maintain a polyploid condition for long. It is not surprising that these animals—some insects, flatworms, roundworms, leeches, and brine shrimp —reproduce by a process whereby an egg can develop into a mature organism without first being fertilized by a sperm.

Polyploidy does not appear to confer any significant advantages on animals, and it strongly limits the reproductive capacity of the individual. It can be considered only as an interesting oddity of little economic or evolutionary significance.

HAPLOID SYNDROME

Higher plants and almost all animals are normally diploid. The fact that diploidy seems intricately linked with structural and physiological complexity suggests a cause and effect relationship. We might conclude that two sets of chromosomes are essential for the existence of complicated organisms. But male bees and wasps are regularly haploid. Haploid plants of cotton, corn, wheat, tobacco, cocklebur, and many others have been found although they are not common. Even haploid tadpoles and mature haploid salamanders are known. Complex haploid organisms are not only possible; they exist. But they are something short of normal.

With the exception of haploid male insects, the haploid individuals of normally diploid organisms show what is known as the haploid syndrome. (A syndrome is a specific combination of symptoms usually associated with a disease or an abnormal condition.) The haploid syndrome is characterized by deviations in development, structural weakness, and smaller size. It is easy to see at first glance that something is drastically wrong with these organisms. A cotton plant with the 1N chromosome number is smaller, has zigzag stems instead of erect ones, and produces no pollen, seeds, or fruit. Haploid frog embryos develop with abnormal nervous tissue, digestive systems, kidneys, and circulatory systems. They are abnormally small and usually die before metamorphosing into adult frogs; it is a wonder they survive as long as they do.

To sum up, these observations support the general correlation between diploidy and complexity. Two complete sets of chromosomes are apparently necessary for the normal development of most higher plants and animals. The genes on one set of chromosomes are not enough; they must be supplemented by complementary genes on the second set. In other cases, strong normal genes on one set of chromosomes may mask or suppress damaging genes present on the other set. Flowering plants appear to be further capitalizing on these benefits; there is some evidence that they are gradually evolving from diploidy to polyploidy. In animals, however, polyploidy is generally a failure because it leads to sterility.

READINGS

Bourne, G. H., *Division of Labor in Cells.* Academic Press, New York, 1962 (paper)

A review of the ultrastructure of cells, including the nucleus and chromosomes. Good photographs and diagrams.

Frankel-Conrat, H., *Design and Function at the Threshold of Life: The Viruses.* Academic Press, New York, 1962 (paper).

The structure of viruses and the function of the chemical components of viral particles. For those interested in a more chemical approach.

Sutton, H. E., *Genes, Enzymes and Inherited Diseases.* Holt, Rinehart and Winston, New York, 1961

How DNA governs protein synthesis and, ultimately, the production of human genetic diseases.

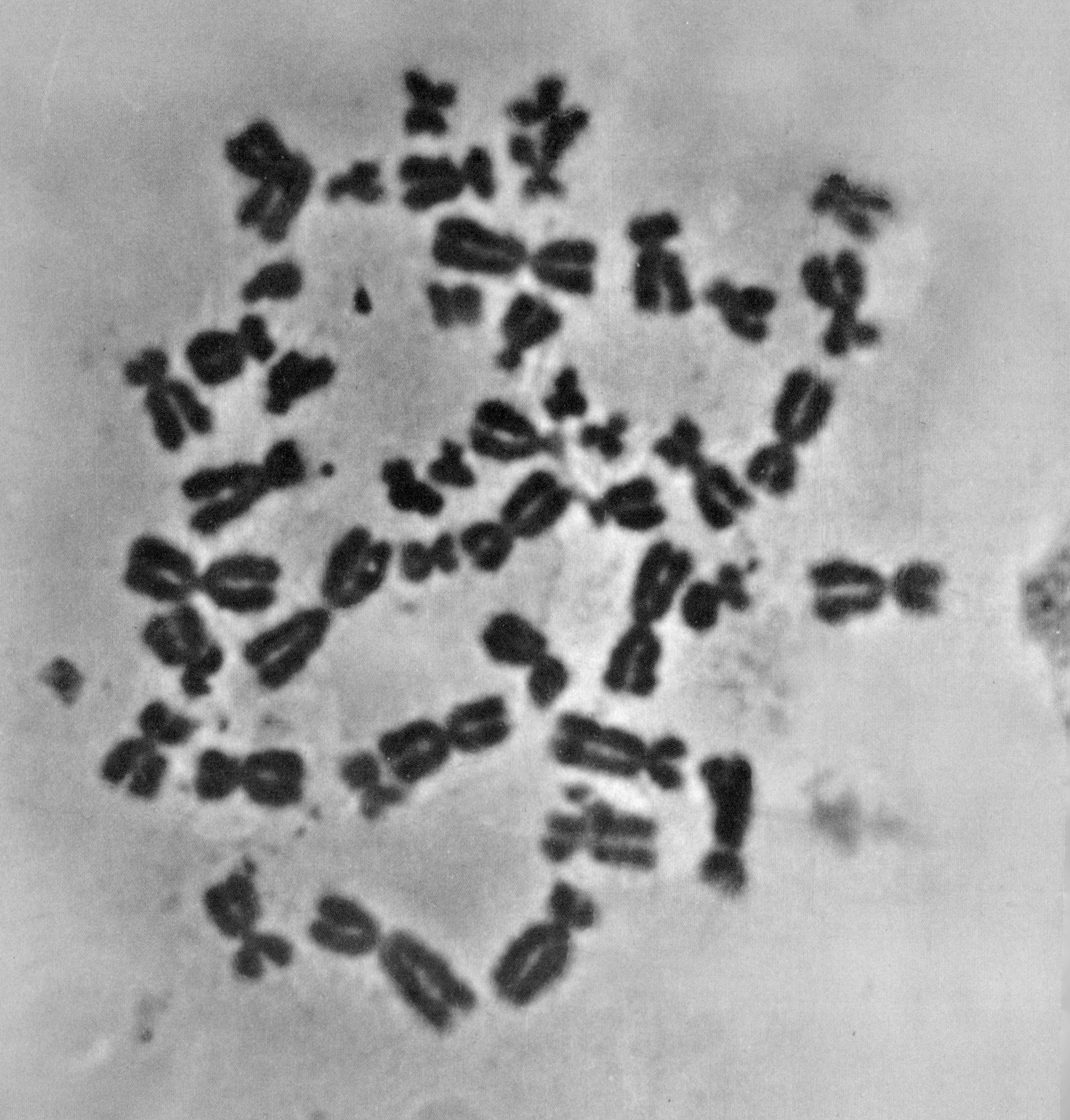

8. CLASSICAL GENETICS

The remarkably high level of organization that characterizes living matter is governed by genes, tiny units of hereditary information. Genes transmit from one generation to the next the information that directs the chemical machinery of the cell. In a biological sense, genes are the only things that an organism inherits from its parents. The information of genes literally enables the new organism to create itself, that is, to organize the chemicals in its environment into its own living matter.

Genetics is easier to understand if we separate the two functions of genes: their control of the cell and their role in heredity. This approach roughly corresponds to the two main avenues of genetic research which we might call microgenetics and macrogenetics. Microgenetics is concerned with the internal architecture of the chromosome, the molecular structure of the gene, and the chemistry of genetic control. Macrogenetics is the study of the inheritance patterns of the whole organism and populations of organisms; it seeks to explain diversity and sameness within populations. In recent years the distinction between molecular genetics and classical or organismic genetics has tended to fade.

Early work at the cellular level identified the nucleus and eventually the chromosome as the control center of the cell (Chapter 5). The electron microscope revealed many details of the fine structure of the chromosome. Research at the molecular level established that the site of the gene is occupied by a nucleoprotein. The nucleic acid component was identified as DNA, and the discovery of its structure by Watson and Crick was the breakthrough that led not only to the cracking of the genetic code, but also to the

◀**Human chromosomes** at metaphase. The two chromatids of each chromosome are evident, as are the locations of the centromeres.

current theory of how genes control the chemical activities of the cell (Chapter 6).

In brief, the theory holds that genes are the primary link in a chain of control. DNA directs the synthesis of RNA, which in turn directs the synthesis of proteins, including enzymes. The enzymes in turn catalyze virtually all the reactions that occur within the cell, including those that synthesize the structural and functional components of the cell, such as other proteins, lipids, and carbohydrates. The presence or absence of the right enzyme effectively determines whether a particular reaction will or will not proceed, and the quantity of the enzyme often influences the rate of the reaction. Because an organism is the product of the structure and functions of its cells, genes ultimately control the anatomy and physiology of each living thing.

CLASSICAL GENETICS

For more than a century, biologists have rightly regarded the mechanism of heredity as one of the crucial mysteries of life. Classical genetics emphasized the role of the gene in transmitting hereditary information from one generation to the next. Loosely speaking, we might say that the classical geneticist is primarily concerned with the heredity of the whole organism. He seeks to explain the similarity and the variability of successive generations of organisms in terms of a few basic ideas and principles, principles that would make it possible to predict the genetic makeup of future generations. Prediction has been an important goal of genetics. The ability to predict heredity has enabled us, to some extent, to control it. Larger crops, disease-resistant plants, faster racehorses, and meatier cattle are among the more obvious results of applied genetics. This is only the beginning. In the future it should be possible not only to breed even finer crops and animals, but also to improve the heredity of humans by altering or eliminating harmful genes.

MENDEL

From the beginning of his agricultural existence, man has practiced a sort of do-it-yourself genetics. He chose seeds from the best plants to sow the following season, sometimes with impressive results, sometimes not. Seeds from the best plants of the year often produced better plants the following year, but sometimes they produced small, misshappen offspring that yielded inferior fruit or grain. It was impossible to predict with any degree of certainty what would happen. Eventually scientists tried to understand this phenomenon, but their early theories concerning heredity likewise had little predictive value. It may have been deduced that a man had blue eyes because his grandmother did, but predicting the eye color of his children and that of succeeding generations was still nothing but glorified guesswork.

An answer came in the 19th century. The science of genetics began with the work of Gregor Johann Mendel, an Austrian monk who in his spare time conducted breeding experiments with common pea plants. He studied such traits as flower color, seed shape, plant height, and pod color and shape. After examining more than 10,000 plants over a period of eight years, he presented his results in two papers that he delivered at meetings of the local natural history society in 1865. His conclusions can be summarized in four statements:

1. Organisms carry factors or particles that govern the expression of inherited traits.

2. These particles are passed from generation to generation in the gametes.

3. Some particles exert a stronger influence than others; they are said to be dominant over the weaker or recessive particles.

4. In each generation the proportion of organisms that show a specific characteristic can be predicted mathematically.

Mendel knew nothing of genes, chromosomes, biochemical pathways, or DNA, but he did know his plants. His observations were critical, his dedication monumental, his conclusions sound, and his predictions astonishingly accurate. Instead of trying to consider the inheritance of many characteristics at the same time, he chose to study individual traits one at a time. By simplifying the problem he succeeded where those before him had failed.

GENES

Mendel's work made it clear that such statements as "He inherited his blond hair from his father" or "She inherited her intelligence from her grandmother," are incorrect. A child's hair and intelligence are his (or her) own. In a legal sense, you can inherit your mother's raccoon coat or the family's

Pair of chromosomes of a corn plant is represented by two vertical bars. Crossbars indicate positions of genes, labeled by the letters P and D. Dark ovals at top indicate position of centromeres.

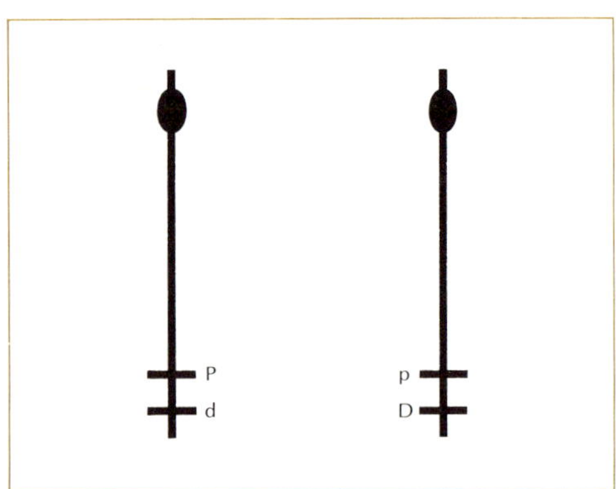

old wicker porch furniture. In a biological sense, you inherit what Mendel called factors or particles; we now call them genes. At present we consider a gene to be a specific portion of a strand of DNA in a chromosome. Possession of a certain gene does not bestow a trait upon an individual; it bestows only the potential to develop that trait. For example, most plants that have the genes for the production of chlorophyll produce this compound only in the presence of light. The plants inherit the genetic potential to produce chlorophyll, but this potential is expressed only in the proper environment. The old argument about which is more important, genes or environment, misses the point: both are essential. "Heredity deals the cards," as the saying goes, "but environment plays the hand."

GENOTYPE

Since the cells of a plant or an animal are the result of the mitotic divisions of cells derived from the fertilized egg, they have the same number and

**

a capital letter represents a dominant gene
a lower-case letter represents a recessive gene
a straight line represents a single chromosome
a box represents a diploid cell
a circle represents a haploid cell
a cross represents a female ♀
an arrow represents a male ♂

The two diagrams below illustrate the use of these symbols. On the left a diploid cell with four chromosomes shows two genes on each chromosome. On the right a haploid cell with two chromosomes shows two genes on each chromosome.

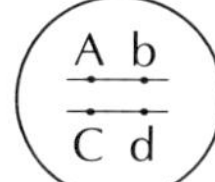

kinds of chromosomes.   They are alike genetically; a geneticist would say that they have the same genotype. By examining the genetic makeup of one cell, we may extrapolate and say that this is the genotype of the entire plant or animal.   A cell of a corn plant, for example, has 20 chromosomes; two of them are depicted schematically in Fig. 1.   Since each cell of the plant is diploid (except for the cells of the pollen grains and the embryo sacs in the flowers), each chromosome has as its partner a homologue that carries genes for the same traits.   Although there are thousands of genes on each chromosome, only two are shown.   The partial genotype of this cell, and of the plant from which it came, is written $\frac{P\ p}{D\ d}$.   What does this mean? In genetics, symbols are commonly used to represent both ideas and structures.   The symbols that we shall use throughout this book are listed in Box A.

The letter chosen to represent a gene is often the first letter of the word describing the unusual or less common expression of the trait. For instance, most plants of this type of corn are about two feet tall after two months of

2
Dwarf and tall corn grow side by side in this demonstration plot. Dwarfism in corn is a recessive trait.

growth. But a few, grown under the same conditions, are only six inches tall (Fig. 2). These plants are genetic dwarfs. The letter d, designating dwarfism, is assigned to the gene that governs the height of these plants. Genes may exist in a dominant and a recessive form. Together, the two are called alleles. In this case, the gene for dwarfism is recessive and accordingly is represented by a lower-case letter. The dominant gene for normal height is designated by a capital D. The line under the letter represents the chromosome that contains the gene.

PHENOTYPE

If we consider just the genes for height, three genotypes are possible: $\boxed{\underline{D\ D}}$ if each cell contains two D genes; $\boxed{\underline{d\ d}}$ if each cell contains two d genes; and $\boxed{\underline{D\ d}}$ if each cell contains one D gene and one d gene. But how do these genotypes affect the physical appearance of the plant? The physical expression of an organism's genetic makeup is called its phenotype. In the case of these corn plants the two possible phenotypes are "dwarf" and "normal." A plant with the genotype $\boxed{\underline{D\ D}}$ or $\boxed{\underline{D\ d}}$ has the normal phenotype because the dominant D gene governs the height of the plant despite the presence or absence of the recessive gene. A plant with a genotype of $\boxed{\underline{d\ d}}$ shows the dwarf phenotype: there is no dominant gene to mask the effect of the recessive gene, and the recessive trait is therefore expressed.

The other genes in our example in Fig. 1 are the dominant P gene for purple stems and the recessive p gene for the normal dark green stems. As the diagram indicates, the genes for coloration and for height are located on the same chromosome; we say that they are linked. For the sake of clarity we diagram them

$$\underline{P\ \ d} \qquad \text{and} \qquad \underline{p\ \ D} \ .$$

The genotype of the cell represented in Fig. 1 may also be written

$$\boxed{\begin{array}{c} \underline{P\ \ d} \\ p\ \ D \end{array}} .$$

This clearly shows that the dominant P gene is linked to the recessive d gene and that the recessive p gene is linked to the dominant D gene.

Let us now consider two more traits: the sugar content of the kernels and the color of the midvein of the leaves. The gene for high sugar content is designated by the letter s. Starchy kernels (dominant gene S) are more common than the mutant sugary kernels. Leaf midveins are either brown (recessive gene b) or green (dominant gene B). The gene for starchiness lies on a separate chromosome from that carrying the gene for sugar content, as well as from the chromosomes carrying the genes for leaf midvein coloration, and those carrying the genes for stem coloration and height. A plant that is purple, tall, has starchy kernels, and leaves with brown midveins might have the following genotype:

$$\frac{P \quad d}{p \quad D} \qquad \frac{S}{s} \qquad \frac{b}{b} \;.$$

To have purple stems the plant must have at least one P gene; to be tall it must have at least one D gene; to produce starchy kernels it must have at least one S gene; and to produce leaves with brown midveins it must have no dominant B genes. In this example the recessive s gene may be replaced by a dominant S gene without changing the phenotype. Since we cannot actually see the genes, how can we tell whether the genotype of a plant with this phenotype is

$$\frac{S}{s} \qquad \text{or} \qquad \frac{S}{S} \quad ?$$

From the information given, we can't. Both conditions are equally possible. The challenge of genetic analysis often is to devise ways to answer such questions.

The first step in the elementary analysis of a genetic problem is to write down what each symbol represents. Then draw the chromosomes, label the known genes, fill in the linkage partners, and list all possible genotypes that could yield the observed phenotype. A worked-out example of these preliminary steps appears in Box B. As the chapter progresses, we shall show how the techniques of genetic analysis can be applied to solve such problems.

Problem: determine the possible genotypes of a corn plant with a tall purple stem, starchy kernels, and leaves with brown midveins

---

GIVEN

P = purple stem　　　　　　　　　　S = starchy kernels

p = normal green stem　　　　　　　s = sugary kernels

D = height of normal plant (tall)　　　B = green leaf

d = height dwarfed　　　　　　　　　b = leaf with brown midvein

Linkage: gene for stem pigment production lies on the same chromosome as the gene for height; P is linked to d and p is linked to D.

SOLUTION

1. Draw the chromosomes so as to get the right number

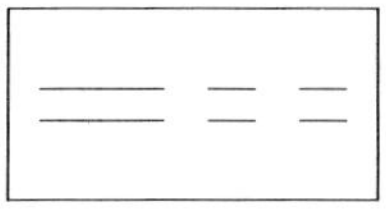

2. Label the known genes

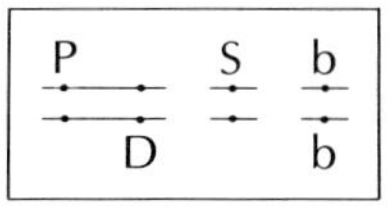

```
P       S   b
═══   ═ ═  ═ ═
    D       b
```

3. Fill in the known linkage partners

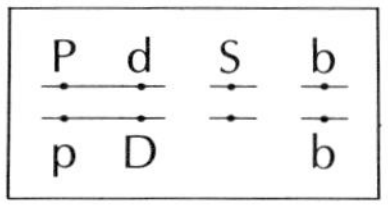

```
P   d   S   b
═ ═   ═ ═  ═ ═
p   D       b
```

4. List all possible alternatives to give the above phenotype

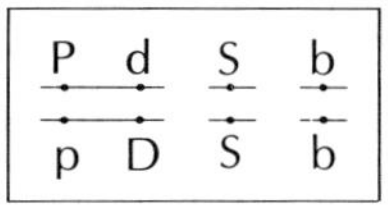

```
P   d   S   b              P   d   S   b
═ ═   ═ ═  ═ ═      or     ═ ═   ═ ═  ═ ═
p   D   S   b              p   D   s   b
```

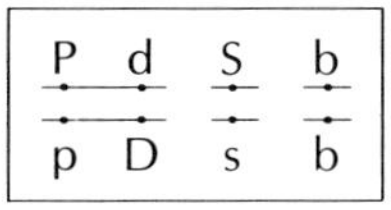

GENETIC PREDICTION

If the aim of genetics were merely to be able to write a genotype for each and every individual of a given population, the task would be formidable but of very little value. The importance of genetics lies in its ability to predict the phenotypes and genotypes of individuals in future generations. In agriculture this allows us to crossbreed certain types of plant with the assurance that the resulting offspring will be superior. In animal breeding stud service assures the production of fine herds. For humans, genetics enables us to predict what the probabilities are that the child of a given man and woman will have a specific genetic defect.

Although most genetic predictions are not 100 percent accurate, a geneticist can predict a certain outcome with a known probability of being right. The information that there is a 50-50 chance that a baby will be a boy is not of much interest to an expectant mother; but the knowledge that any boy born to her will have a 50-50 chance of developing a specific congenital deformity is of critical and immediate importance.

A type of human dwarfism is due to a recessive gene. A person with the genotype $\boxed{d\ d}$ will be a dwarf. The genotype of a normal individual might be $\boxed{D\ D}$ or $\boxed{D\ d}$. The $\boxed{D\ d}$ individual is a carrier. He does not exhibit the characteristic of dwarfism because the dominant gene masks the effect of the recessive one. If one dwarf marries another, all of their children will be dwarfs. But how can we predict the probability that two normal parents, each carrying the gene for dwarfism, will produce a dwarf child? The answer requires a knowledge of the genotypes of the parents' gametes. With this knowledge the geneticist can state the mathematical probability of a given type of sperm uniting with a given type of egg, and thus the probability of producing a child with a given genotype or phenotype. A step-by-step example of this approach is shown in Box C.

Since each parent is a carrier, each has the genotype $\boxed{D\ d}$. The genes for dwarfism are located on a homologous pair of chromosomes. One chromosome contains the dominant D gene and the other contains the recessive d gene. In the process of meiosis, the homologous chromosomes pair and then separate. Then the chromatids of each chromosome separate (Fig. 3). Of all the sperm produced by the father, 50 percent will contain the d gene and 50 percent will contain the D gene. The same is true of the eggs produced by the mother. The genotype and phenotype of an offspring

Box C

Problem: predict the probability that any given child of parents who are both carriers of ateliotic dwarfism, will be a dwarf

GIVEN

Dwarfism is recessive to normal development

D = gene for normal development

d = gene for dwarfism

Father normal but a carrier. Genotype = | D d |

Mother normal but a carrier. Genotype = | D d |

SOLUTION

1. Find the gametes produced by each parent and the percentage of each type

    Sperm: (D) 50% and (d) 50%   *Eggs:* (D) 50% and (d) 50%

    Parent: | D d | *Parent:* | D d |

2. Show all possible matings between sperm and egg types

	(D)	(d)
(D)	D D	D d
(d)	D d	d d

3. Calculate the percentages of each kind of offspring

	(D) (50%)	(d) (50%)
(D) (50%)	D D (25%)	D d (25%)
(d) (50%)	D d (25%)	d d (25%)

    Genotypic frequencies: | D D | (25%), | D d | (50%), and | d d | (25%)

4. Describe the genotypic frequencies as phenotypes

    Offspring: 25% normal, 50% normal-appearing carriers, and 25% dwarf

3

Meiosis involves the synapsis of homologous chromosomes, the subsequent separation of the homologues, and finally the separation of the two chromatids of each chromosome.

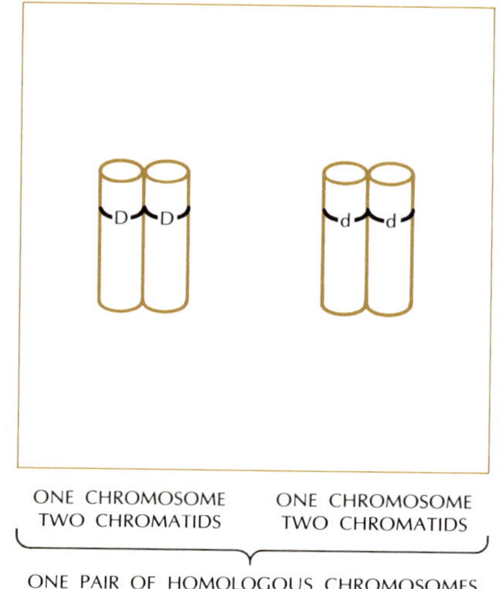

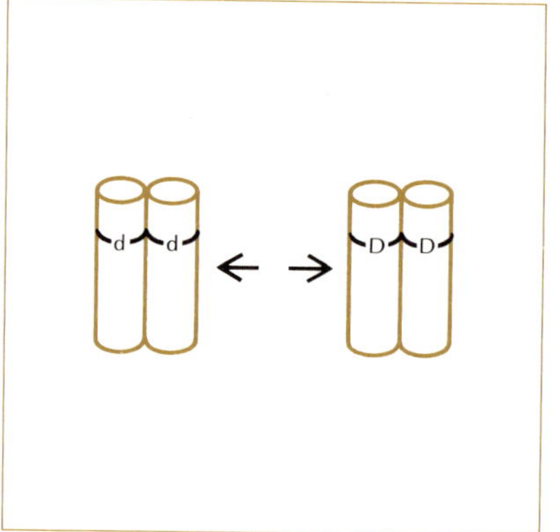

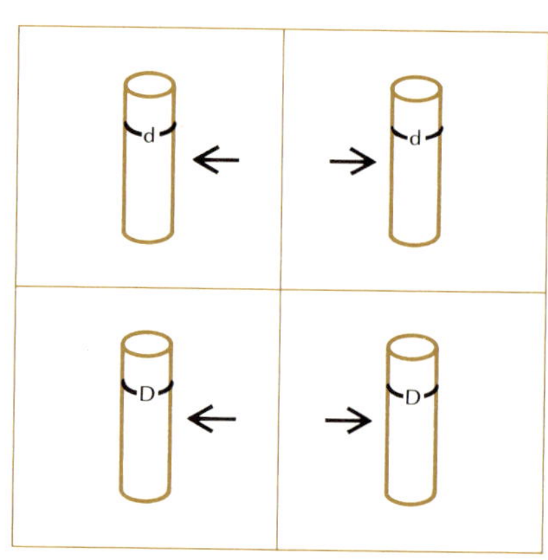

produced by these parents depend upon which type of sperm fertilizes which type of egg.

1. If a D sperm unites with a D egg, a zygote with the genotype $\boxed{D\ D}$ results.   The individual developing from this fertilized egg will be of normal height.

2. If a D sperm unites with a d egg, or if a d sperm unites with a D egg, the genotype will be $\boxed{D\ d}$.   Individuals developing from such zygotes will also be of normal height, but will carry the hidden gene for dwarfism; they will be carriers.

3. If a d sperm unites with a d egg, the genotype of the zygote will be $\boxed{d\ d}$ and the individual developing from it will be a dwarf.

Each of the four combinations is equally probable. These parents can produce three kinds of offspring: normal individuals, carriers, and dwarfs. The chance their child will be a dwarf is 1 out of 4, or 25 percent. The chance that it will be of normal height is 3 out of 4, or 75 percent. The chance that it will be of normal height but a carrier is 2 out of 4, or 50 percent.

The knowledge that there is a 25 percent risk that their child might be a dwarf would cause most couples to consider the advisability of not producing their own children but of adopting children instead. It is a difficult decision. The laws of chance cannot predict any single event with certainty. The laws of probability do not say (in this case) that one out of every four children produced by such a couple will be a dwarf. Instead they say that there is a 25 percent chance that any one child will be a dwarf. Think about the difference in meaning of the two statements. This couple might produce a family of eight children, all of whom are of normal size. No geneticist would be surprised. They would know that such an occurrence was unlikely but certainly not impossible. They also might produce eight children, all of whom would be dwarfs. The laws of probability are based on the analysis of very large populations. A family of eight children might seem like a very large population, but the statistician thinks in terms of thousands or millions. If this couple were to have one million children, about 250,000 (25 percent) would be dwarfs. To the parents this statement is meaningless, but when they are told that it really means that there is a 25 percent chance that each and every child that is born to them will be a dwarf, they know what it means.

DETERMINATION OF GENOTYPES

The problem in Box B illustrates the fact that the genotype of an individual cannot necessarily be deduced from its phenotype. A person showing the dominant phenotype may be either homozygous or heterozygous; that is, in his cells the two genes for this trait may be identical (*homo*) or they may be dissimilar (*hetero*). Some geneticists prefer the terms pure line (similar) and hybrid (dissimilar).

Even if two parents show a dominant phenotype for a trait, their offspring will not always show the dominant phenotype. Again let us use an example of a genetic human disease. Extreme myopia (nearsightedness) is due to a recessive gene m. Persons who are homozygous for the dominant gene $\boxed{M\ M}$ or heterozygous for this characteristic $\boxed{M\ m}$ have normal vision. Let us suppose that a man with extreme myopia marries a woman with normal vision. They would like to know whether their children will be myopic. With only this amount of information it is impossible to tell. Since the father carries only recessive genes for this characteristic, all his children will inherit one gene for the disease. But they will develop the disease only if they inherit another recessive gene from their mother. Since she has normal vision she carries at least one dominant gene for this trait. But if she is a carrier, she can pass the recessive gene to her offspring. How can she tell whether she is a carrier? One possible solution is to examine her family tree. Her mother, she recalled, had normal vision but her father was myopic. Because her father showed the recessive phenotype, he must have been pure line for this gene. All his sperm cells would carry the recessive gene. The woman would have inherited a recessive gene from her father and a dominant gene from her mother, and therefore be a carrier. Any child conceived by this couple would have a 50-50 chance of developing extreme nearsightedness (Box D).

Box D

Problem: determine the probability that a child conceived by a myopic man and a woman with normal vision will be myopic

GIVEN

M = gene for normal vision

m = gene for myopia

Father myopic. Genotype | m m |

Mother normal. Genotype | M ? |

SOLUTION

1. If possible first find out the complete genotype of the mother

    a. List the phenotypes and genotypes of her parents

        Given: her mother has normal vision. Genotype | M ? |

            her father has myopia. Genotype | m m |

    b. Diagram the genetic relationship between the woman and her parents

        ♀(M) Eggs (?)   (m)♂ Sperm   (m)♂

        | M ? | + | m m | =   ♀(M) | M m |

                                       (?)♀ | ? m |

    c. Deduce the genotype of the woman
       We know she possesses at least one M gene since she has normal vision.
       She must possess one recessive gene which she inherited from her
       mother. Therefore her genotype must be
       | M m |

2. Determine the possible types of children she and her husband may produce

   ♀(M) (50%) (m)♀ (50%) (m)♂ (100%)   (m)♂ (100%)

   | M m | + | m m | =   ♀(M) (50%) | M m (50%) |

                                 (m)♀ (50%) | m m (50%) |

3. Express the genotypes as phenotypes

   There is a 50% chance that the child will have normal vision but will be a
   carrier. There is also a 50% chance that the child will develop myopia.

Achondroplasia (dominant	Dwarfism due to faulty development of the skeleton especially the cartilage and bone of arms and legs.
Albinism (recessive)	Almost complete lack of pigment production in skin, hair and eyes.
Alkaptonuria (recessive)	A defect in metabolism that results in the production of urine that blackens upon exposure to air.
Amaurotic Idiocy (recessive)	Mental deficiency, progressive blindness and paralysis.
Anonychia (dominant)	Failure of development of the nails of the fingers and toes.
Epiloia (dominant)	Complex conditions including tumors of brain, heart and kidneys, mental deficiency, epilepsy, production of red and yellow pimples on the face.
Gout (dominant)	Faulty protein metabolism resulting in deposition of crystals of a uric acid derivative in the joints that cause extreme pain.
Huntington's Chorea (dominant)	Uncontrollable muscular spasms and mental deterioration.
Pancreatic Fibrocystosis (recessive)	Excessively salty (sodium chloride) sweat. Also general diarrhoea and stunting of growth.
Phalangeal Synostosis (dominant)	Partial fusion (sometimes complete of two or more fingers or toes.
Phenylketonuria (recessive)	Severe mental deficiency and epileptic seizures.
Polydactyly (dominant)	More than 5 digits on hands or feet.
Porphyria (dominant)	Metabolic disorder characterized by reddish coloration of teeth, sensitivity of skin to abrasion and sunlight. Faulty healing of wounds leaving scars.
Tylosis (dominant)	Callouslike thickening on the palms of the hands and feet.

By studying the family tree it is often possible to deduce a person's genotype for a given characteristic. Table I lists some human genetic traits, and indicates whether the genes that govern these traits are dominant or recessive.

DIHYBRIDS

There are times when it is important to study the inheritance of two or more characteristics at the same time. In wheat breeding, one objective is to develop plants that produce high yields. But high yield will not result if these plants are attacked by a fungus disease known as wheat rust. The overall objective is to develop high-yield plants with resistance to this disease. Simplifying a very complex situation, let us assume that high-yield potential is controlled by a gene L and low-yield by the recessive gene l; disease resistance by a recessive gene r, and susceptibility to this disease by the dominant gene R. These genes lie on separate chromosomes. To predict the possible types of plants that can be produced by crossbreeding, it is necessary to know (or to infer in some way) the genotypes of the parent plants.

A plant with low-yield potential and resistance to disease must have the genotype $\boxed{l\ l\ \ r\ r}$ and can produce only one type of sperm or egg. Each gamete would contain one l and one r gene. But if the parent plant had a genotype of $\boxed{L\ l\ \ r\ r}$ the story would be quite different. Now two types of sperm and egg are possible: those carrying one L and one r gene, and those carrying an l and an r gene. If the parent's genotype were $\boxed{L\ l\ \ R\ r}$, four different types of sperm and egg would be possible

$(L\ R)$, $(L\ r)$, $(l\ R)$, and $(l\ r)$.

It is not possible to have a gamete with both an L and an l gene because these genes are on homologous chromosomes that separate during meiosis. Any given gamete receives only one chromatid from each pair of homologues. For the same reason it is impossible to have gametes with the genotypes of

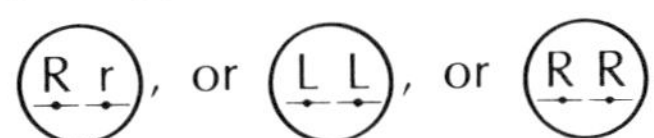

$(R\ r)$, or $(L\ L)$, or $(R\ R)$.

Problem: Predict the plants that can be produced by crossing two plants of the genotype | L l R r | and predict the frequency of each type

GIVEN

Both parents have the genotype | L l R r |

L = high-yield potential   R = susceptible to rust
l = low-yield potential   r = resistant to rust

SOLUTION

1. Determine the possible gametes and their frequency

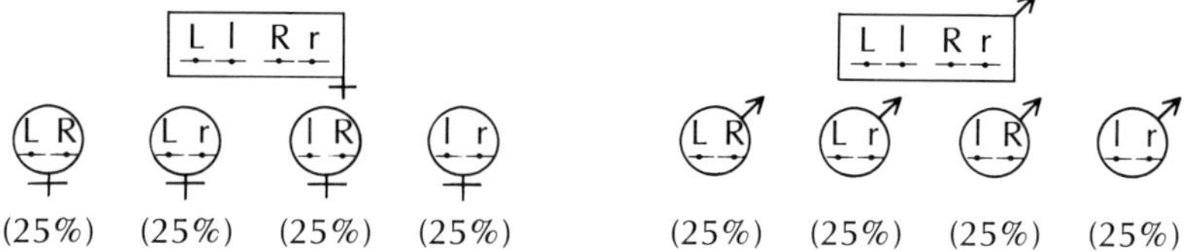

2. Determine the frequency of all possible mating combinations among these gametes

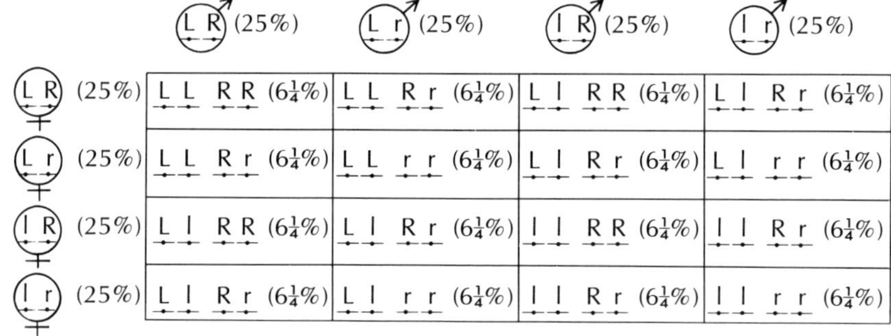

A thorough understanding of the meiotic process will prevent any confusion in determining the possible kinds of gametes produced by an organism of known genotype.

If two plants with the genotype | L l R r | are crossed, the result is an interesting assortment of progeny (Box E). Both parents should produce high yields, but they are susceptible to rust. The male or pollen parent could produce four different kinds of sperm. The female parent could likewise produce four different kinds of egg. Progeny with nine different genotypes

3. List the different genotypes and their frequencies

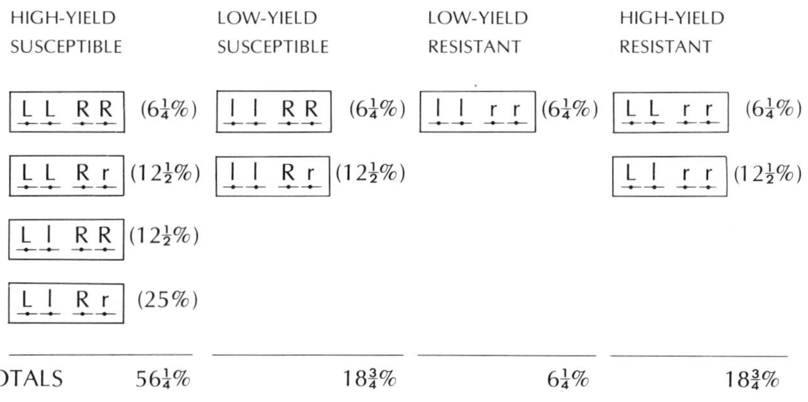

4. Convert the genotypic results to phenotypes

HIGH-YIELD SUSCEPTIBLE	LOW-YIELD SUSCEPTIBLE	LOW-YIELD RESISTANT	HIGH-YIELD RESISTANT
L L R R ($6\frac{1}{4}$%)	I I R R ($6\frac{1}{4}$%)	I I r r ($6\frac{1}{4}$%)	L L r r ($6\frac{1}{4}$%)
L L R r ($12\frac{1}{2}$%)	I I R r ($12\frac{1}{2}$%)		L I r r ($12\frac{1}{2}$%)
L I R R ($12\frac{1}{2}$%)			
L I R r (25%)			
TOTALS $56\frac{1}{4}$%	$18\frac{3}{4}$%	$6\frac{1}{4}$%	$18\frac{3}{4}$%

would be produced but they would show only four different phenotypes:
(1) high-yield/susceptible, (2) low-yield/susceptible, (3) low-yield/resistant,
and (4) high-yield/resistant. Then, by writing down the percentage of each
type of gamete produced by each parent it is possible to predict the percentage of each type of progeny. This type of analysis is much longer and
more complicated than that involved in the inheritance of only one trait.
Most of the plants resulting from such a cross would have high-yield potential but would be susceptible to rust. Less than 20 percent would have both
desirable characteristics, high-yield potential and rust resistance. If a seed

company marketed seed produced from such a cross, they would soon go out of business, and any farmer who bought the seed might well find himself bankrupt.

To guarantee the highest proportion of plants with both a rust-resistant trait and a high-yield potential, the seed sold to the farmer should be produced by crossing plants of genotype $\boxed{L\ L\ \ r\ r}$ with resistant plants. This would guarantee that all plants produced from such seeds would have the genotype $\boxed{L\ ?\ \ r\ r}$. As long as one dominant gene for high-yield potential is present the wheat will have the potential for producing high yield. The homozygous recessive condition of the rust-resistance gene assures that the plants will be resistant to this disease.

TRIHYBRIDS AND MORE

The inheritance pattern of three characteristics at the same time increases the problems of analysis even further. By crossing two parents with a genotype of $\boxed{A\ a\ \ B\ b\ \ C\ c}$, we end up with 8 possible phenotypes in the progeny and 27 different genotypes. By increasing the number of characteristics studied to 10, we get 1024 possible phenotypes and 59,049 possible different genotypes in the second generation. This situation suggests the reasons for the enormous amount of diversity found in any population of individuals of a single species. But this is only part of the story.

INCOMPLETE DOMINANCE

Mendel's ideas about dominance have by now been considerably modified. Modern geneticists believe that very few, if any, gene pairs exhibit complete dominant-recessive characteristics. No gene acts alone. The expression of every gene is influenced by one or more other genes. One of the simplest examples of such gene interaction has been named blending or incomplete dominance. As the name implies, instead of alleles that consist of a dominant and a recessive gene, the alleles consist of two genes which are both active when present in the same cell. Table II contrasts the phenotypes and genotypes of the dominant-recessive alleles and of the incompletely dominant alleles. The power or potency of one gene over its allele may vary

II

Incomplete dominance often shows up as a blend of dominant and recessive traits. Column at left lists genotype and phenotype in cases which follow classic Mendelian rules. Column at right shows results in cases of incomplete dominance (blending).

Dominant-Recessive Inheritance Pattern	Blending Inheritance Pattern
W = dominant gene for red flowers	W^a = gene for red flowers
w = recessive trait for white flowers	W^b = gene for white flowers
| W W | = dominant phenotype (red flowers)	| W^a W^a | = red flowers
| W w | = dominant phenotype (red flowers)	| W^a W^b | = pink flowers
| w w | = recessive phenotype (white flowers)	| W^b W^b | = white flowers

from almost complete dominance to virtual equality. In some cases the effect of the hybrid condition for some genes may be quite unexpected compared with the effects of both pure-line conditions. A case in point is the human disease known as sickle cell anemia. This disease is characterized by the presence of abnormal hemoglobin in the blood which changes the shape of the red blood cells. Instead of being disk-shaped, they become curved or sickle-shaped (Fig. 4). These cells may block the very small blood vessels called capillaries. The result is extreme pain and swelling, sometimes leading to gangrene and even death.

The genotype of a normal individual may be represented as $Hb^A Hb^A$. A person with sickle cell disease has the genotype $Hb^S Hb^S$. Persons who are heterozygous for these genes $Hb^A Hb^S$ are carriers who do not develop the disease. Curiously, the carriers have a resistance to the malarial parasite much greater than that exhibited by either normal individuals or individuals with sickle cell anemia. The reason for this increased resistance to malaria is unknown, and it is quite unpredictable from our knowledge of the effects of the genes in pure-line individuals.

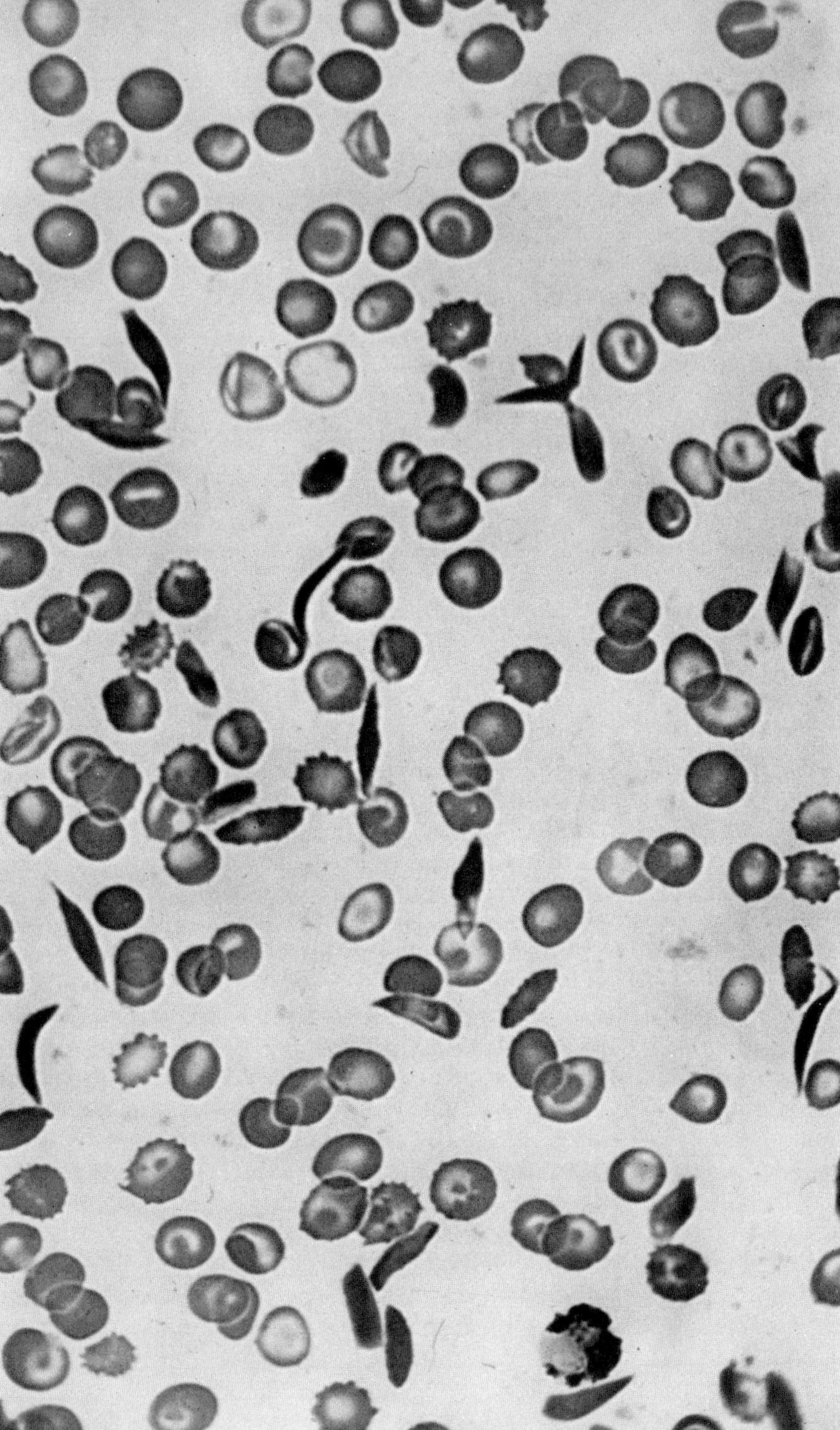

4

Sickle cell anemia, a genetic blood disease common among blacks, is characterized by the presence of sickle-shaped and withered red cells. Normal red cells are disk-shaped.

(Courtesy of Carolina Biological Supply Company.)

5
Coat color of these mice is determined by at least three different forms of a single gene (S^l, s^l, and s). Piebald mouse at left is homozygous s s. Nearly white mouse at center is heterozygous S^l s^l. Black and white mouse at right is heterozygous s^l s.

MULTIPLE ALLELES

Another mechanism contributing to genetic variability was discovered early in the 20th century. It was found that a gene could exist in more than two alternate forms; in other words, that it could have more than two alleles. This phenomenon is now called multiple allelism. Mendel's ideas about dominance had to be modified even further. Workers studying the inheritance of coat color in mice found that yellow, black, and barred coats were due to three different forms of the same gene. The three forms were labeled A^Y, A, and a. In this example of multiple allelism, various combinations of the three alleles give rise to the following coat colors $\boxed{A^Y\ A^Y}$ and $\boxed{A^Y\ a}$ produce yellow coat color, $\boxed{A\ A}$ and $\boxed{A\ a}$ produce a barred coat, $\boxed{a\ a}$ produces a black coat. A fourth combination $\boxed{A^Y\ A^Y}$ produces individuals which do not survive past the embryonic state of development. Another multiple allelic series for coat pattern in mice is illustrated in Fig. 5.

The cells of any one mouse contain only two of these three types of gene. The pair of genes may be identical or may be composed of any combination of two of the three genes. By studying the resulting phenotypes we can deduce that two doses of the A^Y gene are lethal; that the A^Y gene is dominant over both the A gene and the a gene; that the A gene is dominant only over the a gene; and that the a gene is recessive to both the A and the A^Y gene.

III

Blood type in humans is partially governed by three alleles: L^A, L^B, and I. Combinations of these alleles yield the four basic blood types: A, B, AB, and O.

Genes	Blood Type
$L^A\ L^A$ or $L^A\ I$	A
$L^B\ L^B$ or $L^B\ I$	B
$L^A\ L^B$	AB
I I	O

A more familiar example of multiple allelism is the inheritance pattern of blood type in humans. Each of us can be classified as one of four blood types: A, B, AB, or O. The three alleles involved are L^A, L^B, and I. L^A and L^B are dominant over the I gene but neither is dominant over the other. The relationships between the genes and the phenotypes are shown in Table III. Combinations of these alleles can produce six different genotypes and four distinct phenotypes. The general formula for calculating the number of different combinations of X number of alleles taken two at a time is $X(X + 1)/2$. In the fruit fly, a gene affecting eye color has 20 different alleles. Using the formula above, it can be seen that 210 different genotypes are possible for this one characteristic.

POLYGENIC INHERITANCE

To make the situation even more complicated, most characteristics are not governed by just a single pair of genes. In the common fruit fly there are more than 90 different *pairs* of genes, all present in the same cell, that influence the color of the eyes. The condition in which more than one pair of genes in a cell affects a given characteristic is called polygenic inheritance.

Knowledge of genic interaction helps explain the inheritance patterns of many variable characteristics, such as height and skin color in humans, crop yield in plants, and milk and egg production in animals. For decades such characteristics defied genetic analysis because their complicated inheritance patterns did not seem to fit the classical Mendelian laws. The skin color of the offspring of marriages between blacks and whites, for example, varies over a wide range of intermediate shades. It does not follow a simple Mendelian ratio of, say, one white child to three black ones.

With newer and more sophisticated mathematical techniques, including the use of computers, geneticists are now better able to analyze the gradations of inherited characteristics. No longer do geneticists think of a single gene that controls skin color or plant vigor; they think in terms of polygenic systems in which many alleles interact with one another and yield a wide range of phenotypes.

LINKAGE

Early geneticists noted that certain traits tend to be inherited together. This observation led the American geneticist Thomas Hunt Morgan to the idea

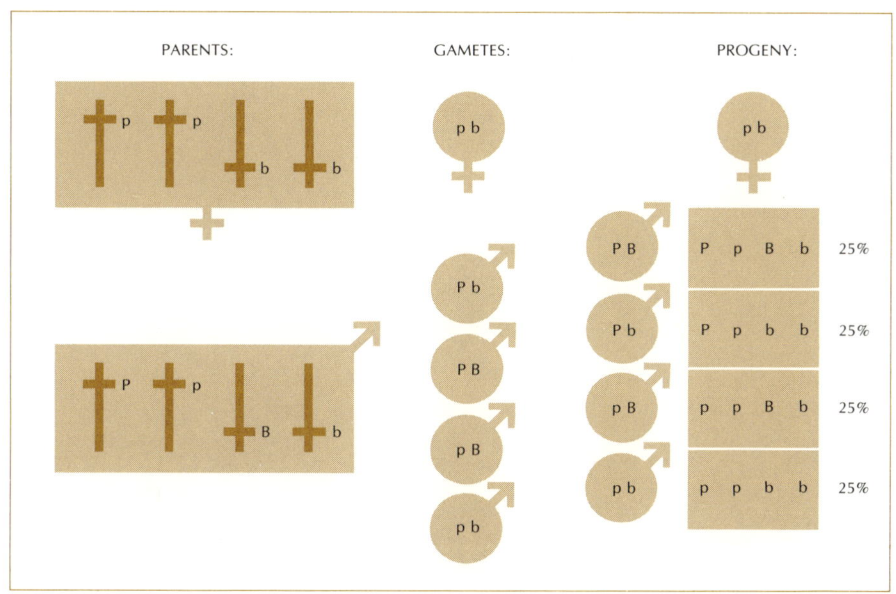

6
If genes for two characteristics lie on separate chromosomes, a parent heterozygous for both characteristics can produce four genetically different gametes. Progeny from a cross between this parent and a homozygous recessive will be produced in a ratio of approximately 1:1:1:1.

of linkage, that genes for associated traits lie on the same chromosome. Moreover, he reasoned that genes that lie close together on one chromosome will be inherited together more often than genes located farther apart. This insight eventually led to the mapping of the sequence of genes along the chromosome; it also led to a Nobel prize for Morgan in 1933.

The construction of a chromosome map depends on a knowledge of the mechanism that Morgan named crossing over. Crossing over is the process by which a part of one chromosome may be exchanged with a similar part of a homologous chromosome (Chapter 7). This exchange produces new combinations of genes on a chromosome. Crossing over serves to maintain the genetic diversity of a species, and thereby enhances its potential to evolve and adapt to long-term changes in the environment.

The first step in mapping a chromosome is to decide whether two given genes lie on the same chromosome. This can often be judged from breeding experiments. Figures 6 and 7 show the predicted results of crossbreeding two corn plants, one heterozygous for two traits, the other homozygous-recessive for both. If the genes lie on separate pairs of chromosomes, the

7

Linkage of genes results in a different progeny ratio from that of Fig. 6. In this case a parent heterozygous for two characteristics can theoretically produce only two genetically different kinds of gametes. The crossing of this parent with a homozygous recessive parent produces progeny in a ratio of 1:1.

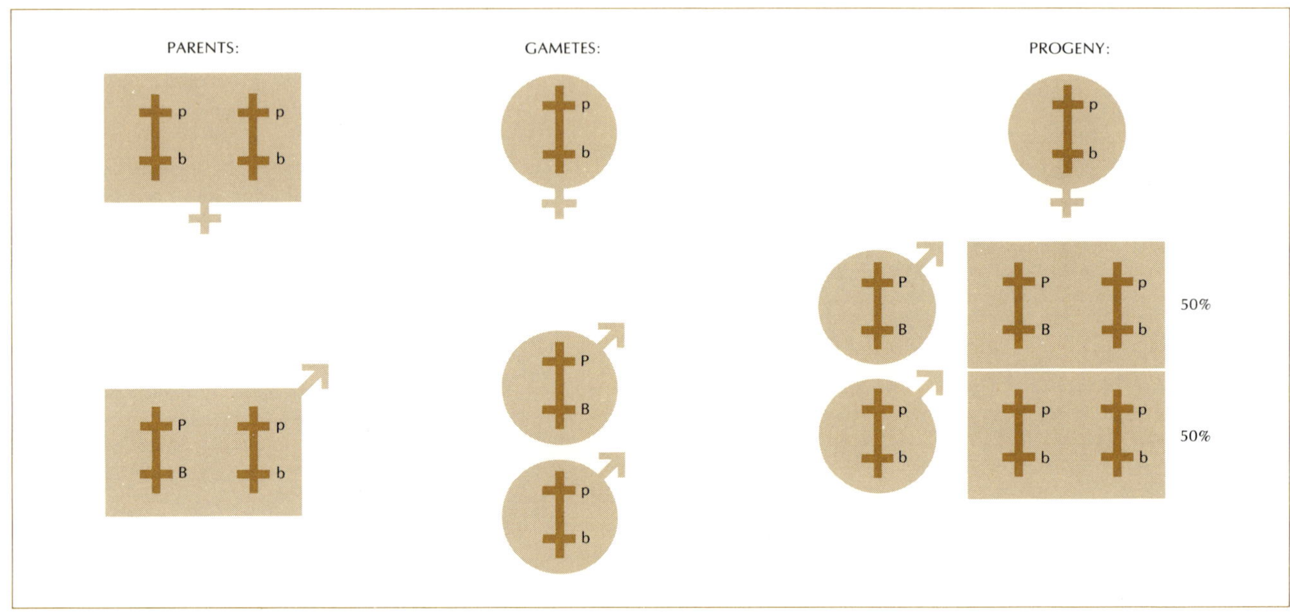

offspring will have four different phenotypes in a ratio of 1:1:1:1, as shown in Fig. 6. But if the genes are linked, we would expect the offspring to show only two different phenotypes in a ratio of 1:1, as illustrated in Fig. 7. In an actual experiment the results lie somewhere between. About 40 percent of the offspring exhibit both dominant characteristics; 40 percent exhibit both recessive characteristics; 10 percent have one dominant and one recessive trait, and 10 percent show the other dominant and recessive traits. It can be inferred from the observed results that the dominant genes are not always inherited together, and also that they are not inherited completely independently. The same can be said for the recessive genes. Since most of the offspring, 80 percent, resemble one or the other of the parents and only 20 percent resemble neither of the parents, it is apparent that the genes are linked most of the time but not all of the time. The true explanation of this phenomenon hinges on an understanding of the process of crossing over.

In Fig. 8 the chromatids are drawn to illustrate diagrammatically what happens during the cross over. In 40 percent of all the cells undergoing

8

Crossing over of chromatids during meiosis may give rise to chromosomes with new combinations of genes. In this case a small percent of the progeny will contain chromosomes with a complement of genes different from that of either parent.

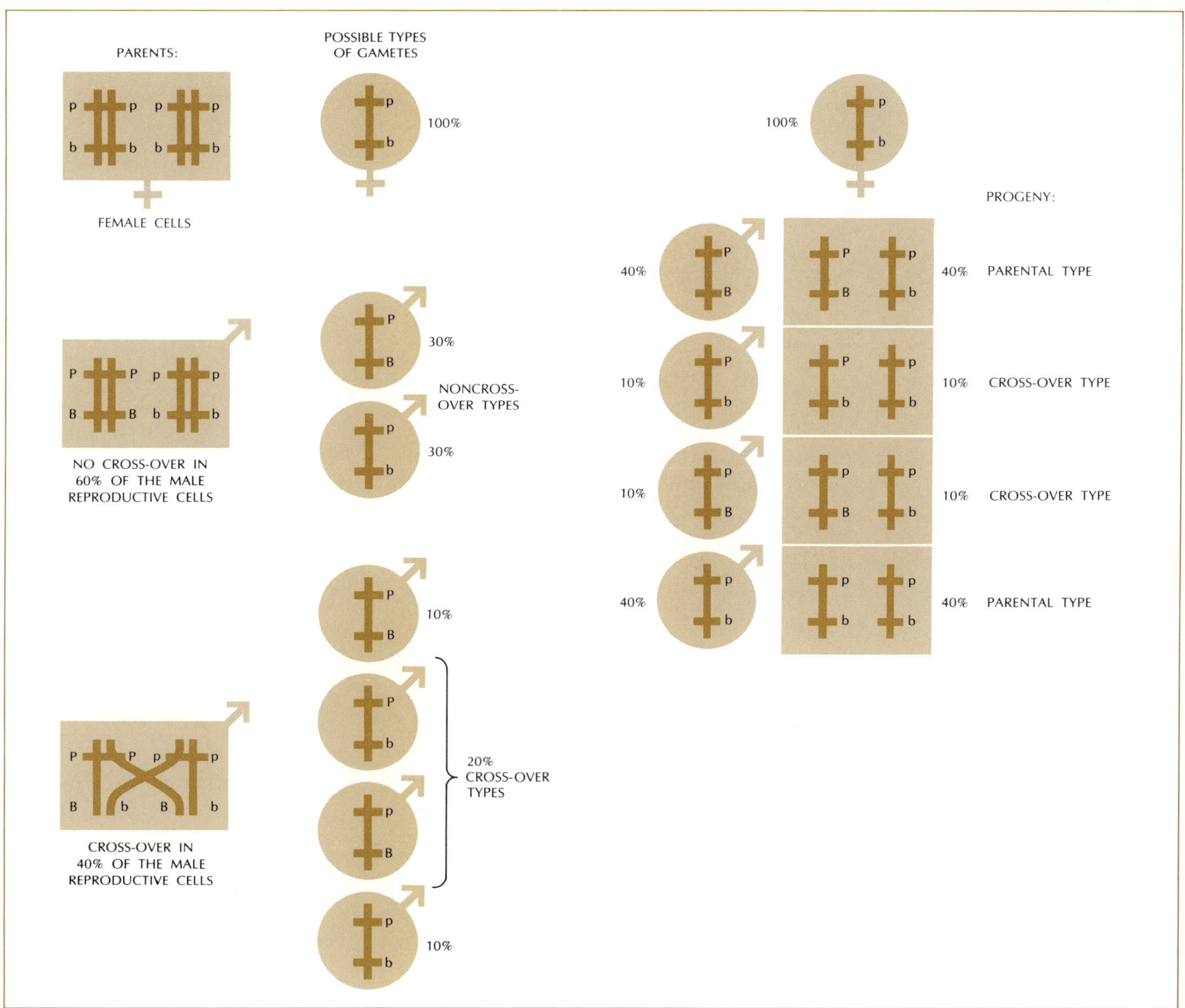

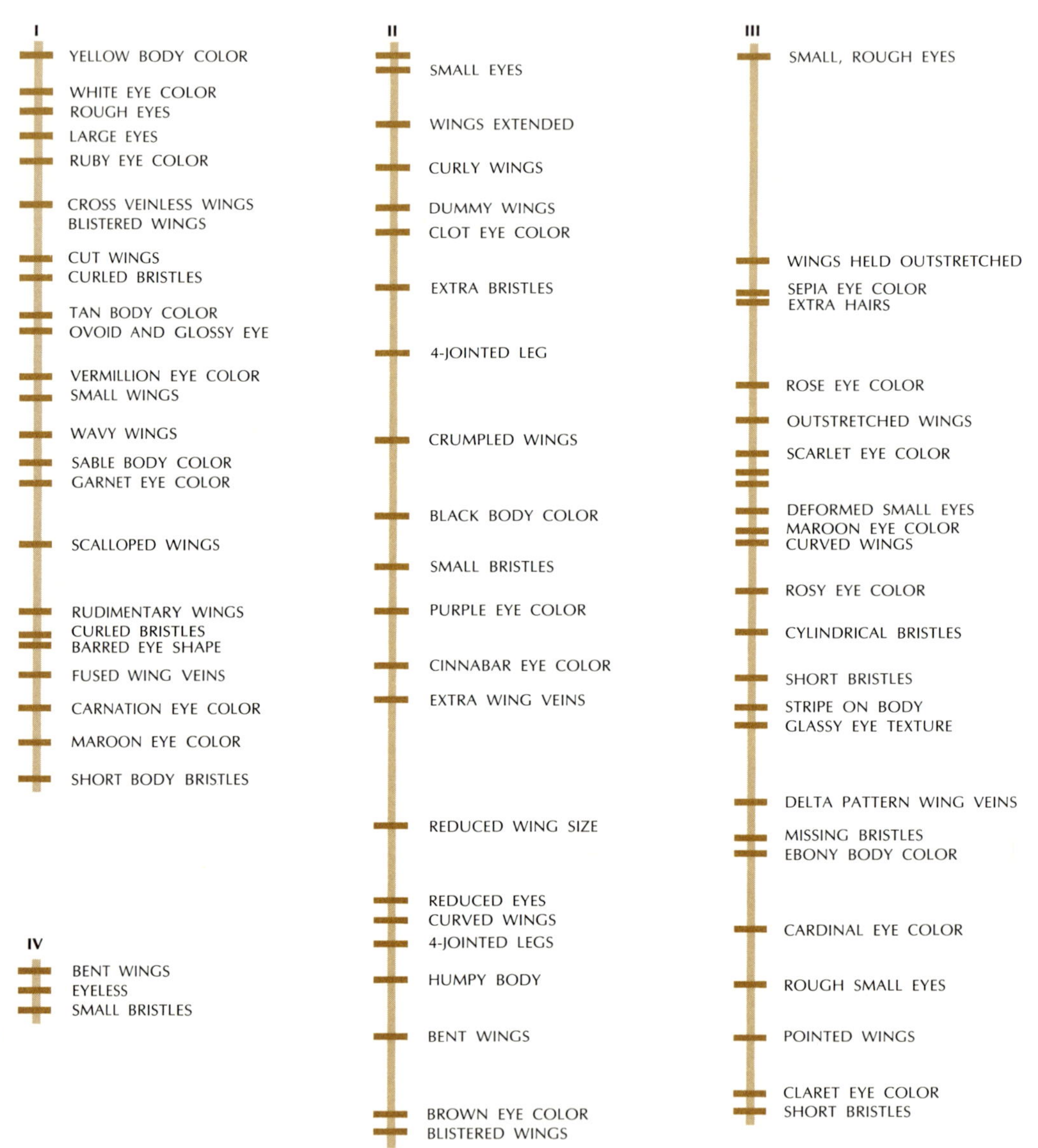

I
YELLOW BODY COLOR
WHITE EYE COLOR
ROUGH EYES
LARGE EYES
RUBY EYE COLOR
CROSS VEINLESS WINGS
BLISTERED WINGS
CUT WINGS
CURLED BRISTLES
TAN BODY COLOR
OVOID AND GLOSSY EYE
VERMILLION EYE COLOR
SMALL WINGS
WAVY WINGS
SABLE BODY COLOR
GARNET EYE COLOR
SCALLOPED WINGS
RUDIMENTARY WINGS
CURLED BRISTLES
BARRED EYE SHAPE
FUSED WING VEINS
CARNATION EYE COLOR
MAROON EYE COLOR
SHORT BODY BRISTLES
IV
BENT WINGS
EYELESS
SMALL BRISTLES
II
SMALL EYES
WINGS EXTENDED
CURLY WINGS
DUMMY WINGS
CLOT EYE COLOR
EXTRA BRISTLES
4-JOINTED LEG
CRUMPLED WINGS
BLACK BODY COLOR
SMALL BRISTLES
PURPLE EYE COLOR
CINNABAR EYE COLOR
EXTRA WING VEINS
REDUCED WING SIZE
REDUCED EYES
CURVED WINGS
4-JOINTED LEGS
HUMPY BODY
BENT WINGS
BROWN EYE COLOR
BLISTERED WINGS
III
SMALL, ROUGH EYES
WINGS HELD OUTSTRETCHED
SEPIA EYE COLOR
EXTRA HAIRS
ROSE EYE COLOR
OUTSTRETCHED WINGS
SCARLET EYE COLOR
DEFORMED SMALL EYES
MAROON EYE COLOR
CURVED WINGS
ROSY EYE COLOR
CYLINDRICAL BRISTLES
SHORT BRISTLES
STRIPE ON BODY
GLASSY EYE TEXTURE
DELTA PATTERN WING VEINS
MISSING BRISTLES
EBONY BODY COLOR
CARDINAL EYE COLOR
ROUGH SMALL EYES
POINTED WINGS
CLARET EYE COLOR
SHORT BRISTLES

◀**Chromosome map** for a small number of the genes on the four chromosomes of a gamete from the fruit fly *Drosophila melanogaster.*

meiosis in the pollen grains of the male parent, crossing over occurs between the gene for height and the gene for purple pigment production; this yields four genetically different types of gamete. The other 60 percent of the meiocytes, in which no cross over occurs between these two genes, can produce only two different kinds of sperm. The relative number of sperm containing each type of chromosome can then be calculated. Forty percent will have the genotype (P B) (10 percent from the meiocytes in which cross overs occurred and 30 percent from those in which no cross over occurred). Forty percent will be of the genotype (p b), ten percent will be (P b), and ten percent will be (p B). Since the female parent is pure-line recessive for both genes, all of her eggs will have the same genetic complement, (p b). In an individual homozygous for each of two genes on a chromosome, crossing over will not produce new genetic combinations.

The cross-over frequency illustrated here is rather high, and suggests that the genes for height and purple pigment lie toward the opposite ends of the chromosome. Two genes that lie very close to each other would show a cross-over rate of almost zero. By performing many crosses, and by comparing the cross-over frequencies of many linked genes, geneticists can arrange them on a chromosome map. Figure 9 is a partial map of the chromosomes of the fruit fly. Although the genetics of *Drosophila* is better known than that of any other organism, thousands of its genes remain unmapped. The far more complex task of mapping the thousands of genes on each of the human chromosomes has only begun.

Chromosome maps aid in the correlation of evidence from three lines of research: classical genetics, biochemistry, and electron microscopy. This correlation has already yielded important insights about the structure and activity of chromosomes. For the future, the precise mapping of the location of each gene opens up the possibility of "genetic surgery," the alteration or elimination of harmful genes.

MUTATION

Hereditary constancy depends on the accurate duplication of genes. Strictly speaking, we carry not the genes of our parents but copies of those genes. Although the copying process is remarkably precise, once in a while it produces a mutation, a gene different from the original. Mutation and sexual reproduction are the mainsprings of evolution; mutation because it is the source of entirely new genes, and sexual reproduction because it creates new combinations of genes.

A certain number of mutations arise spontaneously in each generation. The mutation of any one gene is a rare event, but each chromosome contains thousands of genes; in a population of thousands or millions of organisms, the overall chance of mutation occurring is considerable. Experimental studies suggest that in each generation of *Drosophila,* for example, roughly five percent of the gametes contain one or more mutations. Mutations can arise in any cell, but mutations in the gametes are of primary interest, because these are passed along to the offspring.

Most mutations are decidedly harmful to the organism that inherits them. In fact, many are lethal: the embryo either fails to develop or the organism dies young. For this reason public health officials have become concerned in recent years about the exposure of the human population to agents that increase the mutation rate. Ionizing radiation—X-rays, gamma rays, cosmic rays, α- and β-particles—is the most potent of these agents. In terms of its effect on genes, there is no safe dose of radiation; even a small exposure can produce a mutation, and the greater the dosage, the greater the chance and frequency of mutation. This effect explains the worldwide concern about the danger of fallout from nuclear explosions, and about the disposal of radioactive industrial wastes; it also partially explains the sharp curtailment during the last decade of the use of medical and dental X-rays, and the increased concern with shielding patients' ovaries and testes from ionizing radiation during medical treatment. A number of chemicals can also increase the mutation rate. The most notable is mustard gas, but the list grows longer each year, and the pollution of air and water with these compounds has become a public health hazard. Many research workers believe that mutation-inducing agents are somehow implicated in the production of certain types of cancer, but so far the connection has not been proved. At the very least, many mutagens are also carcinogens and vice versa.

READINGS

Barish, N., *The Gene Concept.* Reinhold, New York, 1965 (paper)
> Brief, easy-to-read discussions of Mendelian genetics, multiple genetic systems, and the structure of the gene.

Cook, S. A., *Reproduction, Heredity and Sexuality.* Wadsworth, Belmont, Calif., 1964 (paper)
> A rather hasty survey of classical genetics, the role of the gene, and the effect of evolution on the process of reproduction.

Dunn, L. C., *A Short History of Genetics.* McGraw-Hill, New York, 1965
> A short and clearly written book, with emphasis on classical genetics.

Medvedev, Z. A., *The Rise and Fall of T. D. Lysenko.* Columbia University, New York, 1969
> The story of the political suppression of free scientific thought in the Soviet Union in the 20th century. The publication of this book in the West led Soviet authorities to declare the author "insane" and to confine him in a mental hospital. Medvedev is a well-known Soviet biologist; Lysenko was a charlatan whose doctrines on the inheritance of acquired characteristics found favor with Stalinist authorities.

Roslansky, J. D. (ed.), *Genetics and the Future of Man.* Appleton-Century-Crofts, New York, 1966
> Rather weighty discussions of the implications of the new knowledge of genetics and man's ability to alter his own genetic future.

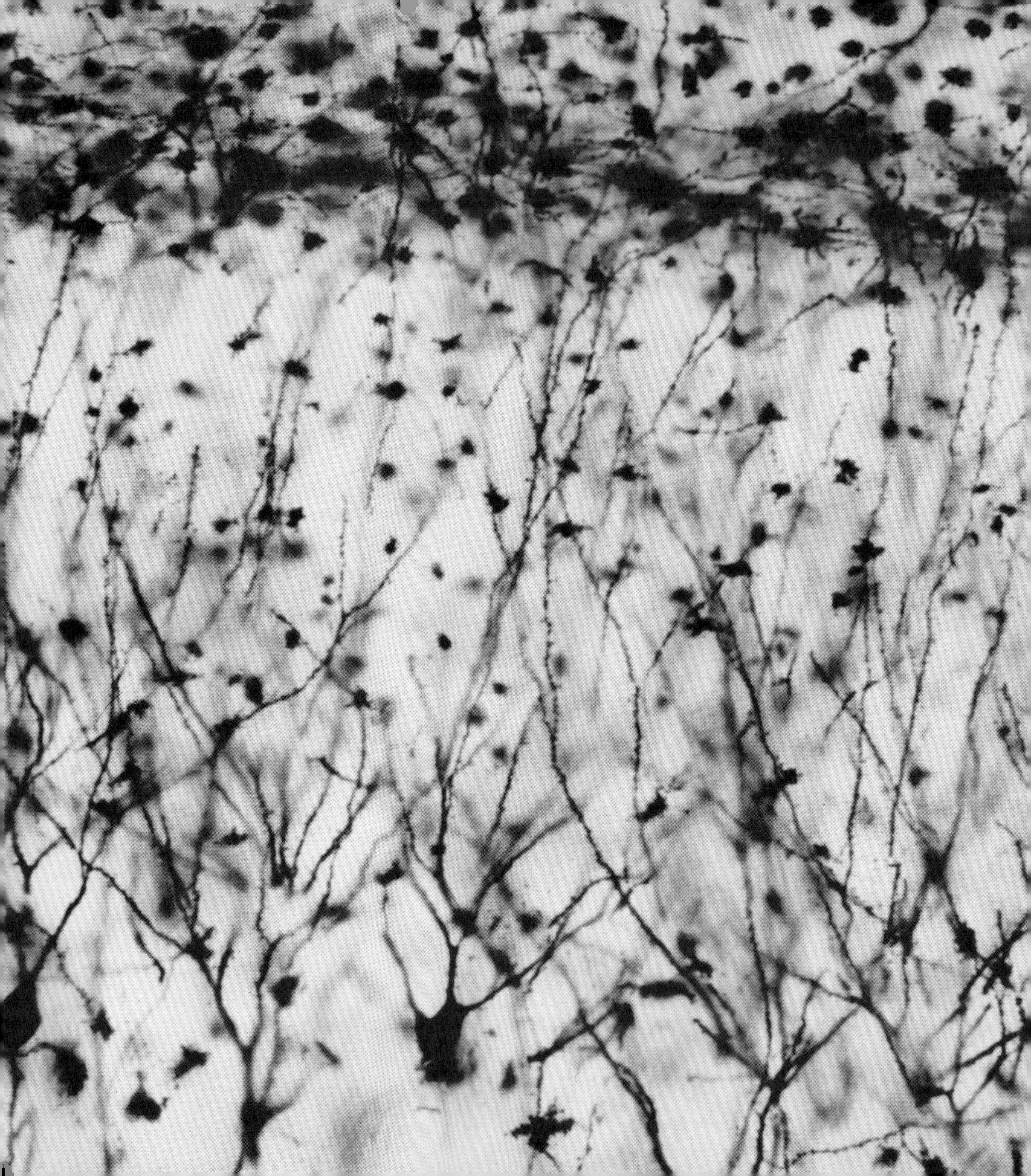

9. CONTROL MECHANISMS

The assertion of the poet John Donne that no man is an island applies equally to all other living organisms. Dependent on outside sources of energy and raw materials, beset by natural enemies, threatened by toxic chemicals and physical dangers, the living organism grows, develops and survives by regulating its activities, both internal and external, in response to the environment.

Encoded in the genes of each organism lies a whole repertory of possible responses which are selected and coordinated by a variety of control mechanisms. In most multicellular organisms control involves some means of communication between different parts of the body. The activity of certain plant and animal cells is often regulated by chemical messages carried by small molecules called hormones. The nervous system of animals is a control mechanism that relies on rapid electrochemical impulses for the transmission of information. Specialized sensors monitor the environment and convert physical and chemical information into coded pulses, which are relayed by nerve fibers at speeds of up to 70 meters per second. Integrated and processed by the switching circuits of the higher nerve centers, the information is fed to various effector organs, sometimes as an electrical signal, sometimes as a chemical message, triggering an appropriate response. In most systems sensors continuously monitor the results and feed back to the effectors information that automatically corrects their future performance. Biological control systems

◀**Brain cells** form a control network of incredible complexity. This photomicrograph depicts cells from the cortex of the rat stained by the Golgi method, which reveals some of the connections among cells. Black "blots" are cell bodies; thin lines are nerve fibers.

have inspired the development of man-made feedback systems, such as the navigation systems that guide space vehicles. Present knowledge of biological control mechanisms is far from complete, but each month journal articles report new discoveries about both the molecular and electrical control mechanisms of living organisms.

CONTROL OF DEVELOPMENT

How does a complex organism composed of many different kinds of cell develop from a single-celled egg? No problem in biology has provoked more conflicting theories, or stimulated more research. Early in the lives of plant and animal embryos many cells begin to become specialized, a process that biologists call differentiation. As these cells in the embryo divide they begin to change: structurally and functionally they become different from the parent cell. In the course of development, successive waves of cell division and differentiation eventually give rise to the many types of cell and tissue found in the adult.

Differentiation is a crucial process in the development of all multicellular organisms because it creates the diversity of specialized cells necessary to perform specific complex functions. The process of differentiation continues throughout the life of the individual. Worn out and damaged cells and tissues are continually replaced. Appropriate cells differentiate as new structures are formed. For example, at a precise point in the growing season of a plant, the cells of the stem tip differentiate and flowers are formed; in animals, new blood cells continually arise from less differentiated cells.

Several generations of biologists have found development and differentiation easier to describe than to explain. The main difficulty has arisen from the apparent conflict between the fact of extreme diversity of cell types and one of the basic axioms of biology: that, as a result of mitosis, every cell in a multicellular organism contains a complete and identical set of genes and therefore contains the same hereditary instructions. If the axiom is true, any theory of development must propose some mechanism by which genes can be switched on and off at specific times in the differentiation process. The nature of this mechanism is the key not only to the mystery of development but probably also to the day-to-day function of genes throughout the life of every organism. It is no wonder that a problem of this magnitude has remained unsolved for so long.

TOTIPOTENCY

For many years biologists thought the zygote to be the only all-powerful or totipotent cell in a diploid organism's life history. They thought that the zygote alone possessed the genetic potential to produce a complete, mature diploid organism. However, a few examples from the plant kindgom seemed to refute this idea. For example, how can one explain the ability of some plants to produce small intact plantlets along the margins of their leaves? The cells of the parent plant that give rise to these new plants must also be totipotent. But they seem to be totipotent only under certain conditions. The expression of their totipotency is inhibited when they are removed from the parent plant. If these cells are taken from the leaves and are placed in test tubes or in petri dishes, they will grow and divide but only in a very disorganized way. Masses of cells are produced; recognizable plants are not formed. This supports the suggestion that the expression of a cell's genetic potential is controlled by the physical and chemical environment of that cell.

This theory was supported by a series of experiments carried out in the late 1950's by F. C. Steward at Cornell University. He took cells from the roots of mature carrot plants and cultivated them in a special culture medium containing coconut milk, the fluid that bathes the developing embryo within the coconut fruit. By this technique Steward was able to grow intact, fully formed carrot plants, flowers and all, from cells of a mature carrot root. It was elegant proof that the zygote is not unique in its ability to produce a fully functional organism. Further work has established that cells from plants as different as ferns and orchids can be stimulated to develop into new plants if they are placed in the correct culture medium. It now seems possible that certain cells of all plants are totipotent. A newly developed technique has enabled Steward and his colleagues to culture thousands of genetically identical plants from a single parent. The availability of adequate numbers of genetically identical organisms should make it possible to settle many of the classical arguments of "nature versus nurture": to what extent various inborn characteristics can be altered by the environment.

A similar breakthrough in the field of animal biology was reported in 1968 by J. B. Gurdon of Oxford University. Gurdon removed the nucleus from a cell from the intestine of a South African clawed toad and inserted it into an unfertilized egg from which the nucleus had been removed. The

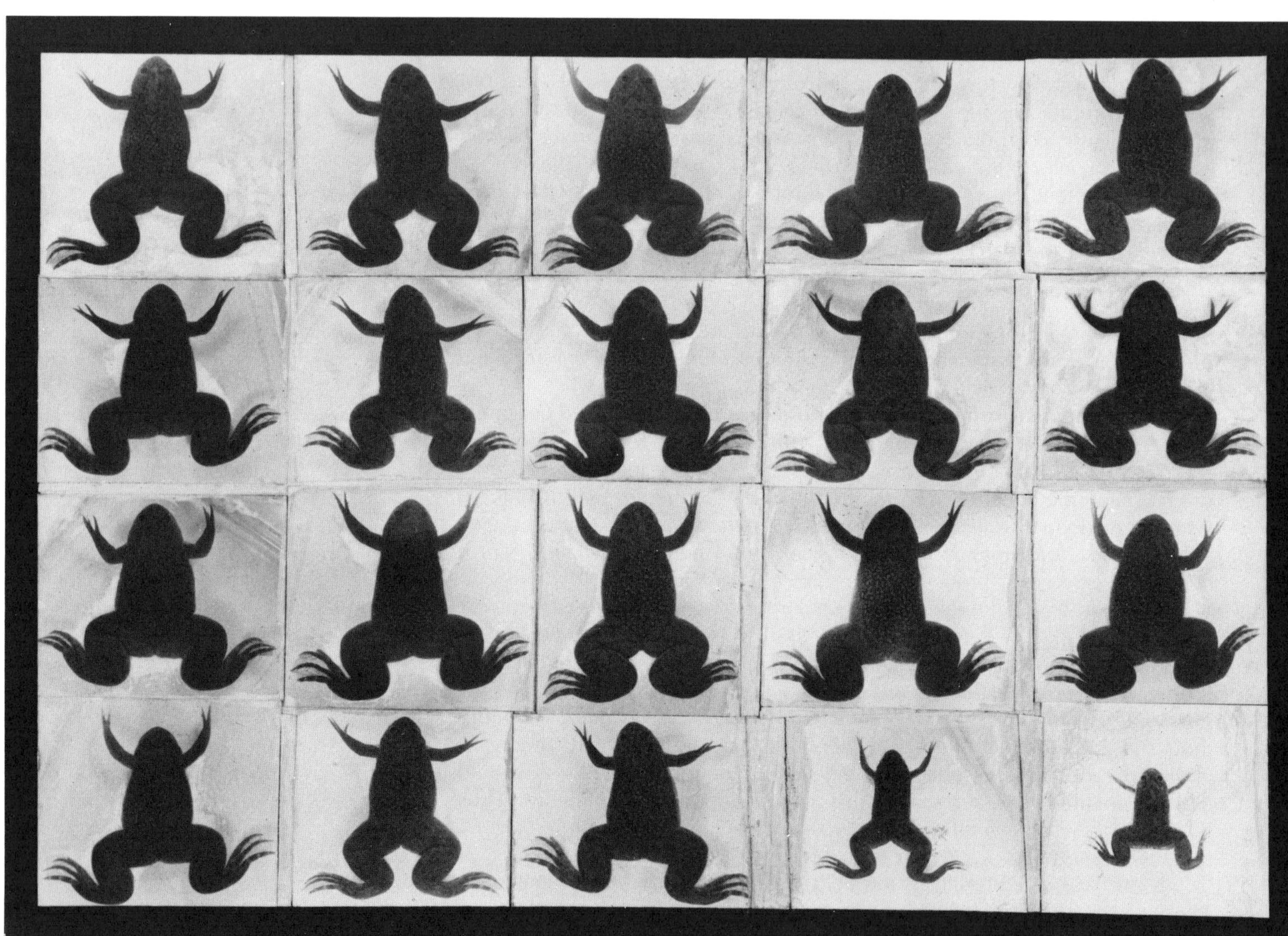

1

egg developed into a normal adult frog, genetically identical to the "parent"
frog which supplied the nucleus (Fig. 1). This was the first demonstration of
the totipotency of a nucleus from a mature, differentiated animal cell, and it
led Gurdon to conclude that cell differentiation depends not only upon
external environmental conditions but also upon an interaction between the
nucleus and the cytoplasm. In the case of the experiments with the South
African toads, genes that had been switched off while the nucleus was in
the intestinal cell were apparently switched back on by some chemical
substance in the egg cytoplasm when the nucleus was embedded in the
enucleated egg. The nucleus directs what the cytoplasm will do, and the
cytoplasm exerts a definite influence on the activities of the nucleus.

INDUCTION AND REPRESSION

A theoretical model of how genes might be turned on and off was proposed
in the early 1960's by Francois Jacob and Jacques Monod of the Pasteur
Institute in Paris. Their work involved the bacterium *E. coli*.

E. coli cells grow well on a medium containing the simple sugar glucose. They contain the enzymes necessary to transport glucose into the cell and to use it as a source of energy. However, if *E. coli* cells are transferred from a glucose medium to a medium lacking glucose but rich in another sugar, lactose (milk sugar), they stop growing. They cannot use lactose as a food. But after a short time they begin to grow again. The cells have begun to synthesize large quantities of two new enzymes: one to transport lactose into the cell and one to break down the lactose molecule into smaller sugar molecules which can be used as a source of energy. When the bacteria are transferred back to the glucose medium, the synthesis of these enzymes virtually stops. Evidently the presence of lactose molecules switches on the genes that control the synthesis of the enzymes necessary for its utilization by the cell. This "switch-on" or induction effect was observed as early as 1900. More than 50 years later researchers discovered a "switch-off" effect, which they named gene repression. The classic example involves the synthesis of amino acids such as arginine, tryptophan, and histidine. The presence of these substances within a cell effectively prevents their manufacture by the cell.

It was not until 1961 that Jacob and Monod concluded that cells control enzyme synthesis by the turning on and off of the activity of specific portions of the DNA in the cell nucleus. Working with the lactose system, they reasoned that because two different enzymes involved in the utilization of lactose were turned on and off together, the genes for the synthesis of these two enzymes might lie next to one another on the chromosome. Moreover, they proposed that another kind of gene—one that controls both of these two genes—must lie next to them. Jacob and Monod called this control or switch-on gene the operator gene. They grouped the operator and the genes that it controls into a functional unit that they called an "operon." Later they proposed the existence of a third type of gene, a regulator gene, which influences the activity of the operator gene.

According to their theory, the regulator gene directs the production of a repressor molecule that switches off the operator. For *E. coli* cells growing on a medium containing glucose but not lactose, one such repressor molecule combines in some way with the lactose operator gene, inactivating it. As a result, the whole operon of genes for lactose utilization is blocked. When the bacteria are transferred to the lactose medium, molecules of lactose enter the cell and combine with the repressor molecule, preventing it from combining with the operator gene. This unlocks the operator and activates the entire operon. A synthesizer molecule begins "reading" the information of the structural genes, resulting in the synthesis of the lactose enzymes. So long as lactose is present in the cell it will combine with the repressor. In effect, lactose represses the repressor.

An *E. coli* cell contains about 2500 genes. At any one time most of them are undoubtedly switched off. The mechanism of gene induction, depicted schematically in Fig. 2, explains how the presence of a chemical substance in the environment can actually result in the activation of one or more of these genes.

Jacob and Monod used a similar model to explain how a chemical can turn off the activity of certain genes. The regulation of the production of the amino acid histidine is an example of gene repression. The operon governing the synthesis of this compound contains genes for the production of at least ten different enzymes. Each is involved in one of the steps of the process leading up to the final synthesis of a molecule of histidine. If a cell containing this genetic system is placed in a medium containing a high concentration of histidine, the cell's machinery for the synthesis of histidine shuts down, and the cell uses the external source of histidine.

The molecular control of the repression mechanism is less well understood than the mechanics of induction. Apparently, in the case of gene repression, certain chemical repressor molecules, normally produced in the cell, are by themselves incapable of blocking the operator genes. First they must be activated. In the case of histidine synthesis, for example, activation apparently depends on the concentration of histidine molecules within the cell. At a certain concentration, histidine combines with the repressor molecule and activates it. The repressor then blocks the operator gene, and thus turns off the entire operon (Fig. 3). This sequence of events ensures that the cell will make the best use of its supply of molecular raw materials by preventing it from synthesizing compounds that are freely available in the environment.

The difference between induction and repression depends directly upon the characteristics of the specific repressor molecule produced under the control of the regulator gene. In the case of induction the repressors are inactivated, switching the operons on; in the case of repression the repressors are activated, switching the operons off. Some of the details of the operon theory, such as the existence of a repressor molecule, have been

Gene induction involves the activation of genes on a strand of DNA. Regulator gene produces a molecule of messenger RNA that triggers the synthesis of a repressor protein. Repressor attaches to the starter gene, blocking the function of an entire operon (1). An inducer molecule can combine with the repressor

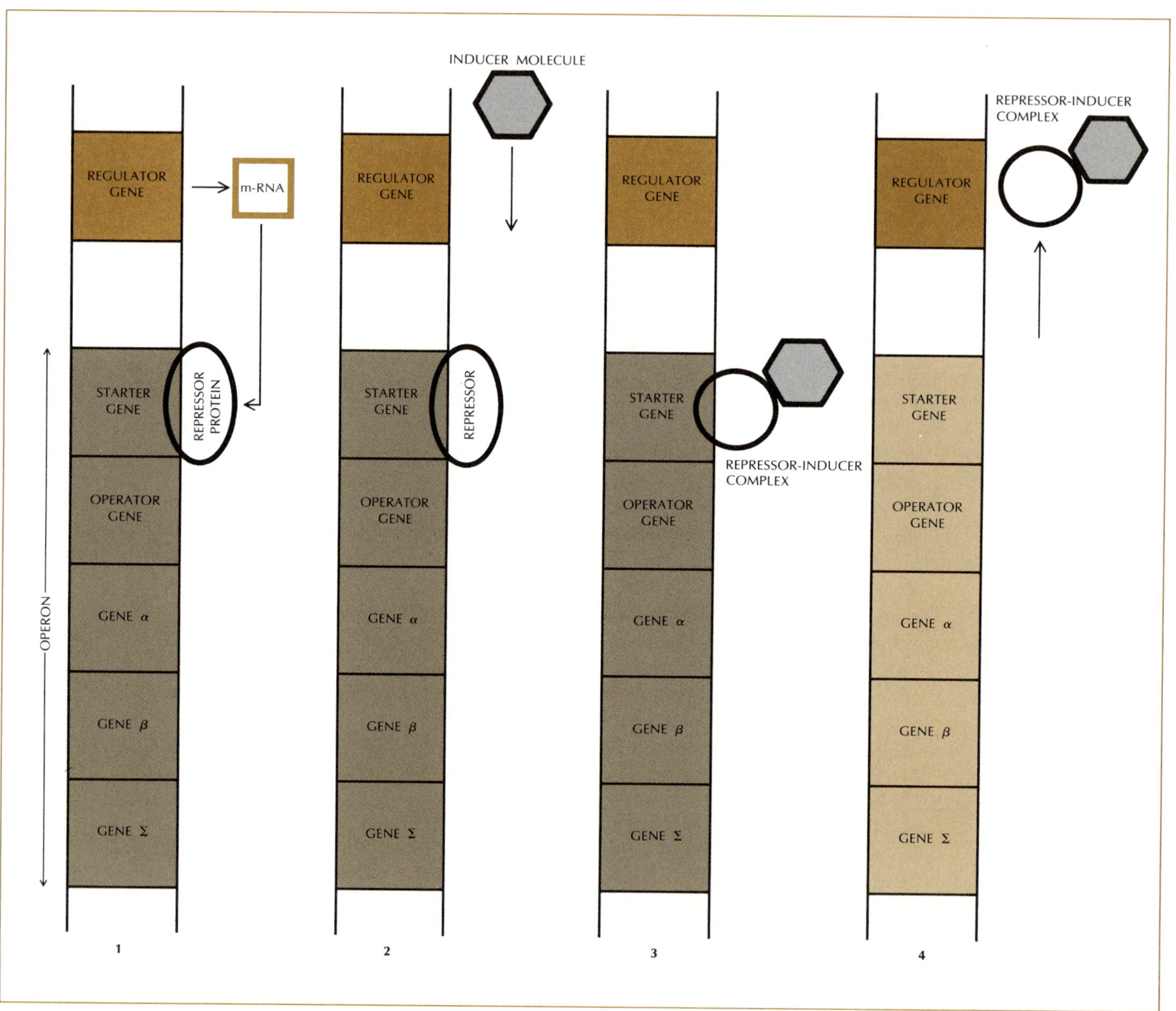

and change it sufficiently to release it from the starter gene (2, 3 and 4). Then a
synthesizer molecule can begin to read the information encoded in the genes
of the operon (5). Functional enzymes are formed as the synthesizer travels along
the DNA chain (6 and 7).

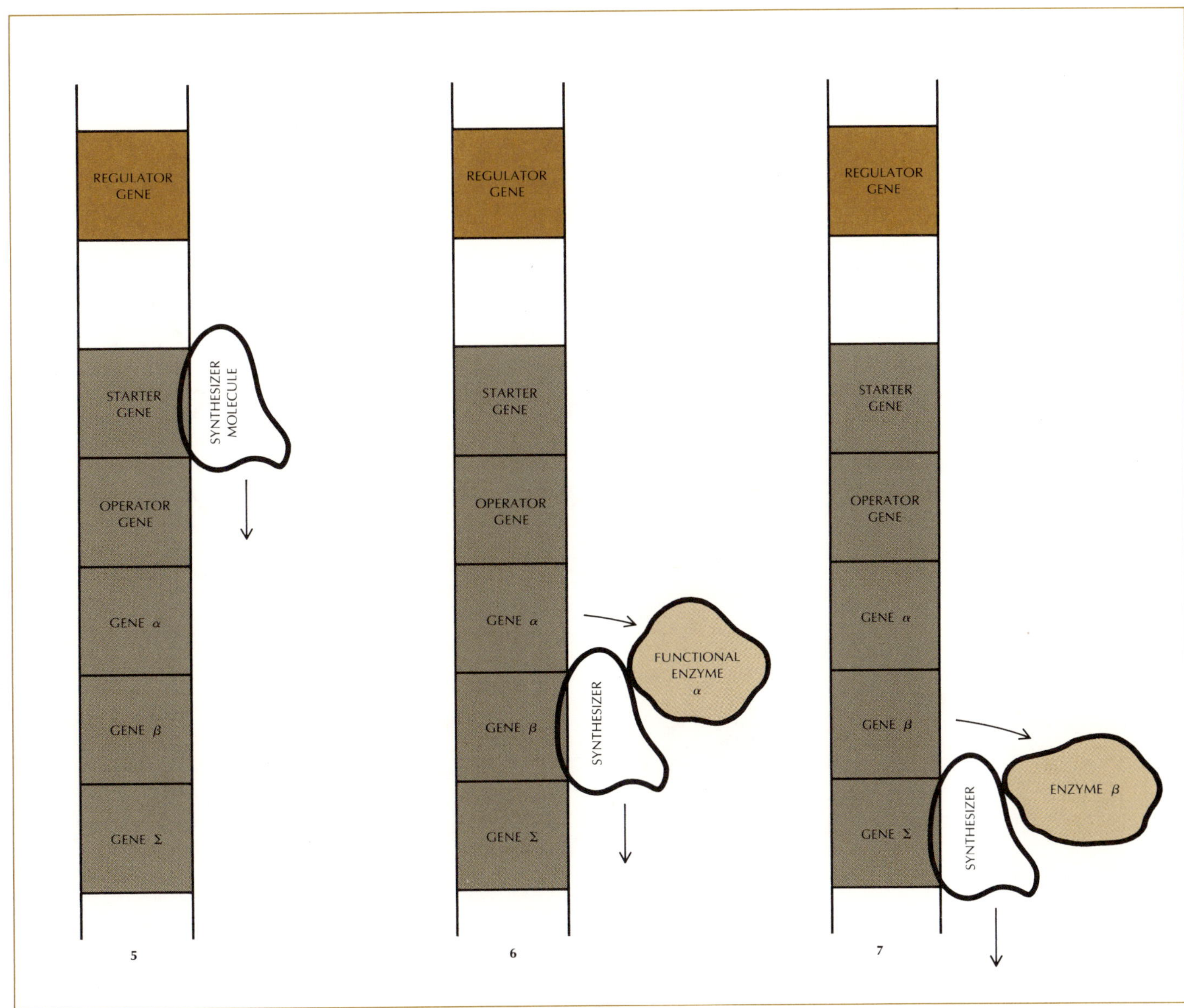

3

Gene repression involves the blocking of previously active genes. Regulator gene initiates the production of a repressor molecule (left). When repressor combines with an activator molecule it can attach itself to the operon (center) and prevent a synthesizer molecule from reading the information on the strand of DNA (right).

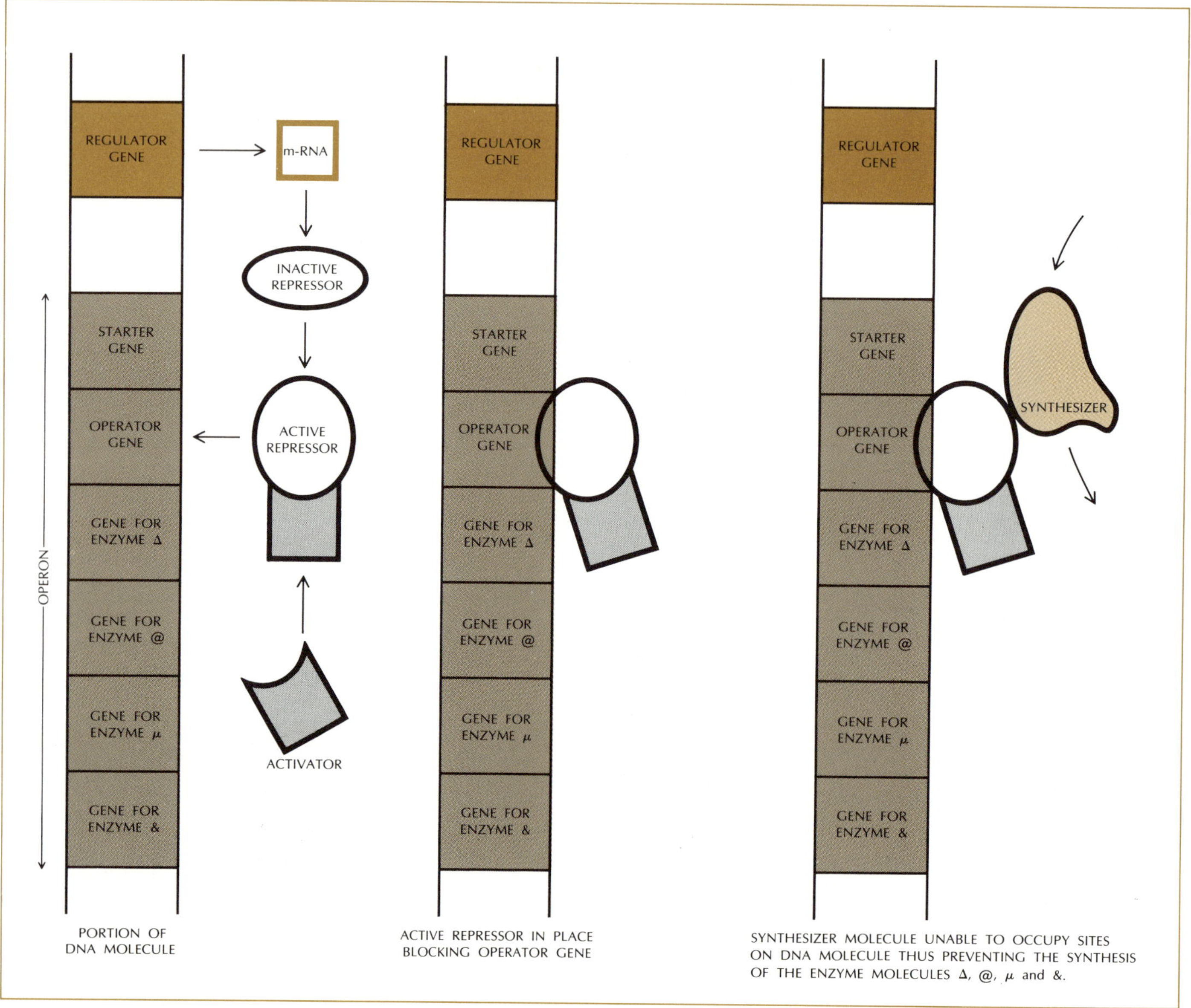

confirmed by experiment. Others, such as the proposed mechanism of repression, remain hypothetical. The main gaps in the theory will probably be filled in the near future; our knowledge of these mechanisms is already considerable, and it is expanding rapidly.

ORGANIZERS

The complete development of a higher organism obviously requires a long and vastly complicated program of sequential instructions. Cells divide, aggregate, migrate, differentiate, and form elaborate structures. In the formation of tissues, organs, and organ systems, the genetic activity of masses of cells must be precisely integrated. What coordinates this swirl of cellular activity? A partial answer suggests that groups of embryonic cells apparently communicate with each other by means of unknown chemical substances. Experimental embryologists have demonstrated this by means of tissue transplantation techniques. In frog embryos, cells that normally lie beneath the neural tube (the structure that develops into the nervous system) were transplanted to some other region of the embryo. These cells induced those nearby to form a neural tube, even though it would not form there normally. Material extracted from killed cells caused the same results. The unknown chemical substance that causes the adjacent cells to differentiate into neural tissue was called an "organizer." Other chemical organizers appear to be involved in the formation of the eyes, the pancreas, the kidneys, the salivary glands, and the skin. Future work may reveal that the entire pathway of development in all organisms is governed by trace quantities of chemical substances as yet unknown.

HORMONAL CONTROL

The chemical and biological activities of cells are controlled by a variety of molecules. The ways in which these affect specialized cells are exceedingly diverse. One area in which researchers have made considerable progress involves the investigation of the production and effects of special chemicals known as hormones. Hormones are chemicals that are produced in very small quantities in one part of an organism, but which exert a profound effect at some distant location from their formation. For example, in man, cortisone produced by the adrenal glands located above the kidneys

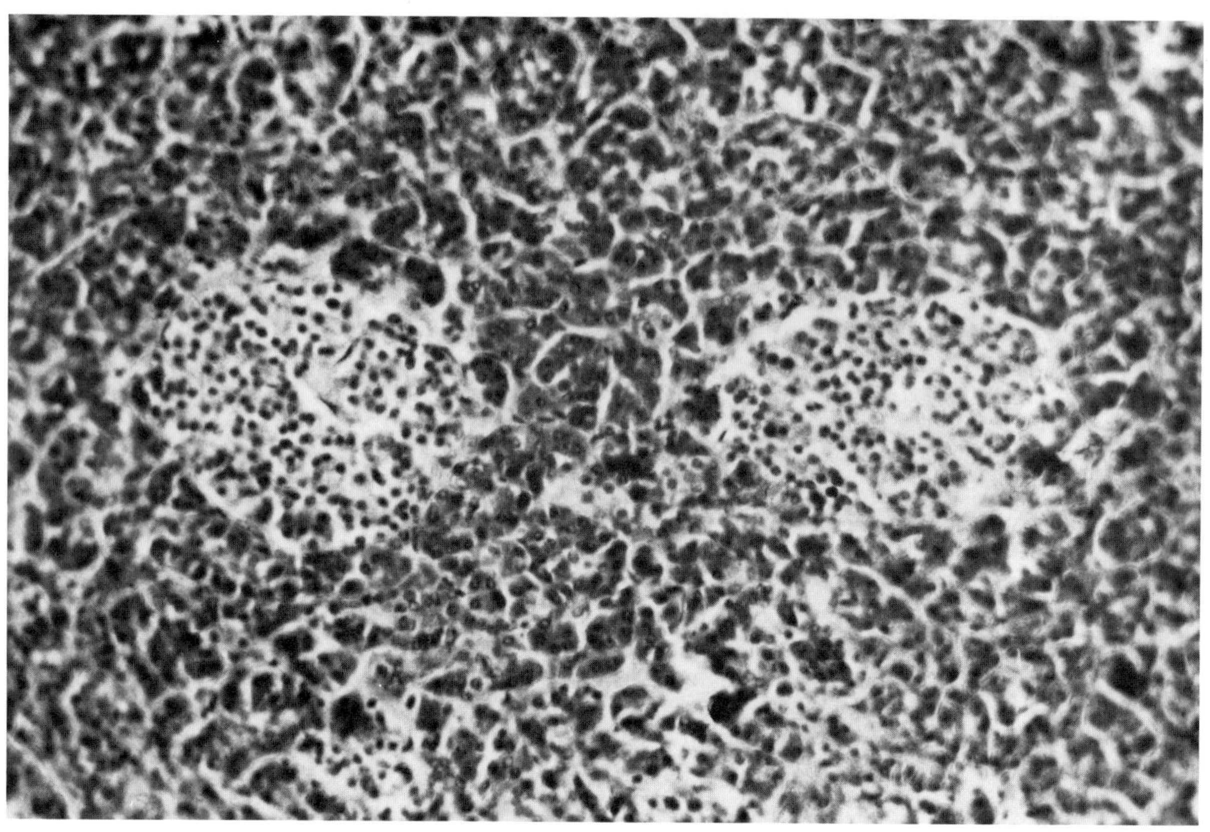

4

Islets of Langerhans are groups of cells in the pancreas that secrete insulin.
In this photo they appear as light-toned areas at left and right center.
Magnification about 200 ×.

causes cells in the liver to produce enzymes necessary for the metabolism
of glucose and amino acids.

In a normal human the concentration of sugar in the blood is regulated
by several hormones. After the digestion of a large carbohydrate meal, the
blood contains a high concentration of glucose. The blood-sugar is monitored by cells in the Islets of Langerhans, tiny glands embedded in the
pancreas (Fig. 4). When they detect a high glucose level, the glands secrete
insulin, a hormone that facilitates the entry of glucose into the cells of the
body. Insulin also stimulates cells of the liver to convert glucose into
glycogen which is then stored in these cells.

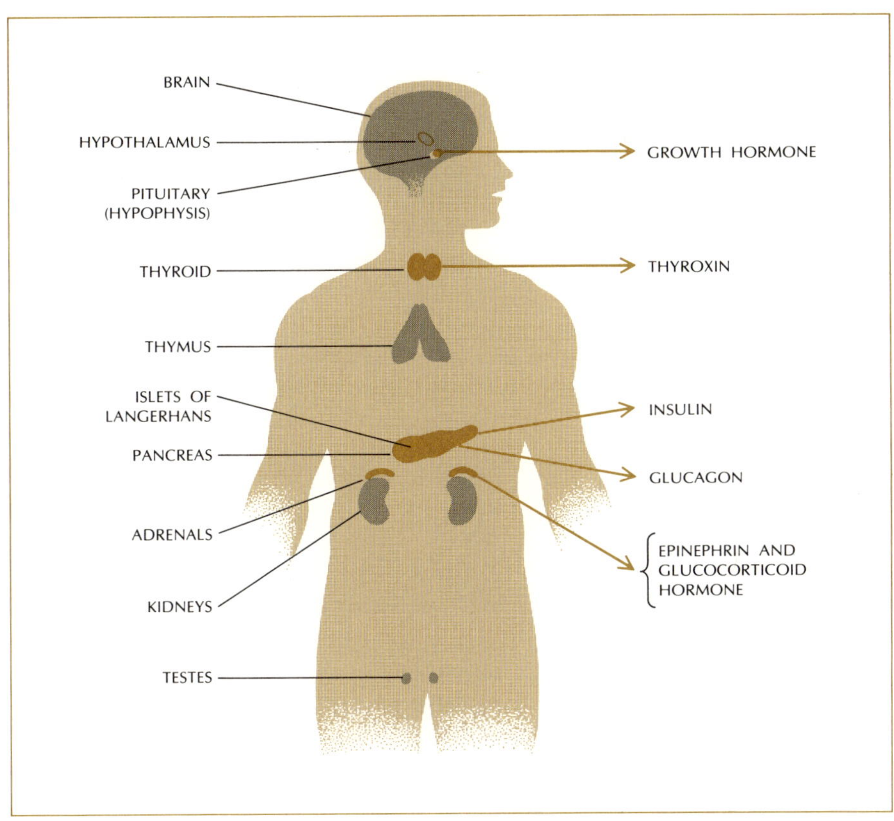

5
Endocrine glands in this diagram secrete hormones, chemical signals that carry information from one part of the body to another. Glands shown in solid color regulate the concentration of sugar in the blood.

Conversely, a low blood-sugar level triggers the formation of a battery of hormones, including glucagon from the pancreas, epinephrine and glucocorticoid hormone from the adrenal glands, and growth hormone from the hypophysis (Fig. 5). In a variety of ways too complex to outline here, they all cause an increase in the concentration of blood sugar.

The interaction of these hormone systems enables the body to maintain a relatively stable blood-sugar level under varying environmental conditions. Surplus glucose is converted to glycogen and stored in the liver. A shortage of glucose in the blood is alleviated by drawing on these reserves. Together, these hormone systems comprise what physiologists call a homeo-

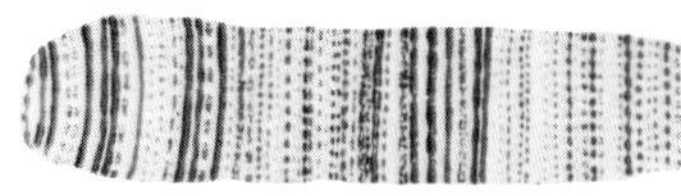

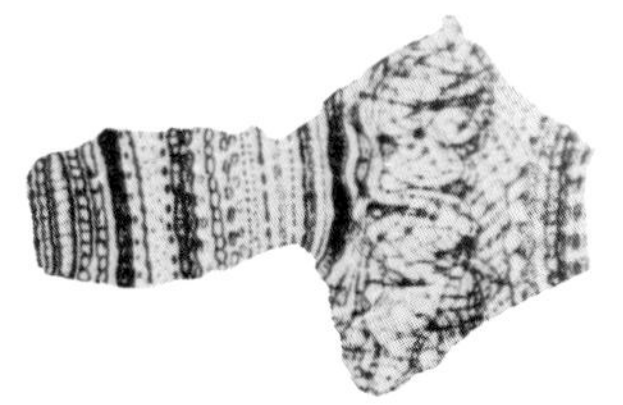

6
Chromosome puffs in the giant chromosomes of certain insects give visual evidence of the activity of specific portions of the chromosome. Special compounds are formed only when a certain region of the chromosome is enlarged into a puff.

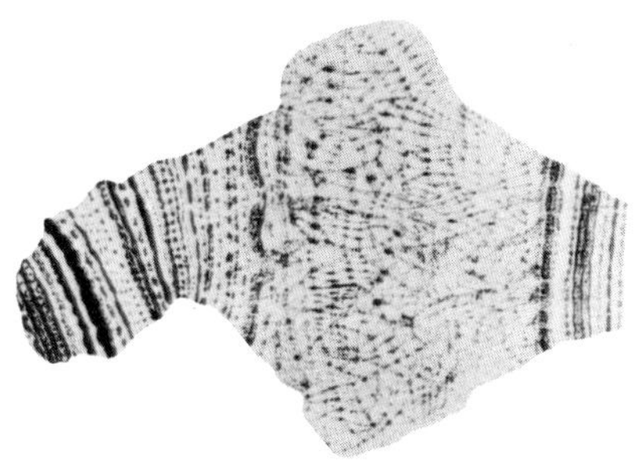

static mechanism: a mechanism that maintains the status quo of a particular system under the stress of changing environmental conditions. Other human homeostatic mechanisms control the water balance within the body, the oxygen content of the blood, and the amount of light that reaches the retina of the eye.

INSECT METAMORPHOSIS

Experimental work with insects has provided insights into how hormones influence growth and development. The life histories of many insects consist of four stages: egg, larva, pupa, and adult. Metamorphosis is controlled by the interaction of several hormonal systems. One of the hormones has been named ecdysone. Artificially injected into a larva, purified ecdysone triggers molting into a larger larva or a pupa. Ecdysone apparently activates the genes that produce enzymes that catalyze the biochemical reactions involved in molting. Photomicrographs of the chromosomes of developing insect larvae have supported this hypothesis. Just before metamorphosis, "puffs" or swellings appear at specific locations on one of the chromosomes, indicating biochemical activity at that location (Fig. 6). The injection of ecdysone into a larva produces puffs at exactly the same locations, leaving little doubt that the effect is caused by the ecdysone. Subsequent biochemical investigations support the hypothesis that the hormone ecdysone causes molting by activating specific portions of the chromosome which initiate the production of the enzymes necessary for molting.

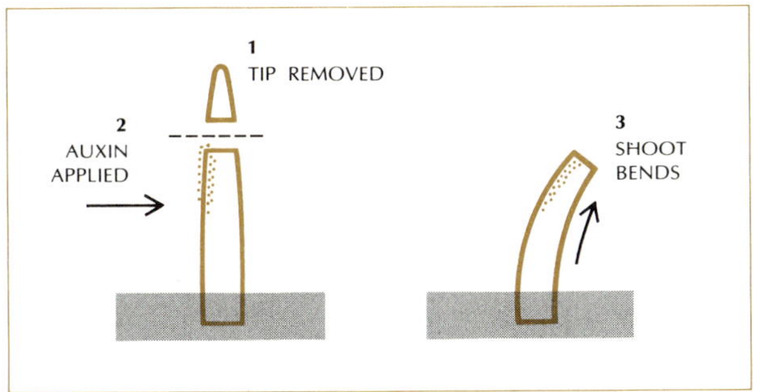

7
Effect of auxin on young plants was demonstrated by severing tip of growing shoot (1). Growth stopped until auxin was applied (2). Then shoot resumed growth, bending away from treated side (3).

PLANT HORMONES

Although Charles Darwin (better known for his work concerning the theory of evolution) suggested the existence of plant hormones as early as 1880, modern work on these substances began about 1930. At that time Fritz Went, working in Holland, proved that oat seedlings produce a hormone called auxin that promotes stem growth.

The auxins are a group of chemicals characterized by their ability to cause plant cells to elongate. Early experiments showed that the normal source of the hormone necessary for stem elongation was the actively-dividing region near the stem tip. Removal of the tip of a plant stem stops its upward growth; if auxin is applied to the cut surface of the stem, growth resumes. The implication of auxin in the bending of plants toward a light source came from similar experiments. In these the tips were removed from plants growing in the dark. This eliminated the natural source of auxin. Then artificially prepared auxin was placed on one side of the cut shoot. The plant began to grow. But instead of growing straight up, it bent to one side. Furthermore, it always bent away from the side to which the auxin had been applied. The interpretation of these results went as follows: auxin moved into the shoot causing the cells on that side of the shoot to elongate more than the unstimulated cells on the other side. This differential growth resulted in the bending of the shoot (Fig. 7). Similar reactions occur when light shines on one side of a plant. It appears that light destroys or inactivates auxin, resulting in an unequal distribution of this hormone in the stem. More auxin persists in the shaded side of the stem, causing this side to grow faster; as a result, the plant bends toward the light (Fig. 8).

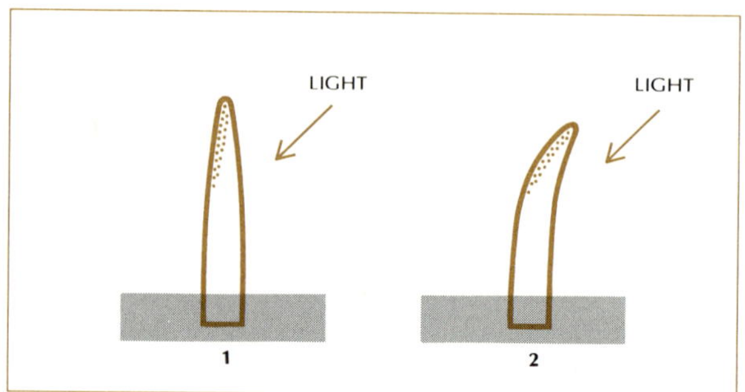

8
Effect of light on auxin causes shoot to grow toward the light. Light inactivates auxin on one side of plant, enhancing growth on the shady side.

Auxins have many practical applications. Commercial rooting compounds, sold under various trade names, contain auxin. When they are smeared on the cut end of a stem, the cells begin to divide and differentiate, forming roots. This works even with woody plants. By spraying fruit trees with appropriate auxin preparations, the unripened fruit that might fall prematurely, is retained on the tree until ripening is complete. Pineapple farmers spray their plants with auxin preparations to ensure that crops will mature at well-separated but preplanned times. Thus, harvesting becomes a year-round activity, allowing for more economical use of automatic equipment, packing, and shipping facilities. It also provides year-round fresh produce for the consumer.

Many artificial plant hormones have been synthesized in the laboratory. One of these, 2,4-D, stimulates plants to outgrow their food supply, resulting in their untimely death. Because broad-leaved plants are much more sensitive to it than are narrow-leaved grasses, 2,4-D is an excellent weed killer for lawns and golf courses.

Like most hormones, auxins are potent in surprisingly minute concentrations. IAA (indole-3-acetic acid), a naturally occurring auxin, is present in plants at a concentration of about six parts per billion. One researcher has calculated that this is about the same ratio as the weight of one needle in 22 tons of hay.

GIBBERELLINS

In the early 1900's a disease called *bakanae* or "foolish seedling" disease
caused drastic reductions in the annual rice crop yield in Japan. This
disease was characterized by plants that were taller, thinner, and paler than
normal. In 1926 E. Kurosawa discovered that a fungus growing in the rice
fields secretes a chemical which causes these changes. The chemical was
isolated in 1938 and named gibberellin. By now other workers have discovered many other gibberellin compounds. Although the chemical differences among these compounds are small, the diversity in their biological
effects is enormous. One type of gibberellin (GA_5) controls the growth of
pea seedlings. Treated with GA_5, genetically dwarf seedlings grow as tall
as normal plants. Another gibberellin (GA_7) controls the formation of the
male sex organs on the gametophytes of ferns. Still another (GA_1) initiates
flower formation in forget-me-not plants. Gibberellins are also essential
in the germination of seeds and in breaking the dormant state in a wide
range of plants.

CYTOKININS

Another group of hormones was discovered in 1955 by the Swedish-born
botanist, Folke Skoog, and his colleagues at the University of Wisconsin.
These compounds are now called cytokinins. They are essential for normal
cell division and differentiation in plants. They enhance flowering and seed
germination, regulate the transport and mobilization of stored foods, control
branching, regulate organ formation, and are involved in the synthesis of
DNA, RNA, and proteins. They have even been implicated in the prevention of the aging process in plants. Unfortunately, cytokinins can only post-
pone senescence, not prevent it.

PLANT DEVELOPMENT

Like animal development, plant development involves a complex sequence
of intricately timed reactions, and workers have barely begun to unravel the
mechanisms involved. One example illustrating the importance of hormonal
control in plant development concerns the germination of seeds and the
subsequent growth of the seedling.

Seed consists of plant embryo (dark color) and food material covered by a protective coat. Embryo survives almost in suspended animation, carrying out life processes at very slow rates.

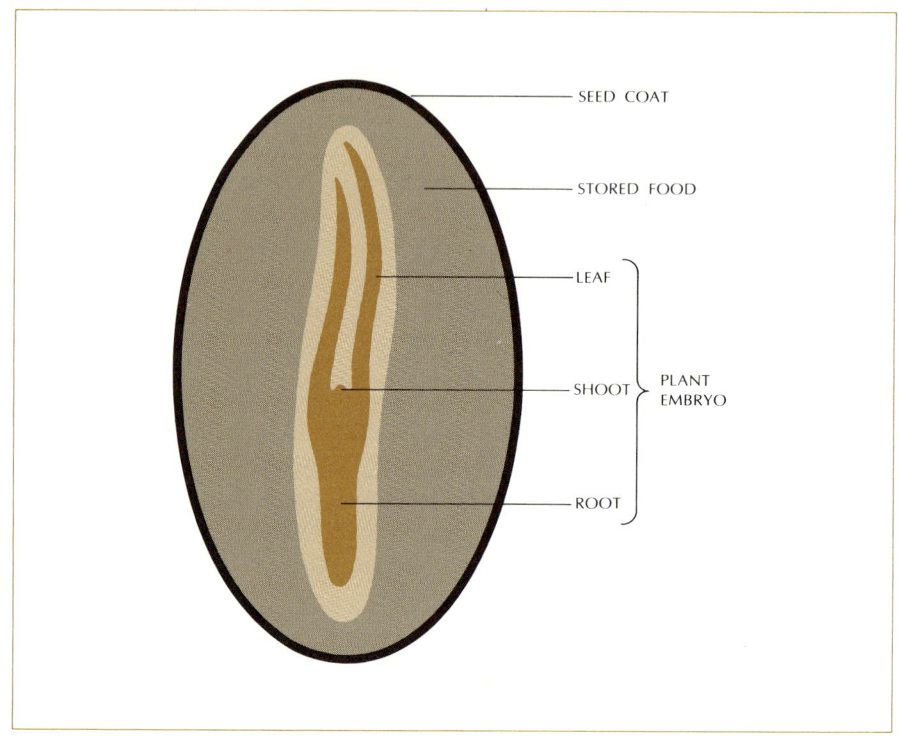

A seed consists of a small plant embryo and a mass of concentrated food material surrounded by a protective coat (Fig. 9). A dry seed is almost in a state of suspended animation. Although it may appear to be dead, its metabolic processes continue at extremely slow rates. Lotus seeds hundreds of years old, when exposed to the right environmental conditions, have germinated and produced healthy seedlings.

Germination of most seeds is triggered by moisture. When water enters the seed it activates the small embryo to begin producing gibberellin. This hormone passes out of the embryo into the cells of the seed surrounding it. It causes them to produce special enzymes that digest the complex stored food into simpler compounds that the embryo can use as an energy source in the growth process. These enzymes also attack the seed coat, weakening it. At the same time the gibberellin has an effect on the embryo itself. It causes the root to begin to grow. The root pushes its way through the

10
Seedling development
includes elongation of stem,
expansion of leaves, and
growth of roots.

weakened seed coat producing external evidence that the embryo is alive
and well on its way to becoming a seedling. Gibberellin has yet another
function. It stimulates the production of enzymes that digest the protein
of the stored food materials by releasing the compound tryptophan which
moves into the embryo where it causes the cells of the shoot tip to produce
IAA. Weakened by the auxin, the cells of the shoot tip begin to take up
water. As the cells swell and grow, the shoot elongates and soon emerges
from the seed coat. The seedling is well on its way toward becoming a
mature plant (Fig. 10).

NERVE CONTROL

Hormonal control mechanisms coordinate much of the growth and differentiation of plants. In complex multicellular animals a network of nerves

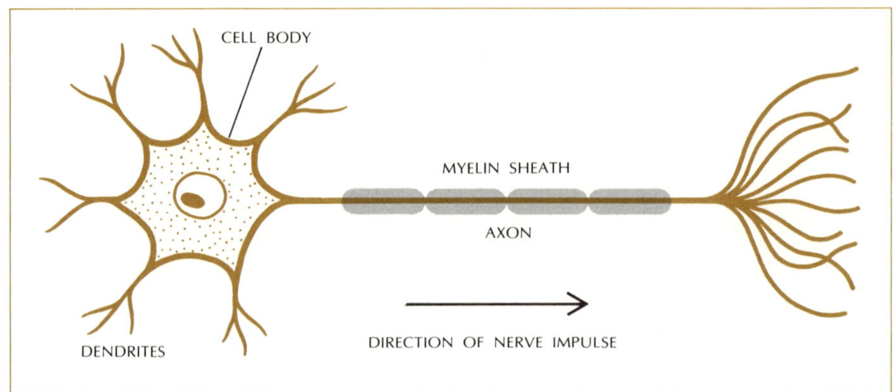

11

Neuron is the basic component of the nervous system. Dendrites conduct signals toward cell body; axon conducts signals away from it. Axons that carry fastest impulses are insulated by a myelin sheath.

coordinates most of the rapid activities of muscles, glands, and organs. In man, and to some extent in other animals, the nervous system includes a complex reasoning mechanism, an evolutionary innovation of enormous adaptive value.

Functionally the nervous system is a control mechanism that gathers information through sensory receptors. Integrated by nerve networks, the information ultimately reaches the brain, where it is sorted, stored, and selected. The selected information forms the basis for commands to effector organs, particularly muscles, which execute the commands. Sensory receptors within the muscles inform the higher nerve centers of the position and movements of the muscles. Other receptors feed back to the brain information about the effectiveness of its commands. A continuous feedback of information is one of the crucial principles of a sophisticated control system; combined with information previously stored, it enables the system to adjust its future performance on the basis of past experience, in other words, to learn.

The nervous system is essentially a network of switching circuits. Basically it consists of variants of a single type of component: the neuron or nerve cell (Fig. 11). Two types of long threadlike filaments extend from the cell body: the dendrites, branching fibers that conduct electrical signals toward the cell body; and the axon, usually a long single fiber that carries signals away from the cell body.

Each of the numerous types of neuron is adapted to a specific function. The dendrites of sensory neurons connect to a variety of receptors specialized to detect a particular type of stimulus. The rods and cones of the eye respond to light; end bulbs in the skin respond to cold, and so on. Uncomplicated bipolar neurons merely relay impulses from one neuron to another; such cells are abundant in the nerve pathways leading to and from the brain. The Purkinje cells in the brain of mammals have a "tree" of branching dendrites; this type of cell can integrate information received from many other neurons (Fig. 12).

12
Nerve cells vary in size and shape. Ganglion cell from an invertebrate (top)
is much simpler than neurons from a mammalian brain (bottom left and right).

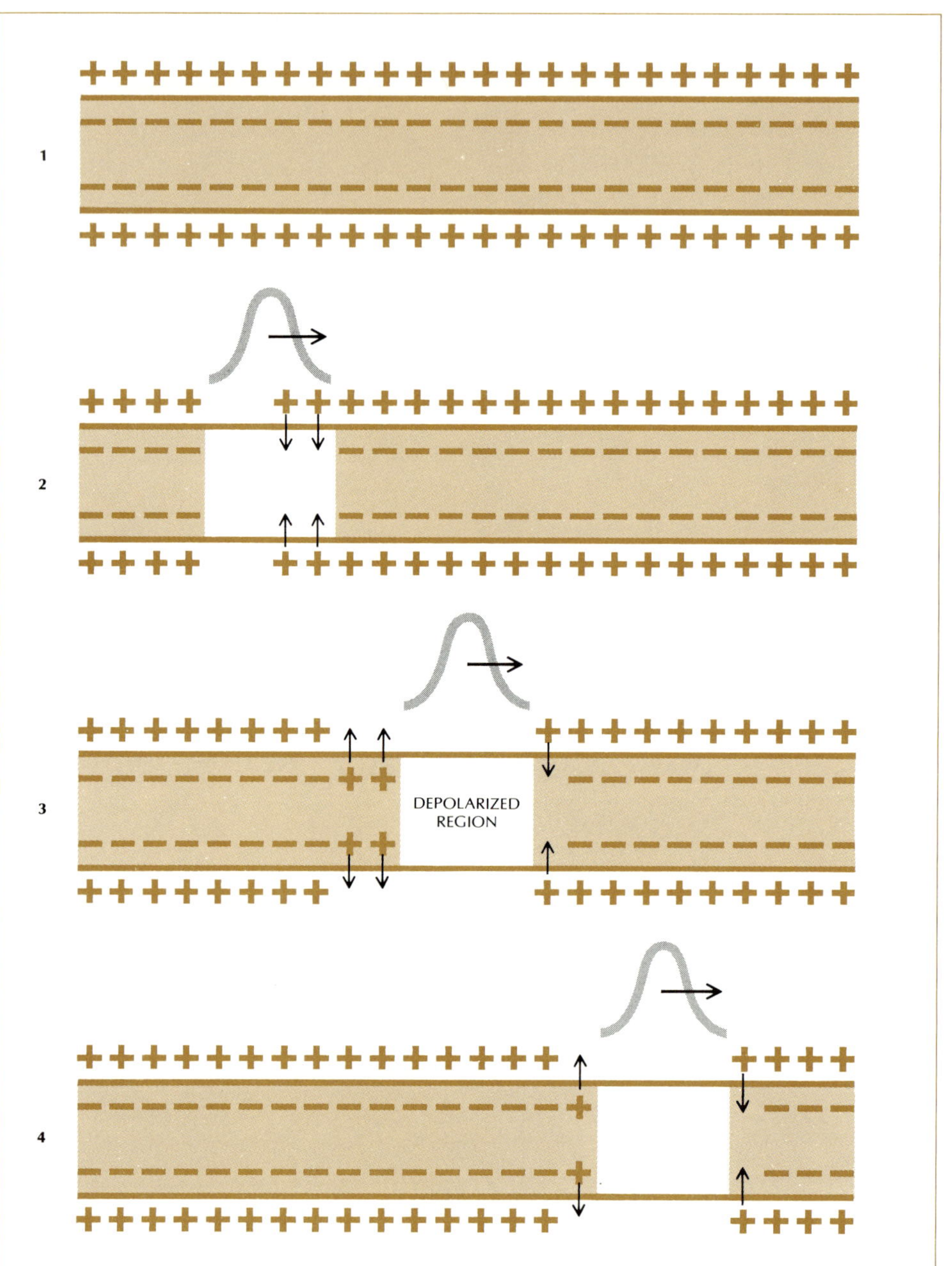

13

Electrical signal is propagated along a nerve fiber by very rapid chemical reactions. Drawing at top depicts fiber at rest, with sodium ions (+) outside the cell membrane. Nerve impulse travels along the fiber accompanied by a wave of depolarization (2). Sodium ions rush into the cell across the briefly depolarized membrane. As the impulse passes, they are pumped back out (3), restoring the original polarity (4).

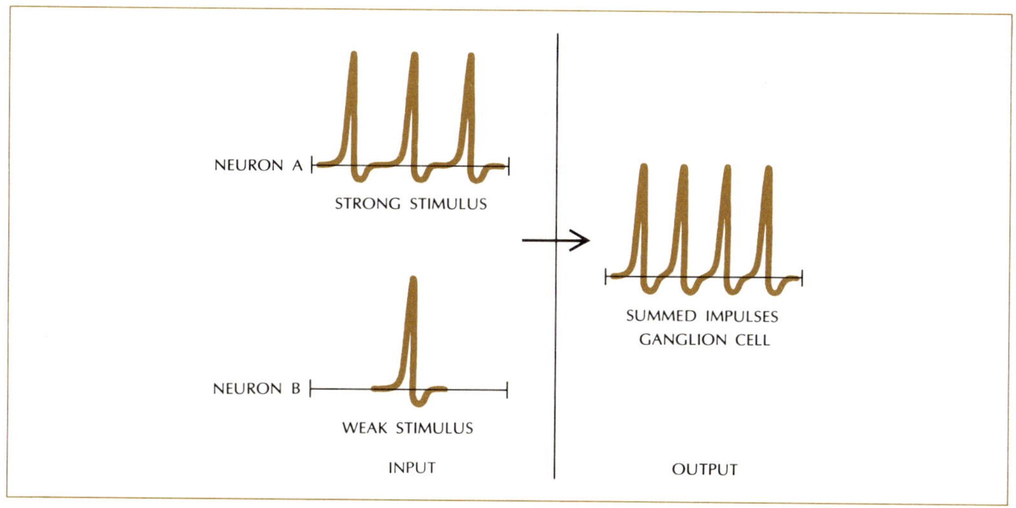

THE NERVE IMPULSE

The electrical signal that travels along a nerve cell results from a series of swift electrochemical reactions across the cell membrane. When the cell is in the resting state (turned off), the membrane pumps sodium ions out of the cell and retains potassium ions inside; this establishes a positive electrical charge along the outside of the membrane. When the cell is stimulated (turned on) the membrane is briefly depolarized and sodium ions rush into the neuron (Fig. 13). A wave of depolarization races along the nerve fiber. Immediately behind the wave, the membrane reestablishes its normal polarity in a few milliseconds (thousandths of a second) as the potassium ions surge outward.

The neuron behaves rather like a telegraph line. In response to stimuli such as light, heat, pressure, or impulses from other neurons, it propagates a series of silent electrical signals. It fires on an all-or-nothing principle: any stimulus strong enough to trigger it will result in a full-strength impulse; a stimulus twice as strong will do no more. Then how can a neuron convey information about the strength of a stimulus? Engineers would say that nerve impulses are frequency modulated (Fig. 14): the stronger the stimulus, the more frequent the impulses, up to a limit of about 1000 per second. This upper limit is imposed by the recovery time of the depolarized membrane.

14
A neuron fires more rapidly in response to strong simuli than to weak ones. Impulses from individual neurons are summed by ganglion cells.

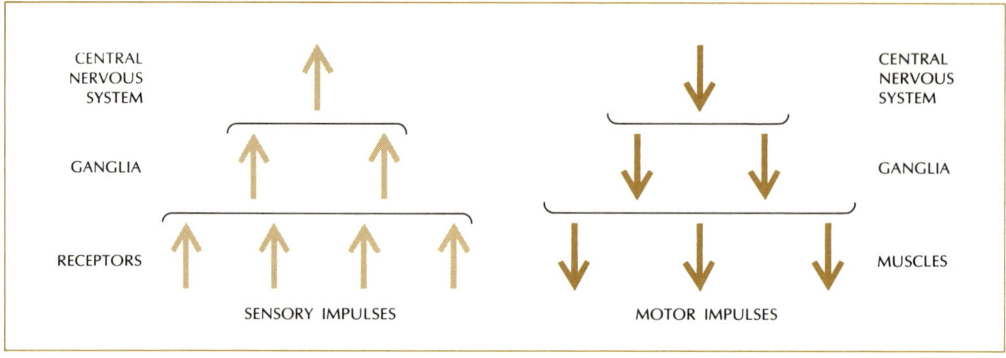

15
Chain of command governs the transmission of nerve impulses. Fibers conduct signals in only one direction: one group carries sensory impulses toward the brain, another group carries motor impulses from the brain. Ganglion cells act as switches, integrating the responses of hundreds of individual neurons. Like the logic gates in a computer, ganglion cells can "decide" whether or not to relay a signal to other neurons.

Neurons are connected to one another by junctions known as synapses. When an impulse reaches the end of an axon it triggers the release of a a chemical compound that quickly diffuses across the synapse, stimulating the next neuron. Since this chemical is produced and released only from the presynaptic fiber and not from the postsynaptic region, the excitative impulse can proceed across the synapse in only one direction. Therefore, a train of impulses normally travels along a neuron in only one direction. Sensory impulses traveling from the arm or leg toward the brain, for example, traverse separate nerve pathways from the motor impulses returning from the brain (Fig. 15).

THE REFLEX ARC

Information gathered by sensory neurons does not necessarily have to pass to the brain for interpretation before the initiation of an appropriate response. Some messages are interpreted and switched directly within the spinal cord itself. The knee jerk test performed by physicians during a routine physical examination is a classic demonstration of one such control mechanism, called a stretch reflex. A tap with a rubber mallet on the tendon below the kneecap suddenly stretches the large extensor muscle on the upper surface of the thigh. Within the muscle the stretch is detected by sensory neurons whose axons terminate in the spinal cord where they synapse directly with the motor neurons that innervate that muscle. Incoming sensory impulses jump the synapse, triggering motor impulses that swiftly cause the muscle to contract, kicking the lower leg upward (Fig. 16).

Stretch reflex. Synapses in spinal cord make possible a few very fast automatic responses, such as the knee jerk reflex shown here. A tap on knee abruptly stretches extensor muscle on top of leg (dark color in diagrams at top). Warning signal from receptors embedded in muscle triggers the firing of motor neuron (dark color in diagram at bottom). Muscle suddenly contracts, kicking leg upward (top right).

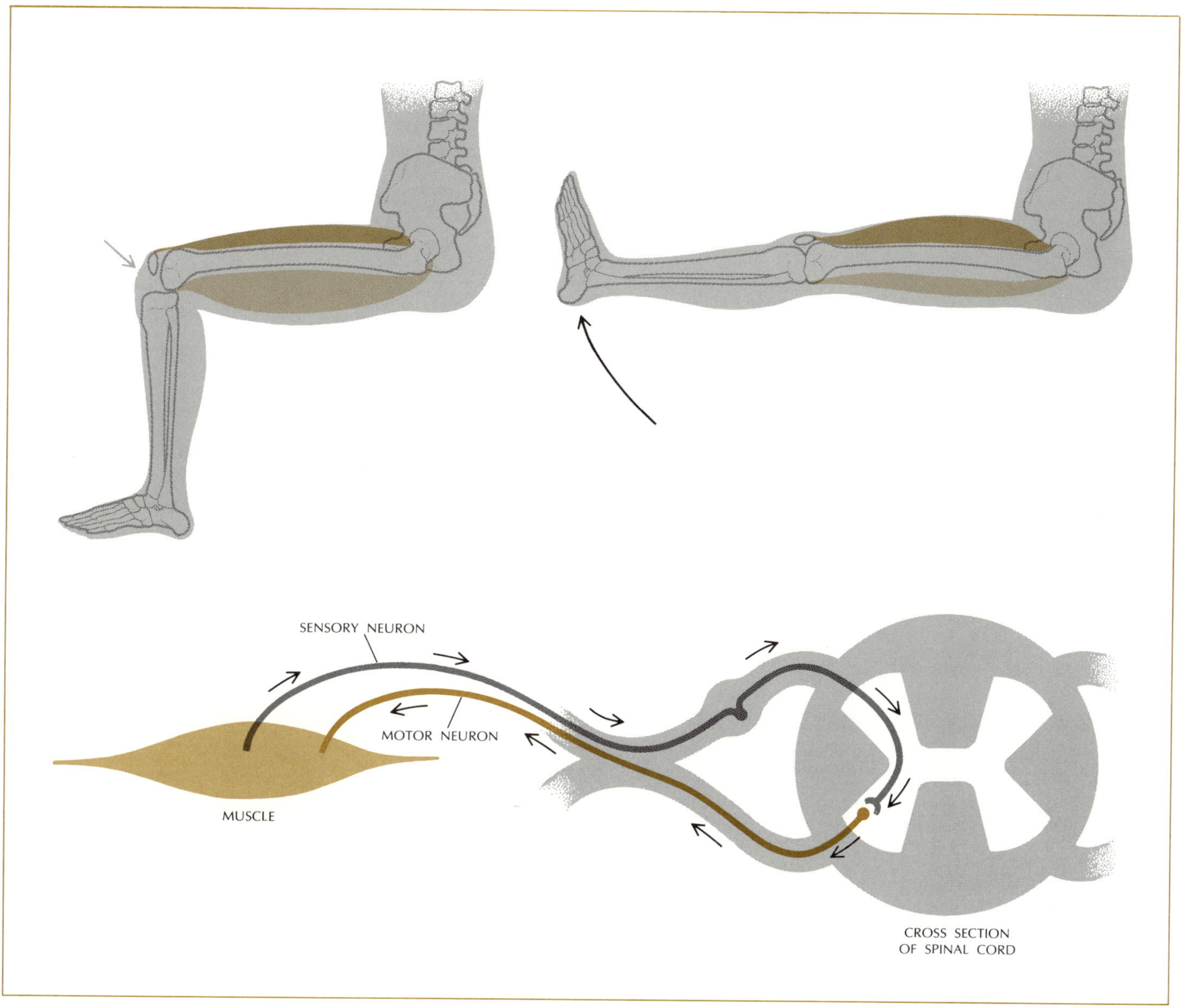

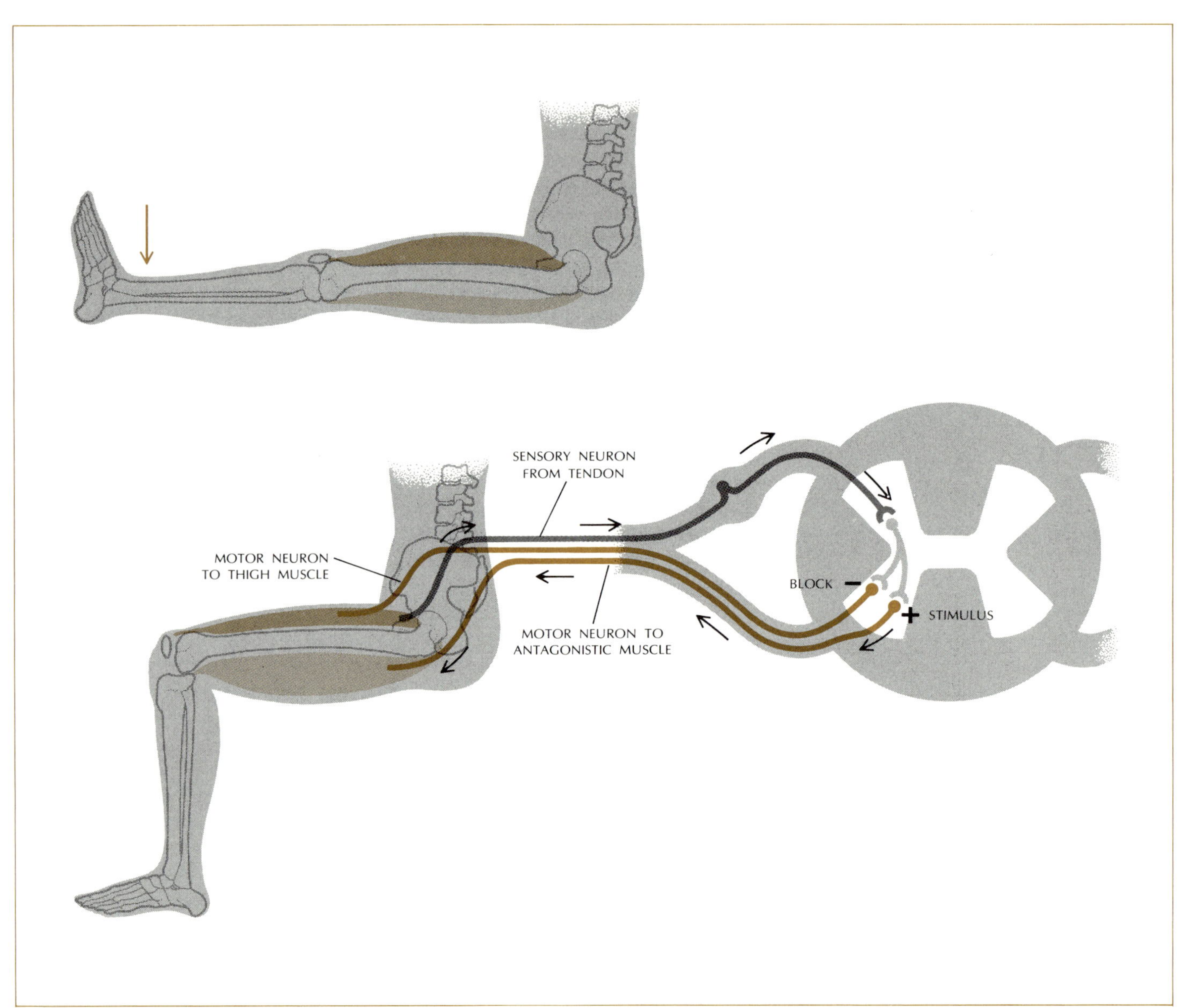

SENSORY NEURON
FROM TENDON
MOTOR NEURON
TO THIGH MUSCLE
MOTOR NEURON TO
ANTAGONISTIC MUSCLE
BLOCK
STIMULUS

◄ **Another reflex arc** prevents damage to muscle from overstretching. If contracted
upper thigh muscle is stretched by trying to force lower leg down (upper
diagram), sensory neurons in the tendon send signals to the spinal cord. In the
cord the message is received by an association neuron (lower diagram).
This neuron stimulates the motor neuron of the antagonistic muscle while blocking any stimulus to the motor neuron of the stretched muscle. The stretched
muscle relaxes to avoid damage.

The response is automatic; no message to the brain is necessary, although
relay neurons in the spinal cord do inform the brain of both the stretch and
the reflex contraction.

Although the stretch reflex itself involves only two neurons, the triggering of the reflex activates other safety mechanisms. Another set of
synapses within the spinal cord blocks the innervation of the powerful
flexor muscles at the back of the thigh. This prevents the simultaneous
contraction of antagonistic sets of muscles, a disaster that could injure both
muscles.

Another system protects muscles from damage by gradual overstretching. When a muscle is stretched slowly, at a certain point of tension a
set of sensory neurons embedded in the tendon of that muscle send a volley
of impulses to the spinal cord. These impulses inhibit the firing of the motor
neurons of the stretched muscle. This reaction prevents excessive contraction that could tear the muscle or injure the tendon attached to it (Fig. 17).
These nerve hookups or interlocks are automatic safety devices that ensure
a fast response to potentially damaging situations.

Model of nervous system could be assembled from electronic hardware. This ▶ block diagram shows flow of information through the system. Key feature of both manmade and biological control system is feedback: system monitors its own performance and adjusts its output accordingly.

OTHER CONTROL FUNCTIONS

Like the reflex arc, other control functions of the nervous system are fully automatic. The smooth muscles that line the arteries and many internal organs are controlled by two sets of nerves, one stimulatory and the other inhibitory. Sensors continuously monitor the performance of these organs and feed information to the brain, which automatically adjusts the nerve impulses that control heart rate, blood pressure, breathing, stomach contractions, and many other functions.

In many animals the nervous system is simply an automatic control system that operates according to a program—a package of instructions—built into the nerve network. To a great extent, even the organisms' patterns of behavior are programmed (see Chapter 14). In the higher animals the nervous system can exert some degree of voluntary control over otherwise automatic functions.

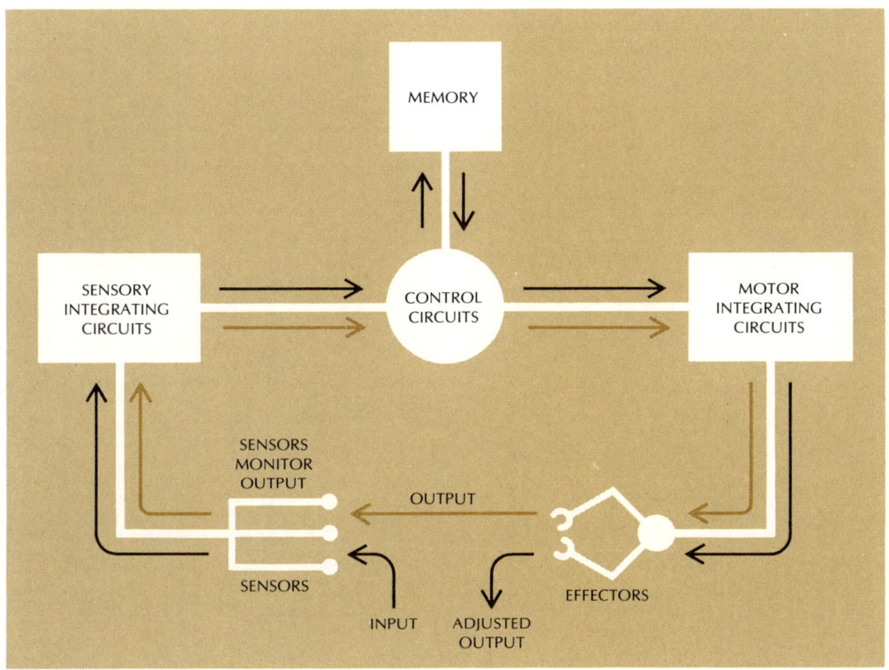

BIOLOGICAL VERSUS ARTIFICIAL SYSTEMS

A control-system designer could assemble a model of a simple nervous system from five types of components: input sensors, conducting wires, integrating circuits, memory units, and output effectors (Fig. 18). In fact many such systems have already been built to control industrial processes. A low-budget model might roughly approximate the performance of the nervous system of a starfish; an elaborate and expensive model could incorporate many of the functions of the nervous system of an organism as complex as an insect. Biological control systems excel man-made ones in compactness and versatility. At the present state of the art, the artificial insect system would be thousands of times bulkier than the natural prototype.

Computers can carry out calculations and certain other repetitive operations much faster than the brain. Asked to find the product of 52,319,418 × 12,369 a human would multiply the numbers in a minute or

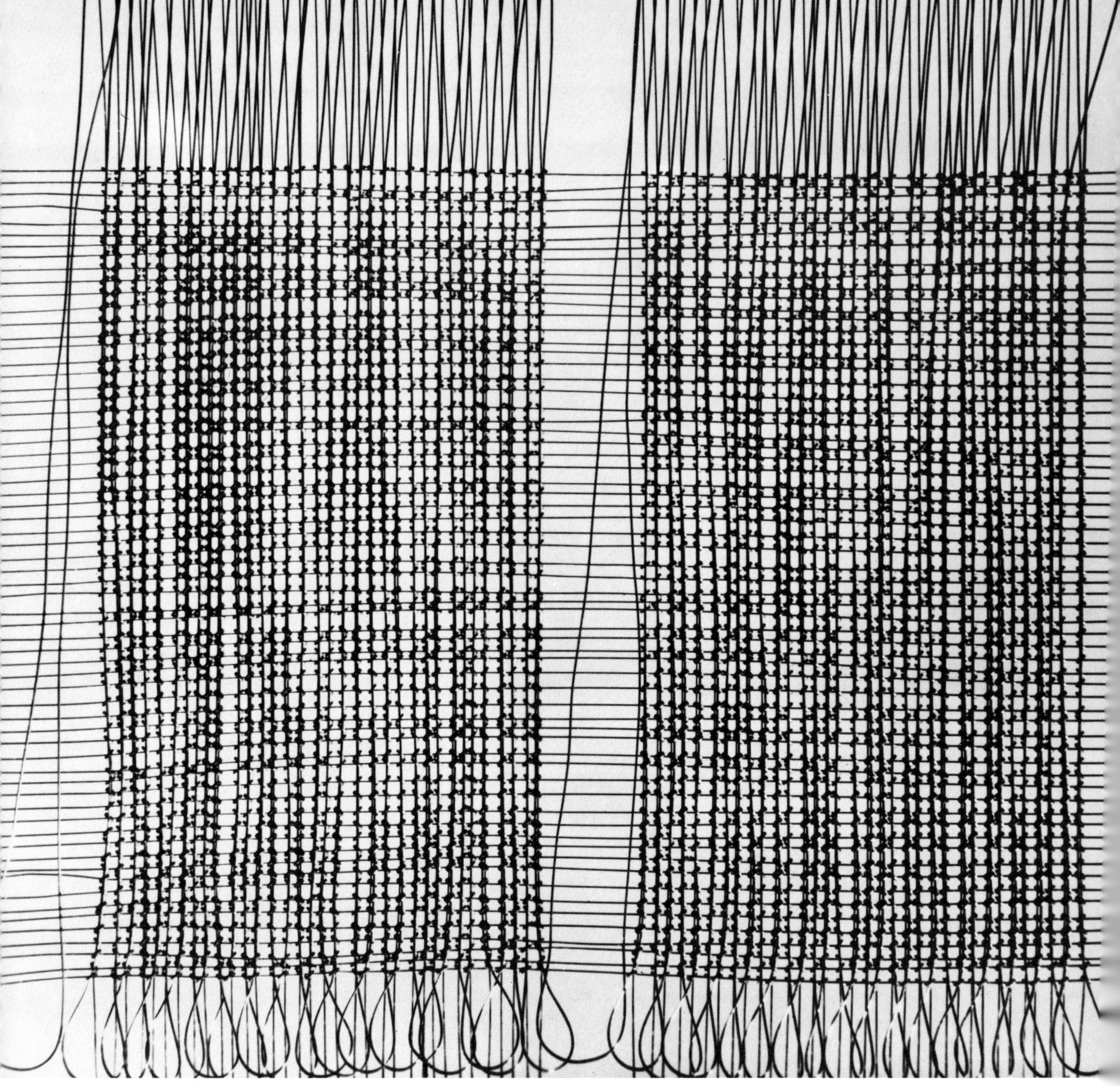

◄**Computer memory** consists of tiny magnetic rings threaded on a grid of wires. Each ring can record one bit of information, such as the digit 1. They are reproduced here larger than actual size. Such memories are being replaced in newer computers by much more compact integrated circuits.

two; a digital computer could solve the problem by repeated additions in less than a millisecond. The short-term memory of a computer (involving information that is used frequently) is usually stored on grids of tiny magnetic or optical switches (Fig. 19). Recent evidence suggests that the neurons in the brain somehow store information in molecules of RNA, a fact that makes the brain thousands of times more compact than a computer memory of equivalent capacity. The microscopic size of each neuron is also crucial to the versatility of the brain. Synapses among millions of neurons in the brain make possible a range of functions that dwarfs the capacity of the most sophisticated computer.

''To err is human'' could well be a motto engraved on the chassis of a computer. A correctly programmed computer rarely makes a mistake and rarely forgets a bit of information. Of course, like any machine it may break down because of the failure of an electrical or mechanical component. As a logic machine, the human brain is erratic by comparison. So far, neurophysiologists have been unable to explain the comparatively high incidence of errors in the performance of a healthy human brain. The mechanisms involved in memory and recall are also the subject of intensive research.

The debate about the relative merits of human and artificial intelligence is renewed with each wave of refinements in computer technology. The subject is of special interest to workers in the branch of mathematics called the theory of games. It is easy to incorporate into a computer an invincible strategy for playing tic-tac-toe; a human opponent can gain a draw but never a win. At checkers, too, a computer will never lose. But at chess, a game that demands an intuitive but rigorous long-range strategy, a player of master rank can consistently beat a computer. About 20 years ago John von Neumann of the Institute for Advanced Study at Princeton, who was one of the pioneers of game theory, estimated that a computer capable of defeating a master would be larger than the Empire State Building. Computers are much more compact now, but they still play dull and amateurish chess.

Nonhuman control systems can outperform humans at many kinds of task. Any man doing uninteresting routine work in an office or factory is probably keeping a good machine out of a job: a temporary situation at best, and one that has already alarmed many labor unions. The one field in which computers will probably never outperform humans is imagination. The computer is a robot incapable of creative or imaginative thought. It is safe to say that although the computer might replace an unimaginative secretary or a plodding bureaucrat, it will never replace a good artist or a dedicated scientist.

READINGS

Butler, J. A. V., *Inside the Living Cell: Some Secrets of Life*. Science Editions, New York, 1962 (paper)

> A semipopular account of cell biology. Covers topics from vitamins and genes to cancer and the functioning of the brain.

Gottlieb, F. J., *Developmental Genetics*. Reinhold, New York, 1966 (paper)

> A brief presentation of topics including genes, embryology, differentiation, and organization.

Harrison, R. J., and W. Montagna, *Man*. Appleton-Century-Crofts, New York, 1969 (paper)

> An engrossing and fact-filled treatise on man. The advantages and disadvantages of being human are presented with humor and scientific accuracy.

Langley, L. L., *Homeostasis*. Reinhold, New York, 1965 (paper)

> A very readable account of some of the many control mechanisms in man.

Nourse, A. E., *The Body*. Life Science Library, Time-Life Books, New York, 1964
> A layman's guide to human anatomy and physiology. Profusely illustrated with beautiful and informative photographs and drawings.

Schmidt-Nielson, K., *Animal Physiology*. Prentice-Hall, Englewood Cliffs, N.J., 1964 (paper)
> The organism as a functioning entity: its need for food, oxygen, water, and energy.

Sussman, M., *Animal Growth and Development*. Prentice-Hall, Englewood Cliffs, N.J., 1964 (paper)
> An introduction to cell differentiation, the embryology of higher organisms, and the morphogenesis of microorganisms.

Tanner, J. M., and G. R. Taylor, *Growth*. Life Science Library, Time-Life Books, New York, 1965
> Superbly illustrated introduction to human embryology, growth, and development.

Twitty, V. C., *Of Scientists and Salamanders*. Freeman, San Francisco, 1966
> A mixture of philosophy, personal anecdotes, and biological investigations on morphogenesis. Based on a lifetime career of studying salamanders. Entertaining as well as informative.

von Neumann, J., *The Computer and the Brain*. Yale University, New Haven, 1958
> This short book, little more than an essay, was written by one of the great mathematicians of this century. It compares the living human brain to digital and analog computers, and analyzes the similarities and differences in their operation.

Wiener, N., *The Human Use of Human Beings: Cybernetics and Society*. Avon, New York, 1967 (paper)
> Cybernetics is the study of feedback control systems and the author was one of the giants in the field. No mathematics, but rigorously reasoned. By now a classic, this book discusses the relationship between entropy and information in terms of both human and nonhuman control systems. Strongly recommended.

Wilson, J. R., *The Mind*. Life Science Library, Time-Life Books, New York, 1964
> From physiology and psychiatry to drugs and disease, the mind is explored in text and illustration.

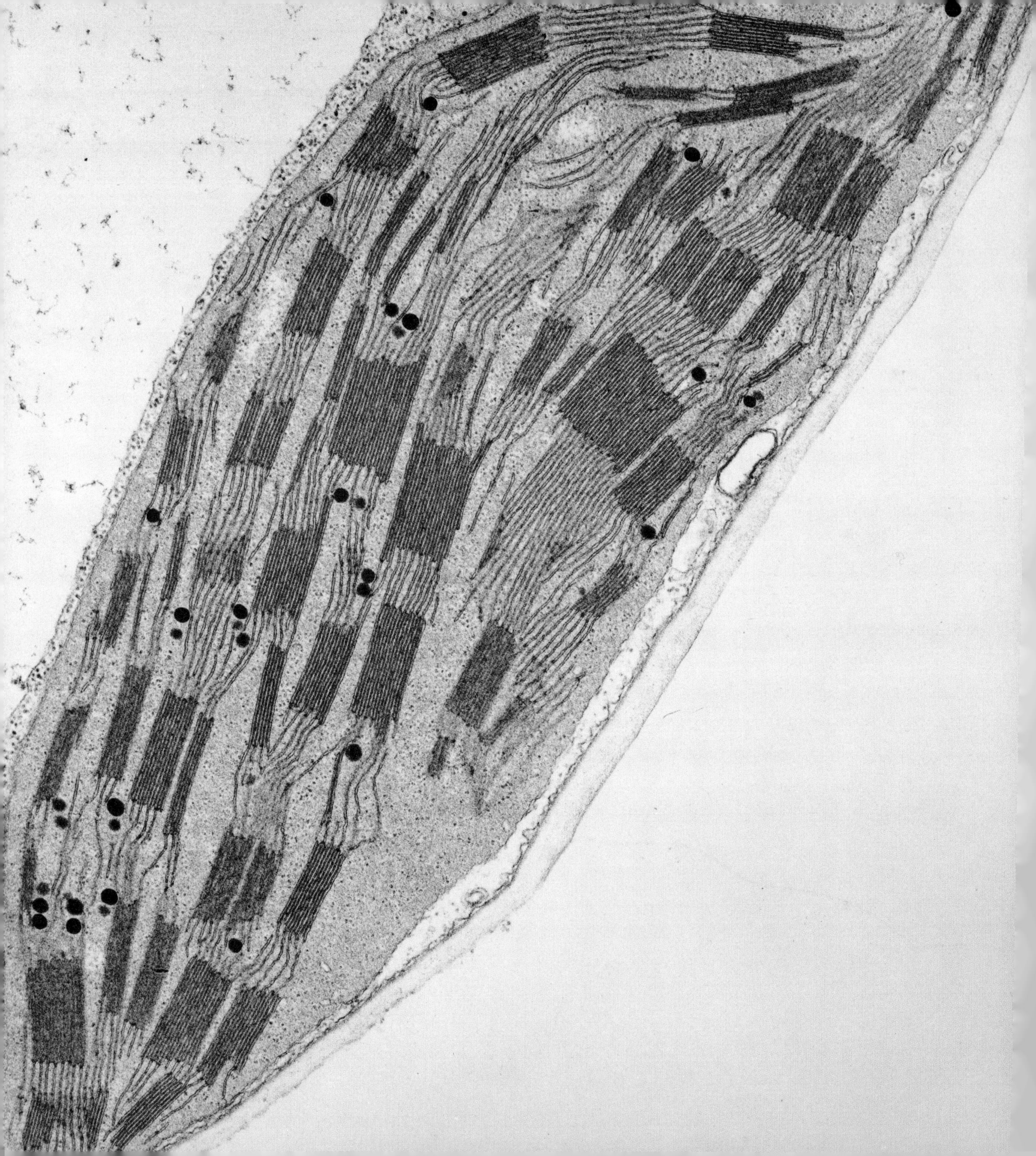

10. PHOTOSYNTHESIS

The sun is the ultimate source of all biological energy and the living world traps solar energy by the process of photosynthesis. Green plants, most algae, and some microorganisms can absorb light energy, which they use in the transformation of simple raw materials into the complex, energy-rich molecules of life. Other organisms—particularly animals, which cannot use raw solar energy directly—derive their energy and raw materials from feeding either on photosynthetic organisms or on each other.

Photosynthesis has powerfully shaped the history of life on this planet and in the future it may impose a limit on the size of the human population. In many parts of the earth, the human population has already outgrown its food supply. To millions of people, malnutrition and famine are familiar causes of death. Starvation will inevitably limit any population that cannot or will not limit itself. The spectre of mass starvation can be banished only by human control of the human birth rate.

Anyone who has seen a tropical rain forest at close hand cannot have failed to have been amazed by the sheer tonnage of living matter created by photosynthesis. An even richer harvest lies in the surface waters of the seas, where tiny photosynthetic organisms produce a total bulk of living matter that dwarfs that of all land plants. But intense cultivation of land and the farming of the oceans can only buy time, perhaps several decades, perhaps a century or two. Today we must increase not only our total output of food-stuffs but also the efficiency with which we use the earth's resources of energy. Further investigation of photosynthesis is important not only be-

◄**Chloroplast** carries out the reactions of photosynthesis in higher plants. This electron micrograph depicts a chloroplast from a leaf cell of timothy grass, magnified about 36,000 ×. Stacks of photosynthetic membranes are called grana.

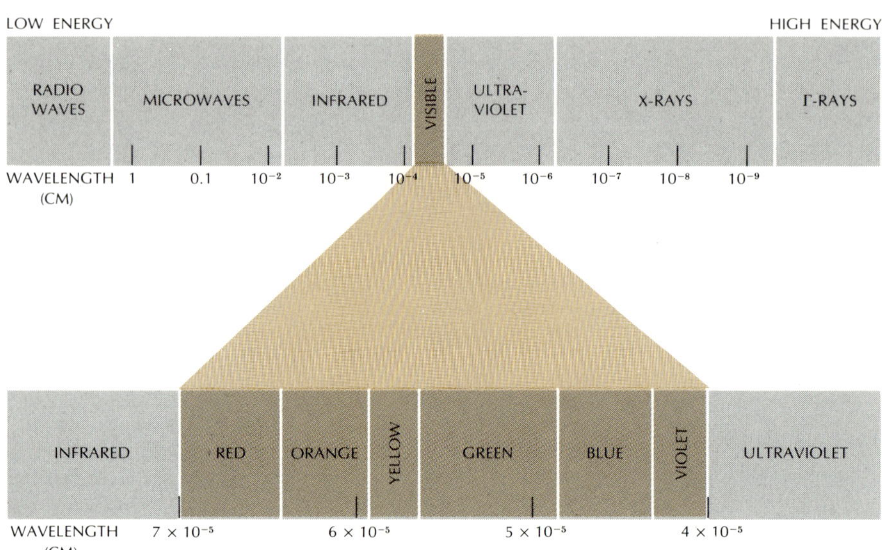

1
Solar radiation is composed of a wide spectrum of electromagnetic energy, from radio waves to gamma rays (top). Visible light comprises only a small band of this spectrum (bottom).

cause it is one of the most basic processes of life on earth, but also because the design of the photosynthetic machinery, refined by billions of years of evolution, may hold valuable lessons for advances in human technology. Understanding of the photosynthetic process may enable us to develop a technology to trap the energy of sunlight economically on a worldwide scale.

SOLAR ENERGY

Like most stars, the sun emits a broad spectrum of radiation: cosmic, gamma, and X-rays; ultraviolet, visible light, infrared, and radio waves (Fig. 1). The wavelength of these radiations varies enormously, from

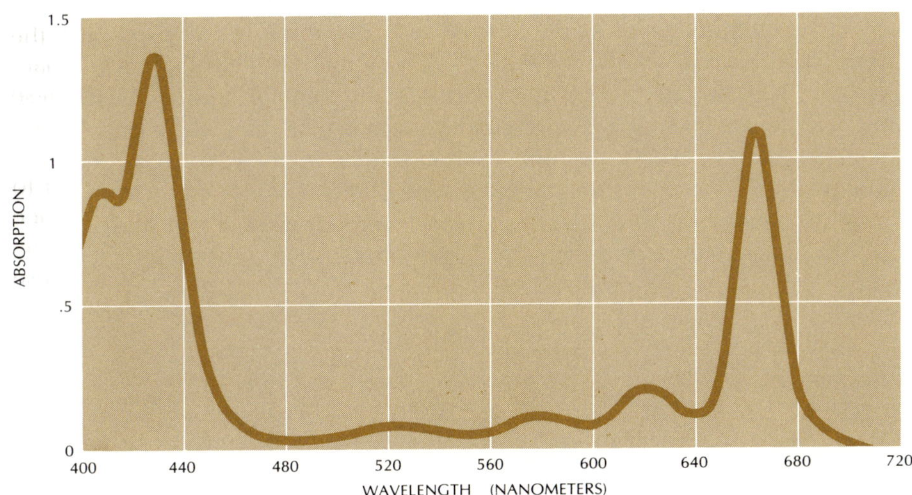

2
Absorption spectrum of chlorophyll indicates that it strongly absorbs both blue light (left) and red light (right). Green, yellow, and orange wavelengths between the peaks are reflected, giving plants their characteristic green color.

roughly 10^{-10} cm (0.0000000004 inch) for gamma rays, to 10^8 cm (over 600 miles) for long radio waves. The eye can detect a relatively narrow band of wavelengths; we call this radiation visible light. The color that light appears to the eye depends on its wavelength. Sunlight, although it appears "white" to the eye, is actually a mixture of wavelengths corresponding to red, yellow, blue, and green light. Dyes and pigments appear to be colored because they absorb some of the wavelengths that strike them and reflect others.

Plants appear green to the eye because their pigment molecules absorb most of the blue and red light and reflect much of the green and yellow (Fig. 2). Although a green landscape is beautiful to behold, the reflected green light is actually being "discarded" by the plants. They can use only the energy of the wavelengths that they absorb.

GENERALIZED EQUATIONS OF PHOTOSYNTHESIS

Photosynthesis begins with the absorption of light energy. Of all the photosynthetic organisms now found on earth, the green and purple bacteria are the most primitive (Fig. 3). The light-trapping pigment in these organisms is slightly different from the chlorophylls found in algae and higher plants; it is called bacteriochlorophyll (Fig. 4). Some of the light energy absorbed by this molecule is lost as heat; the rest is converted to chemical energy. This energy enables the purple sulfur bacteria to produce carbohydrates from only two raw materials: hydrogen sulfide (the stinking gas produced by rotten eggs) and carbon dioxide. Photosynthesis in these organisms can be summarized by the equation

$$2H_2S + CO_2 \xrightarrow[\text{bacteriochlorophyll}]{\text{light}} (CH_2O) + H_2O + 2S.$$

$$\begin{array}{ccccccc} \text{hydrogen} & \text{carbon} & & \text{carbo-} & & \text{water} & \text{sulfur} \\ \text{sulfide} & \text{dioxide} & & \text{hydrate} & & & \end{array}$$

The sulfur may be deposited as small granules inside the bacterial cell or it may be secreted. The simple carbohydrate is the basic building block which the bacterium assembles into complex carbohydrates or uses in the synthesis of lipids, proteins, and nucleic acids.

Other bacteria use gaseous hydrogen as a raw material for photosynthesis

$$2H_2 + CO_2 \xrightarrow[\text{bacteriochlorophyll}]{\text{light}} (CH_2O) + H_2O.$$

Any biological advantages conferred by hydrogen utilization are limited by the scarcity of the raw material. Hydrogen gas comprises only about 0.00005 percent of the earth's atmosphere, a concentration inadequate to support large populations of hydrogen-utilizing organisms.

Photosynthetic algae and higher green plants can also use hydrogen in photosynthesis under certain conditions, but they have evolved a much more efficient chemical alternative. As a source of hydrogen, these organisms can use water. There is no shortage of water on earth. The overall reaction of plant photosynthesis can be summarized by the equation

$$2H_2O + CO_2 \xrightarrow[\text{chlorophyll}]{\text{light}} (CH_2O) + H_2O + O_2.$$

The waste product of this reaction, molecular oxygen, has created an environment favorable to the evolution of the enormous number of animal

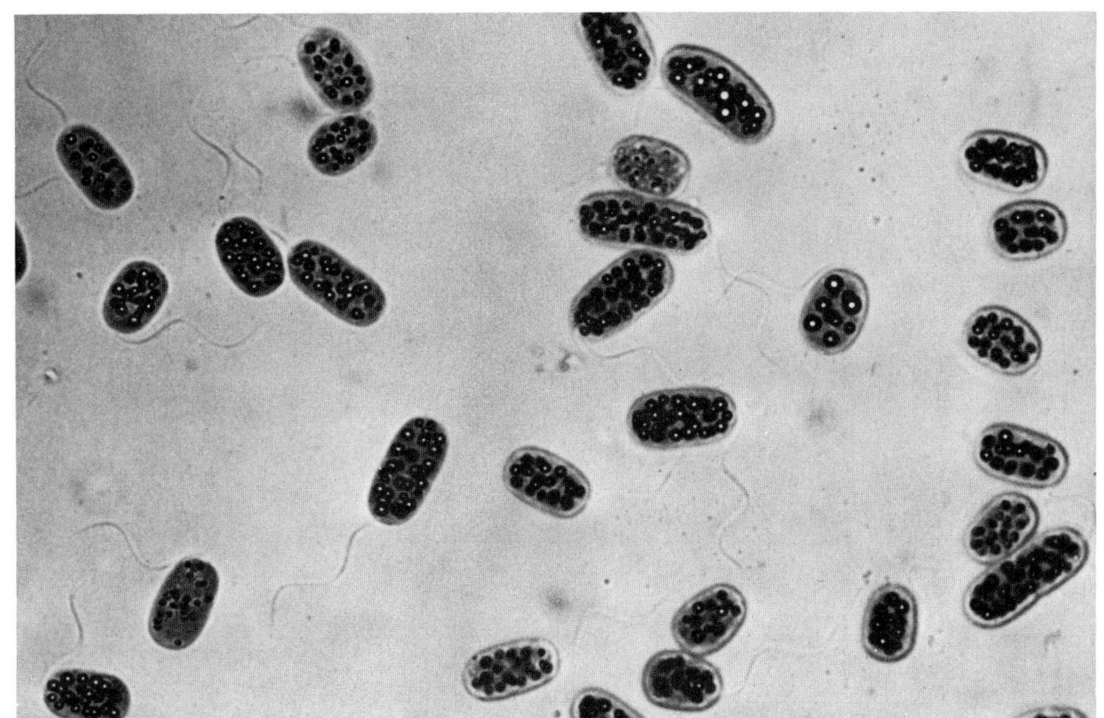

3
Purple sulfur bacteria depicted here belong to the genus *Chromatium*. Globules of sulfur are visible within the cells.

4
Bacteriochlorophyll (left) differs slightly in structure from chlorophyll *a*, one of the forms of the pigment found in higher plants (right). One difference is the double bond between the two carbons at the upper right part of the structure. Another is the presence of an oxygen atom in the upper left portion of the bacteriochlorophyll molecule.

species now present on earth. The role of molecular oxygen in evolution will be discussed in more detail in Chapter 12.

A REFINEMENT OF THE EQUATIONS

The overall equation for photosynthesis was discovered early in the 19th century, but until recently very little was really known about how the process actually worked. Early experimenters mixed carbon dioxide, chlorophyll, and water in glass bottles and exposed them to sunlight. The result was a rather tasteless, pale green soda pop; no sugars formed, and no oxygen bubbled off. Something was missing.

Investigators gradually realized that green plants carry out the process in a complex series of reactions. One of the first questions they asked was how the water and carbon dioxide react in the formation of a molecule of carbohydrate. From the overall equation it seemed logical to assume that the carbon dioxide molecule first splits into carbon and oxygen. The carbon could then unite with the water, forming a carbohydrate molecule, and the oxygen would be given off as a gas. But this line of reasoning did not seem to apply to the photosynthetic process of the sulfur bacteria. If the carbon dioxide were split, and the resulting carbon fragment combined with the hydrogen sulfide, the photosynthetic sulfur bacteria should produce molecular oxygen and a compound with the formula (CH_2S). They don't. For decades the study of photosynthesis was stalled by the question of whether or not the carbon dioxide molecule really splits in the process. Designing an experiment to answer the question was not easy until radioactive tracers became available.

An answer came in 1941. Samuel Reuben, Merle Randall, Martin Kamen, and James Logan Hyde of the University of California at Berkeley designed an experiment using the green alga *Chlorella*. Photosynthesis in this organism is similar to that in higher plants. The raw materials used are water and carbon dioxide. The scientists reasoned that the oxygen given off during photosynthesis must come from whatever compound is split in the process. This could be either the carbon dioxide or the water. Their experiments involved the use of two isotopes of oxygen. Most oxygen atoms have an atomic weight of 16, but some have an atomic weight of 18 (Fig. 5). With an instrument called a mass spectrometer it is possible to tell whether a particular compound contains heavy or normal (light) oxygen

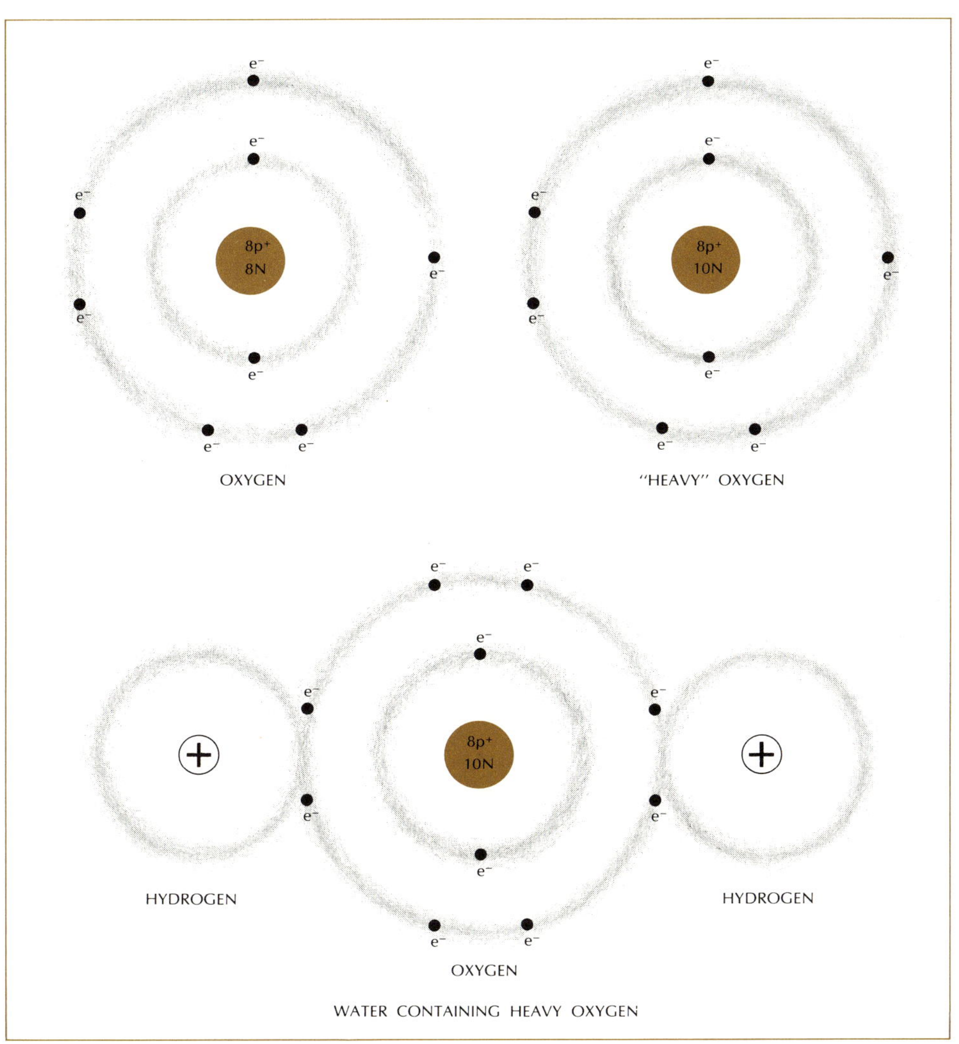

5
Isotopes of oxygen include the most common form, ^{16}O (upper left), and heavy oxygen, ^{18}O (upper right), whose nucleus contains two additional neutrons. Experimenters used water labeled with heavy oxygen (bottom).

Plant cell in this electron micrograph contains three football-shaped chloroplasts (bottom). Large granules within chloroplasts are starch grains.

atoms. Reuben and his colleagues suspended algae in water labeled with heavy oxygen and bubbled in carbon dioxide containing normal oxygen. The heavy oxygen showed up as a gas, suggesting that it was the water molecule and not the carbon dioxide that was split.

By this simple experiment, bacterial photosynthesis and that in higher plants as illustrated by *Chlorella,* were seen to be only modifications of the same overall process. In higher plants water is split and the oxygen is released as molecular oxygen. In bacteria the hydrogen sulfide is split and the sulfur is deposited as sulfur granules. Once again a link was formed showing the close relationships between the functioning of such diverse forms as bacteria and true plants.

PHOTOCHEMISTRY

In the past 30 years many more sophisticated questions have been asked and answered. Photosynthesis has been found to involve two discrete series of reactions. Light is directly required only for the first series, the photochemical reactions that capture the energy of light and convert it into

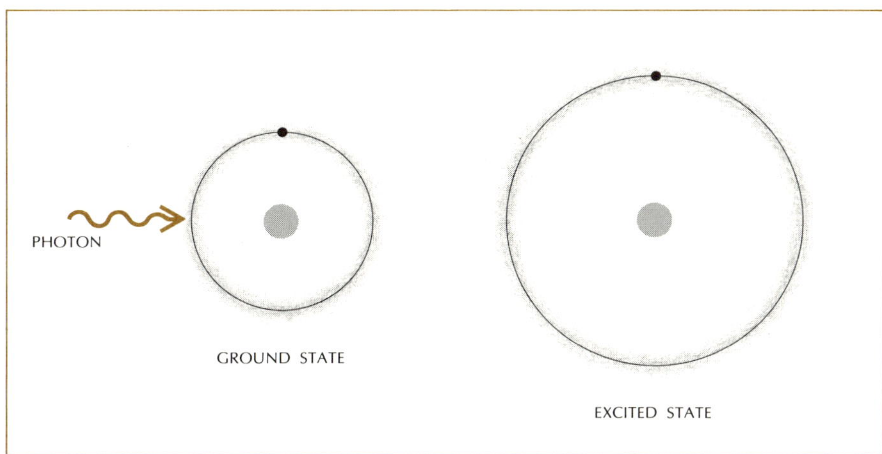

7
Photon of light excites electron from a lower to a higher energy state; that is, to an orbit farther from the nucleus (right).

chemical energy.   These are called the "light" reactions.   The second series, which builds organic molecules from raw materials, can proceed in the absence of light; these are the "dark" reactions. The light reactions occur only in the cells of photosynthetic organisms; the dark reactions occur in most types of cell, including those of animals.

Light will not split the hydrogen from water or from hydrogen sulfide in a test tube. Evidently the living cell contains some remarkable chemical machinery for harnessing the energy of light. The cells of most photosynthetic organisms contain tiny organelles called chloroplasts (Fig. 6) and these are the sites of the photochemical reactions. Photochemistry involves the interaction between light and matter.   Light waves consist of tiny "quanta" of energy called photons. The capacity of a substance to absorb light depends on its atomic structure, particularly on the arrangement of its electrons. Light may excite an electron to a higher energy level, that is, to an orbit that carries the electron farther from the nucleus (Fig. 7). A molecule whose electrons are in the lowest-energy orbits is said to be in the ground state; a molecule with one or more electrons in higher-energy orbits is in an excited state. Excited molecules are highly unstable and tend to

Absorbed energy of a photon can excite the outer electrons of the chlorophyll molecule to such high energy levels (3) that one electron is pushed out of the molecule. Electron is drawn away via the cell's electron transport system (4).

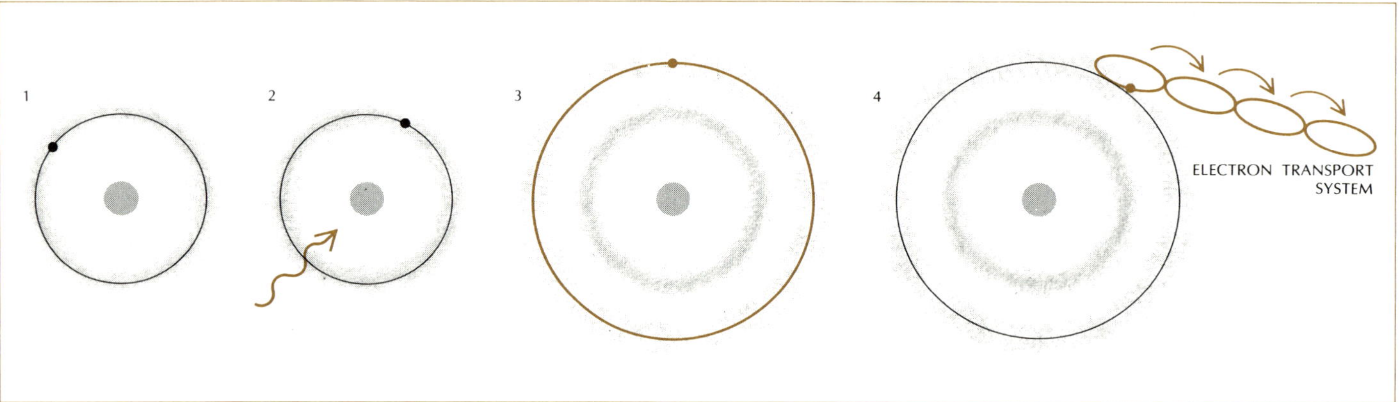

return to the ground state, usually by losing the extra energy as heat or by reemitting it as light.

The chlorophyll molecules in the chloroplasts act like the miniature photocells used in photographic exposure meters. The exposure meter and the plant chloroplast contain mechanisms to conserve light energy and put it to work, instead of letting it leak away. Both the photocell and the chloroplast are transducers: devices that transform one form of energy into another. The photocell converts light energy to electrical energy; the chloroplast converts light energy to chemical energy. In an exposure meter, photons are absorbed by the atoms in a wafer of selenium. Some of the selenium electrons acquire enough energy to separate them completely from the nucleus; they can be drawn off through a circuit, creating an electric current that shows up as the swing of an indicator needle. In the chloroplast, photons are absorbed by the electrons of the chlorophyll molecule. The absorbed energy virtually pushes one of the electrons out of the molecule (Fig. 8). In an intact plant the high-energy electrons are drawn away not

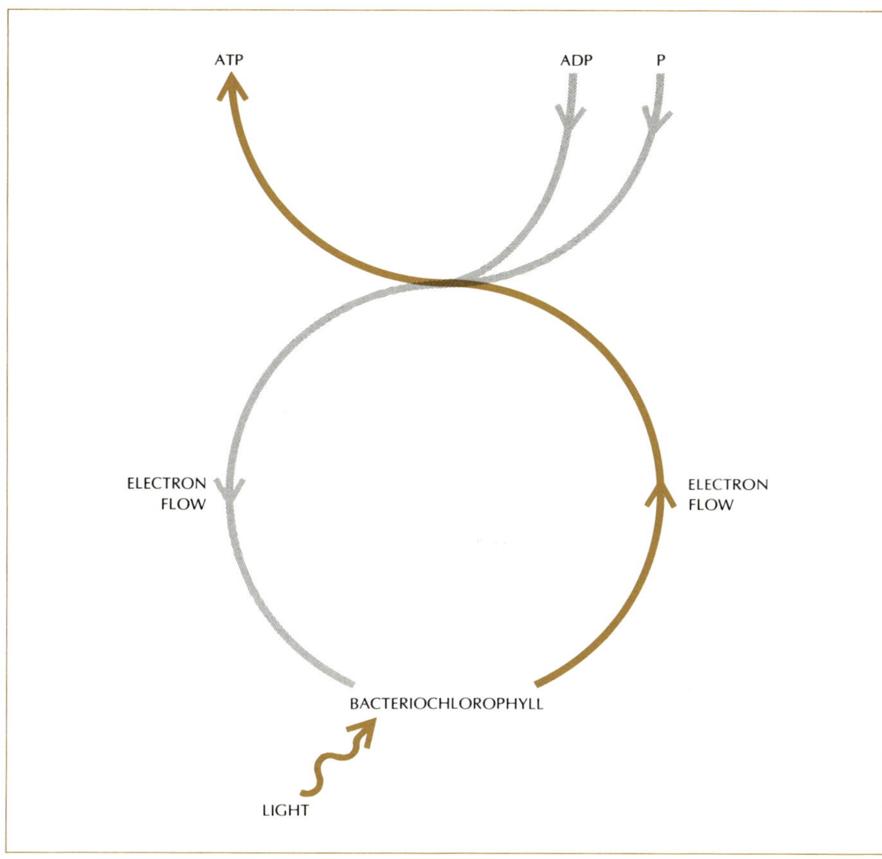

9
Light reactions of bacterial photosynthesis transform the energy of light into the energy of ATP.

by a wire, as in the photocell, but by a circuit of electron-carrying molecules called an electron transport system. The system is a mechanism for the controlled, step-by-step release of energy. The electron transport system enables the cell to harness much of this energy to do chemical work.

BACTERIAL PHOTOSYNTHESIS

In the hydrogen sulfide bacteria the raw materials for the light reactions are light, which comes from the external environment, ADP, and inorganic phosphate, which are present within the cell. The end product is ATP (Fig. 9). The bacteriochlorophyll is neither a raw material nor an end product; it is a molecular transducer that initiates the conversion of light energy into the chemical energy of ATP.

Most of the ATP produced in the light reactions supplies the energy that drives the dark reactions. It is in these dark reactions that carbon dioxide is used in the production of molecules of carbohydrate. In a process

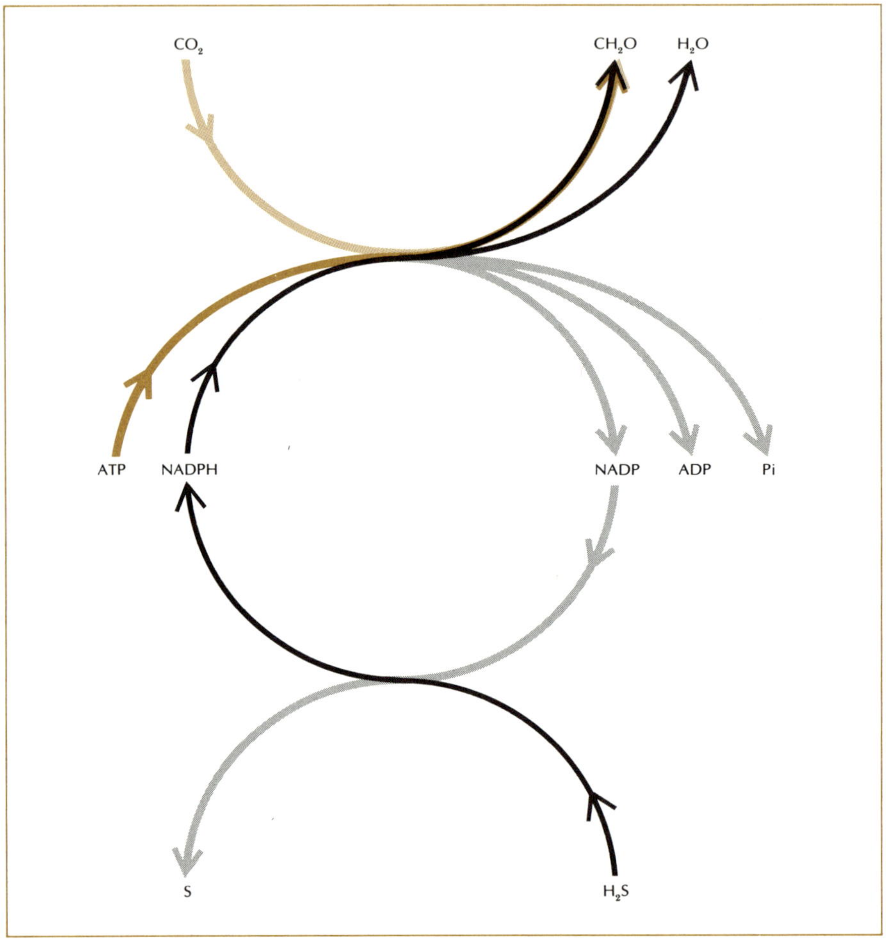

10

Dark reactions of bacterial photosynthesis combine the hydrogen from hydrogen sulfide with carbon dioxide, forming food (CH_2O). Energy-rich intermediaries include ATP and NADPH.

requiring the energy from ATP, the H from the H_2S combines with carbon dioxide, forming a molecule of carbohydrate and a molecule of water. This process regenerates ADP, inorganic phosphate, and NADP (a hydrogen acceptor, nicotinamide adenine dinucleotide phosphate—see Fig. 10), making them available once again for use.

By combining the diagrams for the light and the dark reactions occurring in the photosynthetic sulfur bacteria (Fig. 11), we can see the inter-

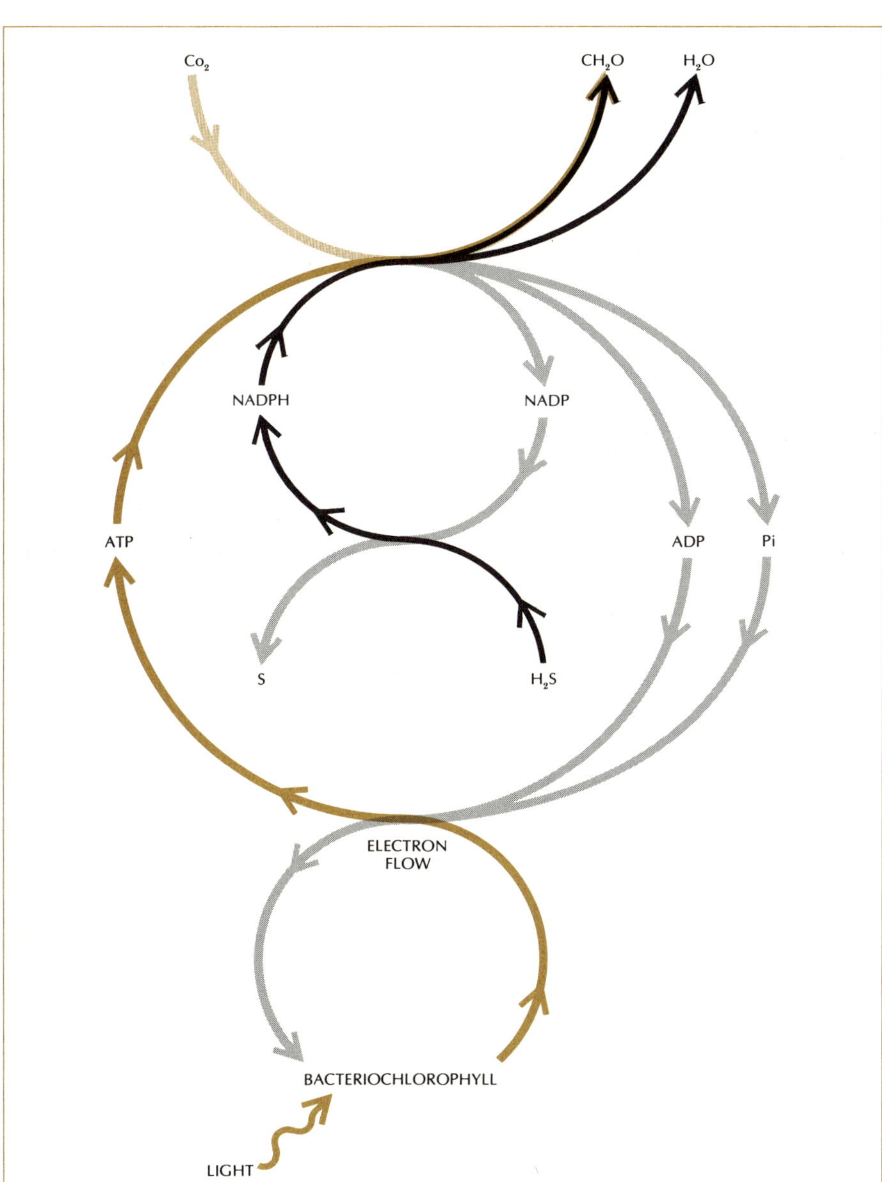

11

Complete photosynthetic cycle of bacteria illustrates the flow of energy from the light that is absorbed by the bacterio-chlorophyll molecule (bottom) to the eventual synthesis of food (CH_2O, top).

actions between the two systems and the ways in which the end products are formed. This diagram is extremely schematic; for simplicity we have omitted many steps. But it is now obvious why the mixing of hydrogen sulfide and carbon dioxide in a bacteriochlorophyll solution exposed to sunlight would not result in the production of carbohydrates. Many intermediate compounds are necessary before the reactions will occur. Furthermore, since every chemical reaction taking place within a living cell is mediated by a specific enzyme, a large number of these specialized compounds must be present before the process can occur. The early experimenters were ignorant of such facts and their efforts to understand photosynthesis were doomed to failure.

PHOTOSYNTHESIS IN HIGHER PLANTS

Photosynthesis in green algae and in higher plants is a little more complicated than that in photosynthetic bacteria. Plants have evolved the ability to use the hydrogen derived from water molecules in the synthesis of carbohydrates. This involves two different light-trapping mechanisms which researchers have designated as photosystem I and photosystem II, and a series of dark reactions.

Photosystem I contains chlorophyll *a* molecules; the absorbed energy reduces NADP to NADPH using the hydrogen ions derived from photosystem II. Photosystem II consists of molecules of both chlorophyll *a* and chlorophyll *b*. Energy collected in photosystem II is used to split water molecules into molecular oxygen and hydrogen ions. The oxygen is released and the hydrogen is used. ATP is formed from ADP and inorganic phosphate in a reaction sequence that links the two photosystems (Fig. 12). If photosystem II is experimentally shut down, photosystem I acts alone and behaves like a bacterial photosystem. This evidence supports the idea that the photosynthetic system in higher plants has evolved directly from the more primitive system found in photosynthetic bacteria.

The end products of the "photo" part of photosynthesis in normal higher plants are ATP and NADPH. Oxygen is a waste product. ATP, NADPH, and carbon dioxide are the raw materials for the dark reactions, the "synthesis" part of photosynthesis.

The details of the synthesis reactions, which lead to the formation of carbohydrates in algae and higher plants, were first thoroughly studied by Melvin Calvin of the University of California at Berkeley. Calvin knew that

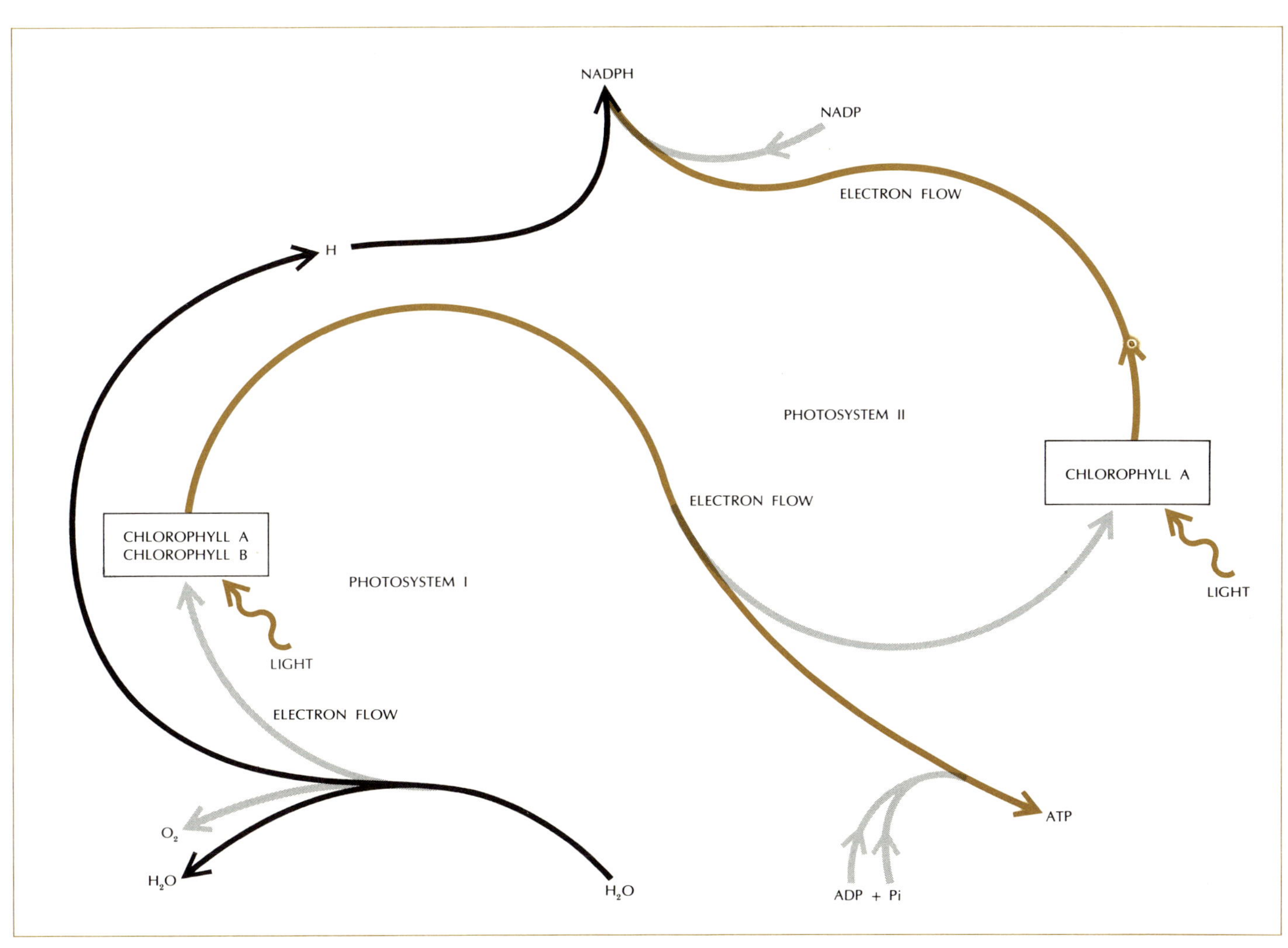

12

Photosynthesis in green plants involves two photosystems. In this process, hydrogen from water is used to produce an energy-rich compound, NADPH, which is later used in the synthesis of organic molecules from carbon dioxide. ATP is also produced as an end product of these light reactions.

Photosynthesis experiment was performed with lollipop-shaped culture vessel (left center). Floodlamps were light source for *Chlorella* organisms in vessel. Samples were drawn off into flask at bottom.

it was inconceivable that a compound as complex as glucose could be formed from carbon dioxide and hydrogen in a single reaction. He wanted to work out the intermediate reactions step by step, beginning with the first compound formed when CO_2 enters the system. As his experimental organism he also chose the one-celled green alga *Chlorella* which is small, easy to grow, and simple in structure. In a liquid culture medium, carbon dioxide can pass directly from the medium into the cells. The products of photosynthesis can be recovered by simple techniques because they remain within the cell; a single-celled alga, unlike a higher plant, has no vascular system through which the newly synthesized compounds are transported to other cells and organs.

Calvin grew the *Chlorella* in a culture vessel shaped like a lollipop (Fig. 13). Into the culture medium he bubbled carbon dioxide labeled with atoms of radioactive carbon (^{14}C). He illuminated the culture for ten minutes, then removed the algae and analyzed the contents of the cells. The radioactive carbon showed up in many different compounds: starch, sugars, amino acids, proteins, and fats. In this short period of photosynthesis the labeled carbon had undergone dozens of reactions. Ten minutes of illu-

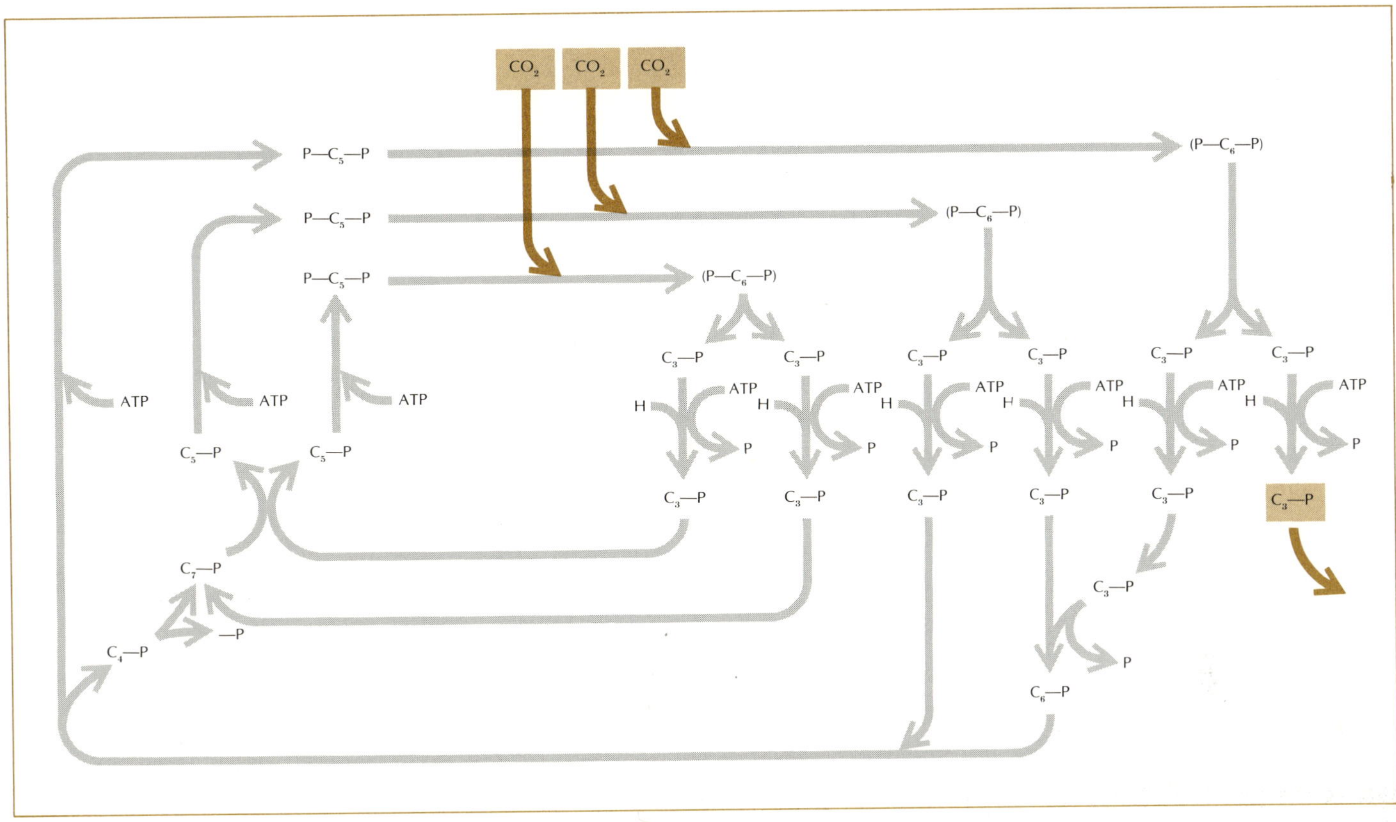

mination was much too long. He shortened the illumination time to one minute, and found that the radioactive carbon showed up primarily in only two compounds: glucose phosphate and fructose phosphate, two six-carbon sugars, each with an attached phosphate group. But even these compounds were too complicated to be formed directly from carbon dioxide in the first step of the photosynthetic process. Calvin reduced the illumination time to fifteen seconds, then to one second. In the one-second sample, most of the radioactive tracer appeared in a single three-carbon compound: phosphoglyceric acid (PGA). He concluded that this compound is the first stable product of the dark reactions of photosynthesis.

Next, by analyzing the contents of cells illuminated for progressively longer time intervals, Calvin was able to trace an entire sequence of reactions following the synthesis of PGA. He found that the ultimate formation of a glucose molecule involves a cycle of at least 14 different reactions. These are summarized schematically in Figs. 14 and 15.

The first step is the combination of carbon dioxide with a five-carbon sugar, ribulose diphosphate, which is normally present in the cell. The result is an unstable six-carbon compound that splits into two molecules

14
Carbon cycle of photosynthesis combines carbon atoms from gaseous carbon dioxide into phosphorylated organic compounds. Energy from ATP is used to drive the reactions.

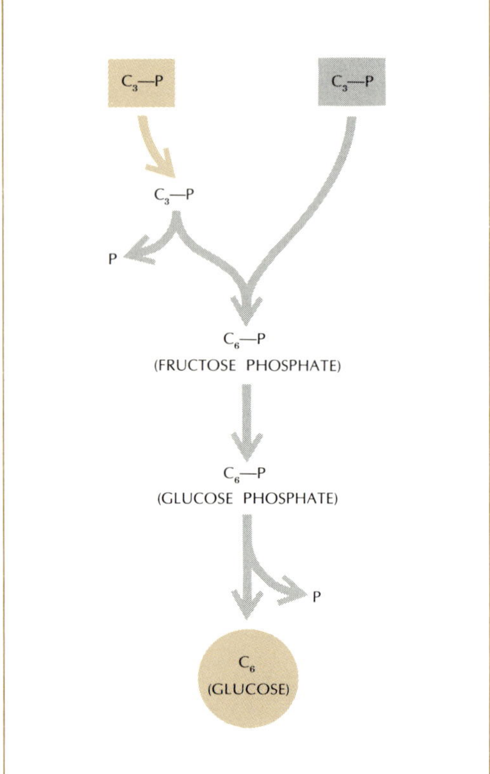

15
Pair of three-carbon molecules from earlier cyclic reactions (Fig. 14) combine, forming a sugar phosphate. This is transformed into glucose phosphate which may be converted into a molecule of glucose.

of PGA. Next the hydrogen from two molecules of NADPH combines with the PGA, forming two three-carbon molecules of phosphogylceraldehyde. This reaction is driven by the energy supplied by two molecules of ATP formed in the light reactions. After a complex series of reactions, two three-carbon molecules link to form a six-carbon sugar with a phosphate group attached. Further reactions remove the phosphate group and convert the sugar to glucose.

Although the overall reaction cycle seems to be relatively simple, the details are much more complex. In particular, the five-carbon sugar ribulose diphosphate is regenerated at each revolution of the cycle by an intricate series of enzymatic reactions. The cycle must be repeated six times for each molecule of glucose produced. For unraveling the complicated cycle of reactions that now bears his name, Melvin Calvin was awarded the Nobel Prize in Chemistry in 1961.

A MODEL OF THE CHLOROPLAST

How efficient is photosynthesis? In the field, the overall efficiency is not very high. According to one estimate, only one or two percent of the energy of sunlight falling on a cornfield during the growing season is recovered in the synthesis of plant material. But in the laboratory alga cells and isolated chloroplasts may show an efficiency as high as 50 percent. The real difficulty in measuring the overall chemical efficiency of photosynthesis lies in the remaining uncertainties about intermediate reactions and the interplay between the light and dark reactions.

Researchers are now trying to relate the chemistry of photosynthesis to the structure of the chloroplast as seen by using the electron microscope. Knowledge of the exact positioning of the pigment and enzyme molecules

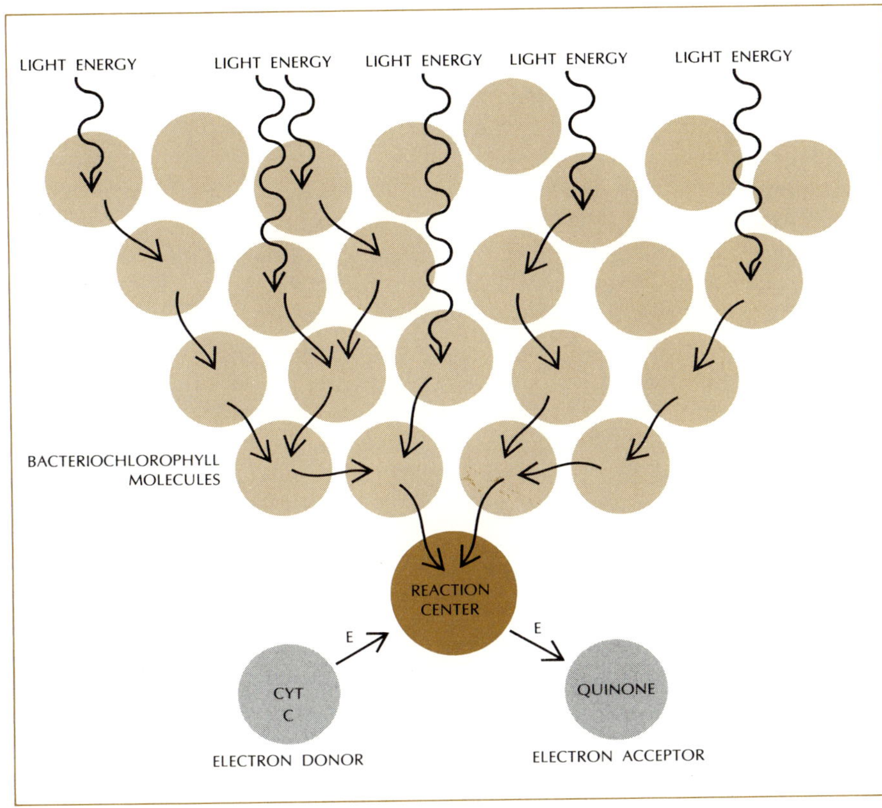

16
Photosynthetic unit collects light, channeling the energy to the reaction center. There it is concentrated and passed on to an electron from a donor compound. The energized electron is led off by an acceptor compound to be used in the chemical reactions of photosynthesis.

in the cell may eventually enable physical chemists to design a model that will fully explain the mechanism of photosynthesis.

The simplest photosynthetic structures appear in bacteria. Photosynthetic bacteria have no chloroplasts; photosynthesis in these primitive organisms is carried on in two types of particle. In the sulfur bacterium *Chromatium* one type of particle functions in the absorption of light and the transfer of energy. Such particles contain aggregates of pigment molecules called photosynthetic units. Each unit contains about 50 molecules of bacteriochlorophyll (Fig. 16). These units absorb photons and concentrate their energy in the electrons of a single molecule. This molecule is the reaction center of the photosynthetic unit. The other particle, which appears to be attached to the bacterial cell membrane, is smaller and contains compounds involved in the production of ATP and NADH. The physical separation of the particles involved in different aspects of bacterial photosynthesis may be expected to soon reveal even more minute details of the energy conversion process of these organisms.

In algae and higher plants all of the light reactions and most of the dark reactions occur in chloroplasts. The photosynthetic units in these structures contain about 300 molecules of chlorophyll and lie within a series of internal membranes called lamellae. Although in algae the lamellae are simply arranged in parallel bands, in higher plants they contain disklike piles called grana. The grana may be compared to the plates of a battery; surrounding them is an amorphous, jellylike material called the stroma, which contains ribosomes, strands of DNA, and the enzymes that catalyze the formation of organic compounds.

There is considerable debate about the ultrastructure of the grana. One technique for preparing them for examination under the electron microscope is called freeze-etching. Isolated chloroplasts are frozen solid in glycerol, dried in a strong vacuum, fractured into chips, and coated with a very thin layer of metal. This metal replica of the chip is then examined under an electron microscope.

As often happens with new techniques, any two workers may not agree on the interpretation of the results. Thus K. Mühlethaler and his team in Zurich maintain that the lamellae are smooth layers in which small globular units are partially embedded. But Daniel Branton and R. B. Park of the University of California at Berkeley argue that the lamellae are accidentally split in the freeze-etching process, and that in nature the globular units are completely buried within the lamellae. This may seem a very minor difference of opinion, but the internal geometry of the chloroplast must be known accurately before we can understand the remarkable machinery of these miniature solar power plants. Right now there is a log-jam of theories about the arrangement of the pigments, enzymes, and associated compounds within the chloroplast. It will be a long time before they are evaluated, tested, modified, accepted, or discarded. Eventually, researchers hope to manufacture intact functioning chloroplasts in the laboratory, a breakthrough that would be as important a step toward the ultimate goal of synthesizing a living plant as the invention of the internal combustion engine was to the development of the airplane.

READINGS

Fogg, G. E., *The Growth of Plants.* Penguin, Baltimore, 1963 (paper)
> A readable treatment of the physiology of plants, stressing the importance of photosynthesis to all other aspects of plant growth.

Galston, A. W., *The Life of the Green Plant.* Prentice-Hall, Englewood Cliffs, N.J., 1964 (paper)
> Brief account of topics including nutrition, growth, differentiation, and the role of green plants in the biosphere.

Steward, F. C., *About Plants: Topics in Plant Biology.* Addison-Wesley, Reading, Mass., 1966 (paper)
> An introductory text in plant physiology with good chapters on photosynthesis and the inorganic nutrition of plants.

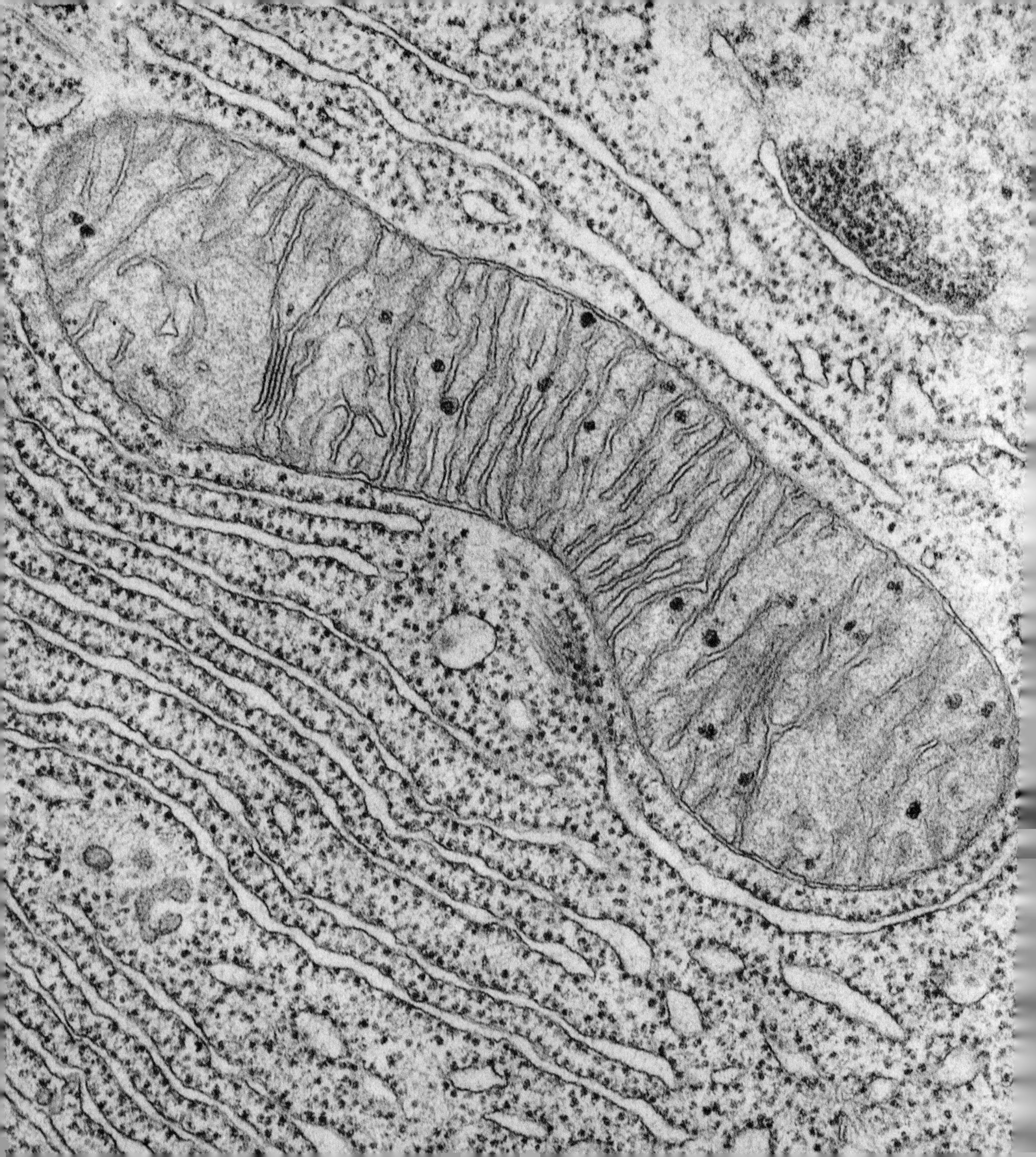

11. NUTRITION AND METABOLISM

Living organisms are basically unstable. The act of living necessitates a continuous supply of matter and energy. New raw materials must be available to build new structures and to replace worn-out components; energy is required to assemble these components and to keep them functioning. In the language of the physicist, the living cell is an open system; that is, matter passes back and forth between the cell and its environment. Gases, water, other inorganic materials, and organic compounds continually enter and leave the cell.

Most photosynthetic plants can manufacture all of their organic compounds from an external supply of carbon dioxide, water, and a few inorganic minerals. Nonphotosynthetic organisms—most bacteria, many protists, nongreen plants, and all animals—require an external supply of preformed carbon compounds, and they obtain these by digesting dead plants, by eating living ones, or by feeding on organisms that feed on plants. Floating in the atmosphere are uncountable billions of carbon dioxide molecules whose carbon atoms have spent the past few eons shuttling between the worlds of living and nonliving matter. Unchanging, virtually indestructible, any given carbon atom may participate in reactions that may build it into the cells of several organisms before returning to the atmosphere in the form of carbon dioxide.

The maintenance of life on earth depends upon five basic processes: photosynthesis, biosynthesis, digestion, fermentation, and respiration. Photosynthesis provides the basic organic building blocks for the synthesis

◄**Mitochondrion** contains the respiratory machinery of the cell. Cristae runs across the short axis of this mitochondrion from a bat pancreas cell. Magnification about 60,000 ×.

of all other organic compounds. Biosynthesis includes thousands of reactions by which complex organic molecules are assembled from simpler precursors. Digestion is a chemical splitting of complex organic compounds into simpler soluble components. Fermentation and respiration break down soluble organic compounds, transferring the energy of their carbon-to-carbon bonds to the more immediately usable energy of the phosphate bonds of ATP molecules. Collectively, these five processes are called metabolism, and to understand them is to understand the chemical machinery of life. Many of the details, and even some of the most basic processes, remain to be discovered, but the overall mechanism is relatively clear and can be described in terms of a few basic concepts.

NUTRITION

Nutrients are substances used in metabolism. Nutritional requirements divide all organisms into two large classes: autotrophs and heterotrophs. Heterotrophs require that energy be in the form of preformed organic compounds, while autotrophs use light or inorganic minerals as their external source of energy. Photosynthetic organisms and chemosynthetic bacteria are autotrophs; all other living things are heterotrophs. Heterotrophs differ sharply from one another in the types of organic compounds they must ingest in order to survive. Some need only a single organic compound; others require a much more varied diet. All derive biologically useful energy by breaking the carbon-to-carbon bonds of the food molecules and most derive further energy by combining hydrogen from the food molecules with molecular oxygen from the environment. Some organisms, both heterotrophic and autotrophic, require additional types of organic compounds. These compounds are not used directly as a source of energy in the production of ATP, but are used as building blocks in the formation of structural macromolecules and enzyme molecules. For example the diet of many organisms, including man, must supply specific amino acids and vitamins.

About 20 amino acids are essential for the synthesis of proteins by all cells. Some organisms can synthesize all they need from a pool of inorganic precursors, but others cannot make certain specific amino acids. Human cells can synthesize sufficient quantities of all but eight amino acids. By comparison, rats require ten amino acids in their food, and the microorganism *Leuconostoc mesenteroides,* a relative of the bacterium that gives

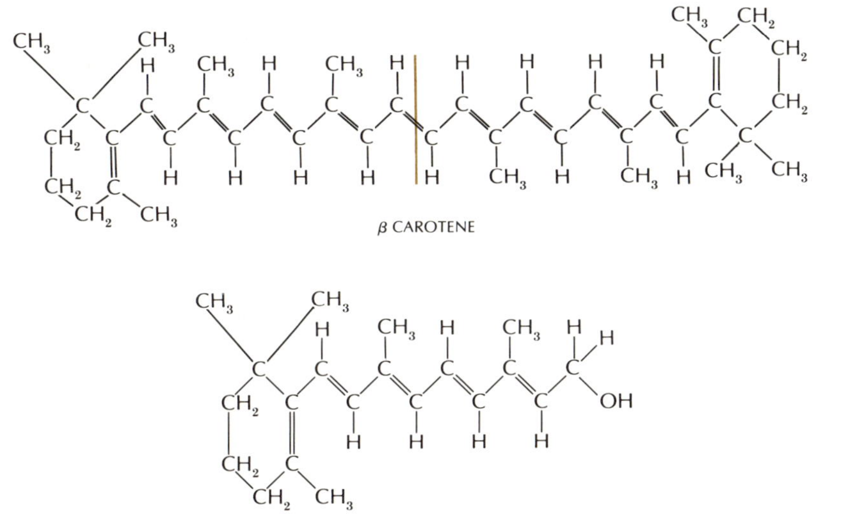

1
Vitamin A is a fat-soluble alcohol. Animals obtain it by breaking down β-carotene, an orange pigment found in plants. Colored line shows point of cleavage.

buttermilk its flavor, requires an external source of 17 different preformed amino acids. If any one of these essential amino acids is withheld from its food for long, the organism dies.

Vitamins are another class of compounds that cannot be made by some organisms. The credit for first isolating a vitamin goes to the Polish chemist Casimir Funk. His work in the early 1900's proved that a specific amine could cure beriberi in pigeons. He suggested that many amines were crucial to life and he named them "vitamines" (thus the modern term "vitamin"). Later work, however, has shown that many of these compounds are not amines at all. Today about 21 different vitamins are believed to be essential for human health; a few are listed in Table I.

The study of the effects of vitamins and their derivatives on plants and animals is an active field of research that is full of surprises. Vitamin A is a good example. It is produced within the bodies of animals by the breakdown of carotene, an orange pigment produced in green plants and used as an accessory energy trapper in photosynthesis (Fig. 1). In humans and other vertebrates this vitamin is important in the prevention of night blindness. George Wald of Harvard University was awarded a Nobel Prize in

I

Five vitamins essential to human health. As their structures indicate, vitamins may be unrelated to one another chemically. Functionally many of them (including four of the five shown here) act as important parts of enzymes. Role of large amounts of Vitamin C in conferring resistance to the common cold is controversial. Excess doses of some vitamins, notably Vitamin A, are toxic. Before their chemical composition was known, vitamins were commonly designated by letters; current practice is to refer to them by their chemical names.

Structure	Sources	Symptons of Deficiency
Vitamin A 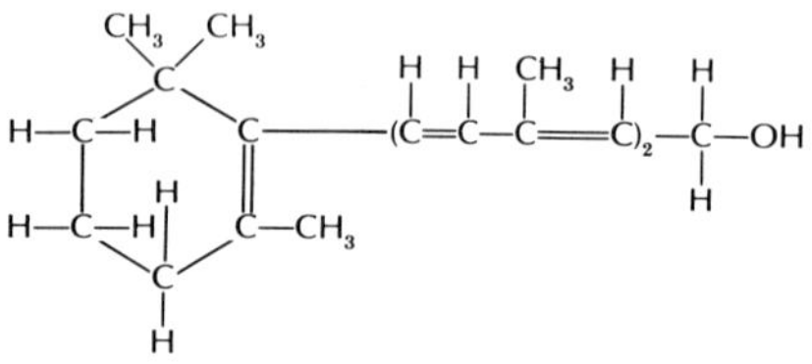	cod liver oil yellow vegetables milk products	night blindness dry rough skin and skin infections
ascorbic acid (Vitamin C) 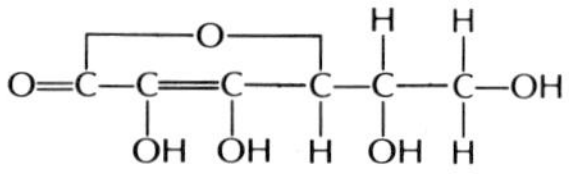	citrus fruit salad greens tomatoes	scurvy: spongy gums, loose teeth swollen joints, internal bleeding susceptiblity to colds?
pantothenic acid 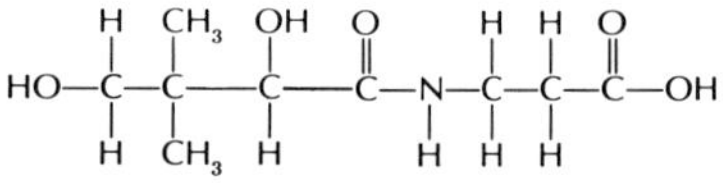	meat: liver, kidneys peanuts eggs	nerve degeneration irritable disposition diarrhea
pyridoxine (Vitamin B$_6$) 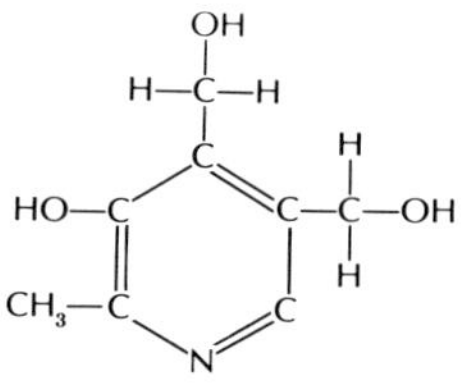	blackstrap molasses wheat bran soybeans	tingling in hands and feet loss of appetite drowsiness inflammation of tongue
thiamine (Vitamin B$_1$) 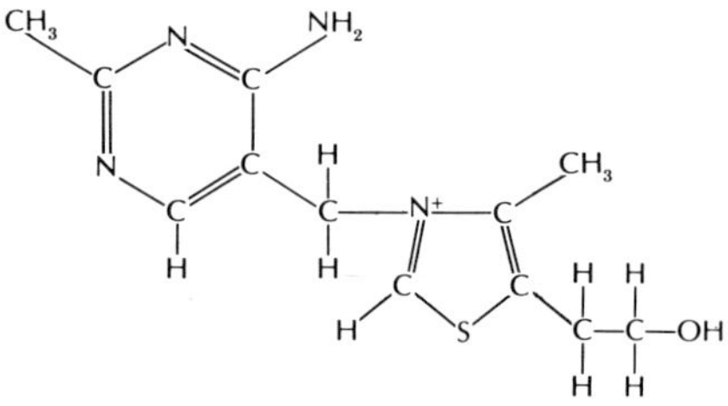	meat: liver, heart, kidney enriched flour yeast	retarded growth (children) beriberi brain damage and convulsions

2
Visual cycle involves the action of light on rhodopsin, a pigment in the eye. As it is broken into
its component parts, retinene and opsin, a message is sent to the brain which is interpreted as
light. The retinene undergoes chemical changes, finally recombining with opsin to regenerate
the active pigment molecule. Green vegetables in the diet supply carotenes that can be
converted into Vitamin A and finally into retinene.

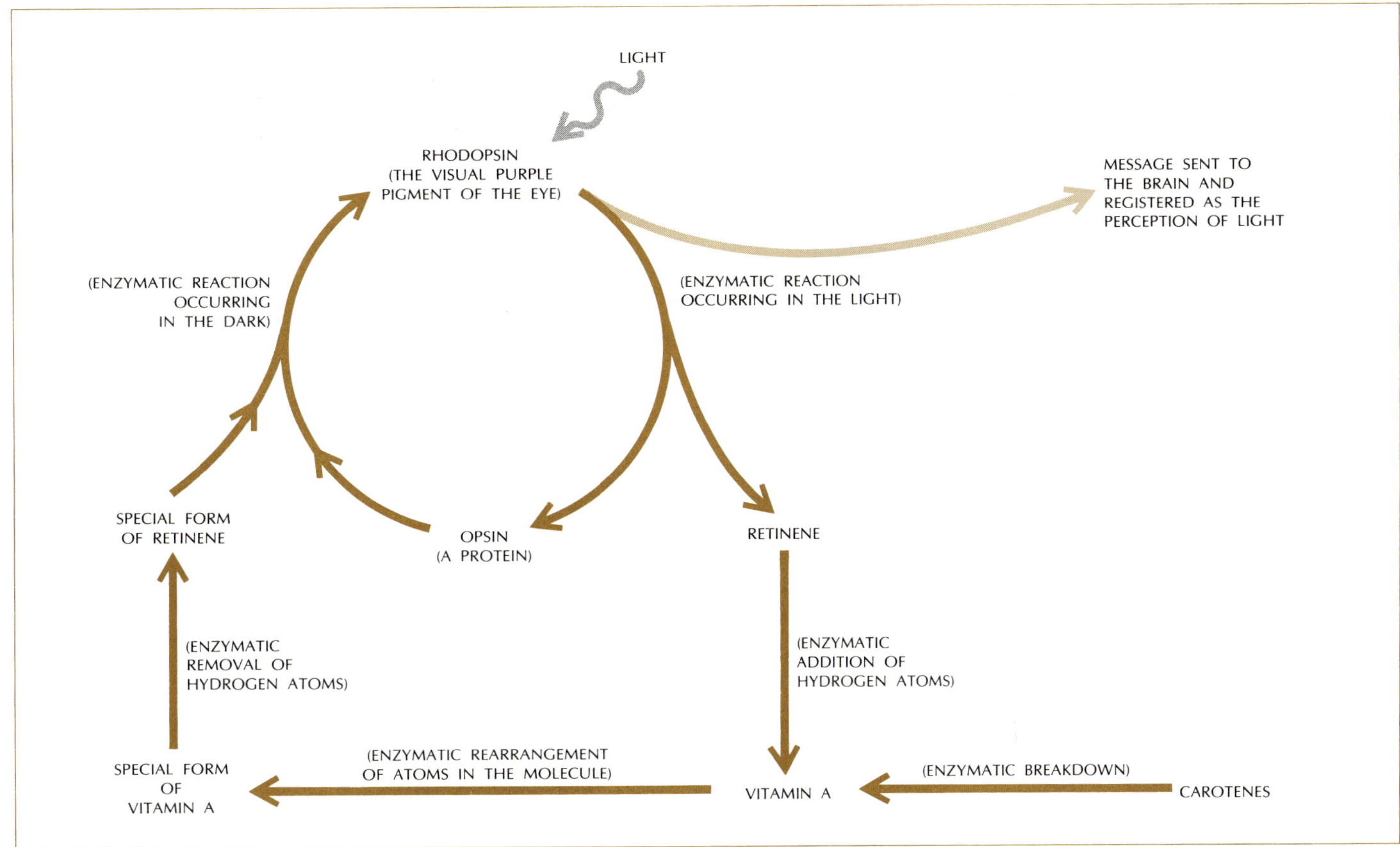

1967 for his discovery of the chemical steps involved in its utilization. He
found that to be active in the visual cycle, the Vitamin A resulting from the
breakdown of carotenes must first be transformed by the removal of two
hydrogen atoms into a special form of the compound retinene. Retinene
combines with the protein opsin to form rhodopsin, the eye pigment known
as visual purple. In special cells in the eye light rays absorbed by rhodopsin
split the pigment molecule into a molecule of retinene and a molecule of
opsin. This process somehow triggers an electrical message to the brain,
which registers it as the perception of light (Fig. 2). The retinene molecule

3

Digestion of sucrose begins with hydrolysis, the chemical splitting of a molecule by the addition of water (color) across one of its bonds. Here sucrose (1) is broken down into the simpler sugars, glucose and fructose (2).

SUCROSE

1

CH_2OH ... $HOCH_2$

HOH

2

CH_2OH ... $HOCH_2$

GLUCOSE   FRUCTOSE

can be enzymatically transformed into Vitamin A which again becomes available to enter the visual cycle.

The other effects of Vitamin A are less well understood. In some way this vitamin also helps prevent the drying of moist surface membranes such as those of the eyes, nose, throat, and lungs. Experiments with rats have shown that a deficiency of Vitamin A somehow results in abnormal sexual processes, general loss in weight, and the development of abnormal bone structure. Obviously, this versatile and potent compound serves extremely diverse functions in the metabolism of organisms.

DIGESTION

To pass through the plasma membrane for use within the cell, nutrients must be soluble. Simple sugars, amino acids, vitamins, and other small molecules can easily cross the cell membrane; but larger molecules must first be split

4
Protists engulf particles of organic matter and seal them in
baglike vacuoles. This sequence of drawings depicts an
amoeba engulfing a flagellated protist.

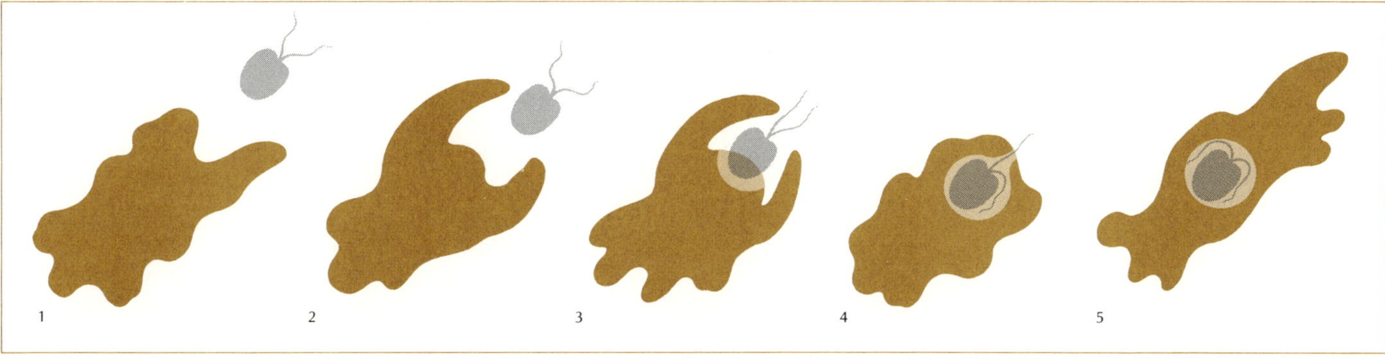

into small fragments.   Digestion is a process by which complex food is
chemically split into simpler soluble components.   It is the enzymatic
hydrolysis (splitting by water) of a food. Water is the agent used in digestion
to fracture a chemical compound into two simpler molecules. The digestion
of sucrose to fructose and glucose is illustrated in Fig. 3.

Bacteria and fungi digest complex food outside their cells by secreting
into the environment digestive enzymes that break down the large insoluble organic molecules. The resulting soluble molecular fragments are
readily absorbed by the cell. Nonphotosynthetic protists like the amoeba
and the paramecium also feed on particles of organic matter. They engulf
the particles and seal them in baglike structures called vacuoles (Fig. 4).
Food sealed in the vacuoles is not immediately available for use because the
chemical machinery for respiration and biosynthesis lies in the surrounding
cytoplasm. Large molecules of protein, fat, and polysaccharides must be
digested by enzymes before they can pass through the vacuole membrane

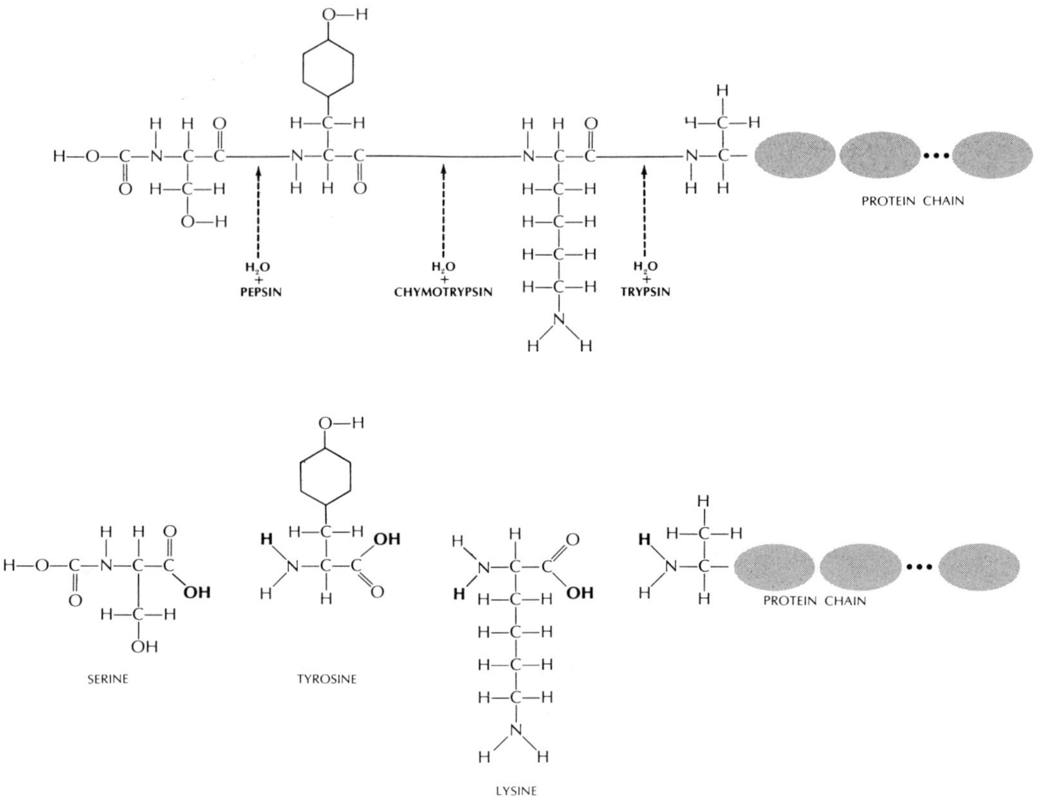

PROTEINS

5

Digestion of proteins, lipids, and carbohydrates is diagrammed on this page and on pages 317 and 318. Digestive enzymes catalyze the splitting of large molecules by hydrolysis: addition of a water molecule across a chemical bond. A hydrogen atom from water attaches to one fragment of the large molecule and an OH group to the other.

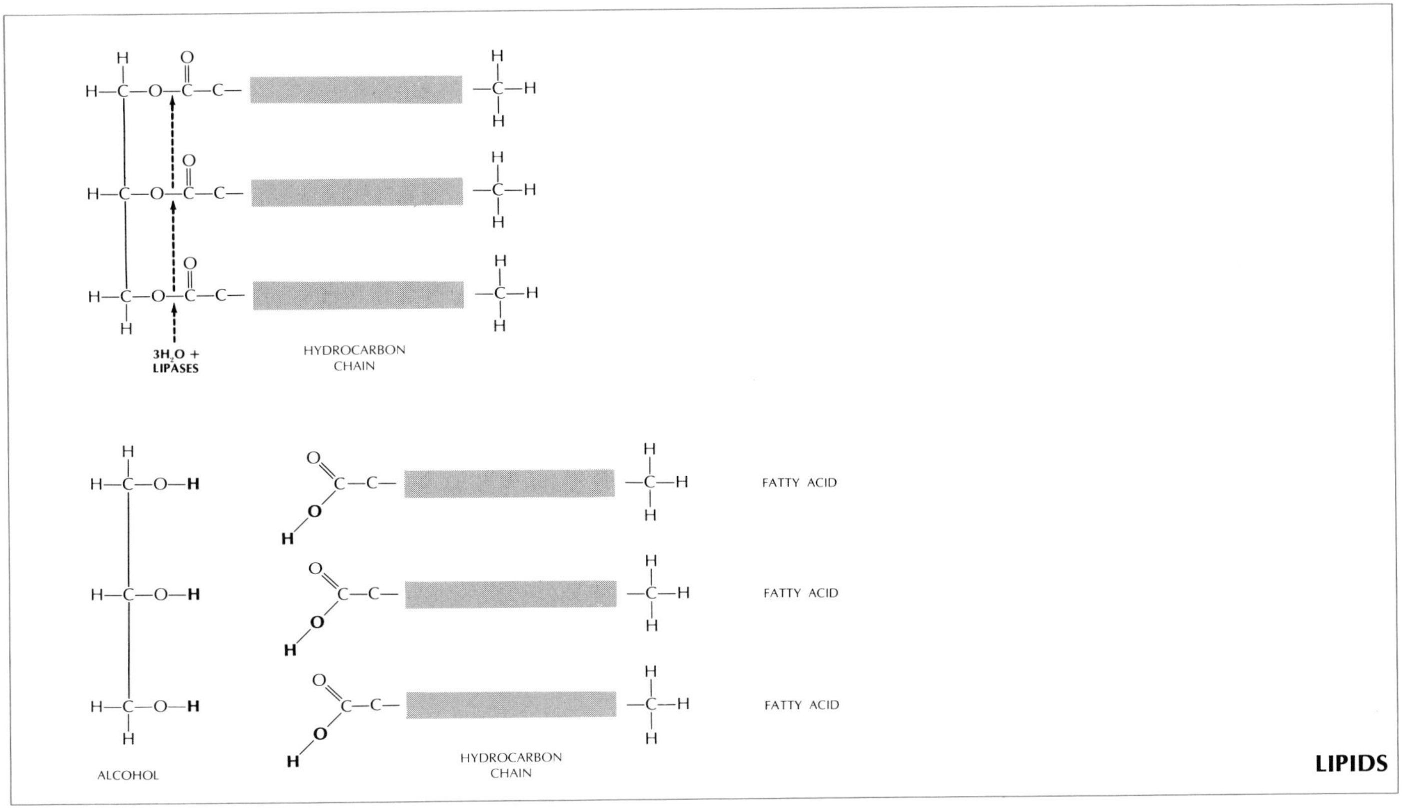

3H$_2$O +
LIPASES

HYDROCARBON
CHAIN

FATTY ACID

FATTY ACID

FATTY ACID

ALCOHOL

HYDROCARBON
CHAIN

LIPIDS

into the cell. Lipases split fats into fatty acids and alcohols; proteases digest proteins into peptides and amino acids; carbohydrases transform giant polysaccharides into small sugar units (Fig. 5).

Many animals possess teeth or other mechanical appendages that chop or shred bulk foods, and specialized glands that secrete into the gut enzymes that digest the chopped pieces into a soup of ions and simple molecules. These substances can pass easily from the gut into the bloodstream, and from there into any of the trillions of cells of the animal body.

Digestion occurs within every living cell. If this were not the case, insoluble food material stored within the cell would be completely un-

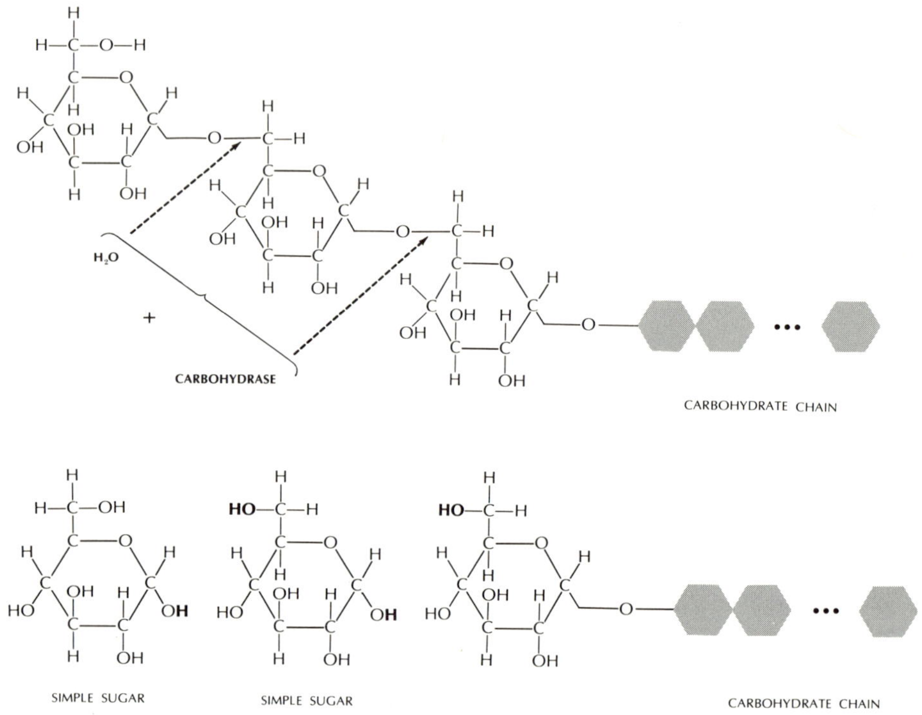

H₂O

+

CARBOHYDRASE

CARBOHYDRATE CHAIN

SIMPLE SUGAR SIMPLE SUGAR CARBOHYDRATE CHAIN

CARBOHYDRATES

available for later use. The glycogen reserves stored in the liver of man could not be called upon to supply a quick source of energy when needed. Similarly, the starch stored in the cells of plants could not be converted to glucose for use as an energy source (Fig. 6).

During periods of starvation, cells literally begin to digest themselves. When an organism stops its intake of food, its cells first digest their reserves of carbohydrates, then fats, and finally the structural and enzymatic proteins. At this stage the life of the organism is near its end.

No part of a cell is really permanent. Even under normal conditions the cell digests and metabolizes many of its own components at a steady rate.

Digestion of starch grains from wheat. Photomicrographs on left show intact grains.
Photomicrographs on right show visual evidence of the beginning of the digestive process.
Pictures on the bottom were taken in normal light; those on the top were taken with polarized light.

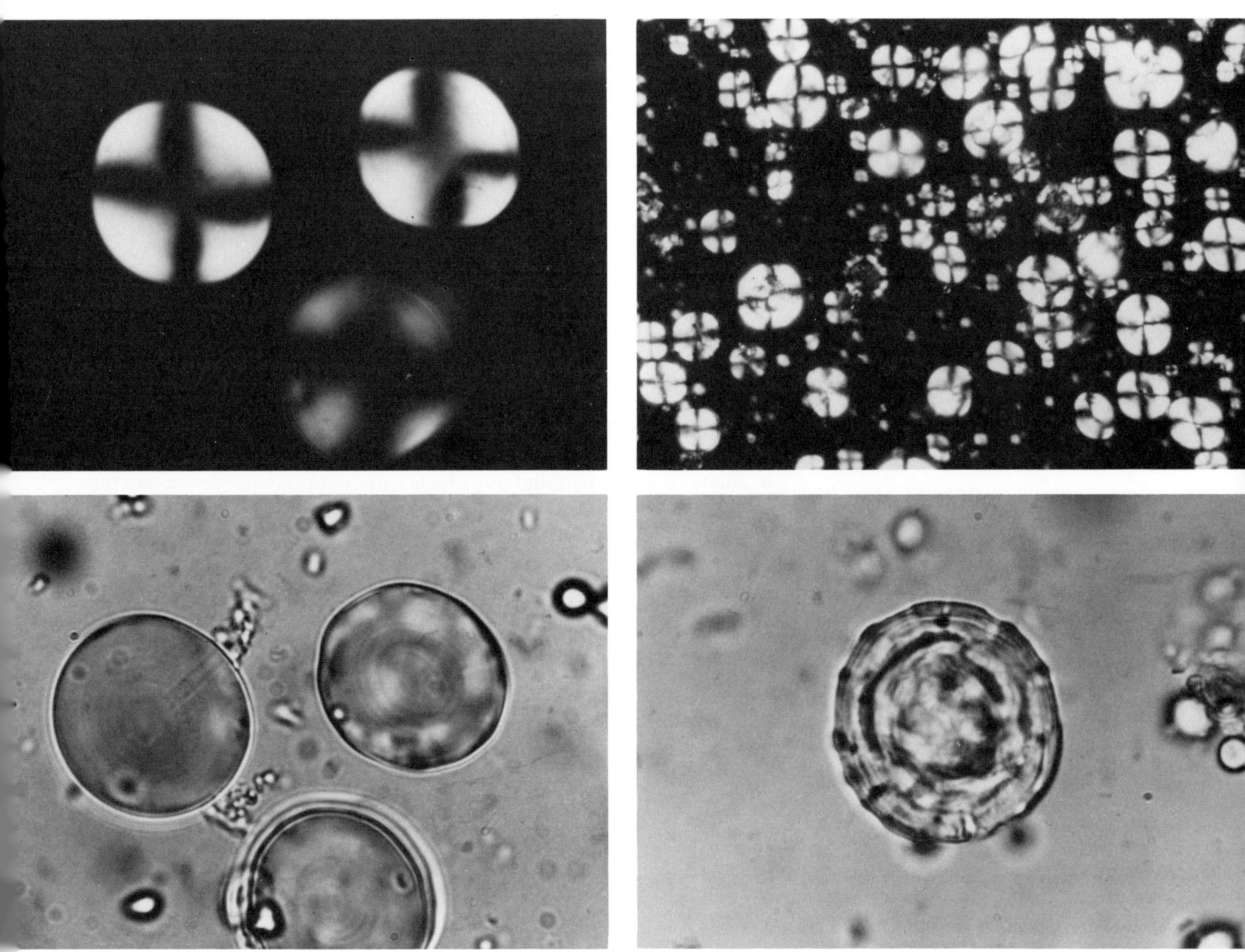

The cell maintains its structure and organization despite this constant turnover by expending energy to synthesize new molecules as fast as it breaks down the old ones. During starvation, the rate of self-digestion simply exceeds the rate of synthesis.

FERMENTATION AND RESPIRATION

Digestion is only a prelude to the utilization of food. It does not produce the useful energy of ATP which powers the metabolic machinery of all biological activity. For this the cell must convert the energy of the carbon-to-carbon bonds of the food molecules into the high-energy bonds of ATP. The actual conversion takes place in the cytoplasm of the cell in the processes of fermentation and respiration. Respiration is an aerobic process (one which requires molecular oxygen) and fermentation is an anaerobic process (one which does not require molecular oxygen). Both create a supply of ATP by the chemical breakdown of food molecules. These processes are essential to all forms of life; fermentation occurs in all cells of all organisms, including green plants as well as microorganisms and animals. In addition, all animals and true plants carry out respiration. A major misconception among laymen is that cells of green plants carry out only photosynthesis, while cells of animals carry out respiration. In fact, respiration occurs in animal cells and both processes occur in the green cells of photosynthetic plants. The general metabolic processes in plant cells are like those in animal cells with the exception that chlorophyll-containing plant cells have the added ability to photosynthesize.

Many types of organisms can use amino acids, fatty acids, alcohols, sugars, and a wide assortment of other organic molecules as a source of energy. Anaerobic organisms cannot break all of the carbon-to-carbon bonds of such food molecules. Upon completion of the fermentation process organic molecules are produced which still contain considerable chemical energy. Aerobic organisms, on the other hand, can break all of the carbon-to-carbon bonds of their food molecules. In addition, they can combine the individual hydrogen atoms with oxygen, forming H_2O, and producing an additional quantity of energy-rich ATP. Thus, more useful energy can be obtained from a given amount of food by aerobic organisms than by anaerobic ones.

The ways in which various organisms metabolize the simple sugar glucose are well known and illustrate some of the differences between the

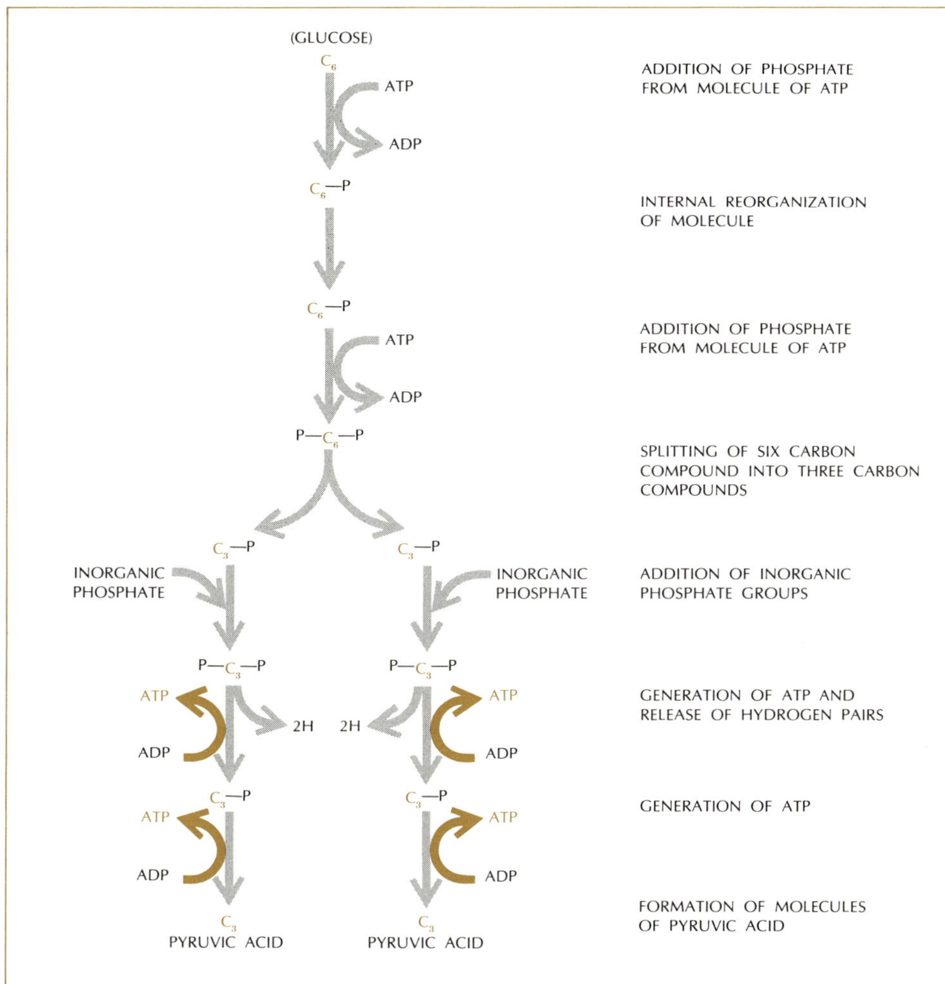

7
Glycolysis converts a molecule of glucose (a 6-carbon compound) into two molecules of pyruvic acid (a 3-carbon compound). Twice as much ATP is generated in the later stages of the process as is consumed in the early reactions.

processes of fermentation and respiration. The fermentation of glucose often proceeds via a series of reactions called glycolysis (from the Greek roots meaning the splitting of a sweet substance). In glycolysis a six-carbon glucose molecule is split, ultimately producing two three-carbon molecules of pyruvic acid. This apparently simple reaction involves at least 11 steps, each catalyzed by a special enzyme. As in photosynthesis, the intermediate reactions enhance the gradual release of energy and the take-up of that energy in the bonds of newly created ATP (Fig. 7).

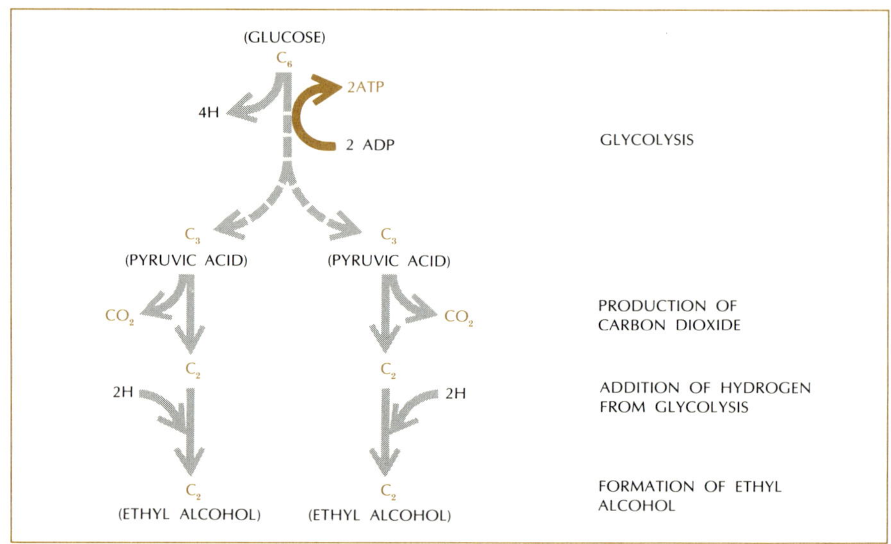

The breakdown of glucose is not spontaneous; to start the process an input of energy is required. In these reactions a molecule of glucose first reacts with a molecule of ATP. The terminal phosphate group of the ATP is transferred to the glucose molecule, yielding molecules of ADP and glucose-phosphate. Second, another ATP molecule transfers its phosphate group to the glucose-phosphate, producing one molecule of ADP and one of a sugar-diphosphate. These two reactions merely "prime the pump." Once the energy-rich sugar-diphosphate molecule has been formed, the remaining reactions "run downhill," splitting the molecule into two molecules of pyruvic acid and producing four molecules of ATP.

Subtracting the two molecules of ATP required to prime the reaction from the four molecules produced by the glycolytic sequence, we have a

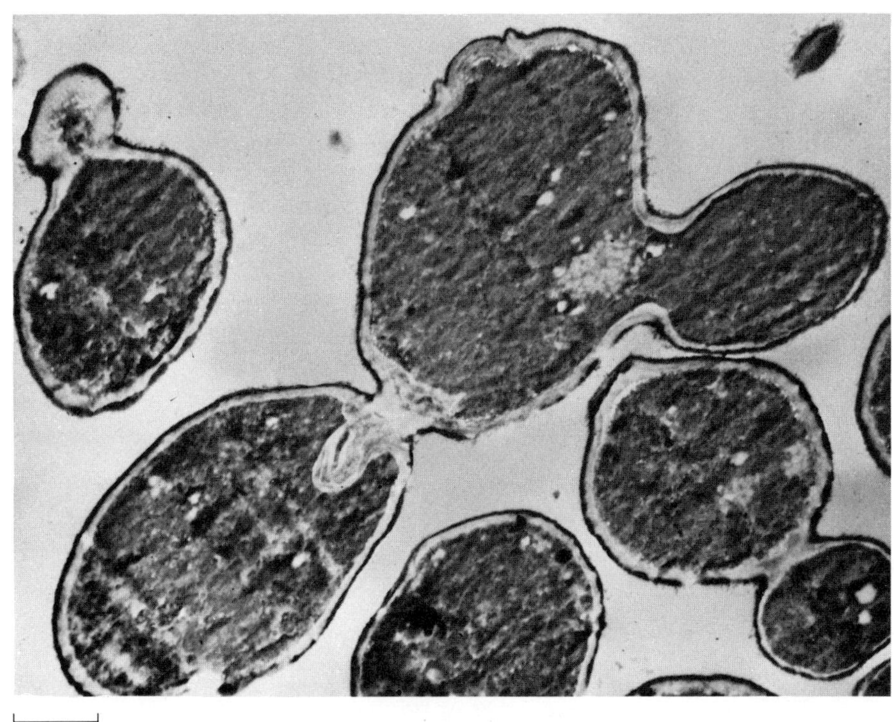

net gain of two ATP molecules produced for every glucose molecule fermented. In the laboratory, the combustion of one gram of glucose (about a teaspoonful) yields about 3800 calories of energy in the form of heat. Heat energy is useless to the cell. In a living cell, the fermentation of one gram of glucose to pyruvic acid converts about two percent of the energy of the glucose, about 76 calories, into the biologically useful energy of the phosphate bonds of ATP.

Once the pyruvic acid has been formed it may undergo additional transformations. By further fermentative processes some microorganisms, including common brewer's yeast, convert this substance into ethyl alcohol and carbon dioxide (Fig. 8). Commercially, this is a very useful reaction: the alcohol gives the kick to whiskey, wine, and beer; the carbon dioxide

9

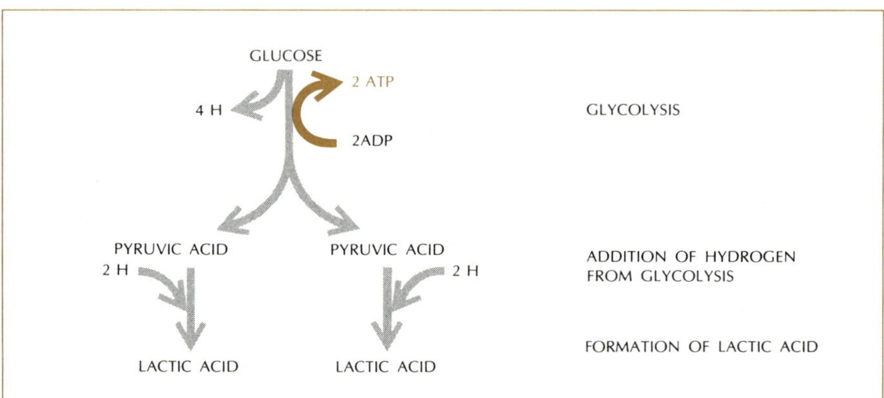

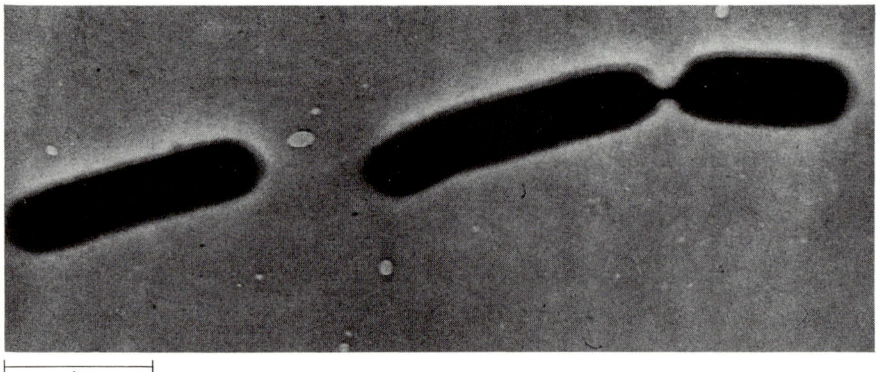

creates the head on beer and causes bread dough to rise. In the souring of milk, lactic acid bacteria ferment glucose to pyruvic acid and then into lactic acid, which leaves the cell as a waste product (Fig. 9). Many economically important organic compounds are produced commercially by fermentative microorganisms; glycerine and Vitamin B_{12} are two familiar examples.

In many bacteria and fungi, fermentation is the main source of energy. Certainly the efficiency of energy conversion is low, but it provides enough energy to keep simple organisms alive. On the other hand, anaerobic reactions alone cannot supply the lavish energy requirements of higher organisms. In the cells of most plants and animals, including man, respira-

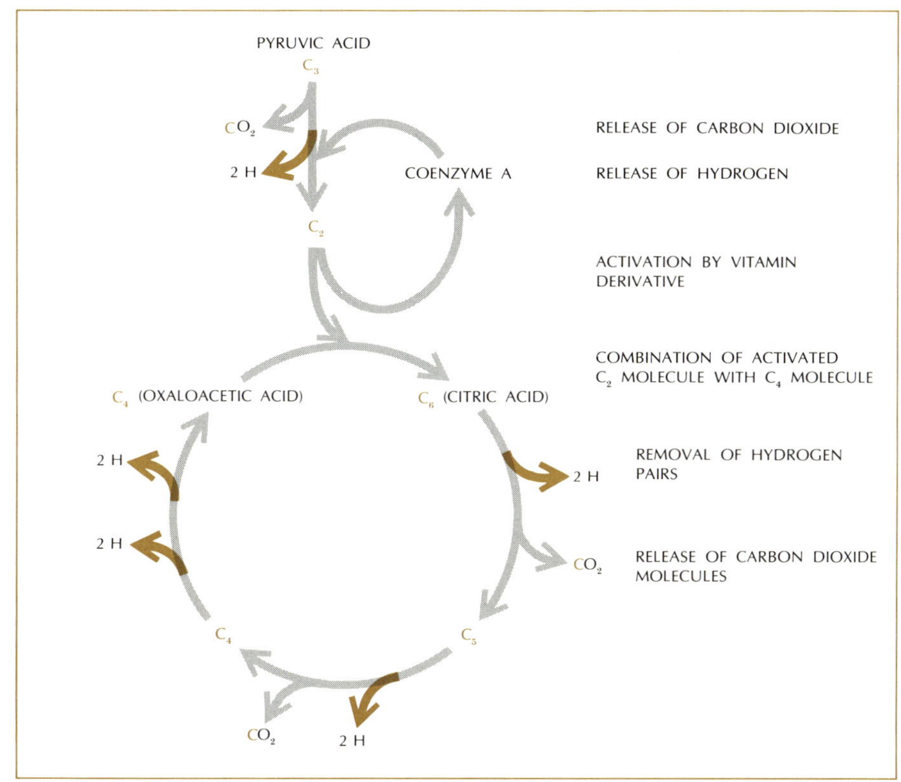

10
Krebs cycle begins where glycolysis ends. Pyruvic acid is altered and fed into a cycle of reactions in which the molecules are progressively broken down, releasing carbon dioxide and pairs of hydrogen atoms.

tion takes up where glycolysis leaves off. Additional reactions oxidize the pyruvic acid from glycolysis into carbon dioxide and water, converting much of its chemical energy into the energy of ATP. First described in 1937 by the biochemist Sir Hans Krebs, the reactions of respiration have been named "the Krebs cycle" in his honor. The Krebs cycle involves an even more complex series of reactions than glycolysis, and the complexity serves the same function: the gradual release of chemical energy and the capture of that energy in the bonds of ATP (Figs. 10 and 11).

First, the three-carbon pyruvic acid molecule is broken into a simpler two-carbon acetic acid compound and a molecule of carbon dioxide. Then, the acetic acid compound, activated by a derivative of the vitamin pantothenic acid, combines with oxaloacetic acid, a four-carbon compound already present in the cell, and forms a six-carbon compound, citric acid. The ensuing reactions break down the citric acid step by step, releasing two molecules of carbon dioxide. The remaining four-carbon compound is a molecule of oxaloacetic acid which is available to accept another two-carbon acetic acid compound beginning the cycle anew.

THE RESPIRATORY CHAIN

During these reactions the hydrogens removed from the different carbon compounds are siphoned off by a transport system somewhat similar to that involved in photosynthesis (Fig. 11). Hydrogens led away from these intermediate compounds are fed into a series of electron carrier molecules collectively called the respiratory chain. At the last link in the chain the hydrogen reacts with molecular oxygen from the atmosphere, forming water.

The breakdown of food in respiration, a process which uses oxygen and produces water, is a reversal of the chemistry of photosynthesis, which uses water and produces molecular oxygen. In fact, many of the electron carriers in the respiratory chain are structurally similar to the corresponding carriers in photosynthesis.

At three points along the respiratory chain the energy of the passing electrons is harnessed to couple an inorganic phosphate group to ADP, forming ATP. Each of the six pairs of hydrogen atoms which are produced during the oxidation of a molecule of pyruvic acid provides the energy for the formation of three molecules of ATP, 18 molecules in all. Because two pyruvic acid molecules are produced from each original glucose molecule, the Krebs cycle produces 36 molecules of ATP for each molecule of glucose oxidized. Adding the two molecules of ATP previously formed by glycolysis, we have an overall yield of 38 molecules of ATP. This represents a thermodynamic efficiency of 40 percent, much higher than the scant 2 percent efficiency of glycolysis alone. In terms of energy this means that respiration makes available for biological work about 1500 of the 3800 calories in a gram of glucose, compared with only 76 useful calories from fermentation. An efficiency of 40 percent is impressive even by the standards of modern engineering. Industrial turbines and engines rarely convert more than a third of the heat of combustion into useful electrical or mechanical energy.

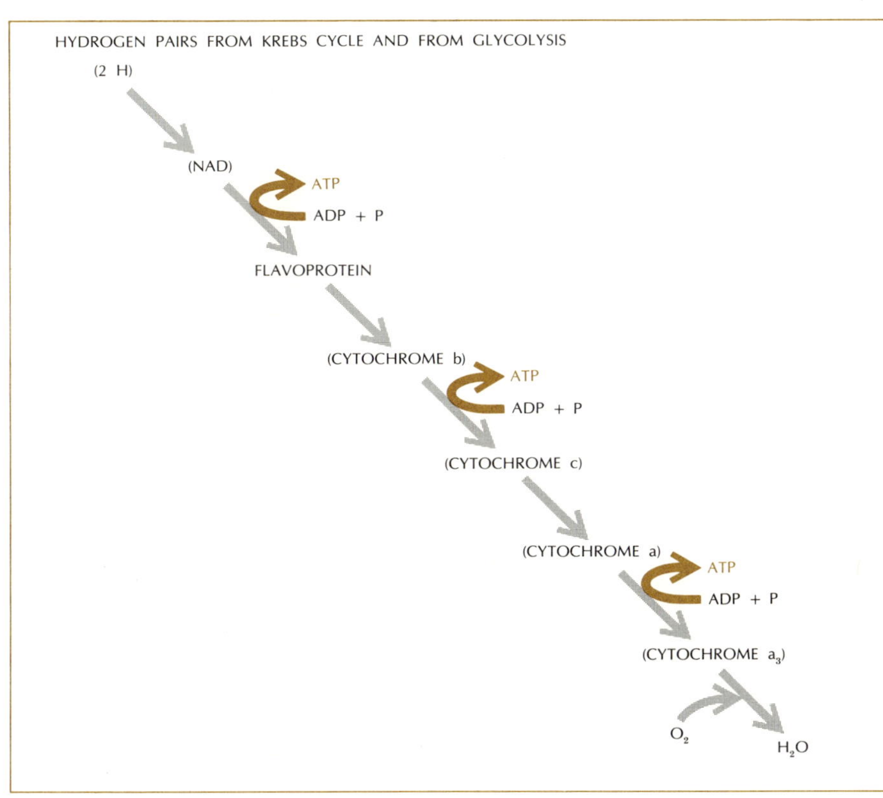

11
Respiratory chain takes hydrogen pairs derived from the Krebs cycle and from the glycolytic reactions and extracts energy from them, using it to produce ATP.

The complexity of respiratory chemistry makes it sensitive to many types of disruption. If one reaction stops, so do the others. A shortage of molecular oxygen inhibits the last step in the respiratory chain. This stops other reactions of this system and in turn shuts down the entire Krebs cycle. The resulting shortage of ATP limits essential processes in the cell and, if prolonged, may result in death. A shortage of the vitamin pantothenic acid prevents pyruvic acid from entering the Krebs cycle and ultimately leads to a shortage of ATP. Iron and some of the B vitamins are necessary for the synthesis of the carrier molecules in the respiratory chain. The absence of any one of the many enzymes that catalyze each step of the Krebs cycle

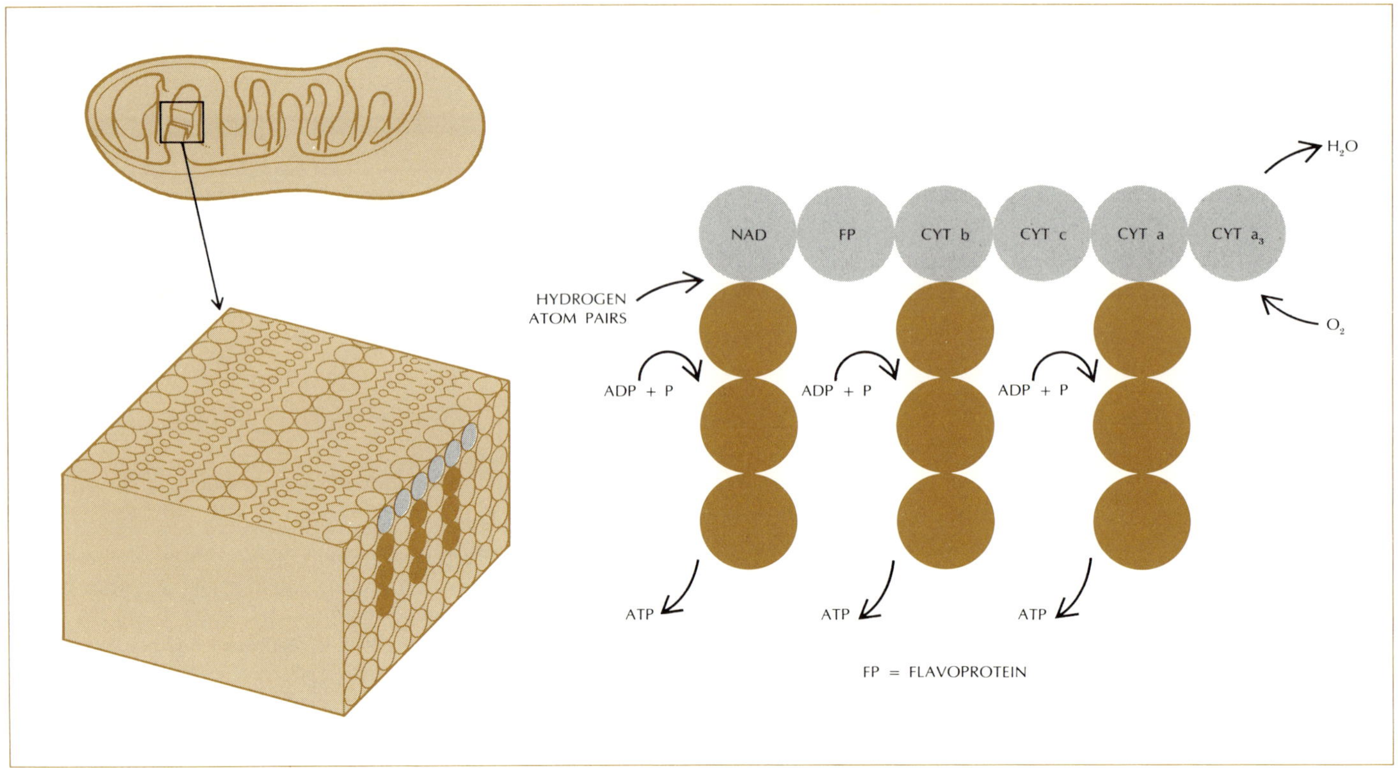

12

Mitochondrion contains numerous internal membranous foldings called cristae (upper left). Cross section of a crista (lower left) shows hypothetical model of arrangement of protein molecules (large circles) and phospholipid molecules (wavy lines and small circles). Definitive arrangement of special molecules within the membrane (right) comprises the respiratory hydrogen transport chain where ATP is produced.

shuts down the entire system. Despite its vulnerability, the process of respiration is remarkably reliable. There may be brief periods in the lives of higher organisms when the Krebs cycle is shut down and glycolysis alone must carry the burden of ATP production, but otherwise both processes continue 24 hours a day for the lifetime of the organism.

THE MITOCHONDRION

In living cells respiratory reactions are extremely fast. Despite the great number of reactions, enzymes, and carriers involved, a glucose molecule is completely oxidized in less than a second. The speed and complexity of these reactions suggest that, like the reactions of photosynthesis, they are not carried out in a random soup of molecules, but in a highly specialized structure. Just as the machinery of photosynthesis is built into the structure of the chloroplast, so the machinery of respiration is built into another subcellular structure, the mitochondrion. A single cell, depending on its type and function, may contain from 50 to 5000 mitochondria.

The chloroplast and the mitochondrion are the two miniature power plants of cellular activity. One converts solar energy to the energy of ATP

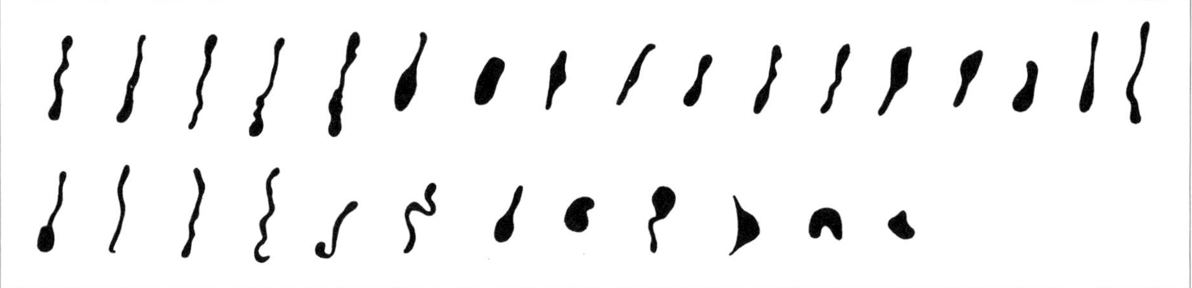

and of organic compounds; the other converts the energy from organic
compounds to the energy of ATP. In both there exists an intimate relationship between structure and function.

Under the electron microscope, the mitochondrion often appears as a
sausage-shaped structure. Cristae, the folds of the inner membrane, divide
the interior of the mitochondrion into a labyrinth (Fig. 12). Built into the
outer membrane are some of the enzymes of glycolysis and some of the
reactions of the Krebs cycle; the respiratory-chain compounds are arranged
along the inner membrane. The remaining Krebs cycle enzymes are dissolved in the fluid matrix that surrounds the cristae. This arrangement
apparently facilitates the transfer of molecules, atoms and electrons to the
correct enzymes and carrier molecules in precisely the correct order for
their efficient utilization.

The relationship between the shape of the mitochondrion and its
metabolic activity is currently under investigation in many laboratories. As
early as 1958 published drawings showed that a single mitochondrion
swells and shrinks during a 15-minute period (Fig. 13). Later experiments
showed that the most striking changes in shape were caused by variations
in the supply of ATP and oxygen. The mitochondrion contracts when it

contains a large supply of ATP and swells when the supply is depleted.

This observation suggests the existence of a physical mechanism for the control of mitochondrial activity. The end products of glycolysis produced on the outer membrane must pass into the interior of the mitochondrion for processing in the Krebs cycle. The increased surface area of a swollen mitochondrion allows these substances to percolate freely into the interior. Once inside, they are used in the production of ATP. As the ATP accumulates it causes the mitochondrial membrane to contract, restricting the input of glycolysis products and slowing down the Krebs cycle. When the concentration of ATP falls, the mitochondrion swells once again; more raw materials enter and the Krebs cycle is accelerated. This mechanism is typical of the built-in control systems that maintain the reactions within the cells at proper rates. The rate at which the mitochondrion synthesizes ATP is apparently governed by the rate at which the cell uses it.

BIOSYNTHESIS

Life processes continuously break down the molecules that comprise the cell. Just to stay alive, the cell must rebuild them at an equal rate; to grow and to reproduce it must synthesize them faster than they are destroyed.

Building and rebuilding molecules is the role of biosynthesis: the most elaborate, most demanding of the energy-consuming processes in the cell. Programmed by the information in the genes, the cell continuously fashions giant molecules from simpler ones at the expense of ATP. Automatic control systems adjust the rate of synthesis to accommodate changes in the environment and changes in the activity of the cell.

In living cells the rate of turnover of specific types of molecule depends on the type of cell. In rat liver cells that are not actively growing or dividing, half of the glycogen is replaced in less than six hours, half of the lipid in seven days, half of the protein in ten days, and half of the DNA in about two months. In many other cells the turnover rates are slower. In the muscle cells of the rat, half the proteins are replaced every three months.

In rapidly growing cells the tempo of biosynthesis is truly amazing, particularly in bacteria, the fastest growing cells on earth. Bacterial cells invest almost all their energy in biosynthesis. From the glucose, ammonium salts and other compounds in a culture medium, a single *E. coli* bacterium can synthesize in one second 12,500 molecules of lipid, or even more incredibly, 1400 complex protein molecules, each containing on the average roughly 11,000 atoms. To put this feat in proper perspective we might note that only one type of large protein molecule has been synthesized in the laboratory despite years of concentrated effort. Albert L. Lehninger of the Johns Hopkins University estimates that a single *E. coli* cell contains roughly 17 million large biological molecules; in 20 minutes its synthetic machinery can reproduce all of them, creating a new organism.

The cells of higher plants and animals grow more slowly, but their rate of biosynthesis is still impressive. In man, the cells that line the intestine reproduce in a few days, liver cells in a few months. The nerve cells in the brain and spinal cord are formed only in the early stages of life and never reproduce.

In all cells biosynthesis follows the same general pattern: the high-energy phosphate group of ATP directly or indirectly activates a building-block molecule, enabling it to combine with the next building block in a series of enzyme-catalyzed reactions. Organisms can take the carbon skeleton of phosphoglyceric acid (PGA) produced in photosynthesis or glycolysis, and from it synthesize all of the carbohydrates necessary for their growth and development. Reacting with ATP, PGA is transformed into the phosphorylated sugar, glucose-phosphate. These molecules may then link, forming long polysaccharide chains such as starch, cellulose, or glycogen. Similarly, fatty acids, alcohols, and fats can also be made from simple building-block molecules derived from PGA (Fig. 14).

The manufacture of amino acids requires additional compounds containing nitrogen. These are originally obtained from the environment by plants, in the form of inorganic ions. In essence, PGA is converted by special enzymes into compounds which can react with the inorganic nitrogen compounds. These reactions produce some of the amino acids. Others are formed from them by reactions that rearrange the parts of the molecules. The synthesis of a few more amino acids (cysteine, cystine, and methionine) requires the incorporation of sulfur atoms into the molecules.

About 90 percent of the cell's biosynthetic energy is spent on the synthesis of proteins. Proteins are the most important organic compounds in the cell; they represent up to 70 percent of the dry weight of an *E. coli*

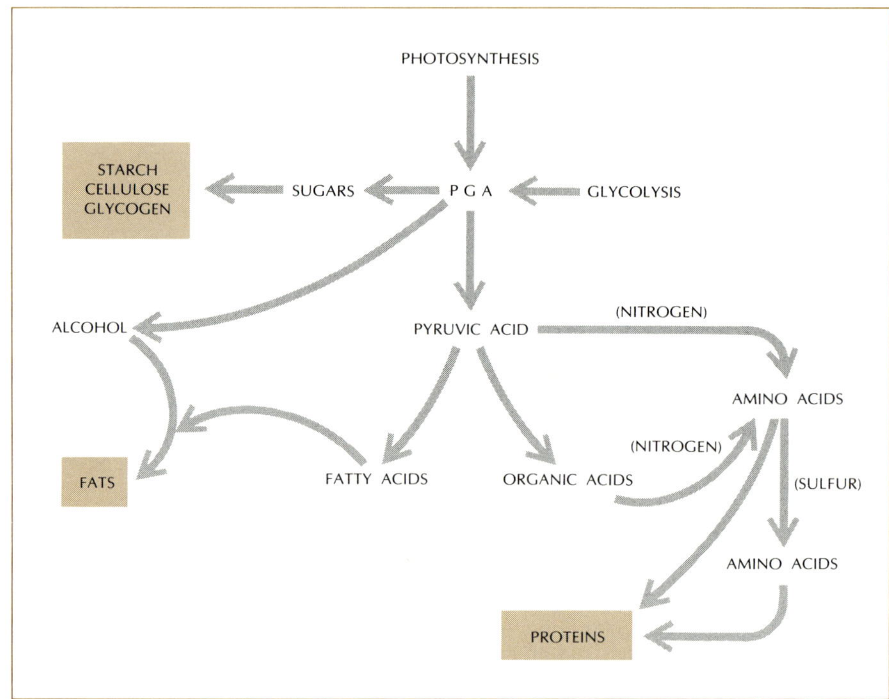

14

Biosynthesis of organic compounds from phosphoglyceric acid (PGA), synthesized in photosynthesis and glycolysis, produces complex carbohydrates and fats and, with the addition of nitrogen and sulfur, proteins.

cell (Table II). We can assume that the average protein molecule contains about 500 amino acid units; to add one of these units to another requires at least two molecules of ATP. When we recall that the cell also uses ATP to synthesize amino acids, we arrive at an estimate that protein synthesis in *E. coli* consumes more than two million ATP molecules per second. In comparison, the synthesis of lipids, nucleic acids, and polysaccharides requires relatively minor amounts of energy; Lehninger estimates that these activities consume less than 300,000 ATP molecules per second.

Lehninger has calculated that the *E. coli* cell contains a total of only one million molecules of ATP, enough for less than half a second of biosynthesis. This means that the cell must continuously recharge the ADP molecules to

II
Chemical composition of a cell. Protein molecules comprise 70 percent of the weight of the bacterium *E. coli,* exclusive of water. The remaining 30 percent consists of lipids, polysaccharides, and nucleic acids. Columns at center and right indicate approximately how many molecules of each type are found in cell, and their average molecular weight. Percentage would vary somewhat in a eukaryotic cell.

ATP at the rate of millions of reactions a second. *E. coli* is an aerobic organism and therefore exploits the efficiency of the Krebs cycle. But many anaerobic bacteria can replenish their supply of ATP and carry out biosynthesis at the same rate as *E. coli* with only the relatively inefficient energy-generating mechanism of fermentation. To sustain their metabolism anaerobic organisms burn 20 times as much glucose as aerobic cells.

15

◄**Hemoglobin molecule** is a globular protein
that contains four polypeptide chains. Heme
groups are shown as planes; sphere embedded in each heme group is atom of iron.

THE ASSEMBLY OF A CELL

In a broad sense the synthesis of structurally uncomplicated molecules is
the first chemical stage in the self-assembly process of a cell. The next stage
is the formation of structurally complex molecules such as proteins. Each
protein molecule has a characteristic three-dimensional structure. Almost
all enzymes are globular proteins, but their peptide chains are coiled together in a highly precise way, not simply tangled like a ball of spaghetti.
Recent work suggests that the three-dimensional coiling and folding of the
protein's polypeptide chain is an automatic consequence of its sequence of
amino acids. For example, the hemoglobin molecule that binds oxygen in
red blood cells consists of four polypeptide chains. The amino acid
sequence in certain portions of these chains evidently creates "sticky"
patches on their surfaces that fit together, holding the four chains in a
specific globular arrangement (Fig. 15).

It is not difficult to extend this line of reasoning to the assembly of
complicated structures such as the cell membrane. Recent research suggests that in the watery interior of living systems layers of lipid molecules of
the right size and shape form automatically for the same reason that
globules of fat form in a pot of stew: the water molecules repel the "greasy"
hydrocarbon chains of the lipids. The lipid layers can then incorporate
those protein molecules whose structure fits the shape of the lipids, resulting
in the formation of a functional membrane (Fig. 16). Biologists now assume
that similar principles govern the assembly of structures like mitochondria
and nuclei, and ultimately the assembly of an entire cell (Fig. 17). This
hypothesis has not yet been proved; the research that could support,
modify, or disprove it has not been done.

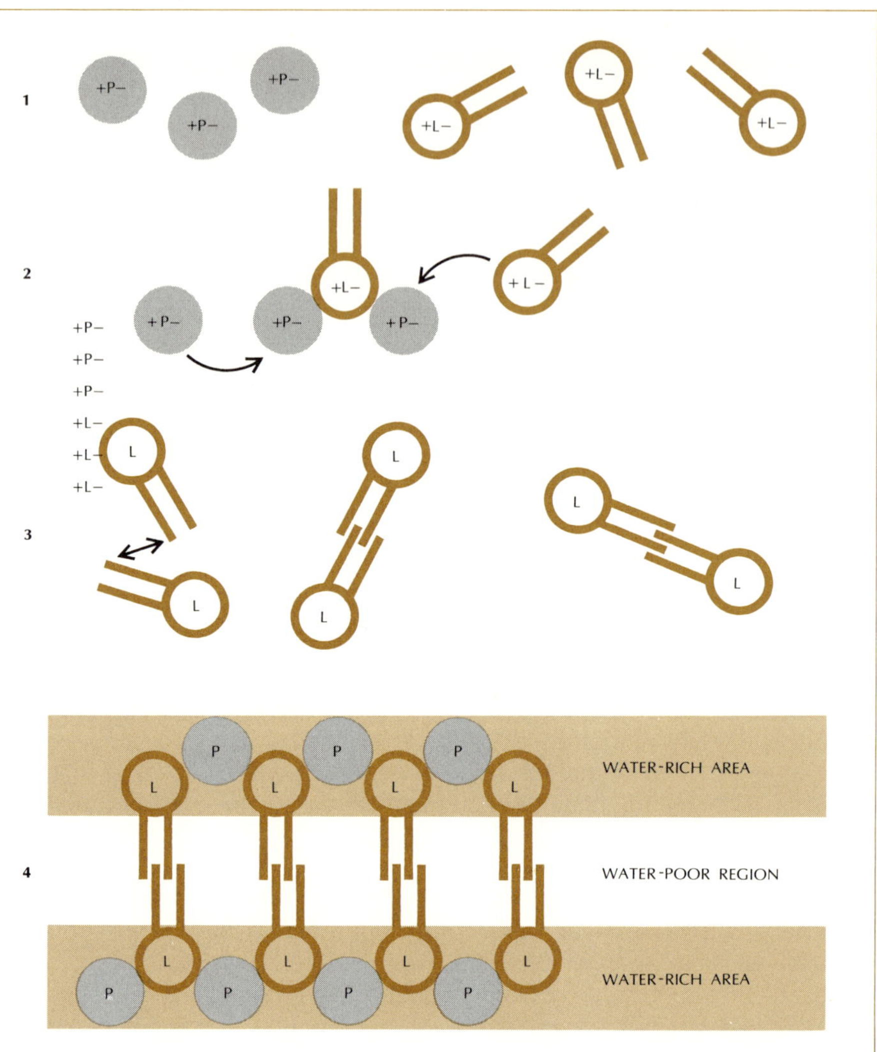

16

Assembly of cell membrane. Protein (P) and phospholipid (L) molecules contain portions which bear electrical charges (1). Unlike charges attract, causing "heads" of phospholipid molecules to align with the protein molecules (2). The noncharged "tails" of the phospholipids attract each other (3), uniting the aggregated molecules into a flat membrane (4).

THE COMPLEXITY OF LIFE

We have discussed digestion, fermentation, respiration, and biosynthesis as separate functions, but in the cell all of these activities proceed simultaneously. Protein may be synthesized while glycogen is being digested; ATP is re-formed in the mitochondria, while in the cytoplasm other ATP molecules are being consumed in the synthesis of large molecules. Certain molecules and ions enter the cell while others diffuse out. And all reactions are under the control of an interlocking hierarchy of control mechanisms.

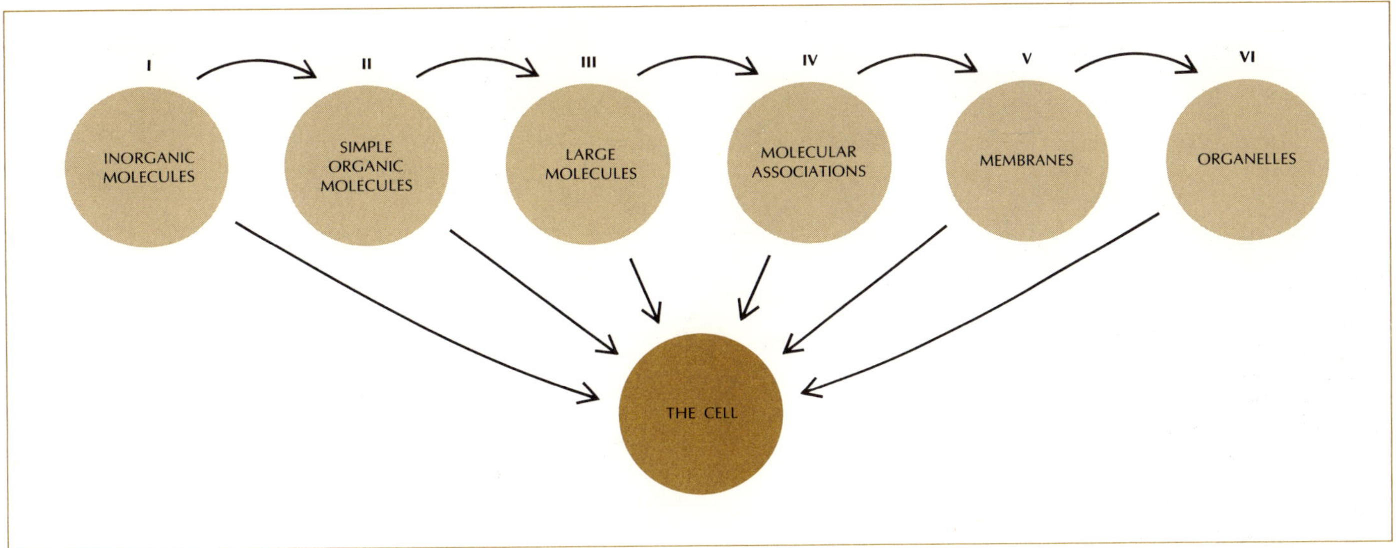

It is no wonder that early biologists were completely baffled by the intricacy of living systems. The chemistry and physics of the day were unable to grapple with such an order of complexity. In despair the early workers concluded that life was beyond explanation. But by tackling problems one at a time, and by reducing very complicated processes to a series of individual chemical reactions, generations of workers have decreased our ignorance and increased our wonder. In the distant future, research should enable man to crack the secret of life. At present the main mystery is just how such a complicated and improbable state of matter came to exist at all.

17
Self-assembly of cell proceeds as simple molecules are changed into ever more complex compounds and finally into structural membranes and organelles.

READINGS

Goldsby, R. A., *Cells and Energy*. Macmillan, New York, 1967 (paper)
> For the introductory student who has a background in chemistry. Presents an overview of the flow of energy in biological processes.

Lehninger, A. L., *Bioenergetics*, 2nd ed. Benjamin, Menlo Park, Calif., 1971
> Difficult for a beginner, but clear and logical discussion of a complex subject.

McElroy, W. D., *Cell Physiology and Biochemistry*. Prentice-Hall, Englewood Cliffs, N.J., 1964 (paper)
> A short introduction to cellular biochemistry; not easy reading.

Pauling, L., *Vitamin C and the Common Cold*. Freeman, San Francisco, 1970 (paper)
> In this controversial little book, a winner of the Nobel Prize in chemistry argues that massive doses of Vitamin C help to prevent the entry of viruses into the cell.

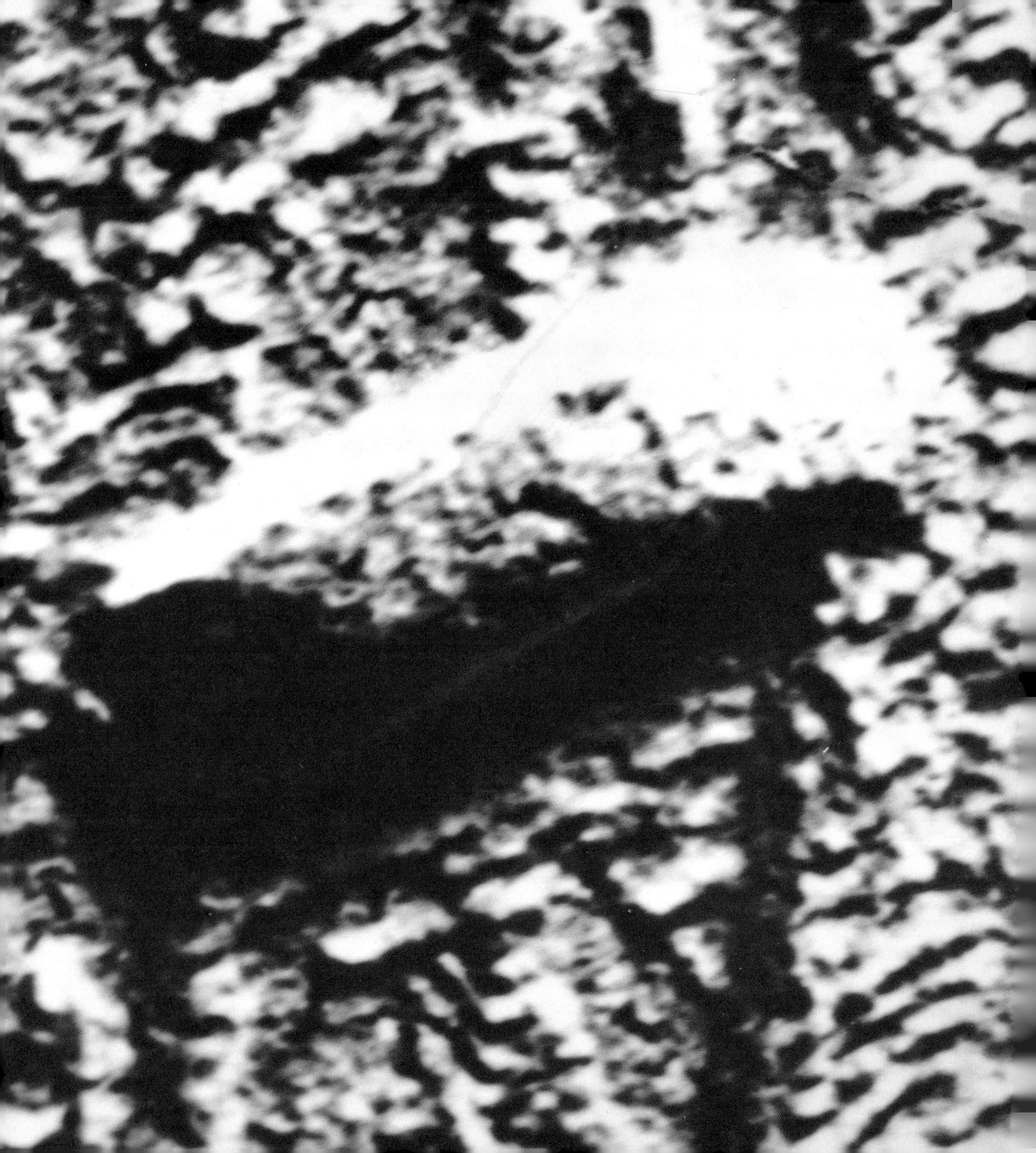

12. THE ORIGIN OF LIFE

How did life on earth begin? Although this question has taunted biologists for centuries, it has only recently become the object of legitimate scientific inquiry. At present we have the beginnings of an hypothesis, but many parts of it are highly speculative.

We recall that until about a century ago most scientists believed that living organisms possessed a Vital Force that enabled their cells to synthesize organic compounds from chemicals in the air, soil, and water (Chapter 3). Belief in Vitalism died slowly because the physical sciences were long unable to explain the complexities of living systems. In 1828, when Wöhler informed his contemporary, the great Swedish chemist Jöns Jacob Berzelius, that he had prepared urea in the laboratory from inorganic precursors, Berzelius refused to accept the implications, acidly remarking that whoever sought fame with urine was bound to end in it.

A worldwide discussion of Vitalism was reopened in 1924 by the publication of *The Origin of Life* by the Soviet biochemist A. I. Oparin. In his brilliant and provocative book Oparin argued that there is no fundamental difference between living and nonliving matter, that both are governed by the same chemical and physical laws. He proposed that conditions on the primitive earth favored the spontaneous synthesis of small organic molecules, followed by a progression of reactions that eventually converted these building blocks into the giant molecules essential to earthly life, including proteins and nucleic acids. In turn, he argued, these lifeless organic molecules somehow assembled themselves by chance into a "protobiont":

◄**Oldest fossil** discovered so far is the rod-shaped bacterium *Eobacterium isolatum.* Found in the Fig-tree chert formation in South Africa, organism is about 3.2 billion years old. Its length is less than one micron.

a precellular organism that might have resembled a modern bacterium.
His theory seemed incredible at the time, but it is now widely accepted as
the basis for current research.

THE GEOLOGIC TIMETABLE

The intellectual difficulties encountered in a search for the origin of life
are enormous. One cannot rule out, for example, the possibility that life
on earth may have resulted from a highly improbable event, such as contamination from a collision with a comet or some other extraterrestrial
source. If life did start in such a way, positive evidence of the earliest forms
will never be found. Much of the evidence written into the early rocks of
the earth's crust has been obliterated by later geological deformation and
heating.

Geologists divide the history of the earth into several eras (Table I). The
most ancient era, the Archeozoic (primal life), stretches from the origin of
the earth to about 2.5 billion years ago. At some time in this remote and
dimly understood era the first living things appeared on earth.

Those primitive organisms have left almost no fossil traces in the
materials of the earth's crust. In fact, the fossil record becomes plentiful
only in the Cambrian period, which opened the Paleozoic era (ancient life).
Precambrian organisms were soft-bodied and few traces of their remains
survive, chiefly because in later geologic ages most Precambrian rocks were
heated to temperatures that destroyed or distorted organic material. A few
Precambrian formations that were not heated provide some important
reference points on the timetable of early life. A formation of flintlike rock
found in South Africa contains microfossils that resemble bacteria and
algae, as well as small amounts of pristane and phytane, compounds produced by the breakdown of chlorophyll (Fig. 1). The age of this formation,

YEARS AGO (MILLIONS)	ERA (AND DOMINANT LIFE FORMS)	PERIOD	EMERGING ORGANISMS
PRESENT	**CENOZOIC ERA** MAMMALS AND MODERN PLANTS	QUATERNARY	FIRST MEN
75		TERTIARY	
230	**MESOZOIC ERA** REPTILES	UPPER CRETACEOUS	FIRST BIRDS
		LOWER CRETACEOUS	
		JURASSIC	FIRST TRUE MAMMALS FIRST FLOWERING PLANTS
		TRIASSIC	
600	**PALEOZOIC ERA** HIGHER INVERTEBRATE ANIMALS	PERMIAN	MODERN INSECTS
		PENNSYLVANIAN	FIRST REPTILES
		MISSISSIPPIAN	
		DEVONIAN	FIRST AMPHIBIANS
1000		SILURIAN	FIRST LAND PLANTS
1900		ORDOVICIAN	FIRST VERTEBRATES
		CAMBRIAN	AQUATIC LIFE ONLY (MARINE INVERTEBRATES)
2500	**PROTEROZOIC ERA** PRIMITIVE PLANTS (ALGAE) AND INVERTEBRATE ANIMALS	KEWEENAWAN	FIRST FOSSIL ANIMALS
		ANIMIKIAN	
3100		HURONIAN	FOSSIL ALGAE & BACTERIA (GUNFLINT CHERT)
		SUDBURIAN	
	ARCHEOZOIC ERA UNICELLULAR LIFE	KEEWATIN	
		GRENVILLE	OLDEST FOSSIL ORGANISMS (FIG-TREE CHERT)
4500	PRIMITIVE CRUST		IGNEOUS ROCKS

Evidence of chlorophyll shows up as two peaks (arrows) in this gas chromatogram of material extracted from the Gunflint chert, which is about 2 billion years old. The peaks on the chromatogram indicate the presence of organic compounds containing from 16 to 24 carbon atoms. Peak A is probably pristane and peak B is phytane, chemical fossils of chlorophyll.

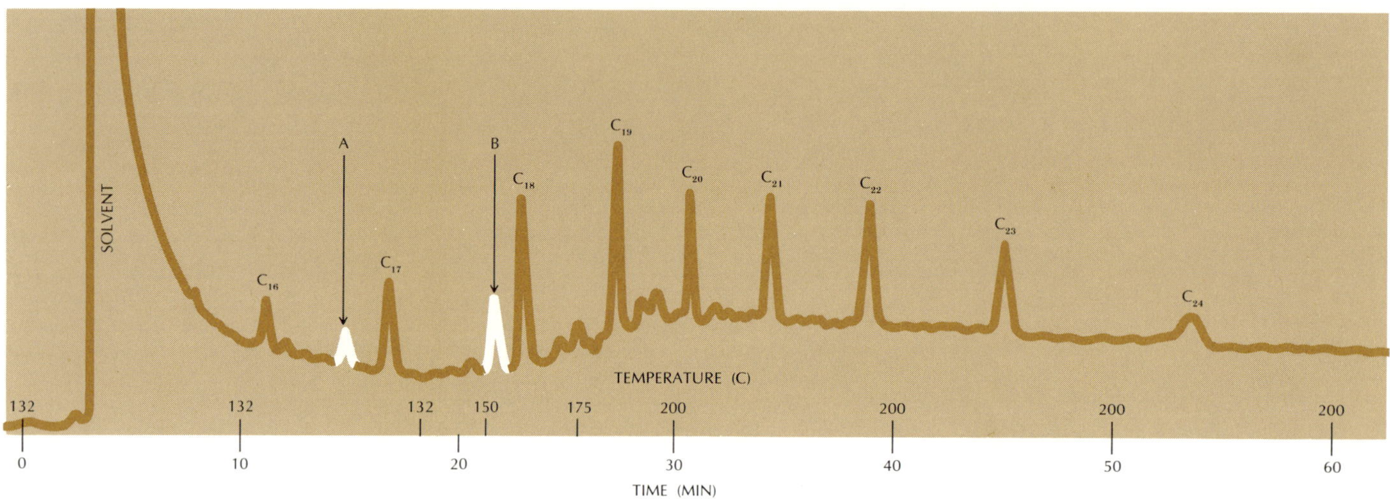

which is known as Fig-tree chert, is put at 3.2 billion years. The Gunflint chert formation in southern Ontario, 1.9 billion years old, contains well-preserved fossils of bacteria and algae (Fig. 2). If the organisms are as old as the rocks in which they lie, and not the result of later contamination, life must have originated much more than three billion years ago, because photosynthetic organisms represent a fairly advanced stage of chemical evolution. Some workers date the origin of life at roughly four billion years ago, only half a billion years after the birth of the planet.

THE BIRTH OF THE PLANET

Current estimates put the age of the Galaxy at 10 to 15 billion years, although the solar system appears to be much younger. According to the most popular theory, about five billion years ago a cloud of matter—the solar nebula—condensed from interstellar gas and dust. The gravitational forces that gave birth to the nebula condensed the cloud still further. The central mass became the sun and the peripheral material gave rise to the

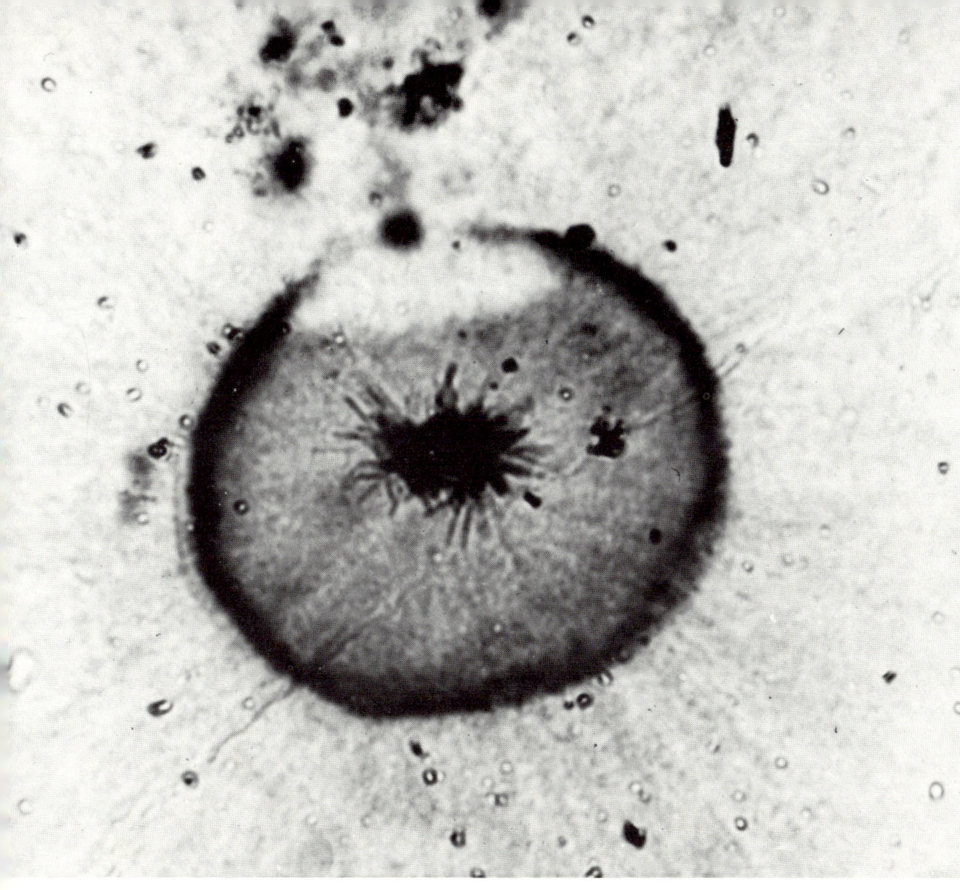

Colony of fossil algae found in Gunflint chert formation in Ontario resembles modern blue-green alga of the genus *Rivularia,* hence its name, *Paleorivularia ontarica.* Magnification about 900 ×.

planets. Several lines of evidence suggest that the earth condensed about 4.5 billion years ago. This was the starting point for molecular evolution, a process that was shaped by the composition of the atmosphere, the temperature, the climate, and the geology of the early earth.

THE PRIMITIVE ATMOSPHERE

Compared to other bodies in the solar system, the earth's atmosphere today is deficient in hydrogen, helium, neon, krypton, xenon, argon, nitrogen, and carbon. During its hot geological infancy, our planet apparently lost large quantities of elements that are gaseous at temperatures of a few hundred degrees. This suggests that the earth's primitive atmosphere is of secondary origin; that is, the atoms that formed the atmosphere were initially combined in solid chemical elements and were later liberated by the heat of the earth.

The exact composition of the early atmosphere is unknown. Almost everyone agrees that free oxygen was absent. One school of thought maintains that the primitive atmosphere consisted mostly of methane, with smaller amounts of ammonia, water vapor, and hydrogen. Other investigators believe the atmosphere was rich in carbon dioxide, with smaller amounts of carbon monoxide, nitrogen, water vapor, and hydrogen. Still others hold that hydrogen, methane and other hydrocarbons, cyanide compounds, ammonia, water vapor, and hydrogen sulfide were all abundant. Recent studies of the atmospheres of our neighboring planets, Mars and Venus, by the Soviet spacecraft *Venera 4* and by the American *Mariner 5* suggest that carbon dioxide might have been the main component, with free nitrogen comprising less than 20 percent, and water vapor and free oxygen being rare.

TEMPERATURE AND CLIMATE

The temperature of the earth depends on the incoming energy from the sun and the energy radiated outward from the atmosphere. But because the composition of the primitive atmosphere is unknown, the heat balance of the early earth cannot be calculated with any degree of certainty.

Most reactions of biological interest occur in aqueous media, and the temperature range of these reactions is bracketed by the freezing and boiling points of pure water: 0°C and 100°C respectively. Freezing virtually suspends biological reactions, and temperatures above 100°C tend to decompose many organic molecules.

At first the earth must have been too hot to have had any liquid water. One theory holds that finally, as the newly formed planet cooled, water vapor began to condense as the first rain. Initially the rain was revaporized before it hit the ground, but soon the surface temperature allowed water to accumulate on land. The primeval rainstorms were nothing like those of today: sheets of water scoured the land clean of all material and swept it into the sea. The water also washed the atmosphere clean of dust and soluble gases and added them to the dissolved matter that was accumulating in the oceans which soon covered most of the globe.

It is with these raw materials and under such physical limitations that the slow process leading to the formation of a living cell began. We shall never know precisely what events or conditions shaped the first organisms. At best we can hope to develop a plausible hypothesis for the origin of life based on studies of modern organisms, scanty historical evidence, and a few assumptions about the environment of the early earth. Following Oparin's theory, we can envision the development of life in three stages: the evolution of organic molecules, the invention of a genetic system, and the origin of cells.

SMALL ORGANIC MOLECULES

Several workers have tried to reproduce in the laboratory the steps that might have led to the creation of small organic molecules early in the Precambrian era. Oparin suggested that the primitive atmosphere consisted of a mixture of gases, particularly hydrogen, but lacked free oxygen. He proposed that energy from the ultraviolet radiation of the sun or from the electric discharge of lightning flashes would facilitate the combination of these chemicals into organic compounds.

The first steps in the experimental verification of Oparin's hypothesis came from the laboratory of Harold Urey and his graduate student Stanley Miller, then at the University of Chicago. In 1953 Miller reported the results of experiments in which he had placed hydrogen, methane, water, and ammonia in an enclosed vessel and had supplied energy in the form of an electric spark. He found many different kinds of amino acids, other organic acids, and even urea in the "soup" that formed in the apparatus. These results were significant not only because a wide variety of organic compounds formed, but also because most of these substances are essential components of the biochemical pathways of modern organisms. Miller's report stimulated similar experiments in laboratories around the world. Some of these procedures are so simple and straightforward that they have since become projects for high school students.

Other workers experimented with somewhat different raw materials and environmental conditions, depending on their ideas about the possible conditions that prevailed on the primitive earth. Some heated their reactants to 1000°C; others performed experiments at room temperature. Some used methane as the only carbon source; others used compounds such as carbon monoxide, carbon dioxide, formaldehyde, and hydrogen cyanide. Some used ultraviolet light as the energy source; others used X-rays, ionizing radiation from cyclotrons, ultrasonic vibrations, or simply sunlight. All of these experiments successfully produced organic compounds: amino acids, sugars, alcohols, and even simple peptides. The

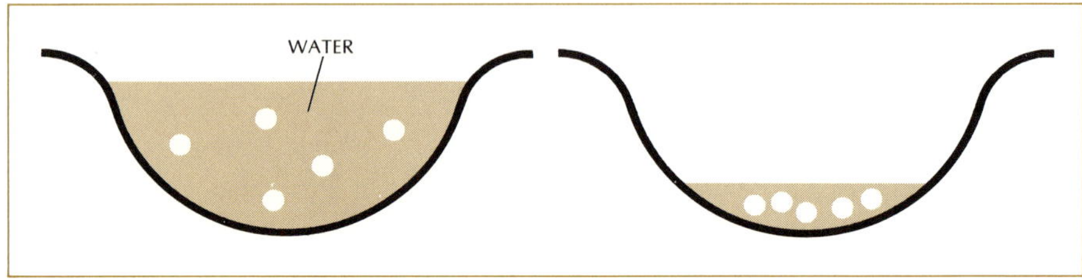

3
Evaporation could increase the concentration
of small molecules (white disks) in small tidal
pools by removing some of the water.

same types of semirandom reactions could have occurred on the primitive
earth. But to produce life, order and direction had to be imposed on the
reactions.

GIANT MOLECULES

The second stage in molecular evolution was the synthesis of the giant
molecules that control the chemistry of life: nucleic acids and proteins.
The spontaneous synthesis of nucleic acids from preexisting components
is chemically and thermodynamically quite understandable. But it is almost
impossible to imagine the random synthesis of specific nucleic acids that
would in turn govern the production of the many enzymes necessary to
catalyze all of the reactions of a living cell. This difficulty is one of the
major stumbling blocks of modern theory. Scientists no longer doubt that
nucleic acids were synthesized spontaneously from nonliving matter, but
the possible steps in this synthesis represent a crucial gap in our under-
standings of the origin of life.

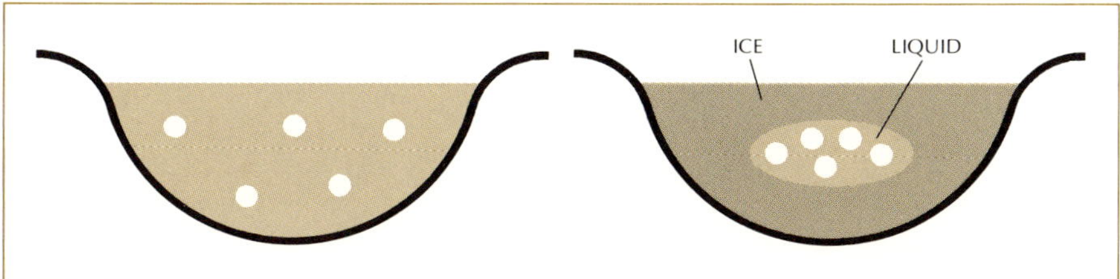

4
Freezing of pools would concentrate organic molecules in small pockets of unfrozen liquid (right).

5
Adsorption on surface of solids such as clay can enhance synthesis of organic compounds by bringing small molecules together under conditions that favor chemical reaction.

THE PRIMORDIAL SOUP

At best, the primitive seas were a very dilute "soup" of small organic molecules created by nonbiological processes. Some concentration mechanism was probably necessary to enhance the speed and probability of chemical reactions. The most likely mechanisms seem to be evaporation, freezing, and adsorption (Figs. 3, 4, 5).

In tidal pools or in volcanic crater lakes evaporation of water could increase the concentration of organic and inorganic substances. Freezing would have a similar effect: when an aqueous mixture freezes, salts and organic molecules are not locked into the crystal structure of the ice but remain in a liquid state. The adsorption of organic molecules on surfaces of clay particles must also have been important in prebiological synthesis. In some cases the clay minerals themselves could have acted as catalysts.

We can imagine that an entire lake or pool functioned more or less like an organism. Inorganic substances entered the system and were transformed into small organic molecules, which in turn were polymerized into larger molecules. Some of the already synthesized polymers might have acted as catalysts, speeding up certain reactions. The polymers themselves

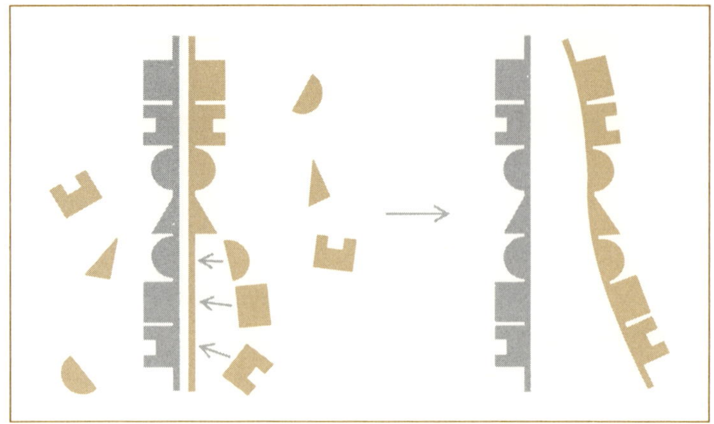

6
Self-duplicating molecule can serve as template for the synthesis of an exact copy (top) or a complementary structure (bottom). Chemical building blocks in this schematic diagram represent small organic molecules.

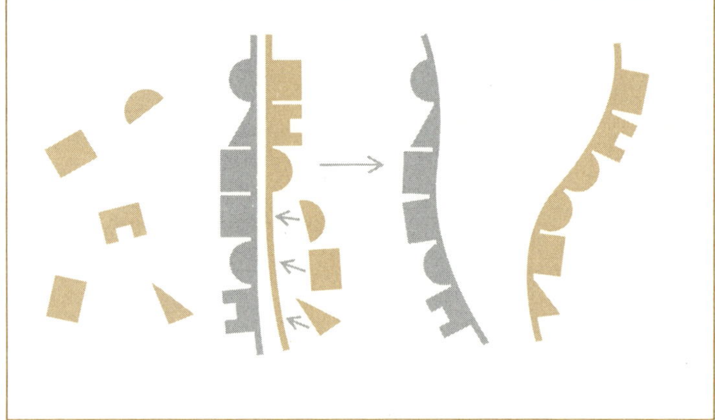

might have incorporated new components into their structure, growing and becoming more complex. At some point one of the giant molecules must have acquired the capacity to trigger the synthesis of other molecules like itself. This was the first major step toward life, and it probably occurred at the molecular rather than at the cellular level. Today we think of organisms as being composed of cells. But it is unrealistic to suppose the simultaneous creation of a genetic system and the packaging of that system in a membrane-bounded structure as complicated as a cell. Membranes and true cells were a much later development.

SELF-DUPLICATING SYSTEMS

In the underpopulated, nutrient-rich waters of the early earth, existence would have involved very little competition. The first self-duplicating systems might have been linear polymers that served as templates for organizing smaller molecules into copies or complements of themselves

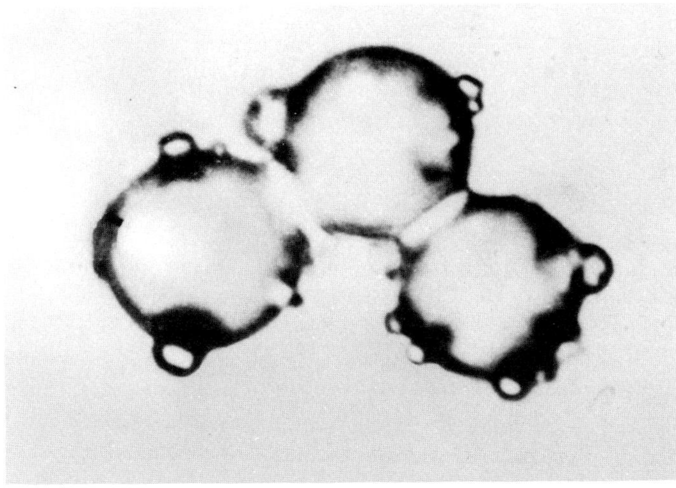

7
Proteinoid microspheres produced in a test tube. The larger microspheres are "reproducing" by the production of small buds. These will separate from the parent spheres, grow, and in turn produce buds of their own.

(Fig. 6). Inorganic ions or small organic molecules might have acted as catalysts: not very efficient catalysts compared with modern enzymes, but there was little competition and an abundance of time. What did it matter that it might have taken centuries to assemble a single replica?

Acting over the course of countless millennia, many random, reproducible "mistakes" in copying could have produced a rich assortment of polymers in the primordial broth. The next step could have been the appearance of new catalysts that enhanced certain "useful" reactions. Perhaps among the original polymers there were polynucleotides that not only served as templates for replicating themselves, but could also catalyze, in a limited way, the organization of amino acids into polypeptide chains. Such a chemical system could easily have evolved into the genetic mechanisms found in modern cells. We now think that the genetic code is similar in all living organisms; in terms of molecular genetics there is only one form of life on earth.

A rival hypothesis holds that the first living systems were aggregates of organic molecules, tiny droplets bounded by a thin film. Known as coacervates, these droplets show some of the properties of cells. Sidney Fox of the University of Miami has demonstrated that coacervates can form under conditions similar to those on the early earth. Fox and his assistants placed dry amino acids in a cavity in a piece of volcanic lava. They heated the lava chunk at 170°C for a few hours, then allowed it to cool and washed it in a hot salt solution. The washings from the lava contained large numbers of droplets, which he called microspheres (Fig. 7).

Microspheres resemble true cells in that they are spherical, relatively stable, and do not spontaneously disintegrate. By adjusting the acidity of the solution in which the microspheres form it is possible to produce spheres with single and double membranes. If the diffusion pressure of the solution is changed the spheres swell or shrink indicating that the membrane is at least to some extent differentially permeable, an essential property of the membranes of true cells.

Although their structure does not even vaguely approach the complexity of true cells, coacervates could have served as containers in which new organic compounds could have been synthesized. Simple chemicals could have remained within the spheres long enough and in high enough concentrations, to have enhanced reactions that might not have occurred in the external environment. When the spheres reached a critical size they might have burst, enriching the environment with the products of these reactions, the raw materials for the synthesis of newer and more complex molecules. So far, the difficulty with this model is that it does not incorporate a genetic system—a mechanism for reproduction and mutation—and this is a crucial weakness.

The origin of cells is the greatest gap in present theories of the origin of living organisms. It will undoubtedly persist for some time because so little is known about the complex molecular organization of cells and their membranes. It is a highly speculative and possibly futile exercise to postulate the origin of a structure that we do not yet understand. Nevertheless, recent experiments on cell reproduction have led to the hypothesis that the cell is a self-assembling system; that the right molecules, synthesized in the right order, will combine spontaneously by purely physical processes to form the membranes and organelles of a new cell. This idea will probably shape the research on the origin of cells during the next decade.

PRIMITIVE CELLULAR ORGANISMS

Most workers believed that the first true organisms were aquatic, unicellular, and that they lacked a true nucleus; in other words they were prokaryotic cells, like those of modern bacteria. And like many modern prokaryotes, they were heterotrophic, deriving metabolically useful energy by breaking down preformed organic molecules. The external soup of raw organic compounds would have bathed these organisms with nutrients.

As the number of organisms increased at an accelerating pace, the storehouse of food in the primordial soup was depleted faster than it could be replenished. Some organisms probably evolved the capacity to synthesize

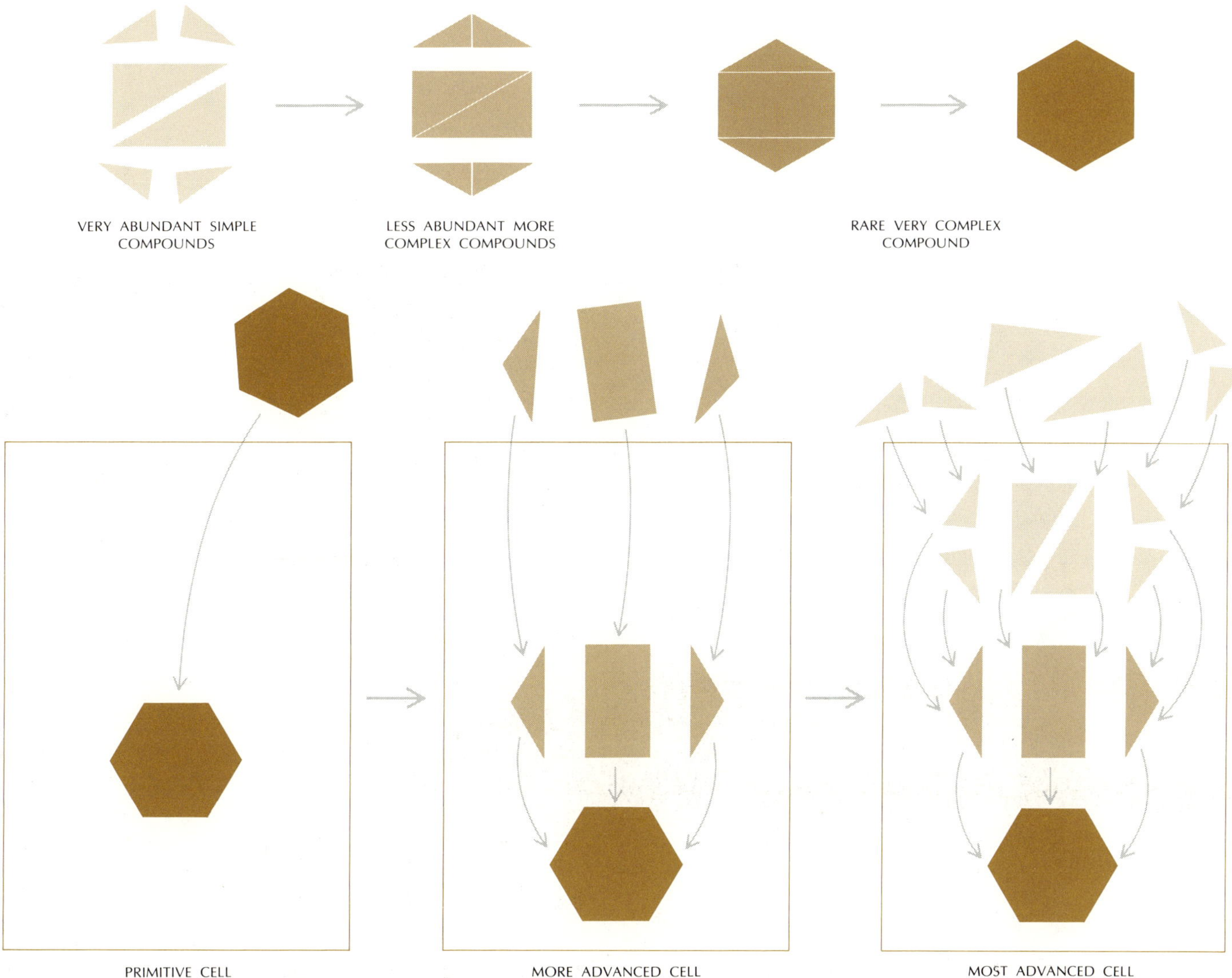

8
Evolution of synthetic pathways. In primordial seas complex compounds were formed by the spontaneous combination of simple compounds (top). The most primitive cells relied on these rare complex compounds as a source of energy and as building blocks for cell components (bottom left). Advanced cells (center and right) eventually evolved the capability to make all the complex compounds they needed from simple molecules that were abundant in the environment.

a few of the organic compounds that less advanced organisms had to obtain from the environment. The capacity for limited biosynthesis reduced their dependence on the external supply of special molecules. When an essential nutrient disappeared from the soup, an organism that could synthesize it from available precursors would gain a decided survival advantage. The synthetic pathways in modern cells probably evolved backwards, one step at a time, beginning with the end product of the pathway (Fig. 8).

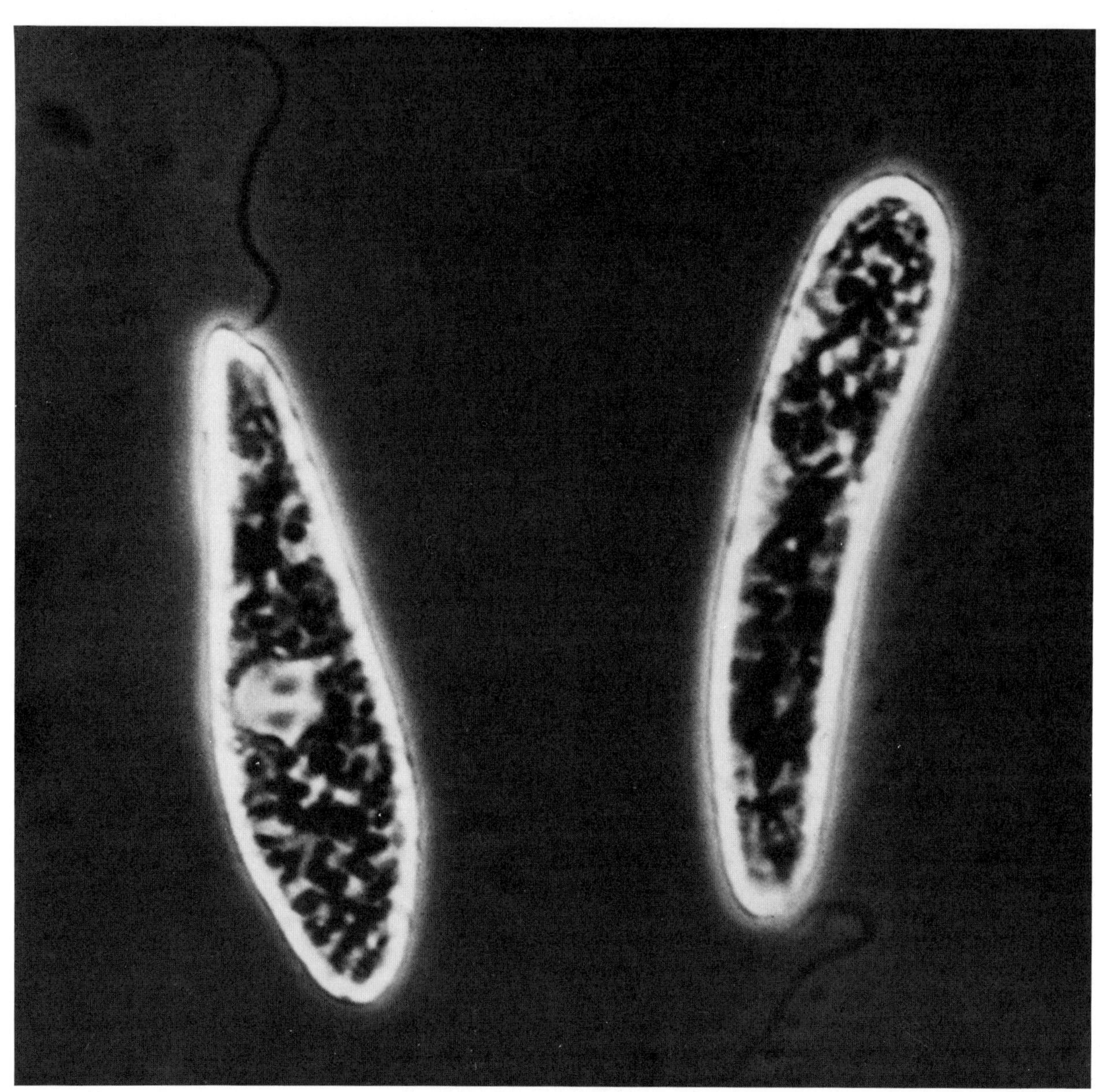

9

PHOTOSYNTHESIS AND THE OXYGEN REVOLUTION

The dependence of organisms upon the nonbiological supply of organic compounds could not persist indefinitely. Unless life could tap some new and abundant source of energy it would reach a dead end. The development of pigment molecules was one of the great advances of evolution. Coupled to energy-transfer molecules, they enabled organisms to trap the energy of the sun and to use it in simple metabolic processes. A primitive photosynthetic organism could function heterotrophically when there was no light, but could supplement this process with photosynthetically produced biological energy when light was available. Living examples of such organisms persist today among the photosynthetic bacteria. Another example—although by no means a primitive organism—is the protozoan *Euglena gracilis,* commonly found in stagnant pools receiving drainage from agricultural land (Fig. 9).

Thus the process of photosynthesis probably appeared rather late in the evolution of single-celled forms because it would have conferred substantially increased survival value only when the nutrients in the primitive seas had been depleted. Subsequent innumerable improvements in the photosynthetic process eventually led to a "modern" system that could split water, combine the hydrogen with carbon to make organic molecules, and release molecular oxygen as a by-product.

The release of molecular oxygen into the environment by advanced photosynthetic organisms initiated a new and spectacular round of evolution. The availability of oxygen opened the way for the development of aerobic respiration (Chapter 11) with its more efficient breakdown of food molecules. Aerobic organisms obtained more useable energy from their food than anaerobic organisms. The increased energy supply contributed to the development and maintenance of more complex structures and control mechanisms.

The production of oxygen by photosynthetic organisms also changed the atmosphere from an open window to a protective screen. Molecular oxygen released by primitive aquatic plants diffused from the water into the atmosphere. Eon by eon its concentration slowly increased. In the upper atmosphere some of the oxygen molecules were transformed by ultraviolet radiation into molecules of ozone (O_3), a gas that is transparent to visible light but opaque to deadly ultraviolet radiation.

In time a protective envelope of ozone 14 to 15 miles thick surrounded the earth. Aerobic organisms, previously restricted to an eternally submerged existence where ultraviolet rays could not penetrate, could spend short periods out of the water. As they developed mechanisms to prevent dessication they could prolong these periods on land. Thus oxygen, one of the by-products of photosynthesis, made it possible for animals to colonize the land, and it undoubtedly set the stage for the explosive evolution of animal life that marked the dawn of the Cambrian period, about 600 million years ago.

LIFE FROM NONLIFE

When did life on earth actually begin? What events changed a nonliving aggregation of chemicals into an organism? We shall probably never know the answers. Some workers believe that, given certain conditions on the primitive earth, the eventual emergence of life was inevitable. They stress that life is a natural consequence of the evolution of matter. Still we do not really know how highly organized a structure must be to be alive, to reproduce, and to evolve.

There is little hope of resolving this problem by extrapolating backwards from living organisms. Even the simplest organisms alive today are highly evolved, complicated, and sophisticated systems. They probably bear no more resemblance to the first living things than the aircraft in Leonardo's sketchbooks bear to modern jetliners. On this planet the record of the early stages of development is forever erased, and the first explorations of the moon have yielded no evidence of early life. Perhaps some primitive stage in the evolution of life is preserved on another planet, or in a remote outpost of our Galaxy—a biological antique awaiting discovery by space explorers of the future.

READINGS

Blum, H. F., *Time's Arrow and Evolution.* Princeton University, Princeton, 1968 (paper)

> The relationship between the second law of thermodynamics and the process of evolution. Difficult but rewarding.

Ehrensvard, G., *Life: Origin and Development.* University of Chicago, Chicago, 1962 (paper)

> Some of the biochemical possibilities that may have led to the origin of life and early evolution. Not easy reading.

Handler, P. (ed.), *Biology and the Future of Man.* Oxford University, New York, 1970

> Chapter 5 is an excellent essay on the origins of life by five research workers in the field, headed by N. H. Horowitz of Caltech.

Keosian, J., *The Origin of Life.* Reinhold, New York, 1964 (paper)

> Brief discussions of the ideas of spontaneous generation, Oparin's notions on the origin of life, and the uncertainties of the basic concepts in this field.

Oparin, A. I., *Genesis and Evolutionary Development of Life.* Academic Press, New York, 1968

> A logical step-by-step presentation of Oparin's arguments on the chemical and physical processes leading to the formation of life from nonliving matter.

Shklovskii, I. S., and C. Sagan, *Intelligent Life in the Universe.* Dell, New York, 1966 (paper)

> Extremely thorough, learned, and fascinating treatise on the fitness of the universe for life, and of man's efforts to discover life on other planets.

13. THE RISE OF MODERN ORGANISMS

The impressive antiquity of the earliest fossils suggests that the immense gulf between molecules and organism was bridged in a primal surge of evolutionary invention. No more than a billion years after the birth of the planet, Archeozoic seas harbored the first signs of life: primitive cells without nuclei and undoubtedly incapable of either mitosis or sexual reproduction. During the next two billion years or so the pace of evolutionary innovation slowed, primarily because of the genetic conservatise of the prokaryotic cell. Many modern species of blue-green algae, for example, are living fossils, structurally almost identical to ancestors that flourished in Precambrian seas. Less than a billion years ago, according to one theory, another great surge of evolution began with the appearance of a new kind of one-celled organism. It was the eukaryotic cell, the basic building block of all higher organisms. Lynn Margulis of Boston University has argued very persuasively that this new type of cell originated as a commune of prokaryotes living symbiotically within the boundaries of a single "cell" membrane. Gradually this association evolved into hereditary symbiosis. Organelles such as mitochondria and chloroplasts might be the distant descendants of bacteria and blue-green algae that once lived independently. The theory has been advanced and discarded more than once in past decades, but it no longer seems unreasonable. Certainly some animal

◀**Primitive bird,** *Archaeopteryx,* dates from the Mesozoic era. It shows clearly reptilian features, particularly the long bony tail. In this fossil impression the head is bent back, so that the jaws and teeth are not clearly visible. Birdlike features include feathers and well-developed wings.

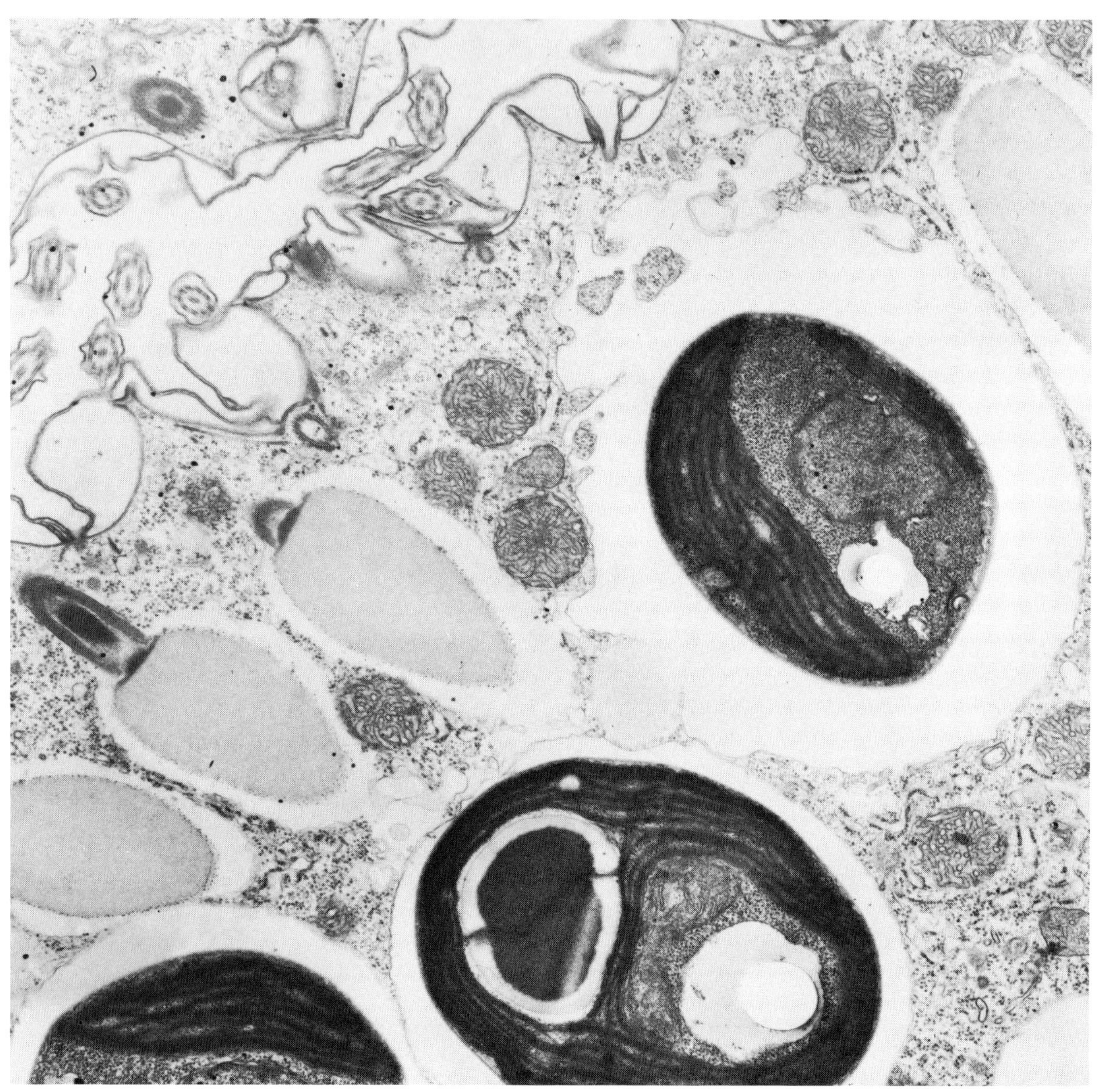

◀**Photosynthetic algae** of the genus *Chlorella* live symbiotically within the modern one-celled *Paramecium bursaria* and supply some of its food. Algae appear as dark ovals in this electron micrograph. Magnification about 19,000 ×.

organisms living today, such as the corals of tropical seas, derive at least part of their nourishment from photosynthetic algae that live within their cells (Fig. 1).

The Bitter Springs chert formation in central Australia, roughly one billion years old, contains the earliest microfossils of eukaryotic cells: the green alga depicted in Fig. 2. Primitive eukaryotic cells capable of sexual reproduction, respiration, and photosynthesis were probably the ancestors of all higher plants and animals. These ancestral cells vanished long ago, but not before they ignited an explosion of diversification. Sexual reproduction, with its recombination of parental genes, eventually gave rise to cells that could aggregate into tissues and organs and perform specialized functions. For almost two billion years all organisms had been more or less alike, but the emergence of the eukaryotic cell changed all that. The thrust of the next surge of evolution was toward organisms of increasing size, complexity, and diversity. By the dawn of the Cambrian period, which opened the Paleozoic era roughly 600 million years ago, the seas teemed with a rich variety of life that included the prototypes of all modern organisms except higher animals and plants. The earliest organisms and many of the intermediate ones have vanished, leaving behind only a few fossils in the rocky crust of the earth. Nevertheless, a rough picture of the early stages of evolution can be assembled from indirect evidence: by extrapolating backward from younger fossils, and by studying the relationships among the modern descendants of ancient organisms.

Such evidence led Lamarck and Darwin to propose rival theories of evolution. Lamarck's ideas have been largely discarded and Darwin's modified, but Darwin's theory stands as the greatest revolution of 19th-century natural philosophy, and as one of the central unifying ideas of modern biology.

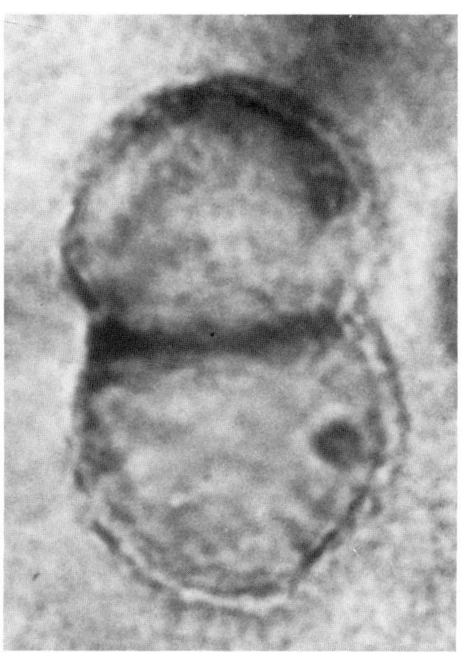

First eukaryotic cells appeared more than a billion years ago. Photomicrograph shows cell division of the green alga *Glenobotrydion*. Dark nuclei are visible at far right.

LAMARCK

The French naturalist Jean Baptiste de Monet, chevalier de Lamarck, proposed a theory of evolution in 1809. Even at that early date he regarded evolution as an established fact and he concentrated on explaining how it occurred. Lamarck classified living organisms like the rungs of a ladder, with the simplest forms at the bottom and the most complex at the top. According to his theory, simple organisms arose spontaneously, and slowly evolved into higher, more specialized creatures by adapting to different environments. Lamarck argued that the mainspring of evolution was the inheritance of acquired characteristics.

He observed that a structure such as a man's arm grows brawnier as a result of frequent exercise, and that muscles shrivel and deteriorate if they are not used, for example, if the arm is in a cast for many weeks. He assumed incorrectly that such characteristics were inheritable; that the son of a blacksmith could inherit the massive arms of his father. By the same logic, a kangaroo might have evolved from some ancestral four-footed animal that began to use its tail and hind legs to jump rather than to walk. While these parts of the body became overdeveloped, the forelimbs would have degenerated from lack of use. The parents would have passed these traits along to their offspring, which would have developed them further and in turn passed them along to their offspring. After hundreds or thousands of generations, this process would have given rise to the modern kangaroo, with its tiny forelimbs and massive hindquarters.

Biologists now know that acquired characteristics do not modify the genetic makeup of the parent; they cannot be transmitted to the next generation and hence can play no role in evolution. Although Lamarck's ideas about the mechanism of evolution have been thoroughly discredited, they continue to mislead some laymen even today, just as they misled Lamarck's intellectual successor, Charles Darwin.

DARWIN

Born the year Lamarck published his theory (1809), Charles Darwin was the son of a physician. Darwin grew up in an atmosphere of learning, books, and wealth. His father sent him to study medicine at Edinburgh University. After two years of learning little and enjoying it less, Charles left medical school and entered Cambridge, where he enjoyed the social life immensely. By intense cramming just before examinations he managed

3
H.M.S. Beagle (center), the sailing vessel that carried young Charles Darwin on his voyage, was painted in 1841 by Owen Stanley as she lay at anchor in Sydney harbor.

to graduate with his class in 1831. The H.M.S. *Beagle* was about to begin a five-year exploratory voyage around the world and Darwin signed aboard as a naturalist (Fig. 3). In December the ship set sail and Darwin's life work had begun.

The *Beagle* made numerous stops in its voyage around the world (Fig. 4) and from each habitat Darwin collected a wealth of sample orga-

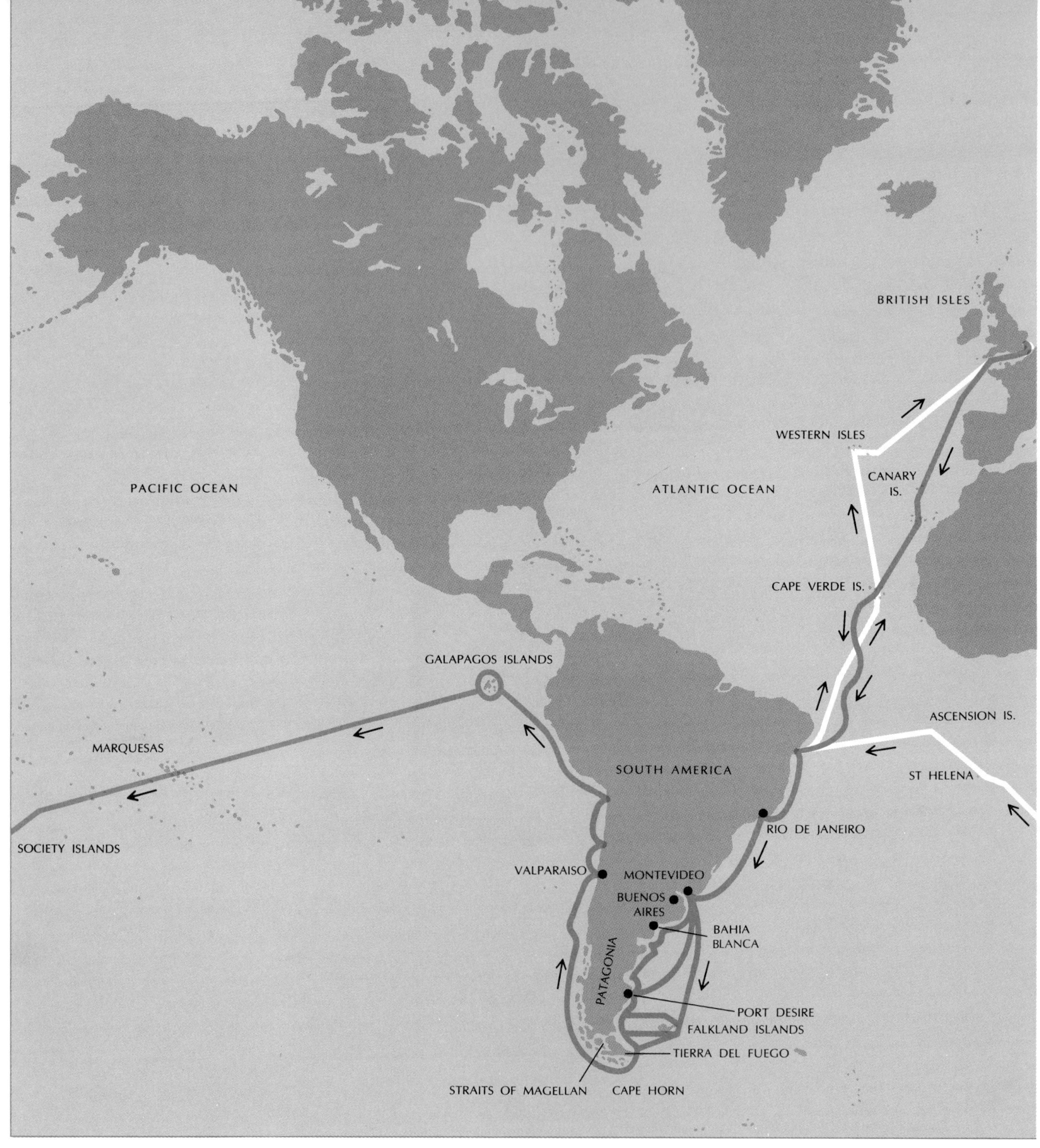

BRITISH ISLES
WESTERN ISLES
CANARY IS.
PACIFIC OCEAN
ATLANTIC OCEAN
CAPE VERDE IS.
ASCENSION IS.
GALAPAGOS ISLANDS
MARQUESAS
SOUTH AMERICA
ST HELENA
SOCIETY ISLANDS
RIO DE JANEIRO
VALPARAISO
MONTEVIDEO
BUENOS AIRES
BAHIA BLANCA
PATAGONIA
PORT DESIRE
FALKLAND ISLANDS
TIERRA DEL FUEGO
STRAITS OF MAGELLAN
CAPE HORN

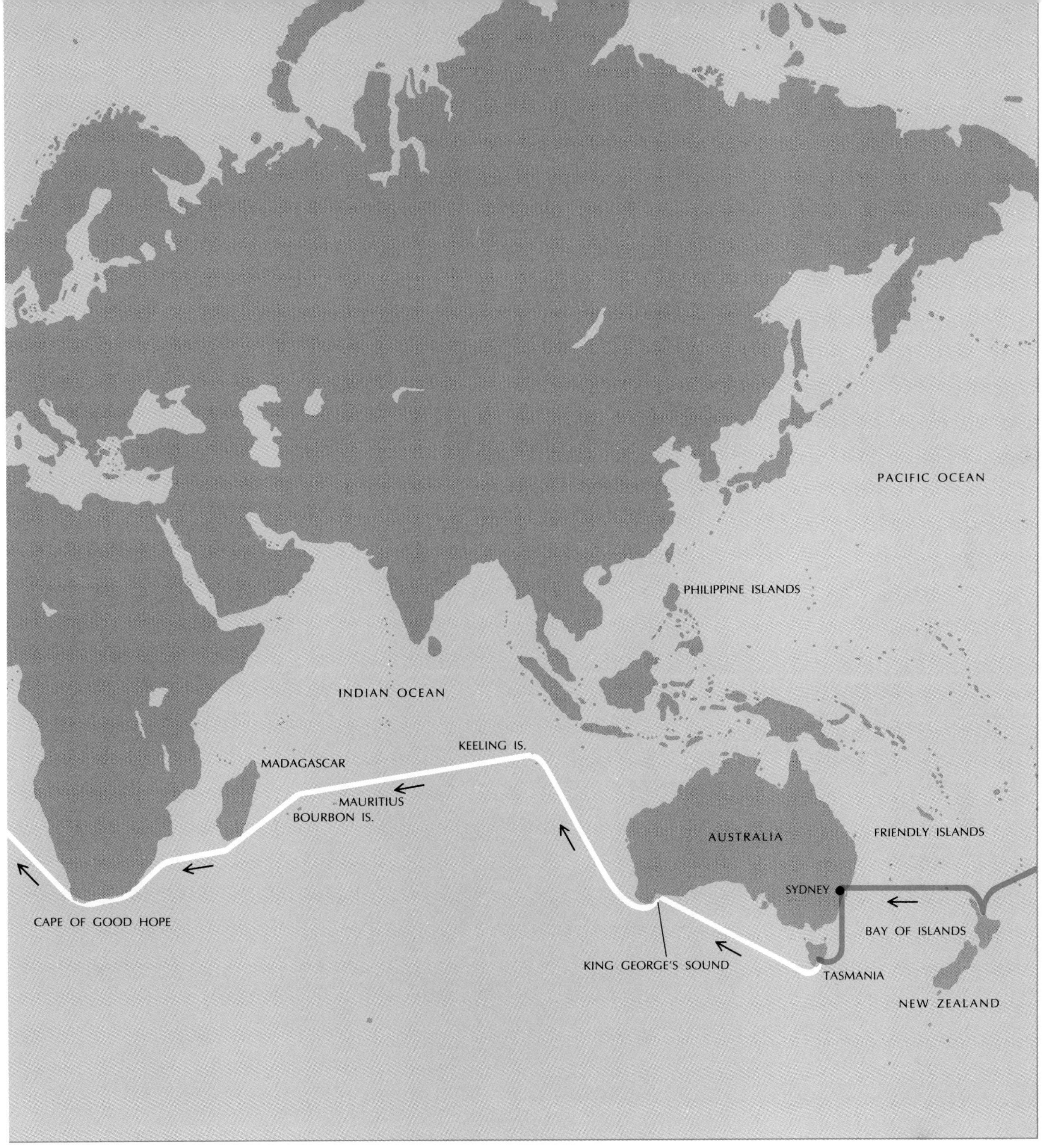

Voyage of the Beagle took Darwin around the world in five years, with long stopovers in South America and the Galapagos Islands. Outbound voyage is shown in color, homeward leg in white.

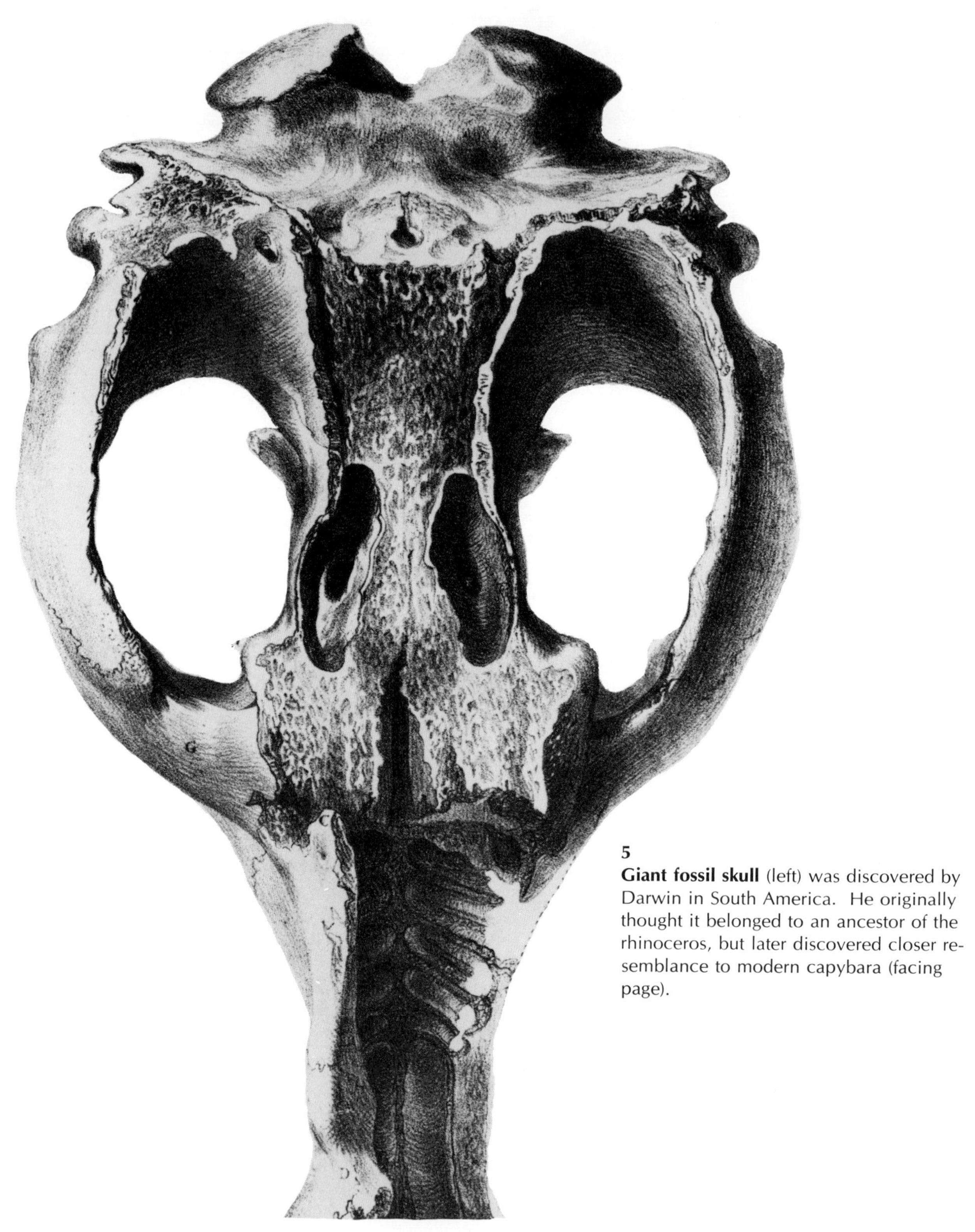

5
Giant fossil skull (left) was discovered by
Darwin in South America. He originally
thought it belonged to an ancestor of the
rhinoceros, but later discovered closer resemblance to modern capybara (facing
page).

nisms. Two localities provided especially significant evidence for Darwin's increasing conviction that different species were the products of thousands of years of evolution. One of these localities was the small village of Bahia Blanca in Argentina. Here Darwin found a bed of exotic fossil skeletons which included the skull of a huge animal that Darwin at first thought was a relative of the rhinoceros. Later, he discovered that the giant skull looked more like the head of the capybara, a small South American rodent (Fig. 5). He also found an enormous jawbone that was similar to, but many times larger than, the jaw of the modern sloth. These discoveries confirmed Darwin's belief that modern animals are descended from extinct prehistoric ancestors.

6

Galapagos finches. Darwin discovered that the Galapagos Islands harbored 14 species of finch, which apparently evolved from a single species that migrated from South America.

DARWIN'S FINCHES

Darwin's second great find came in the Galapagos Islands. Straddling the equator about 600 miles due west of Ecuador, these 13 bleak, sun-baked volcanic islands are unlike any others on earth. Here, Darwin found living exotic animals: 200-pound tortoises, four-foot marine lizards, and cormorants that could not fly. Most important, he found a small group of drab birds which he was able to identify as finches. To his complete surprise, the islands contained 13 different species of finch (Fig. 6). Where did they come from? Darwin believed that they were all descended from a single pair that had somehow been blown from South America in the distant past.

In 1836 the *Beagle* reached England and Darwin was home. With him he brought 24 field notebooks crammed with information that would occupy his mind for the next two decades. During those years Darwin was greatly influenced by the ideas of another biologist: Alfred Russell Wallace had arrived independently at conclusions roughly similar to Darwin's, and together they announced their new theory of evolution. Darwin elaborated on it and eventually, in 1859, published his monumental book, *On the Origin of Species by Means of Natural Selection, or the Preservation of Favoured Races in the Struggle for Life.*

DARWIN'S THEORY

Darwin documented his theory with an overwhelming mass of data, but based it on a few fundamental concepts. First, he observed that the individual organisms of a species differed from one another in small but recognizable ways. For example, some were larger and stronger than others, and some showed a slightly different pattern of instinctive behavior. Second, he knew that a population of organisms invariably produced many more offspring than could possibly survive and grow to adulthood. He concluded that the individuals best adapted to their environment would survive and reproduce; those less well-adapted might not. Any small inheritable variation that would endow an individual with a slight advantage in the struggle for existence would tend to be preserved; variations with little or no survival value would disappear. Thus, each succeeding generation would differ more and more from its original ancestor. In time the ancestor might become completely extinct, and its modern descendants might be different enough to be called a new species. He called this mechanism natural

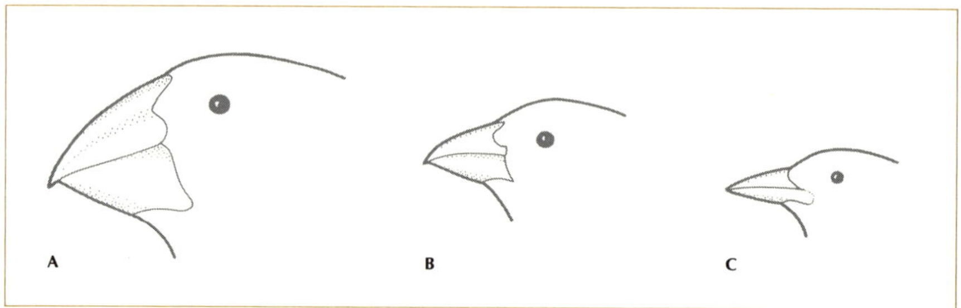

selection, and it became a keystone of his theory. For Darwin, natural selection explained the relationship between the giant South American fossils and the modern capybaras and sloths: the contemporary forms were direct descendants of the ancient ones. It also explained the origin of the 13 species of finch: small random variations in the offspring of the original finches enabled them to invade habitats that were unavailable to their parents. Some of the present species of finch on the Galapagos Islands feed on insects, some on seeds; some feed primarily in the trees, some on the ground; some eat only small food particles, some can ingest large ones (Fig. 7). One species has acquired the complex ability to use ''tools'': with pieces of twig it probes inside tree trunks for insects. Each species is adapted to exploit to the fullest a specific habitat different from those of its relatives.

Darwin recognized that an organism's habitat includes not only the physical factors of heat, light, water, wind, and soil, but also the presence of other organisms. The tool-wielding finch would probably never have evolved on the South American mainland because it would have been thrown into immediate competition with native long-billed birds like the Jacamar and the Scythbill which also probe trees and air plants for insects. To compete successfully for food, the finch would have had to develop and master a complex pattern of instinctive behavior, the use of the twig, in a single abrupt burst of evolution. To Darwin, this seemed highly improbable. He had observed that evolution proceeds in a series of small steps over very long periods of time. On the Galapagos Islands there were no native long-billed birds to compete with the finch during the time that it acquired and

◄**Beaks of Galapagos finches** are adapted to specialized feeding habits. Some of the finches feed on large seeds (A), some on small seeds (B), and others on insects (C).

perfected its new pattern of behavior. The availability of a vacant habitat favored the evolution of a new species adapted to exploit it. Finches that instinctively used twigs would have no rivals in their search for food, thus enhancing the survival value of the new characteristic.

NATURAL SELECTION

The idea that natural selection is the mainspring of evolution was one of the most powerful of Darwin's insights, but from his time to our own it has frequently been misinterpreted. Many writers came to see Darwin's "struggle for life" as a cruel and bloody combat: "nature red in tooth and claw." In the popular mind, natural selection came to be thought of in terms of "survival of the fittest" and "the law of the jungle," notions that have been used to justify some of the deplorable aspects of human conduct, from economic exploitation to genocide. In nature the struggle for existence rarely involves combat. Plants never fight, and animals of the same species rarely kill one another; in fact, animals armed with nature's most lethal weapons—horns, claws, fangs, beaks, venom—have evolved elaborate patterns of instinctive behavior for settling disputes among themselves without killing or injuring their opponents (see Chapter 14).

In the modern view, natural selection is a positive, creative force. The organisms best adapted to their environment usually find food and mates more easily and can take better care of their offspring, ensuring that a large number of them survive to reproductive age. The struggle for existence is really a matter of differential reproduction. The most successful organisms are simply those that leave behind the greatest number of offspring. In other words, nature "selects" inheritable variations that favor the survival of more offspring. During thousands of generations, this trait will become progressively more abundant in the population and the nature of the species will gradually change (Fig. 8).

Man is the only species with the ability purposefully to modify its own evolution. At present the genetic heritage of our species is deteriorating with each generation because our medical competence and our proper humanitarian concern for living people allows humans to live who, without special medical care or social compensation, would die. In other ages, many individuals with serious genetic defects such as diabetes would either not survive to reproductive age, or their children would die in infancy. In either case, the frequency of the harmful genes in the population's gene pool would remain small. But now that insulin enables diabetics to survive and lead relatively normal lives, and hospitals can safely deliver the children of diabetic mothers, the incidence of diabetes in the population is continually increasing. Theodosius Dobzhansky of Rockefeller University foresees a future in which insulin injections may be as common as taking aspirin tablets. The impression that medicine is conquering disease is thus an illusion. We often are merely controlling the symptoms of diseases as we increase the load of genetic defects we transmit to future generations.

Present social policies accelerate the trend toward the accumulation of genetic defects insofar as we help everyone to live according to his need and "to reproduce according to his greed, or lack of foresight, skill or scruple." Those are the words of H. J. Muller, a geneticist whose pessimism about man's evolutionary future is shared by many of his colleagues. In man, as in other species, Darwinian fitness does not refer to the survival of the strongest, healthiest, or most intelligent; it refers only to reproductive success. In modern urban society, the fittest could include a feebleminded woman who produces a dozen feebleminded children.

In order to slow down our genetic deterioration, we must soon control both the quantity and the quality of human reproduction. There are at least three drastically different approaches to genetic control in human populations. One is a direct attack on the harmful gene by physical or chemical techniques that would delete it, cause it to mutate to a normal form, or replace it with a good gene. Among geneticists the consensus

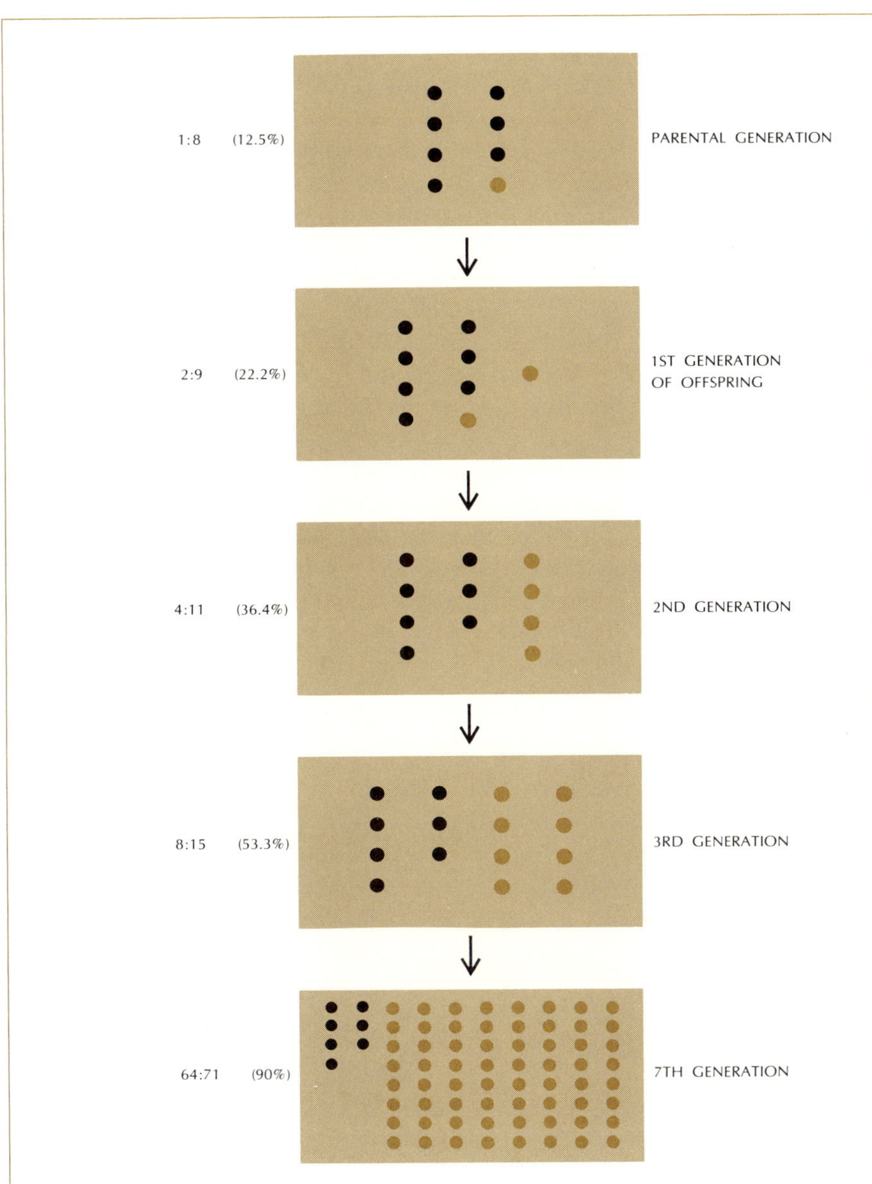

8
Natural selection favors the gene pool of organisms that produce the greatest number of surviving offspring. In this diagram, best adapted individuals (color) leave twice as many offspring. By the seventh generation, the favored type comprises 90 percent of the population.

seems to be that this approach, which is often called "genetic engineering," will not become technically feasible in the near future. Another approach is selective breeding, aimed either at breeding out harmful traits or at breeding in desirable ones. Possible techniques include voluntary birth control under the advice of heredity clinics that would screen potential parents; voluntary sterilization; artificial insemination of a woman by a genetically superior donor other than the husband; and artificial inovulation, sometimes called foster pregnancy, which is the reverse of artificial insemination. According to James F. Bonner of the California Institute of Technology, in another decade or two a third possibility may exist—that of producing infants who are genetically identical to living adults. This would be done by asexual reproduction; that is, by replacing the nucleus of a fertilized egg with the nucleus of a cell from a donor, presumably a genetically outstanding individual, and implanting it to develop normally in the uterus of a volunteer mother. This is the technique used by Gurdon at Oxford to produce a generation of genetically identical toads, as described in Chapter 9. In effect, this method produces an identical twin of the donor, one generation younger. Stanford geneticist Joshua Lederberg has suggested the advantages of applying this procedure to humans: "If a superior individual—and presumably, then, genotype—is identified, why not copy it directly, rather than suffer all the risks . . . involved in the disruptions of recombination and sexual reproduction?"

The subject of eugenics touches some of man's deepest emotions, and it raises issues that transcend the boundaries of science. The central issue, however, is not whether man should tamper with his own evolution—we are already doing that by default—but by what means, and toward what ends. Those who want to explore this question are referred to the readings at the end of this chapter, particularly to the book by Dobzhansky (who tends to be skeptical about eugenics); the article by Lederberg (who isn't); and the paperback by Ramsey (who emphasizes the ethical questions).

THE MODERN THEORY

Darwin's theory itself has evolved considerably in the last hundred years, chiefly as a result of the rise of modern genetics. Darwin's greatest failure was his inability to explain the origin of inheritable variations. For lack of a better explanation, Darwin espoused Lamarck's incorrect idea of the inheritance of acquired characteristics. Curiously enough, the correct ex-

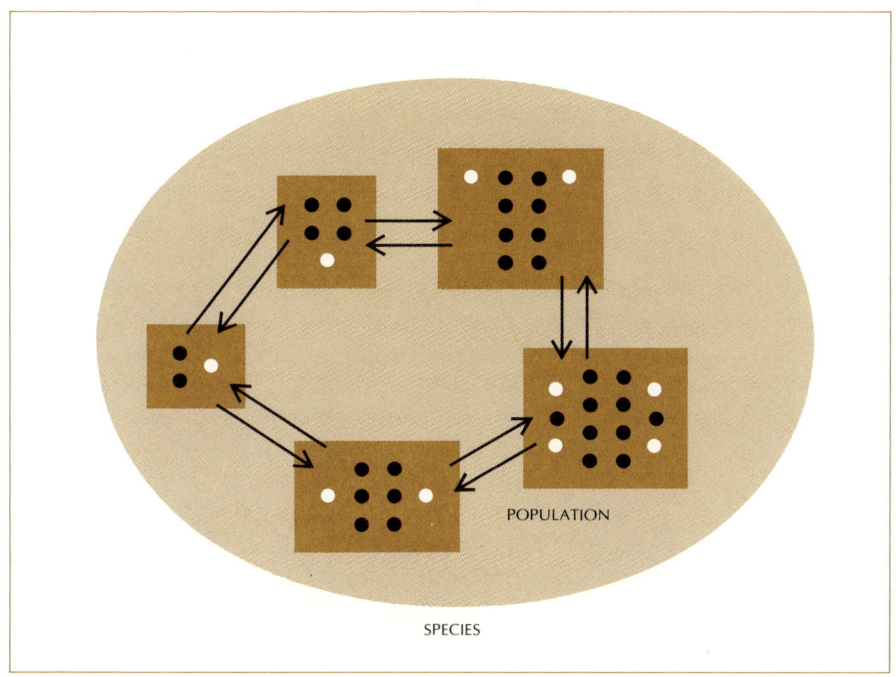

9
Gene flow. Interbreeding among populations reshuffles the genes of a species. Here colored blocks represent populations of individuals and circles represent genotypes. A mutant gene appearing in one population may soon spread to all populations. Arrows indicate occasional interbreeding.

planation depended on the work of one of Darwin's contemporaries, Gregor Mendel; but the significance of Mendel's work was not recognized until years after Darwin's death.

Genes are the raw material of evolution. Basically, evolution is a change in the genetic composition of a population, under the influence of natural selection. In the strict Mendelian sense, a population is a group of interbreeding individuals. Within large populations there are many subpopulations whose members interbreed preferentially. Among humans Americans are one example; the Chinese are another.

All the genes of all the members of a species comprise a "gene pool." Interbreeding continually reshuffles the genes in the pool; any mutant genes that arise in one population eventually may become distributed throughout all the populations of that species as a result of occasional mating between individuals from different populations. Within a species, this gene flow is gradual and, barring any barriers to breeding among subpopulations, inevitable (Fig. 9).

THE HARDY-WEINBERG LAW

The central axiom of modern evolutionary theory is based upon a statistical law worked out independently in 1908 by G. H. Hardy in England and G. Weinberg in Germany. The Hardy-Weinberg law states that within a large Mendelian population, if mating is random and if there are no mutations, the relative frequencies of each gene allele tend to remain constant from generation to generation. In such a case, the rate of evolution is zero.

Almost no population fulfills all the conditions of the law. First, mating patterns are usually not random; the best-adapted members find mates more easily. Second, gene mutations occur constantly and spontaneously, although at a relatively low rate (Chapter 8). Third, some interbreeding populations, such as a flock of sheep, are too small to be governed by the Hardy-Weinberg statistics, just as a coin tossed ten times does not always result in five heads and five tails, although it would come much closer to this ratio if it were tossed a thousand times. Nevertheless, the Hardy-Weinberg Law provides a useful baseline for studies of population genetics and evolution.

THE MECHANISM OF EVOLUTION

Variation in heredity depends upon two processes: mutation and sexual recombination (Chapter 8). Sexual recombination reshuffles existing genes into new patterns. Mutation creates new genes by altering the structure of DNA.

Because it is the only known source of new genes, mutation is essential to evolution. All new mutations are tested in the marketplace of natural selection. Beneficial mutations will become increasingly abundant in the gene pool, as shown in Fig. 8. But an overwhelming majority of mutations are harmful, and sooner or later may be weeded out of the gene pool in several ways. Dominant lethal mutant genes result in the death of the embryos or the immature organisms in the following generation; some result in the birth of sterile offspring. These mutant genes will not be passed on to future generations. Nonlethal mutant genes, however, may be passed from generation to generation. In these cases the harmful gene is eventually eliminated from the population by differential reproduction. If, in each generation, a parent carrying the harmful mutant gene has one less surviving offspring than a parent carrying the normal allele, the frequency of the mutant gene in the population will continually diminish.

In spite of the deleterious effects of most mutant genes, mutation provides a mechanism that enables a population to adapt to gradual changes in the environment. The gene pool of any reproducing population contains variant genes, and the environment exerts a pressure that "selects" or favors the survival of the traits most adaptive to that environment. A change in the environment would favor another set of genes. Almost all genetic traits are constantly being selected for or selected against, often in very subtle ways. In time, even a very slight positive or negative selection pressure can drastically transform the nature of an entire species.

10
Peppered moth exists in two forms, one light, one dark. Photograph at left shows both forms resting on a lichen-covered tree trunk; light-colored moth is almost invisible. View at right shows both forms on the soot-covered trunk of an oak tree.

THE PEPPERED MOTH

A case history of the effect of natural selection on a modern population has been recorded by H. B. D. Kettlewell of Oxford University. Kettlewell studied the peppered moth, a species which, like several others in the insect world, is composed of two distinct forms, one light and one dark (Fig. 10).

Collections of these moths taken before the middle of the 19th century contained only light-colored specimens; a single dark-colored freak was collected near the industrial city of Manchester in 1845. Since that time the number of dark forms observed and collected has steadily increased; and today the population of peppered moths near Manchester is composed of about 99 percent dark forms. Biologists assumed that this drastic shift in color was related to the rise of industry in the Manchester region. Factory chimneys dumped into the air enormous quantities of soot which settled as a grimy film across the landscape. Perhaps caterpillars feeding on soot-covered leaves had ingested some compound that increased the biosynthesis of melanin (the dark pigment) in the adult moths. But dark moths also appeared, although very rarely, in the open countryside far from industrial pollution. Experiments in which caterpillars were fed soot-covered leaves failed to produce any dark-colored moths.

Another suggestion was that the dark color was the result of a genetic mutation. But mutations are relatively rare events and certainly could not occur at a rapid enough rate to explain a complete reversal of the ratio of light to dark moths in only a century.

Kettlewell postulated that the mechanism involved was mutation followed by natural selection. In areas distant from factories and in the Manchester region before the coming of the factories, the trees and rocks upon which the moths would alight were covered with light-colored lichens. Against the lichens, a light-colored moth would be almost invisible to predatory birds, but a mutant dark moth would be a highly conspicuous meal for hungry birds. The chances that a dark moth would reach maturity and reproduce were slim indeed. Thus the mutant gene for dark color suffered genetic death. The coming of the factories changed the appearance of the countryside. Deposits of soot prevented the growth of lichens on rocks and darkened the surfaces of vegetation. Against this background, the light-colored moths would be conspicuous to predators while the dark ones would be well camouflaged, and hence more likely to survive and produce offspring. The mutant gene for the production of melanin conferred a survival and therefore a reproductive advantage. As a result its frequency in the population increased. In only a century, a population of light-colored moths had evolved into a population of dark ones. Kettlewell was able to substantiate his hypothesis by releasing dark and light moths in both sooty and clean habitats. In sooty habitats light ones were rarely recaptured in

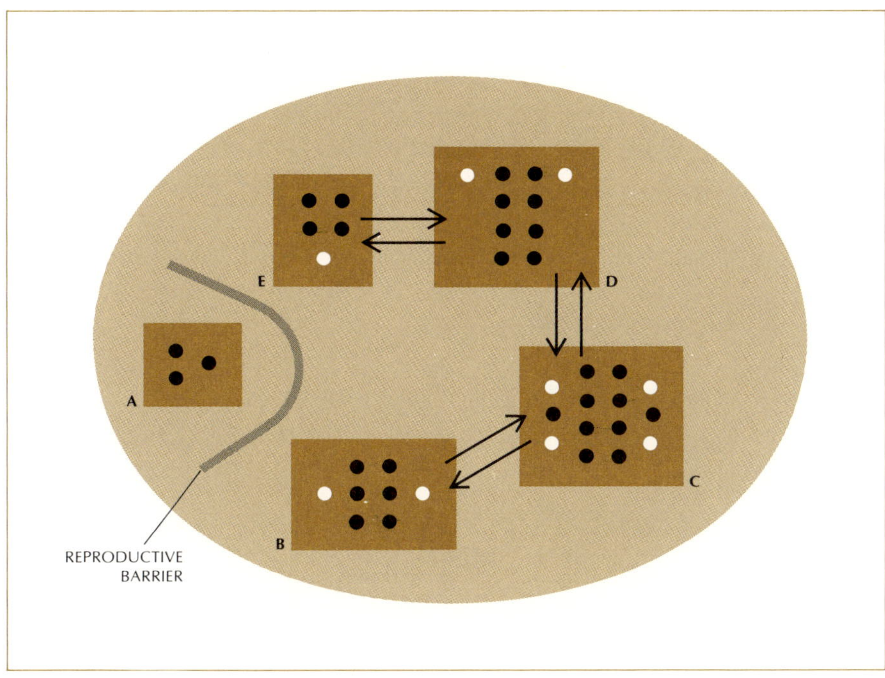

11
Reproductive barrier (curved bar) prevents populations from interbreeding and prevents the flow of genes from one population to another. Establishment of barrier may result in the creation of new species.

traps; presumably most had been eaten. In clean habitats the dark ones were rarely recaptured.

Kettlewell's case history illustrates how a single type of gene mutation may cause an inheritable variation which, in a certain environment, confers a higher survival value than the original gene. The reprodutive rate of the mutant form is higher than that of the original form, and the frequency of the mutant gene increases in the population. In the case of the peppered moth, a change in gene frequency led to the predominance of only a new type of individual; in other cases it can lead to the production of a new species.

THE ORIGIN OF SPECIES

The splitting of a single species into two new species requires reproductive isolation. Otherwise, any mutant genes that might arise in one population would eventually become distributed throughout the entire species by breeding among populations.

Speciation takes place in at least four stages. The first is the establishment of a reproductive barrier which prevents members of one population from breeding with members of another, but allows free breeding within the two populations (Fig. 11). Such a barrier might arise as a result of sudden

Windflowers are plants that grow in Alaska and in the Rocky Mountains of western United States. Botanists believe that these two populations were separated from one another by a continental glacier during the ice ages. Drawings show two forms of same species.

accidental migration, as was the case with the first Galapagos finches. Other barriers result from subtle long-term changes in the environment, as was the case with a North American plant called the windflower (Fig. 12). Populations of these plants now grow in central Alaska and 2000 miles away, in the Rocky Mountains of the western United States. Botanists assume that the original range of these plants extended from Alaska, through Canada, into the western United States. About 20,000 years ago, during the peak of the last Ice Age, a huge continental glacier spread over parts of Alaska, most of Canada, and part of the northern United States. The ice completely erased the vegetation between the unglaciated portions of Alaska and the continental United States, thus separating the Alaskan and Rocky Mountain populations.

When two populations cease to interbreed, all gene flow between them stops. But, although a reproductive barrier is essential for the formation of a new species, it does not make speciation inevitable. The common skunk cabbage (*Symplocarpus foetidus*) grows abundantly in China and also in the northeastern United States. According to geological and fossil evidence these populations have been separated for millions of years, but they have

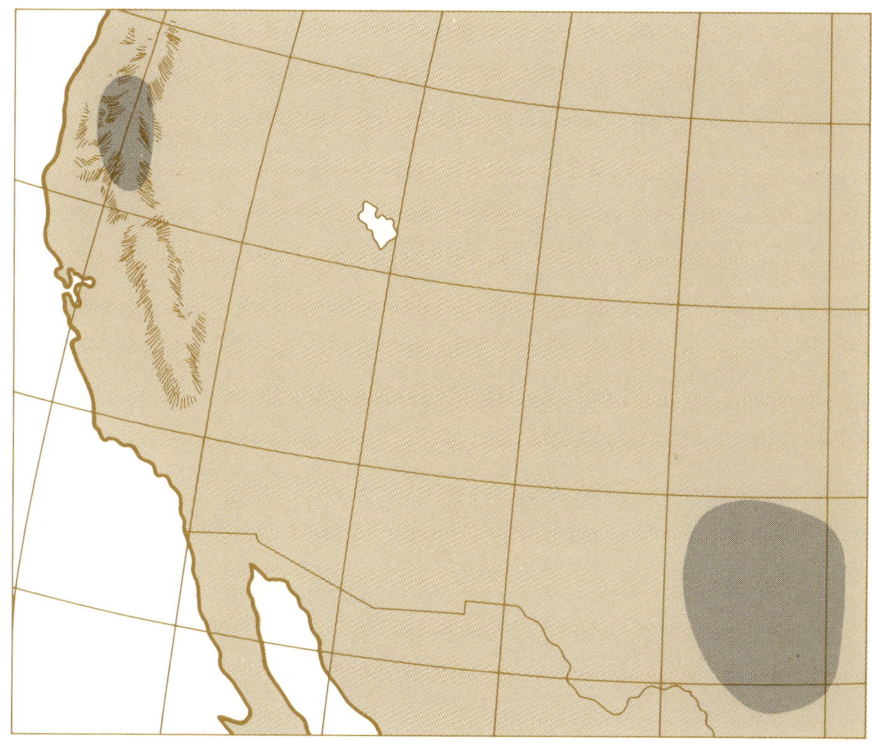

13
Suture zones are regions where separate populations of closely related organisms have recently come in contact. Map shows two such zones (dark tint): the Cascade-Sierra zone in California and Oregon, and the Central Texas zone.

not evolved into separate species. Significant changes in gene frequency occur only when new mutations confer a reproductive advantage on individuals. The habitats of the skunk cabbage in China and the United States are very similar and hence do not tend to select for different mutants. This conclusion is supported by the occurrence of the same distribution pattern of many other plants.

The gene frequencies of two isolated populations will change if they occupy areas where the environmental conditions differ. Over many thousands of years, after the accumulation of favorable mutations or new combinations of genes, the members of two populations can become distinctly dissimilar from each other in anatomy, physiology, and/or behavior.

At an early stage of evolution, if the populations are reunited, their members can still interbreed freely and produce abundant offspring. Today, pairs of closely related "species" are found breeding freely in "suture zones" such as the slopes of the Sierra Nevada Mountains and the Cascade Range (Fig. 13). Biologists believe that a recent warming trend in the climate of this area has allowed the far western species, previously restricted to the lowlands on the western side of the mountains, to move up the mountainsides and to spill through the passes onto the eastern slopes, where they encounter closely related species previously restricted to the Great Basin and the Rocky Mountains. The present interbreeding populations in this suture zone include species of such diverse organisms as poplars (trees), juncos (birds), martens (mammals), and butterflies (insects). Eventually, this interbreeding may eliminate the evolutionary differences between the eastern and western species.

A third stage in the process of speciation occurs when the members of the two populations have become so different that, even if they do mate, the matings yield very few offspring, or offspring poorly adapted to the environment and hence unlikely to reach maturity and produce offspring of their own. Recently the ranges of two species of narrow-mouthed toad, *Microhyla olivacea* (found in Texas, Oklahoma, and northern Mexico) and *Microhyla carolinensis* (found in the entire southeastern United States) have met along a suture zone in western Texas and Oklahoma. Interbreeding between the populations does occur but natural selection does not favor the hybrids. Individuals of the two species in this overlap zone have evolved distinctive differences in mating calls, the call of *Microhyla carolinensis* being much lower and shorter. This change in behavior drastically reduces the number of matings between individuals of the two species and consequently restricts the gene flow between them. The two species of *Microhyla* remain distinct and continue to evolve along separate pathways.

Once two populations reach the third stage, the gene flow between them is so severely limited that the next step is almost inevitable. In this fourth stage the accumulated differences between the two populations are so great that no viable offspring are produced by matings with members of the other population; or in a more extreme case, the members of one population simply will not or cannot mate with any of the members of the other.

It should now be clear why there is only a single species of man on earth today. Man's habit of wandering over the surface of the globe, freely mating with any individuals he may encounter, prevents the establishment of completely isolated populations. No group of men has had the chance to remain separated from the rest of humanity for the time necessary for the evolution of a new species.

BIOLOGY AS HISTORY

Evolution has created the vast diversity of modern organisms as well as countless other species that have long since vanished. Before they disappeared, many of the vanished organisms left their traces, as fossils, in the rocks and minerals of the earth's crust. The fossil record is the history of the interaction of life and earth, a history of the fate of biological inventions like a seed or a wing at the hands of natural selection.

Any trace of an ancient organism is a fossil. Some fossils are skeletons, some are petrified pieces of dead plants. Others are shells or fragments of organisms embedded within the sediment of the ocean floor, or in tar, ice, amber, or quicksand. Still others are footprints or other evidence of activity left in soft mud or sand that later solidified into rock.

Fossils can usually be dated according to the rock strata in which they lie. In sections where the earth's crust has not been subject to extensive geologic disturbance the deepest layers of rock are the oldest. In general the deep layers are inaccessible, but in some areas strata of various ages, including the oldest, have been exposed by upheavals or cut open by rivers, as in the Grand Canyon (Fig. 14).

Rocks contain built-in atomic clocks: radioactive elements like uranium-238 and potassium-40. Radioactive uranium, for example, spontaneously decays into lead at a slow but precisely known rate. By comparing the relative amount of naturally occurring lead with the lead derived from the radioactive disintegration of uranium in a sample of rock, geologists can estimate its age within an error of about 10 percent. Fossils composed of dead organic material not more than 25,000 years old can be dated directly by measuring the concentration of radioactive carbon-14 in the sample. This is the same isotope used in biochemical tracer experiments.

Paleontology, the study of the fossil record, is a whole branch of science in itself. Table I in the last chapter sets forth the five major eras in the history of the earth: the Archeozoic (primal life), Proterozoic (primitive life), Paleozoic (ancient life), Mesozoic (middle life), and Cenozoic (modern life). The last chapter traced the story of life from the origin of the earth, roughly 4.5 billion years ago, to late in the Proterozoic era, about one billion years ago—an interval that stretches across most of the earth's existence.

Numbers in the billions are so large that they have little meaning for most of us. Table I in this chapter (page 384) scales down the history of the

14
Rock layers of the earth's crust are clearly visible in this photograph of the Grand Canyon. In general, deepest layers are the oldest. Tinted areas indicate location of oldest rocks in photograph. These layers, which date to the Proterozoic era, were laid down more than 600 million years ago.

Geologic calendar compresses the entire 4.5 billion years of earth history into a 30-day month. One day equals 150 million years. Life appeared about 500 million years after the earth's origin—that is, sometime during the fourth day on this calendar. Oldest known fossils lived 3.1 billion years ago (tenth day), but life was not abundant until the opening of the Paleozoic era (27th day). The four most recent days are enlarged at bottom. The entire history of civilized man, from ancient Sumer to the present, comprises only 30 seconds.

ORIGIN OF EARTH

1 · 2

3 · 4 FIRST LIVING THINGS? · 5 · 6 · 7 FIRST CELLS? · 8 · 9

10 OLDEST FOSSILS (GUNFLINT CHERT) · 11 · 12 · 13 · 14 · 15 · 16

17 · 18 · 19 · 20 · 21 · 22 · 23

24 FIRST EUKARYOTIC CELLS? · 25 · 26 · 27 ABUNDANT LIFE · 28 · 29 · 30

PRESENT

ARCHEOZOIC ERA   PROTEROZOIC ERA   PALEOZOIC ERA   MESOZOIC ERA   CENOZOIC ERA

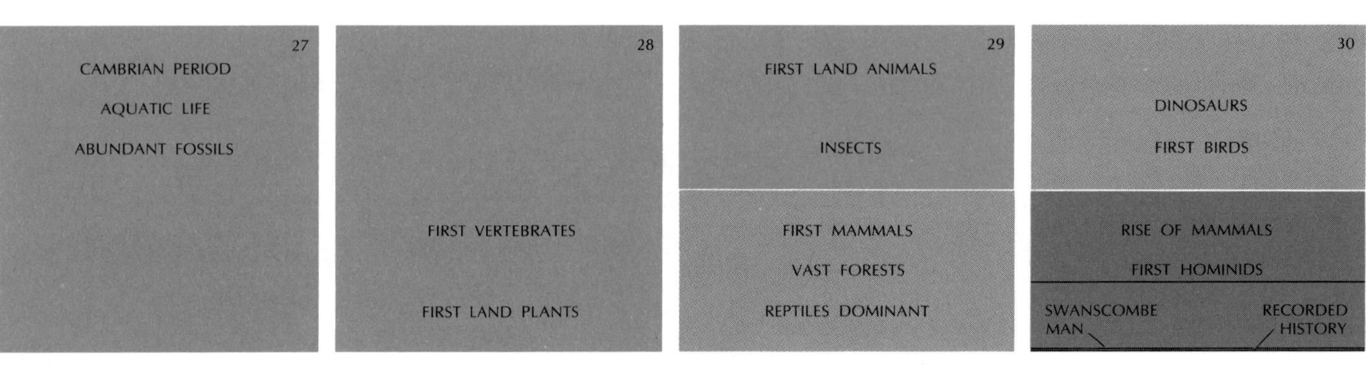

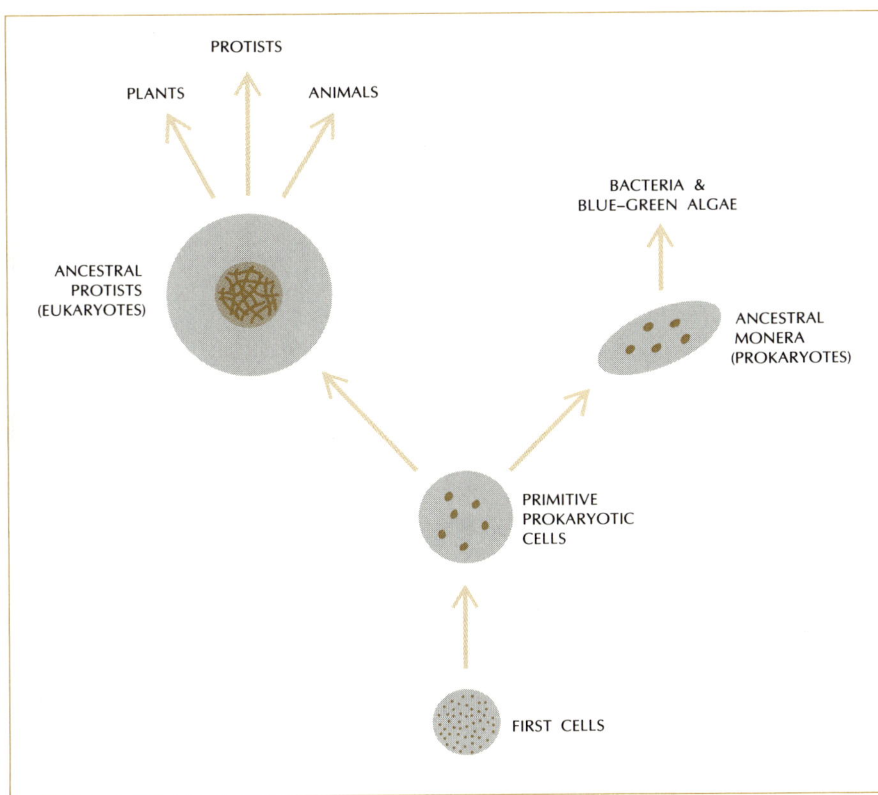

15
Organismic evolution begins with the first cells. They were probably prokaryotic, with genetic material (color) diffused throughout the cell. Primitive line of prokaryotes probably split, giving rise to ancestral Monera and Protista.

earth to a 30-day month. On that scale, the Archeozoic and Proterozoic eras fill 26 days. Fossils from these early eras are notably rare, but they suddenly become abundant in the next era, the Paleozoic, which begins on the twenty-seventh day of our one-month scale.

For lack of fossil evidence, students of ancient life must reconstruct the steps by which the first cells evolved into these prototypes from what they know about the relationships between modern organisms. Present theories hold that the first living things appeared on earth during the first half-billion years of its existence, sometime during the first four days on our one-month scale. The first cells, which probably appeared somewhat later, are believed to have been prokaryotes, the prototype of modern Monera. Some of these ancestral prokaryotic cells evolved into eukaryotic cells, the prototype of the modern Protista. This was the first branching of the tree of evolution (Fig. 15). One of the earliest branches was the ancestor of modern animals, the Metazoa.

Most of these primitive organisms have vanished without a trace. In Precambrian rocks, paleontologists have so far discovered only traces of

16

Precambrian animals in these three photographs are among the oldest ever discovered. Top left is a segmented worm, *Spriggina floundersi;* bottom left is the disk-shaped *Tribrachidium heraldicum;* on the right is the frondlike *Dickinsonia costata.* All lived in the Proterozoic seas. Their anatomy indicates considerable evolution from unknown primitive ancestors.

algae, microfossils of fungi, and impressions of a handful of animals. The animal fossils include corals, segmented worms, and two strange creatures that look like jellyfish (Fig. 16). Discovered in Australia in 1947, they are the oldest known animals.

THE PALEOZOIC ERA

The primitive organisms of the Archeozoic and Proterozoic eras were all aquatic. So were those of the early Paleozoic era which stretched from about 600 million to about 230 million years ago (two and a half days on our one-month scale). Sometime during the Paleozoic era life emerged from the ancient seas and invaded the land. Ancestral protists evolved into green algae, which in turn gave rise to psilopsids, the first land plants.

Fossil psilopsids, which probably lived along the seashore and in swamps, date back more than 300 million years (Fig. 17). By the end of the

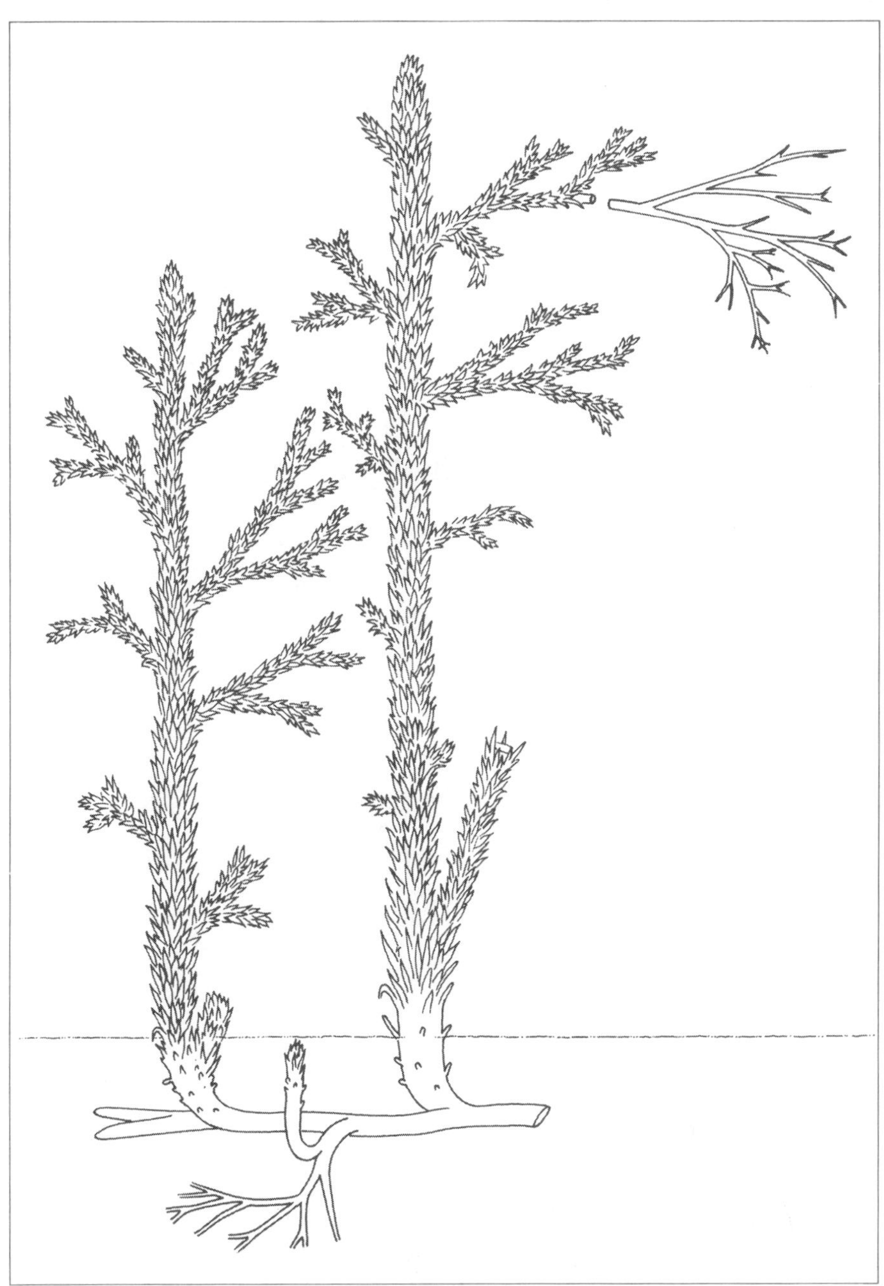

17

Fossil psilopsids from the Paleozoic include many species
of the genus *Asteroxylon*. These primitive land plants were
composed merely of simple shoots clothed with leaflike
scales. Simpler psilopsids lacked even the scales.

◀**Trilobite,** a possible ancestor of modern crabs and lobsters, flourished in Paleozoic seas. This fossil impression dates to the late Cambrian period, about 520 million years ago, and was found in Wisconsin. It is reproduced here almost twice its actual size.

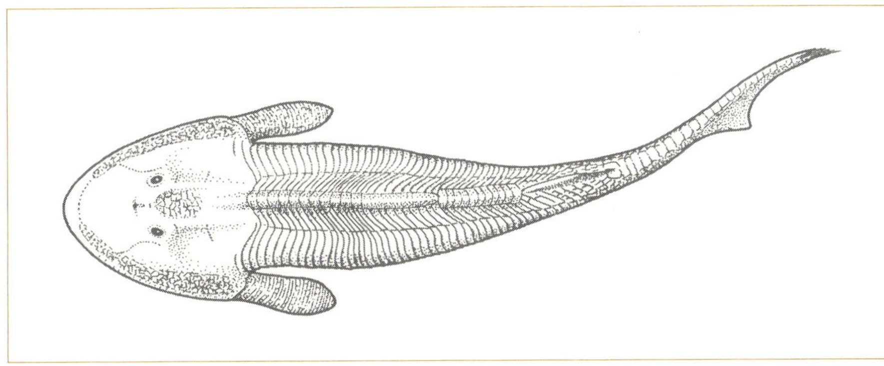

19
Ostracoderm was an armored fish that inhabited fresh water during the Paleozoic era. One of the first vertebrates, it is depicted here in top view.

Paleozoic, almost all psilopsids had become extinct. The survivors and some of their descendants gave rise to the modern Metaphyta (higher plants). The psilopsids of the Paleozoic are the remote ancestors of the trees and green plants we know today.

Land plants provided food for the Paleozoic animals that were soon to follow them out of the sea. The earliest known land animals were scorpions that appeared about 300 million years ago. The first insects appeared shortly thereafter. Trilobites were abundant in the seas of the early Paleozoic (Fig. 18); these extinct organisms may be distant ancestors of modern crustaceans, like lobsters and crabs. *Limulus,* the horsehoe crab, appeared at the end of the Paleozoic and survives today, apparently unchanged by 200 million years of natural selection.

Toward the middle of the Paleozoic the first vertebrates appeared. These included the ostracoderms, jawless armored fish that lived in fresh water (Fig. 19). Indirectly descended from them are the cartilaginous

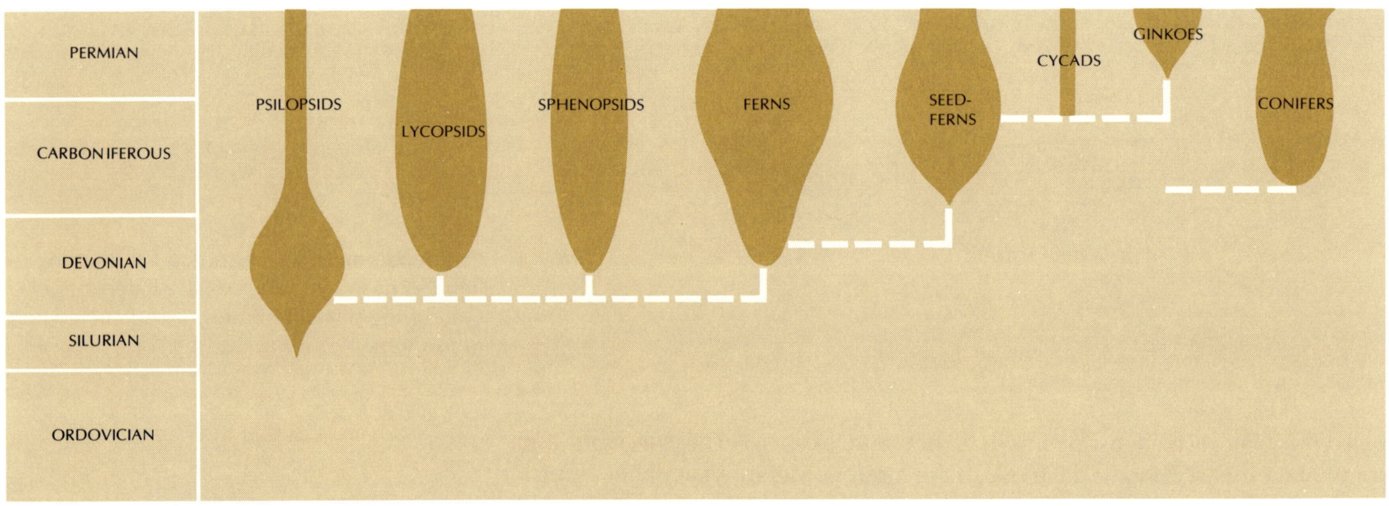

20
Plant evolution during the Paleozoic era is depicted schematically in this simplified diagram. Width of shaded areas indicates relative abundance. Dashed lines show hypothetical relationships among groups.

fishes (sharks, skates, and rays), bony fishes, and amphibians, the first land vertebrates. The evolution of plants and animals during the Paleozoic is summarized in Figs. 20 and 21.

The end of the Paleozoic was marked by the birth of many mountain chains and the wholesale extinction of existing species. The difference between Paleozoic and Mesozoic organisms is so great that early geologists believed that the organisms of the Paleozoic were wiped out by some worldwide catastrophe and replaced by entirely new and different kinds of plants and animals. Certainly some disaster devastated the living world

21
Animal evolution during the Paleozoic. Some groups have become
more abundant with time, while
others have decreased in abundance. All but the placoderms have
descendant forms living today.

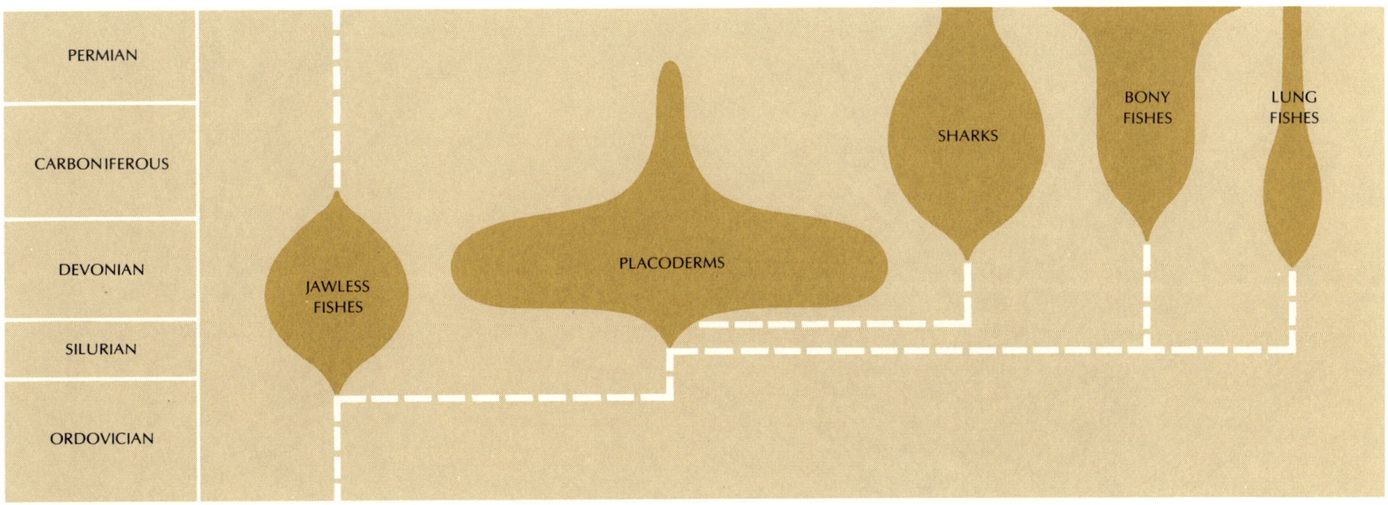

during this period, but among the land-based survivors at the close of the
era were amphibians, reptiles, and insects. In the sea, the dominant survivors were sharks and bony fishes.

THE MESOZOIC ERA

The Mesozoic era began 230 million years ago (a day and a half ago on our
one-month scale). During this era, the species that survived the Paleozoic
began to rebuild the living world. Among plants, the Mesozoic became the

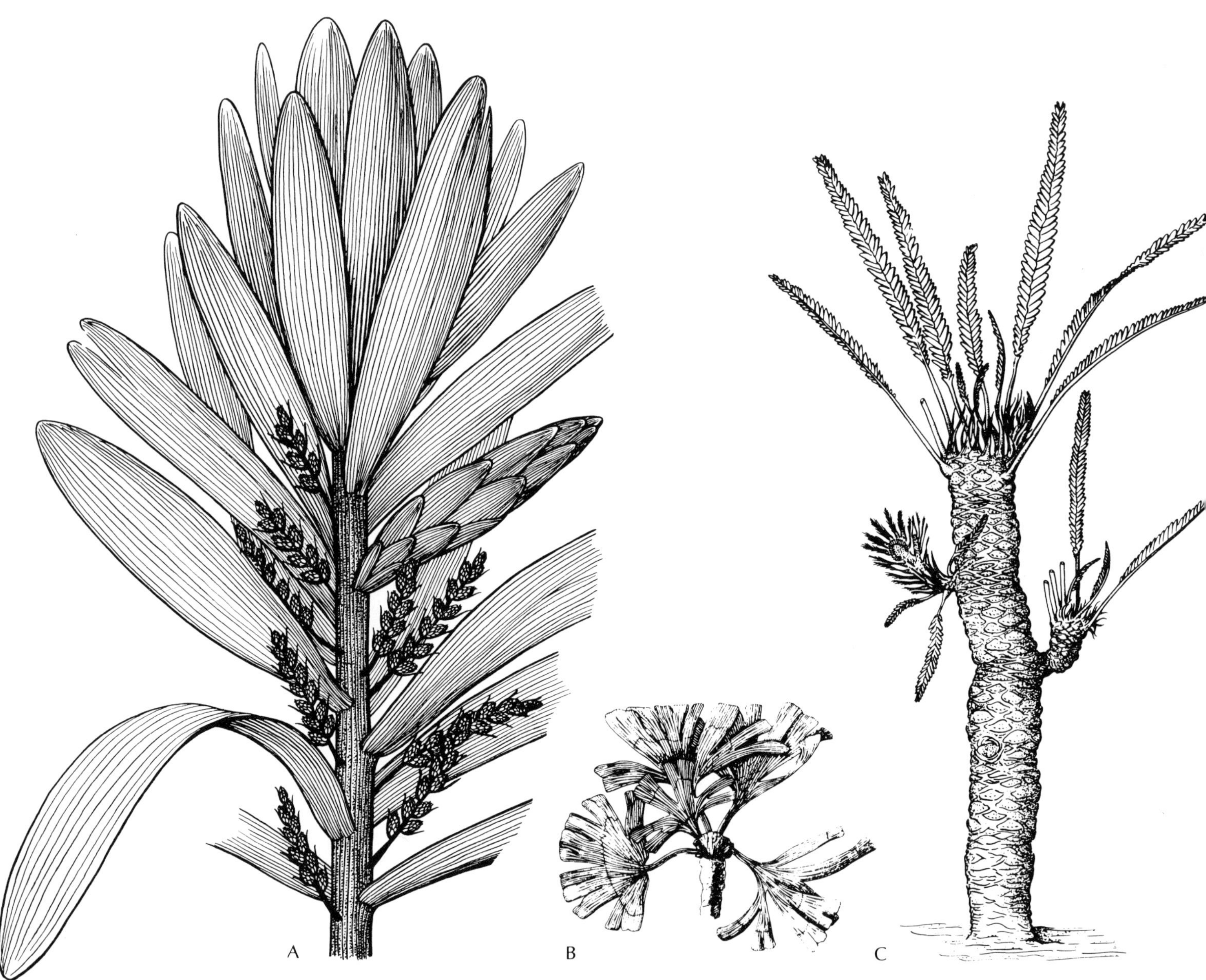

era of gymnospersm: conifers, ginkgoes, and cycads (Fig. 22). All were probably descended from the seed ferns of the Paleozoic: the cycads directly, the conifers and ginkgoes indirectly. Forests of these trees, which dominated the Mesozoic landscape, included many species of conifer that still survive, particularly pines, redwoods, and cypresses. The Mesozoic also saw the rise of angiosperms, the ancestors of modern flowering plants. Their origin is obscure, but forests of tree-forming angiosperms, notably oaks, maples, and palms, flourished during this era.

Among animals, the Mesozoic was the era of reptiles. Derived from the forms of the late Paleozoic, the organisms of the early Mesozoic gave rise, indirectly, to flying reptiles (pterosaurs) and to the legendary dinosaurs that came to dominate the land in the late Mesozoic, about 100 million years

22
Mesozoic plants include conifers (A), ginkgoes (B), and cycads (C). Although species shown here are extinct, many closely related descendants persist today. Photograph at right shows fossil cycad leaves from the Petrified Forest in Arizona. Leaves are embedded in shale that dates to the Triassic period.

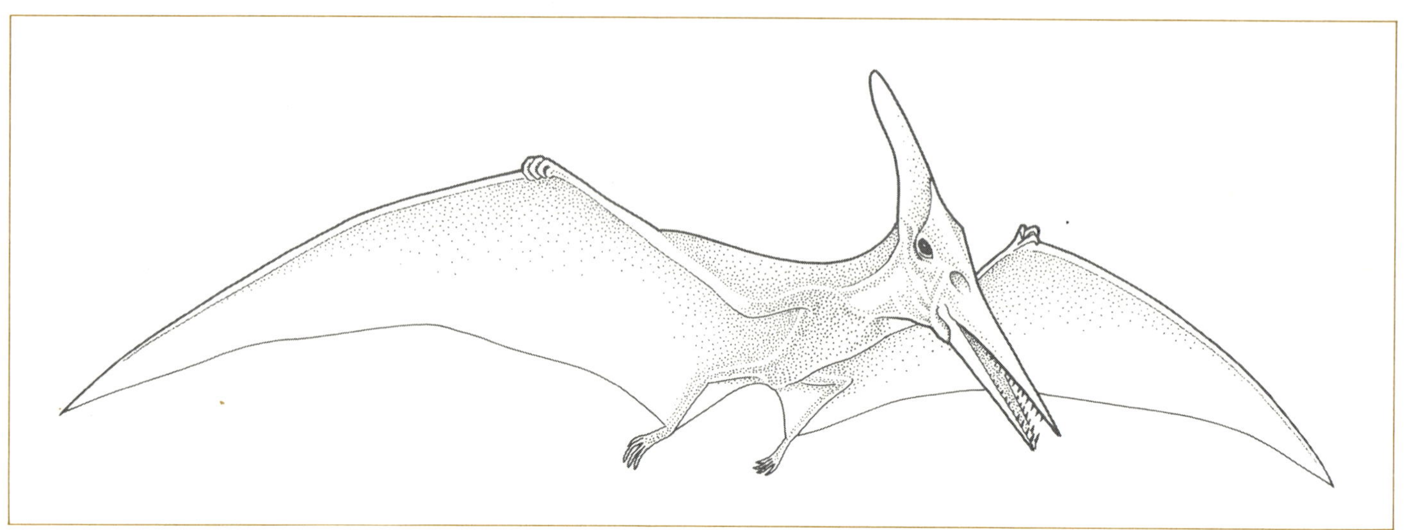

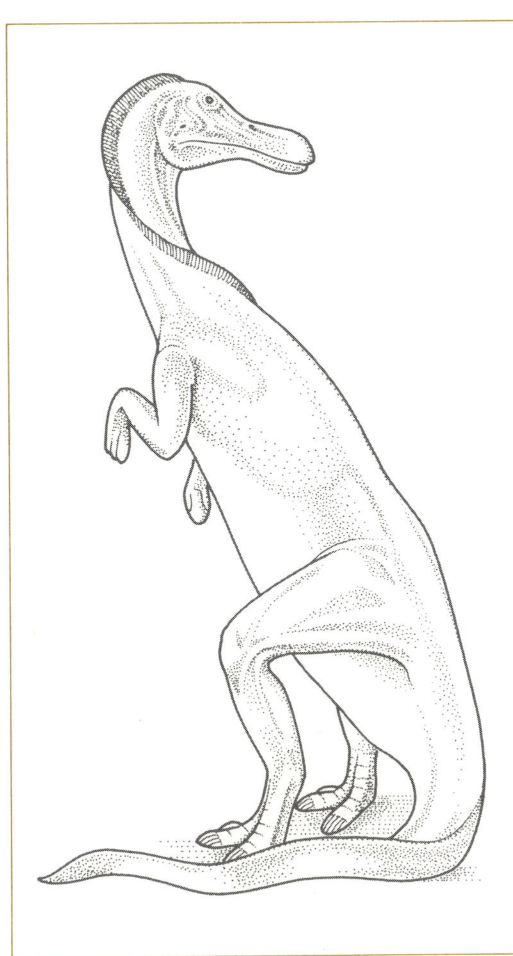

Mesozoic reptiles once dominated the earth. Pterosaurs (facing page, top) were flying reptiles; *Dimetrodon* (facing page, bottom) was a terrestrial reptile; plant-eating dinosaur (this page) walked on its two hind legs.

ago (Fig. 23). The transition from reptiles to birds is clearly illustrated by the fossil of *Archeopteryx,* an animal with reptilelike teeth and tail, and birdlike feathers and wings. During the early Mesozoic, another group of reptiles evolved into the first mammals, but this group of fur-bearing creatures was to remain insignificant for another 80 million years.

The Mesozoic ended with the Laramian revolution, a massive geological upheaval that shaped the highest mountain ranges of today, including the Rocky mountains and the Andes. The climate of the earth began to cool, a trend that was to continue during the Cenozoic era. According to current theories, the earth is normally much warmer than it is today. Every 200 million years, the long warm cycles are interrupted by Ice Ages. The earth is now emerging from the fourth great Ice Age since the Proterozoic era. The current Ice Age has profoundly influenced the environment, and hence the evolution, of the organisms of the modern Cenozoic era, which began 75 million years ago (about 12 hours ago on our one-month scale).

THE CENOZOIC ERA

The plants and animals of the Cenozoic, including man, are creatures of an Ice Age. Among plants, the Cenozoic is the era of angiosperms, particularly flowering plants and trees. Fossil trees of the Mesozoic and early Cenozoic show no annual rings, indicating that they grew at a steady rate throughout the year. The warm year-round growing season of the Mesozoic gradually gave way, in all but the tropical regions, to the annual cycles of climate that prevail today. Faced with harsher climate, some seed plants of the Mesozoic acquired highly effective adaptations. Their wood-making activity was curtailed. They became smaller nonwoody annuals and biennials. This was the origin of the small-bodied flowering plants that now comprise about 80 percent of the flora of temperate regions. The great forests that covered these regions during the early Cenozoic gradually thinned out, with important consequences for animal evolution.

The Cenozoic is called the era of mammals, but it is also the era of modern birds, bony fishes, and insects. The extinction of Mesozoic reptiles opened a variety of habitats to mammals, and dozens of lines evolved to exploit these opportunities. One of these lines, the primates, took to life in the trees.

The first hominids, members of the taxonomic family of man, appeared about 30 million years ago. From their apelike ancestors they inherited gripping ability, the opposable thumb, feet specialized for life in trees, color- and three-dimensional vision, voice communication, and an enlarged brain. Early hominids were probably small and light. They descended from trees almost as soon as they split from the ancestral primate stock, probably because stands of trees were farther apart than in earlier times. Vulnerable to attack by saber-toothed tigers and other carnivores as they dashed across open land, early hominids were subject to powerful

24
Homo africanus, also known as *Australopithecus africanus,* is the oldest human hominid. He lived in Africa about two million years ago. He made and used primitive stone tools.

selective forces that favored the evolution of muscles that would enhance their speed and agility. The human two-footed gait is unique among ground-dwelling animals, and so is the arrangement of powerful muscles in the buttocks, thighs, and calves.

The oldest ''human'' hominid is *Homo africanus,* discovered recently in Tanganyika (Fig. 24). About two million years ago this primitive man fashioned tools for hunting, but was primarily a vegetarian. The oldest

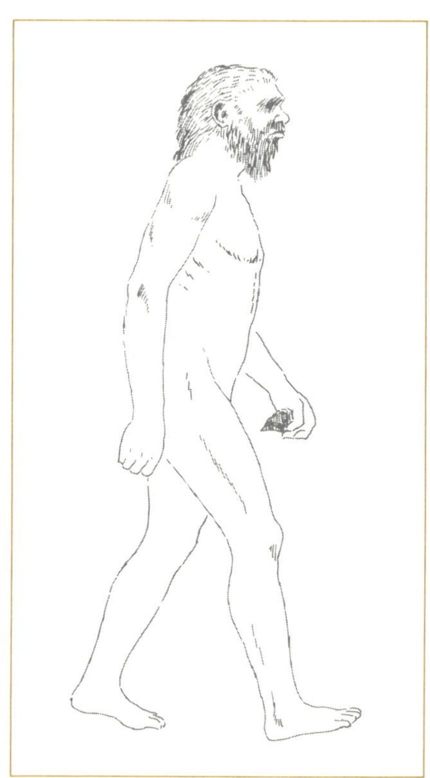

25
Swanscombe man, who lived in England more than 250,000 years ago, is oldest known representative of our species, *Homo sapiens.*

representative of our own species, *Homo sapiens,* appears to be Swanscombe man, who lived in England more than 250,000 years ago, less than 25 minutes ago on our scale (Fig. 25). His brain was apparently about double the size of that of *Homo africanus.* By 20,000 years ago, well into the Stone Age, all human species except man had become extinct. The societies of ancient Mesopotamia and Egypt emerged about 5000 years ago. The entire history of civilized man, from then to now, represents 30 seconds on our scale.

FUTURE EVOLUTION

The emergence of human intelligence and consciousness represents a vast advance in evolution, one that raises profound questions. Is intelligence just a normal stage in the evolutionary progression of life? Evolution goes on, even today; present organisms are not an end point, merely the gateway to the future. But what kind of future? What is the next stage?

In only about 500 generations, man has risen from the Stone Age to his present dominance of the planet. He has acquired the ability to accumulate and store knowledge, endowing each future generation with a solid foundation of information on which to build. Man's power and dominance is multiplied in each generation by an ever-accelerating technology.

Despite such advances, the record so far suggests that in man evolution has created an organism that threatens the survival of all species. Human activities have already sent many species to their doom. Among his own species, man's efforts to solve the problem of overpopulation, starvation, and poverty are halfhearted failures. He has made open sewers of the air, the lakes, and the rivers. His uncontrolled aggressions leave him unmatched among animals for ferocity in slaughtering his brothers. The highest priority of human technology is the manufacture of weapons that can devastate the planet in less than an hour. The same technology cannot protect him from natural catastrophes. His proudest cities lie at the mercy of hurricanes, tidal waves, earthquakes, and droughts. Overpopulation and the demands of technology have depleted in a few generations a reserve of natural resources accumulated over thousands of centuries. On an exhausted and polluted planet, the most pertinent question about the next phase of evolution is, "Will man survive to see it?"

READINGS

Berrill, N. J., *Man's Emerging Mind*. Dodd, Mead, New York, 1965 (paper)
> Extremely readable account of the evolution of man. Spiked with subtle humor and interesting insights.

Darwin, C., *The Origin of Species by Means of Natural Selection*. Crowell-Collier, New York, 1962 (paper)
> A book to be sampled if not completely devoured by all students of biology.

Darwin, C., *The Origin of Species.* Penguin, Baltimore, 1969 (paper)
   An abridged version of Darwin's classic work.

Dobzhansky, T., *Mankind Evolving.* Yale University, New Haven, 1962 (paper)
   The biological and cultural components of human evolution, from the point of view of a geneticist. Relevant to many current problems, including racism, uncontrolled population growth, genetic diseases, and intelligence.

Eiseley, L. C., *Darwin's Century: Evolution and the Men who Discovered it.* Doubleday, Garden City, N.Y., 1958 (paper)
   The pioneers and early defenders of the theory: Darwin, Huxley, Wallace, and others. Well written.

Howell, F. C., *Early Man.* Life Nature Library, Time-Life Books, New York, 1965
   Splendidly illustrated and well-written discussion of the evolution of man.

Irvine, W., *Apes, Angels, and Victorians.* Meridian, New York, 1955 (paper)
   Another biographical approach to Darwin and his times. Droll humor, clear writing.

Karp, W., *Charles Darwin and the Origin of Species.* Harper & Row, New York, 1968
   Fascinating description of Darwin's life and the discoveries that led to his monumental theory. Richly illustrated.

Lederberg, Joshua, *Experimental Genetics and Human Evolution,* in Bulletin of the Atomic Scientists, October, 1966, pp. 4–11
   A geneticist of world stature discusses eugenics, particularly proposals for improving our species by producing identical twins of genetically outstanding individuals.

Lerner, I. M., *Heredity, Evolution and Society.* Freeman, San Francisco, 1968.
   Interesting and timely examples of the social, political, and psychological implications of organic evolution and the laws of heredity.

Margulis, L., *Origin of Eukaryotic Cells.* Yale University, New Haven, 1970
   Written for professional biologists, but the serious beginning student can follow the line of argument. This book will influence research for the rest of the decade.

Moore, R., *Evolution*. Life Nature Library, Time-Life Books, New York, 1962

> A clearly written look at evolution, from the structure of genes to the emergence of man. The illustrations are up to *Life's* usual standards.

Moorehead, A., *Darwin and the Beagle*. Harper & Row, New York, 1969

> Lavishly illustrated, with a text by a nonbiologist who knows how to write.

Ramsey, Paul, *Fabricated Man*. Yale University, New Haven, 1970 (paper)

> Subtitled "The Ethics of Genetic Control," and written by a renowned professor of ethics at Princeton University, this mind-stretching little book examines both the moral and the biological aspects of eugenics. An excellent overall discussion, and the Notes at the end of the book include further references for the serious student.

Simpson, G. G., *The Meaning of Evolution*. Yale University, New Haven, 1949

> A more sophisticated look at the process of evolution by an expert who is incapable of writing a muddy paragraph.

Stebbins, G. L., *Process of Organic Evolution*. Prentice-Hall, Englewood Cliffs, N.J., 1966 (paper)

> Genetic mechanisms of speciation, trends in evolution, and the process of human evolution. All topics are treated clearly and carefully.

Wallace, B. and A. M. Srb, *Adaptation*. Prentice-Hall, Englewood Cliffs, N.J., 1964 (paper)

> The process of evolution as shown by examination of reproductive behavior, communication patterns, habitat preferences, and structural modifications.

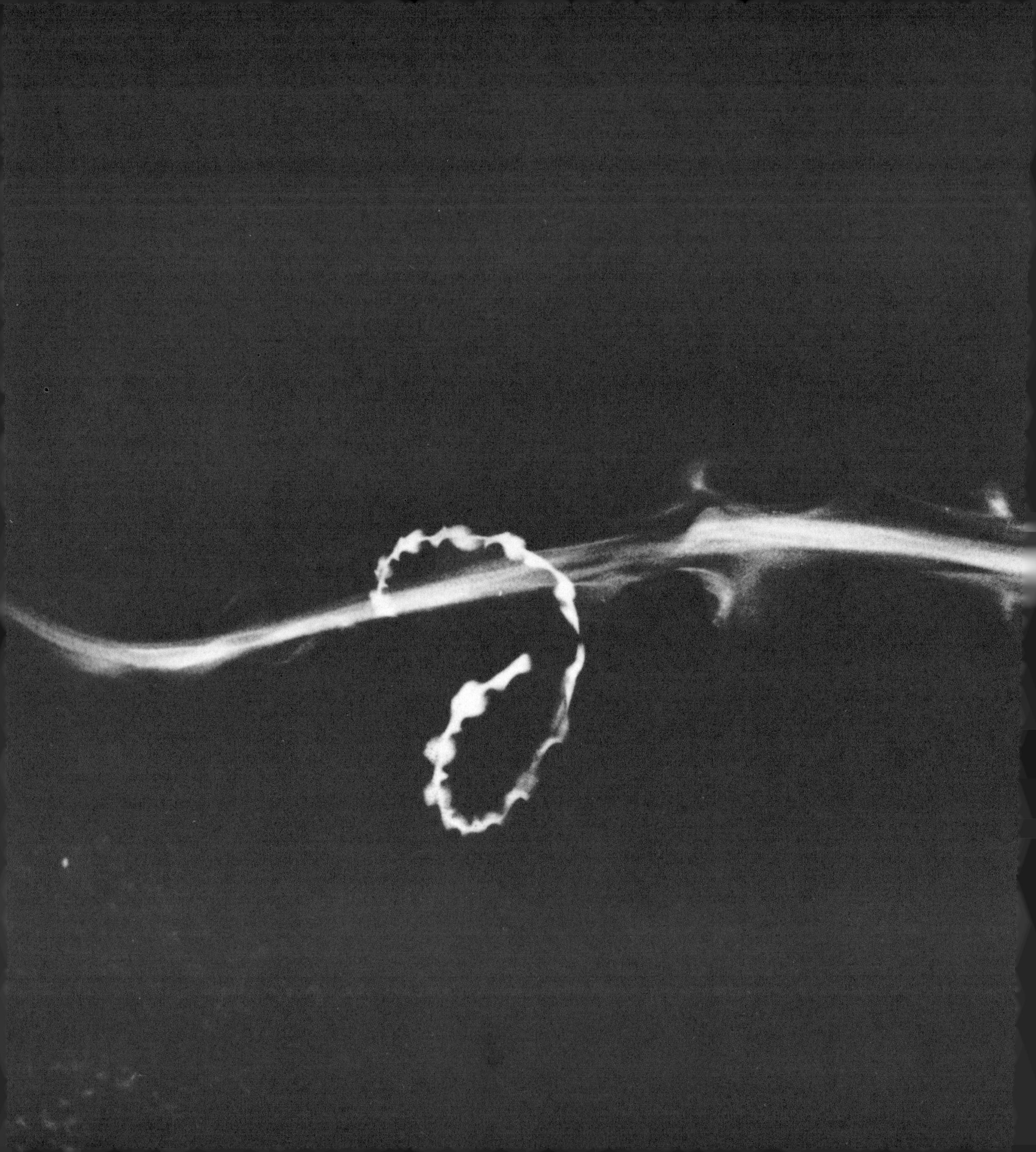

14. THE BIOLOGY OF BEHAVIOR

The first clue that Hoyle's Black Cloud (Chapter 1) might be alive was its behavior. Unlike other celestial objects, its response to natural forces was not passive and uncomplicated. Compared with inanimate matter, a living system embodies a whole new level of structural complexity; it is capable of responding actively to its environment. Taken together, its responses comprise behavior.

In its broadest sense, behavior includes activities as diverse as the diurnal migration of plankton, the growth of plants toward sunlight, and the blinking of a firefly. In recent decades, the most interest has centered on the behavior of higher animals, especially that of mammals and birds. Students of animal behavior—ethologists and animal psychologists—have discarded the old idea that all animal behavior is instinctive, or that it can be completely explained in terms of a few basic drives, such as hunger and sex. They now view behavior as the product of heredity, learning, and development.

The entire repertory of an animal's behavior depends on information coded in its genes: specifically, on the information that determines the structure of the nervous and endocrine systems. Incorporated into the cells of these two systems are the programs for instinctive behavior as well as the memory and logic circuits that enable an animal to expand its repertory of behavior by learning. During the development of these systems in the young animal, its pattern of behavior changes gradually from immature to adult.

◄**Bat captures moth** in midair despite the moth's evasive spiral. Bat enters from right, flying about six meters (20 feet) per second. Moth (center) detects bat sonar and attempts to escape, but bat reacts and changes course in time to intercept.

Behavior helps an animal adapt to its environment, and like all gene-controlled adaptive mechanisms, it is subject to natural selection. Behavior evolves: signals become more sophisticated, courtship activities less ambiguous. Several stages in the evolution of particular traits of behavior survive among species alive today, enabling ethologists to reconstruct some of the "fossil behavior" of the past.

INHERITED BEHAVIOR

Inherited or innate behavior ranges from simple reflexes to complex instincts. In simple, short-lived animals virtually all aspects of behavior are preprogrammed. Each species of spider, for example, weaves a particular type of web. Even individuals reared in complete isolation will fashion a web that is stereotyped in every detail. Learning plays no part in this activity; the spider literally cannot spin a web of another type. The spider's capture of prey, the ceremony of mating, and the rearing of offspring are also innate behavior.

The nervous and endocrine systems coordinate the often complex physiological processes that underlie inherited behavior. In birds the cycle of reproductive behavior is initiated by a seasonal increase in the length of the day. Fed into the central nervous system via the visual pathways, the perception of increasing day length triggers the alarm of some sort of internal biological clock. The nervous system shifts the hormone balance of the bird by stimulating the hypothalamus, a part of the brain located directly above the pituitary, the master gland of the endocrine system (Fig. 1).

As the link between the nervous and endocrine systems in vertebrates, the hypothalamus is involved in sexual behavior, control of hunger and thirst, and the regulation of body temperature. From the higher parts of the brain it receives descending nerve fibers which continue on to the pituitary. It is also linked to the pituitary by a miniature circulatory system. Stimulated by the activities of the hypothalamus, the pituitary releases polypeptide hormones into the bloodstream, which carries them to respective organs, such as the testes and ovaries. In turn, the sex glands release steroid hormones that have a direct effect on behavior: a male will court a female only when his level of testosterone is sufficiently high, and a female will respond to courtship only when she is secreting estrogen.

Experimental evidence suggests that steroid hormones affect behavior by acting directly on the brain. A capon (a castrated rooster) will not usually

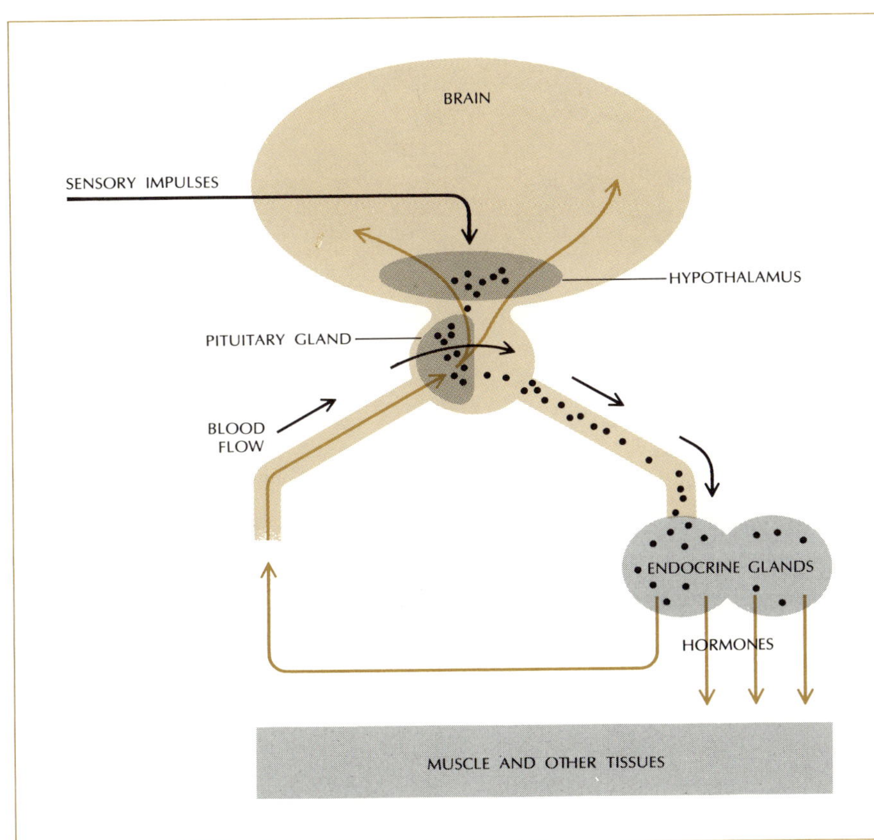

1
Feedback loop regulates cycle of reproductive behavior in birds. Perception of increasing day length, fed into brain via sensory pathways, stimulates hypothalamus and pituitary gland to release polypeptide hormones (dots). These hormones in turn stimulate ovaries and testes to produce steroid sex hormones (colored arrows). Sex hormones affect muscle and other tissues and via the bloodstream are fed back to the brain, thus closing the control loop.

court a hen. A tiny crystal of testosterone surgically implanted in its hypothalamus will restore its courtship behavior, but not, of course, its fertility. Electrical stimulation of the hypothalamus has a similar effect.

Steroid hormones may also influence sexual behavior by altering the response threshold of the sensory system to certain types of stimuli. A bird will not always incubate eggs. It may eat them or ignore them unless it is stimulated to incubate, either physiologically by the secretion of the female hormones estrogen and progesterone, or psychologically by the normal sequence of activities leading from courtship to nest building.

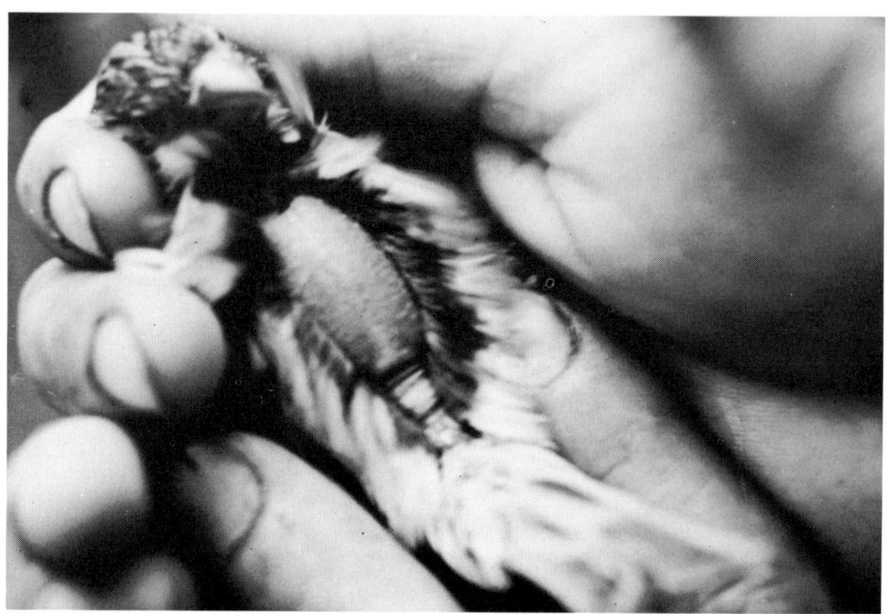

2
Incubation patch appears on the breast of
certain female birds shortly before egg-laying.
Photo shows a female house finch.

The females of some species of bird have an incubation patch on their
breasts (Fig. 2). As a result of hormone stimulation, shortly before egg-
laying the patch loses its feathers, becomes highly vascular and much more
sensitive to touch. These changes apparently improve the heat flow from
parent to eggs. Moreover, tactile stimulation of the patch apparently orients
the bird on the nest. If the patch is locally anesthetized the bird will fail
to sit normally. It will rise and resettle frequently over the eggs, which suggests that tactile feedback from the eggs is necessary for normal incubation.

LEARNING

Most animals can modify their behavior, at least to some extent, on the
basis of experience. Bees can learn the color of the flowers that yield nectar.
If a glass dish containing sugar water and other dishes containing distilled
water are arranged on paper squares of different colors, bees will quickly

learn which color corresponds to the sugar water. They will return to the square of that color even if the dishes are switched.

A very short memory limits the ability of insects to learn. The octopus has a better memory and shows a surprising capacity for learning. Its capabilities represent the zenith of invertebrate intelligence. The best memories, and hence the best learning abilities, are found in vertebrates, particularly mammals. They are more intelligent than any other group of animals.

The classic experiments on animal learning were performed by the Russian psychologist Ivan Pavlov (1849–1936). By ringing a bell immediately before blowing powdered meat into the mouth of a dog, he conditioned the animal to associate the bell with food. Soon the dog's mouth began to water every time he heard the bell whether or not food was present. Using food or escape from punishment to reinforce modifications in behavioral responses, trainers can teach animals to perform a surprising variety of tasks. To get food, a rat can be trained to find its way through a maze; chickens can learn to count or to distinguish between symbols such as a square and a circle; chimpanzees can learn to fit two bamboo sticks together to reach a banana.

Young chicks, quail, and turkeys crouch in the grass and remain immobile when an alarming stimulus appears overhead. This behavior is innate: the crouching response is identical in both wild and incubator-reared chicks. Like pecking, crouching behavior is modified in the first few days after hatching. Incubator chicks will at first crouch in response to almost any objects passing overhead: live hawks and pigeons, cardboard models of flying ducks and hawks, and even cardboard circles. After seeing these objects several times a day the young birds gradually stop crouching; eventually they stop crouching even when they see a live hawk. Learning not to respond to a stimulus is called habituation. Habituation to irrelevant stimuli such as falling leaves or flying insects is adaptive, but habituation to hawks could prove fatal.

Another special type of learning is common among certain birds. In many species, including coots, quail, and ducks, a bird only a few hours old will follow its mother as she leads her brood away from the nest. The young will follow her for a few weeks, then gradually become increasingly independent. The act of following is innate but following the parent is not. The young bird learns to follow any object whose features it recognizes soon after hatching. Incubator-reared hatchlings will follow objects that have little resemblance to the normal parent—a man, a cardboard box, a football, colored balloons, or the adults of another species, including grown hawks.

Known as imprinting, this form of learning can occur only during a short, sensitive period early in the animal's life; ducklings, for example, imprint 5 to 24 hours after hatching. Imprinting can even affect the bird's sexual preferences. Many biologists who have hand-reared young birds have later found themselves the object of the bird's sexual interest. Similarly, chicks imprinted to cardboard boxes when only a few days old will court these boxes when they reach maturity. Although we might expect the social and sexual preferences of birds to be inherited traits, learning is clearly involved.

The relative importance of inheritance, learning, and development in shaping behavior varies from species to species and depends on other features such as the complexity of the nervous system and the life cycle of the animal. In general, learning is dominant in the development of the behavior of animals (like the chimpanzee) which are endowed with a complex nervous system, a long life span, and a prolonged period of parental care of the young. Conversely, inherited behavior is dominant among animals like insects, whose life cycles provide little or no opportunity to learn adult behavior. The female digger wasp lays each of her eggs in a different burrow; after provisioning each burrow with food, she dies. The new generation of digger wasps that emerges from the burrows the next spring possesses a full repertory of wasp behavior. Clearly, this prefabricated behavior is encoded in its chromosomal DNA. The chimpanzee and the digger wasp embody two extremes of learned versus inherited behavior. The behavior of most animals falls somewhere in between, since it is the sum of varying proportions of heredity and learning.

DEVELOPMENT

The development of the nervous system in the young animal also influences behavior. This is well illustrated by an analysis of the behavior of the so-called precocial birds: those species whose young hatch with the ability to peck competently, to feed and drink for themselves, to hide from predators, and to follow the mother off the nest. Their pecking behavior is instinctive; pecking by incubator-reared hatchlings is identical to the pecking of normally reared ones. Soon after hatching, the accuracy of the

3

Effect of learning on pecking accuracy was tested by fitting one group of newly hatched chicks with a hood and goggles that deflected their visual images seven degrees to the right. A control group of chicks was fitted with nondistorting goggles. In a few days both groups became more accurate in pecking at a brass nailhead "seed" (grey disk). Colored spots indicate pecks. Investigators concluded that increased accuracy was due not to learning but to maturation of the nervous system.

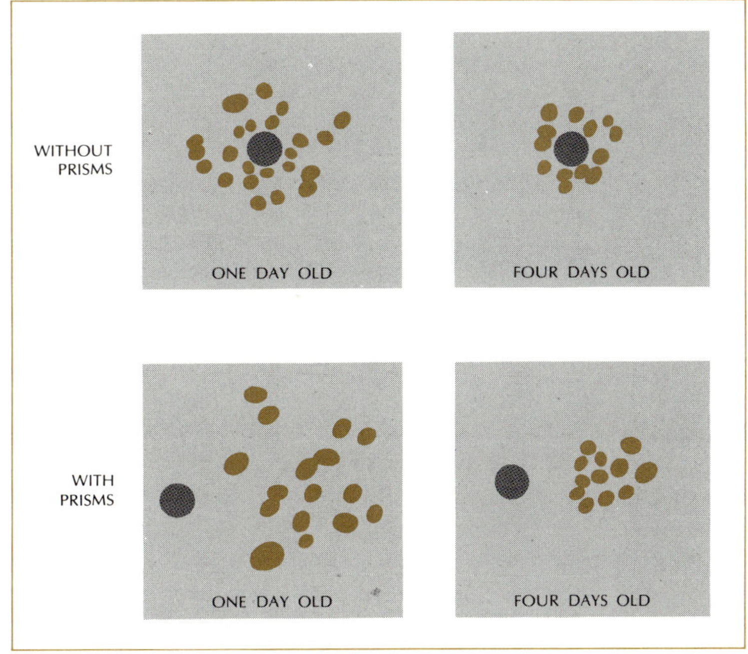

pecking in both groups improves. To determine whether this was due to learning, experimenters fitted newly hatched chicks with prismatic goggles that deflected the visual image several degrees to one side. This prevented the newly hatched chicks from picking up seeds; in other words, their pecking could not be rewarded, and hence the chicks could not learn to improve it. Nevertheless when the experimental chicks were a few days older, they clustered their pecks as accurately as control chicks who were fitted with nonprismatic lenses (Fig. 3). The increased pecking accuracy in the older chicks was possible because of the maturation of the nervous system.

THE EVOLUTION OF BEHAVIOR

Evidence of the behavior of extinct species has vanished almost without a trace, although fossilized termite nests and worm burrows do provide crude evidence of the habits of those organisms. Some concept of the evolution of behavior can be obtained by comparing the behavioral patterns of living species.

Behavior is adaptive: the fleeing impala zigzags wildly, prolonging the chase and exhausting the swifter but less endurant cheetah; a flock of starlings closes ranks at the sight of a hunting hawk, preventing him from isolating and capturing a member of the flock. During its evolutionary history every animal species loses individuals whose behavior patterns are nonadaptive, and the genes that control this behavior are weeded out of the gene pool.

Strictly speaking, an instinct is a complex pattern of inherited behavior. Animal behaviorists recognize six major categories of instinct: nutrition, aggression, reproduction, social behavior, sleep, and grooming. Each of these major instincts involves simpler levels of behavior. The simplest are known as fixed action patterns.

The genetic control of fixed action patterns has been demonstrated with breeding experiments on two related species of African parrot. One species cuts strips of leaves or bark for its nest, tucks several of these strips into its rump feathers, then flies to the nest site. The other species collects sticks or cuts a single strip of bark which it carries to the nest in its bill. Hybrids of the two species show intermediate fixed action patterns that incorporate portions of the behavior of both parents. At first this inherited hybrid behavior is an ineffective compromise; young hybrids do not tuck and carry strips in their plummage effectively, nor do they carry strips regularly in their bill. Eventually, hybrids come to carry the material in the bill, and engage in less and less tucking. After three years the hybrids learn to carry material in the bill almost as effectively as the one parent species, but even then they lapse from time to time into brief episodes of tucking. The carrying behavior of the hybrids reveals a powerful genetic component. The gradual shift from one carrying technique to the other demonstrates that only with the greatest difficulty can an animal learn to override a fixed action pattern, even one that has been genetically altered by interbreeding with another species.

4

Head-bobbing display of lovebird has become progressively more ritualized. Primitive species bobs head with rapid shallow nod. Intermediate species bobs head more slowly, and in a wider arc. Highly evolved species bobs head very slowly, in even wider arc.

THE EVOLUTION OF DISPLAYS

The production of hybrids between two different species is often an evolutionary dead end because the offspring of such matings are usually sterile. Natural selection often favors activities that discourage the "mistake" of mating with another species. The courtship ceremony of many species assures that only members of the same species will mate. Courtship behavior includes instinctive displays which are fixed action patterns that emphasize some highly effective stimulus, such as gaily colored plumage. The spreading of the peacock's conspicuous fan is an example of this.

The courtship displays of lovebirds (parrots of the genus *Agapornis*) show various stages in the evolution of behavior. The genus includes six species. The behavioral or physical characteristics of four species are considered primitive. The characteristics of the fifth are more complex; those of the sixth are considered most advanced. The courtship displays of each of these parrots include head bobbing and a stylized grooming movement. Head bobbing during courtship has clearly evolved from similar movements performed when feeding the young. As in so many other cases, the evolved behavior (head bobbing) and its precursor (feeding movements) are both part of the animal's normal repertory. As new species of *Agapornis* evolved, their head-bobbing displays began to differ from one another. The primitive species bobs its head rapidly and shallowly; the intermediate species, more slowly and deeply; and the most advanced species, very slowly and very deeply (Fig. 4).

Stylized grooming movements have also evolved from earlier behavior. All six species perform a different version of these movements when courting. The four primitive species perform a stylized preening movement; the intermediate species, a stylized scratching movement (scratching the face with the foot); and the advanced species, an extremely stylized scratching movement in addition to a tail-wagging movement. All these stylized movements are highly stereotyped, that is, they are virtually identical whenever they are performed. Conversely, in the actual grooming of their feathers the parrots show considerable variability, devising many different kinds of preening, scratching, and tail-wagging movements.

Grooming behavior is the precursor of the courtship displays, but how did it evolve into courtship signals? The answer lies in the observation that parrots, like our own species, resort to irrelevant grooming behavior when faced with a conflict of motivation, such as the desire to approach and to flee from an object at the same time. When a man faces a critical decision he may suddenly scratch the top of his head. When courting, a male parrot approaches a female for sex with apprehension and even fear, and for good reason: an unresponsive female may nip his toes and many a male has lost one or two. If this conflict between sex and fear is strong at the moment a male parrot lifts his foot to mount the female, he may simply continue the upward movement and scratch his face with the foot. If such scratching conveys to the female information about the male's intentions and thus facilitates courtship, then natural selection might gradually transform scratching into a courtship display.

To be useful, a display must be clearly distinguishable from routine behavior. To breed successfully a female parrot must differentiate display scratching from the scratching used in normal grooming. It is not surprising that in the course of their evolution, displays gradually become more exaggerated and stereotyped, that is, progressively less ambiguous. The fully evolved display is said to be ritualized, and the evolution of a display is called ritualization. Display tail wagging is not markedly different in form from the tail wagging of normal grooming. It is considered to be un-ritualized and probably represents an early stage in the evolution of a display. Display scratching is partly ritualized in some species; the movements are fast and perfunctory and frequently the bird scratches the bill instead of the feathered areas of the head. In other species scratching is even more ritualized, being performed very fast and directed at the bill

Display scratching is part of the courtship ritual of lovebirds. Male at right is courting; female at left is scratching.

(Fig. 5). Display preening, an alternate series of quick nods of the head to the right and left that may only occasionally bring the bill to the feathers, is highly ritualized.

Comparative studies concerning the evolution of fixed action patterns such as displays have been the primary source of our knowledge of the evolution of behavior. However, it must be realized that not only sexual courtship activities, but any type of behavior which is genetically controlled in whole or in part is subject to natural selection and may eventually evolve, including the ability to run, or to swim, or to talk, or even to learn to plan ahead.

COMMUNICATION

Communication facilitates the social interplay among the members of a biological community. The transfer of information is not restricted to members of the same species. Gaudy blossoms attract insects to nectar sources "in return" for the transfer of pollen; the bold coloration of the wasp is a "hands off" signal that benefits both wasp and picnicker; the alarm call of a small bird upon sighting a hawk warns many animals to take cover as well as members of its own species. Of course, when the communication is between members of the same species, the interplay is more obvious, and the signaling systems are likely to be more complex. The signals themselves can involve any of the senses: touch, sight, hearing, smell. The choice seems to depend chiefly on the distance the signal must travel and on its purpose. Obviously, tactile signals are not suitable for long-distance communication. Visual signals are most effective at moderately close range because obstacles between signaler and receiver prevent reception of the signal. Moreover, because most visual signals depend on environmental light, they are largely ineffective at dusk or at night. However, some animals that are active during the day do signal visually, particularly if they live in open environments such as plains, savannas, or the clear waters around a

Social grooming is an example of communication by touch. It plays an important role in friendship, courtship, and dominance. Photos depict grooming behavior in chimpanzee (left) and gorilla (right).

coral reef. Both chemical and acoustical signals carry well over long distances and pass readily around obstacles in the environment. This makes them ideal for communication among nocturnal animals and among animals inhabiting forests or murky waters.

TACTILE SIGNALS

The spawning of a minnow-sized fish known as the three-spined stickleback is coordinated by a tactile display. After the female stickleback enters a nest prepared by the male, he nudges her abdomen with his snout and she responds by depositing eggs. If the male is removed once he has led the female into the nest she can be artifically signalled to lay eggs by tapping her abdomen with a glass rod.

Another example of tactile communication is the social grooming behavior of primates. Beyond its obvious function of cleansing the fur of ticks and sloughed-off flakes of skin, grooming plays an important role in the social behavior of primates, particularly in establishing friendships, pair bonds between male and female, and the dominance of one male over another (Fig. 6).

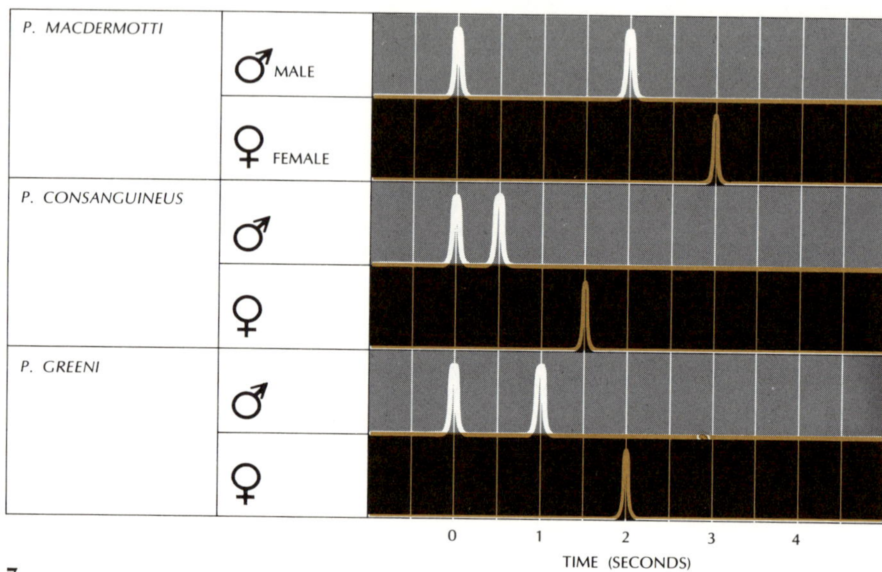

7

Flash patterns of firefly communicate both the sex and species of the sender. Flying male seeking mate emits precisely timed pattern of flashes. Female on ground sends answering flash after a specific delay interval. Male then lands near her and mates. Patterns were supplied by James E. Lloyd of the University of Florida.

VISUAL SIGNALS

Fireflies communicate at night by producing bioluminescence. The flying male of each species emits a specific pattern of flashes and the sedentary female responds after a certain delay. Her response flash and delay time are characteristic of her species (Fig. 7). Male fiddler crabs attract females to their burrows with a claw-waving display. The pattern of waving is highly stereotyped, and each species of fiddler crab has its typical pattern. In mammals facial features are common signals. The fear grimace of monkeys and apes exposes the teeth by pulling back the corners of the mouth; the "open-mouth threat" face takes on the opposite countenance, with the teeth covered and the corners of the mouth pulled forward (Figs. 8 and 9).

8
Fear grimace of monkeys and apes is sometimes called
the fear grin because corners of the mouth are pulled
back, exposing closed teeth.

9
Open-mouth threat display differs from fear grin in that
jaws are open and corners of mouth more or less cover
the teeth.

ACOUSTICAL SIGNALS

Bird songs are male displays that attract females and repel other males.
The songs of several species are now known to be partially inherited and
partially learned (Fig. 10). The learning process is similar to imprinting.
Many insects also communicate by song, each species having its own distinctive, innate tempo. With few exceptions insects produce their songs by
rubbing together parts of the body that are specifically modified for this
purpose. Marine animals broadcast acoustical signals through water, an
excellent sound-transmitting medium. It is well known that whales and
porpoises orient themselves with their own sonar echolocation systems, but
recent evidence suggests that these marine mammals also communicate

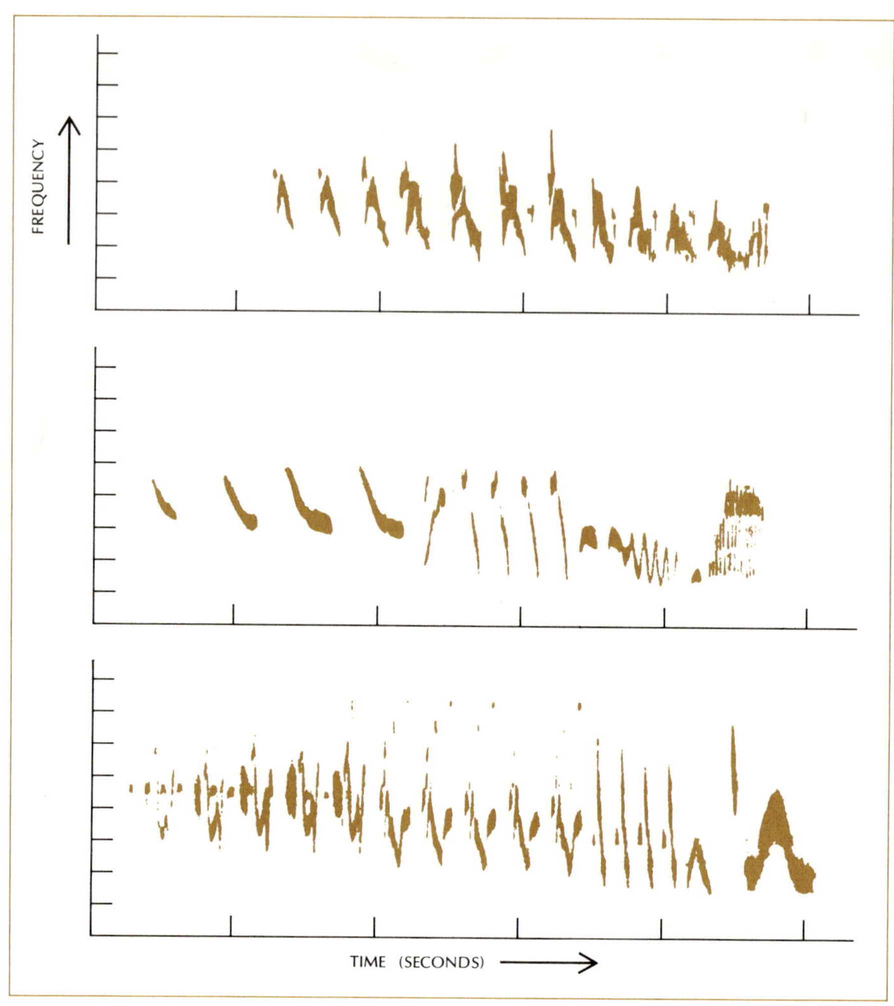

Songbirds learn songs by listening to older birds. Sonagrams of chaffinches show role of learning and instinct. Sonagrams in the top row record the very simple songs of two birds raised in isolation who have never heard another finch. Sonagrams in the middle row show songs of two birds raised in isolation but permitted to hear each other; those in the bottom row are normal songs of two adult wild chaffinches.

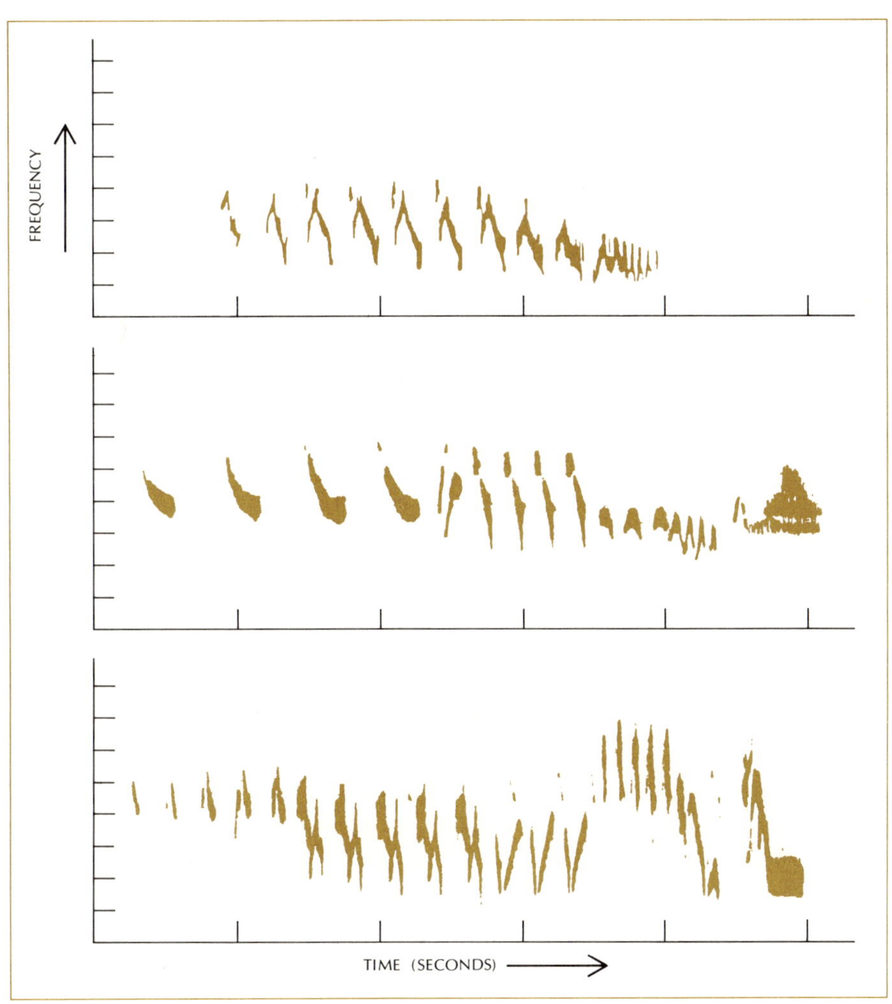

vocally. Some recorded whale vocalizations are surprisingly similar to bird songs.

CHEMICAL SIGNALS

Pheromones are the chemical signals of insects, the compounds that coordinate such social behavior as trail marking and courtship. Ants mark their trails with at least two different chemicals, a highly volatile short-term marker for short trails and a less volatile long-term marker for lengthier trails. Trail marking provides a means of communicating the direction and

11

Scent marking. Maxwell's duiker, a miniature antelope from west Africa, marks its mates and its territory with scent from glands on the sides of its face. Photos show territorial marking by duikers in Bronx Zoological Park.

distance to a goal. A single ant can blaze a chemical trail that will lead a thousand of his colleagues to a picnic basket.

Virgin female silkworm moths assume a calling position and emit a chemical that, when carried by the wind, can be detected by a male moth more than a mile away. Following an upwind course, the male locates the female and mates with her. Mammals also secrete pheromones; they mark the boundaries of their territories by depositing scent, often in the urine, on prominent objects (Fig. 11). Lemurs mark themselves and their mates as well as their environment in this way.

SIGNAL AND LANGUAGE

What do these signals mean? In rare instances a specific signal seems to represent some object in the environment; that is, the signal is the animal's "name" for that class of object. The vervet, a monkey that inhabits the forests of tropical Africa, has several predator warning calls: one for fast-moving ground predators like leopards, one for large snakes, one for eagles. Each call evokes an appropriate response from other vervets. Upon hearing the "leopard call," vervets will climb into the treetops away from thickets; but upon hearing the "eagle alarm" they run or drop into thickets rather than remain in exposed locations such as short grass or the branches of trees.

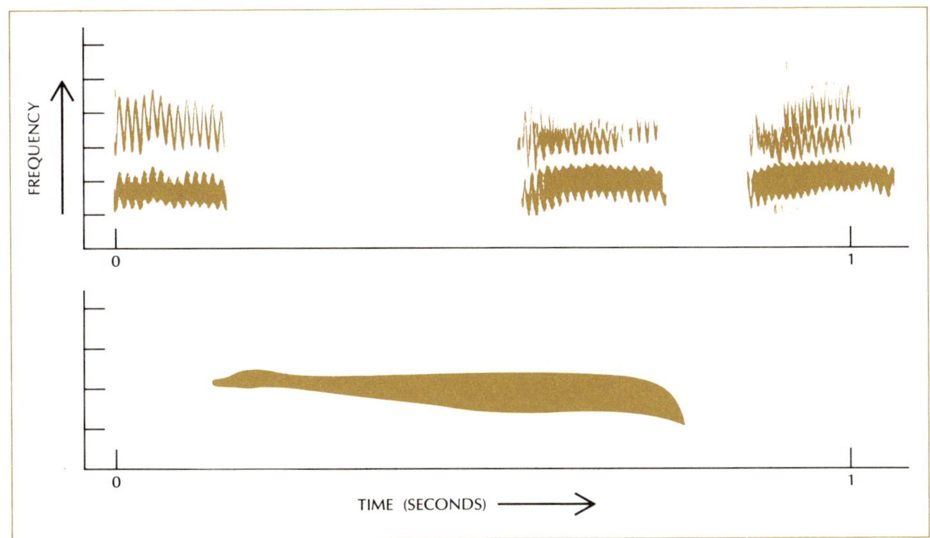

12
Alarm and distress calls of two finches. Fledgling distress call of American goldfinch (top) is rapidly repeated and contains a mixture of frequencies, making the caller easy to find. Hawk alarm call of British chaffinch (bottom) is a pure tone, very difficult to locate.

In many instances we can infer the meaning of the signal from the behavior of the signaler and the receiver. Some signals seem to convey a message such as "Get out!" or "I'm a virgin silkworm moth, located here." More precise information about the content of a signal can often be obtained by examining the signal itself. For example, loud, repeated acoustical signals that contain many frequencies are easily localized, but a pure tone that fades in and out has ventriloquistic properties and its source is not easy to find. The distress calls of newly fledged birds fall into the first category, and the alarm calls of birds sighting a hawk fall into the second, as shown by the sonagrams (voiceprints) reproduced in Fig. 12. The fledgling's distress call transmits to the parent information about the position of the young bird, but locational information is obviously not part of the function of the hawk alarm call. Chemical signals also provide information about the position of the signaler; their evaporation establishes a concentration gradient that enables the receiver to home in on the source. And, of course, visual signals transmit information on position as soon as they are detected.

The complexity of the communication system increases with the complexity of the social system. Frogs are not highly social and their communication is correspondingly simple. In the spring male frogs move to ponds where they sing; their song is an example of long-distance communication, giving only the location, sex, and species of the callers. Sexually mature females are attracted to the ponds by the singing. Male frogs

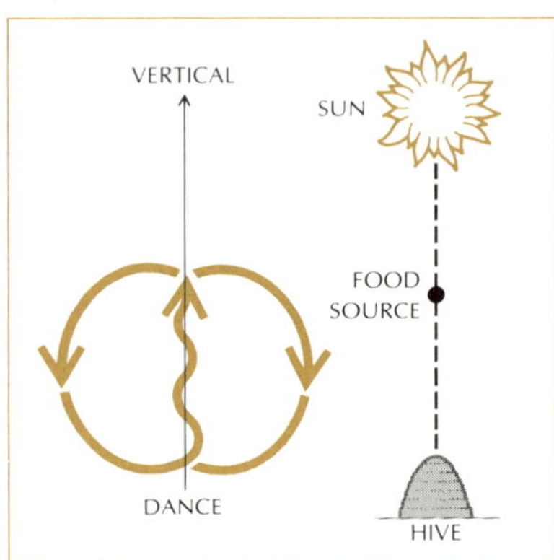

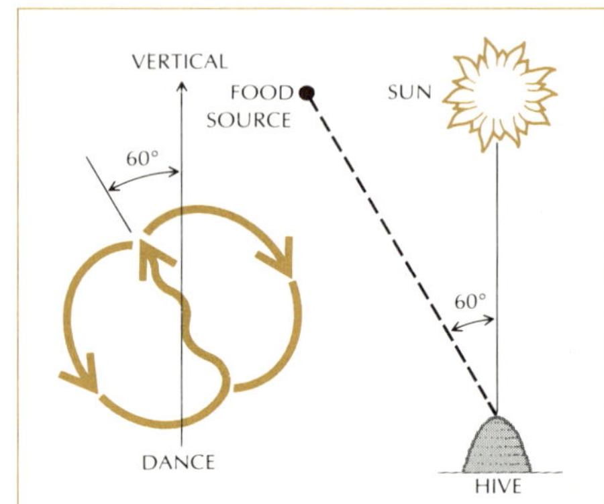

will attempt to mate with just about any nearby object of the appropriate size. If a male attempts to clasp and mate with another male, the second frog gives a harsh call, struggles, and is released; but if a female is clasped she remains still and lets nature take its course.

In contrast, the honeybee has evolved an extremely complex signal system. The classic work of Karl von Frisch of the University of Munich has unraveled part of this remarkable system. His work has revealed that upon finding a rich source of food a scout honeybee returns to her hive and performs a waggle dance on the surface of the comb. Depending on the direction of the source of nectar, the scout performs the dance at a specific angle from the vertical. The length of the waggle run indicates the distance to the source (Fig. 13). Flower scent on the body of the scout indicates which species of flower provides the food. In addition to the visual signals, the scout produces sounds that may transmit the same kind of information. This redundancy of information is also characteristic of advanced man-made communication systems, and of the error-checking codes frequently programmed into computers.

The complexity of some animal communication systems has led some laymen to call them animal language, but there is no true comparison between human language and the communication systems of other animals. Human language is uniquely complex and flexible, a system of culturally learned sound patterns that can be variously combined to generate an infinite variety of words and sentences. New words can be coined to refer to

13

◄**Dance of the honeybee** on the vertical face of the
comb indicates the location of a source of nectar. If
the source is very near the hive, returning scout performs round dance (left). If nectar is farther away,
scout performs waggle dance (center and right).
Direction of waggle run (wavy arrow) relative to the
vertical indicates direction to food relative to the
position of the sun. Waggle dance is actually a ritual
flight during which the scout bee buzzes his wings and
swings abdomen from side to side. Length of waggle
run indicates distance to source.

new concepts. Nothing in the animal kingdom, even among the higher
primates, approaches human language. The origin of human language remains a mystery, a phenomenon unique in the animal kingdom.

ORIENTATION

The ability of animals to orient themselves to their surroundings and to
navigate accurately over long distances stands out among the impressive
feats of animal behavior. Animal guidance systems rely on many of the
same cues as do man-made systems: sonar echoes, the position of the sun
and stars, and the measurement of elapsed time. But at least one animal
technique, navigation by smell, has no parallel in human technology.

Some species of bat broadcast ultrasonic sound through the nose or
mouth, and the returning echoes enable the animal to locate and avoid
obstacles such as tree limbs and telephone wires, and to capture moths in
flight. Echolocation is a surprisingly precise mechanism for orienting movement. Although moths can detect bat sonar and take evasive action, the
bat can follow their spiraling dives and usually snare them.

The list of animals that use sonar includes several birds, particularly
the oil bird of Trinidad, many whales and porpoises, rodents such as the
Norway rat, and possibly some species of fish, which spend part or all of
their lives in dark caves, tunnels, or in the ocean depths.

SUN COMPASS

Von Frisch studied the orientation mechanism of the honeybee by sealing off a hive from all but red light, which bees cannot see but which permits them to be observed by humans. The dances of scout bees became disoriented in the red light, but given a glimpse of the sun, they immediately reoriented their dance pattern. Many other insects, birds, and fish use the sun as a compass to determine the direction of a specific goal; they can then set out in this direction and reach the goal after traveling a prescribed distance.

To maintain its original compass direction, an animal must compensate for changes in the position of the sun throughout the day. This ability depends on some sort of preprogrammed clockwork mechanism built into the animal. The functioning of the time-compensating mechanism in bees is most apparent when bees are preparing to swarm. The scout bees indicate the position of a proposed new hive by the same kind of dance they use to indicate the location of food. The dance goes on for hours and even for days in the dark interior of the hive. Although, in this case, the continuously dancing bees cannot see the sun from inside the hive, they gradually change the angle of their dance as the day wears on, a change that coincides with the movement of the sun during that time.

Biological timekeeping mechanisms occur in many organisms, from protozoa to vertebrates. Beyond the hypothesis that such biological clocks must operate on some electrochemical cycle, the nature of the mechanism remains a mystery and is currently the subject of lively speculation and intensive research.

NAVIGATION

Long-distance seasonal migrations are part of the life cycles of whales, reindeer, sea turtles, salamanders, salmon, and even monarch butterflies, but birds outdistance them all. The bristle-thighed curlew nests along a restricted piece of Alaskan shoreline and migrates 6000 miles over the trackless Pacific to winter in the Polynesian Islands. Some migrants have very specific destinations: Kirtland's warblers nest only in a few counties in Northern Michigan and winter only on a few of the Bahama islands.

To orient themselves, migrants may rely on a compass sense, a time sense, and, for lack of a better term, a map sense that is not yet fully under-

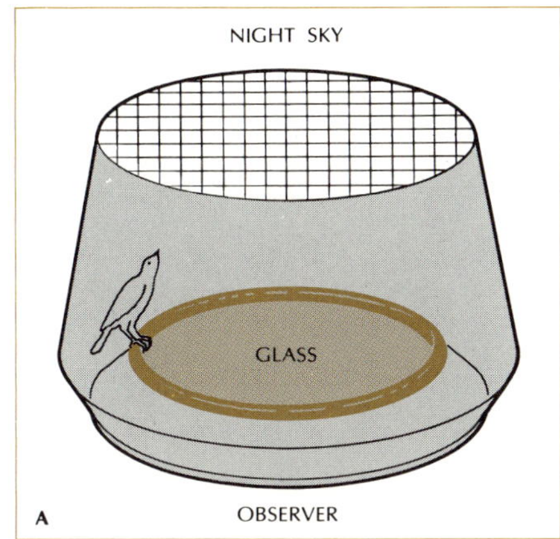

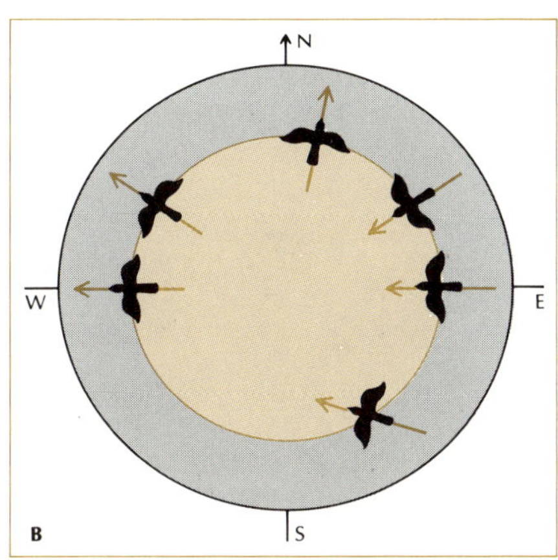

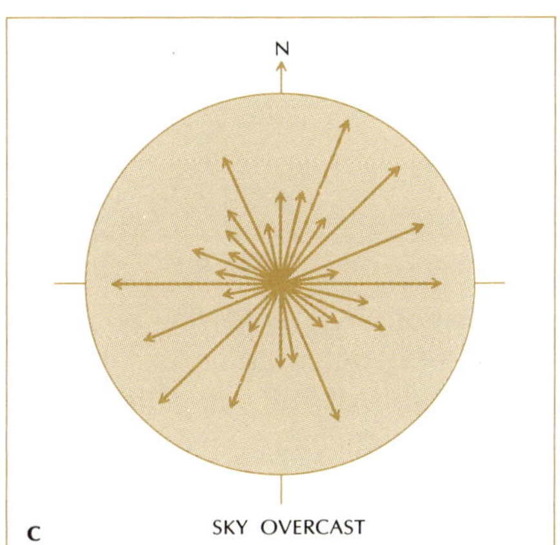

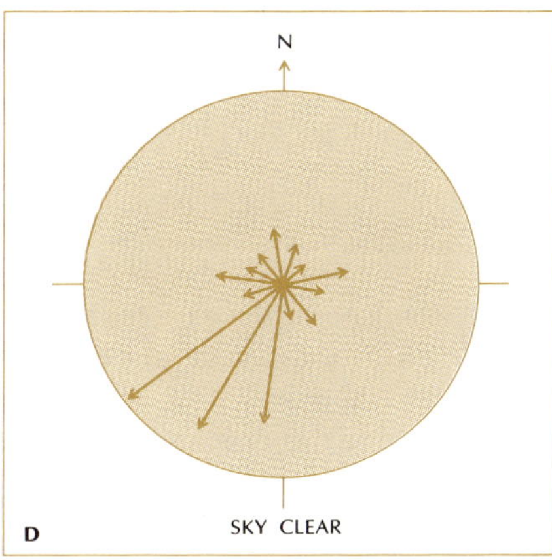

14
Celestial navigation of migrating birds has been demonstrated in several experiments. Here bird perches on a ring in a glass-bottomed cage (A). An observer below the cage records the direction of the bird's flutterings when it can see the night sky (B). Results are summarized in diagrams C and D. Length of arrows indicates number of times that bird fluttered in each direction. During their migratory season (autumn), birds in this study oriented themselves toward the southwest, the direction of their winter quarters, on clear nights when they could see the stars. On overcast nights, birds fluttered randomly.

stood. When caged out-of-doors, starlings flutter in the normal migratory direction in both spring and fall. A series of experiments demonstrated that the direction of their fluttering depends on a knowledge of the position of the sun.

Other birds, especially small songbirds, migrate at night. When caged during the migratory season they flutter at night in the correct migratory direction so long as they can see the natural night sky. If their cages are taken into a planetarium, the direction of their fluttering can be consistently altered by changing the planetarium sky, demonstrating that these birds fly by the stars, not by the sun (Fig. 14).

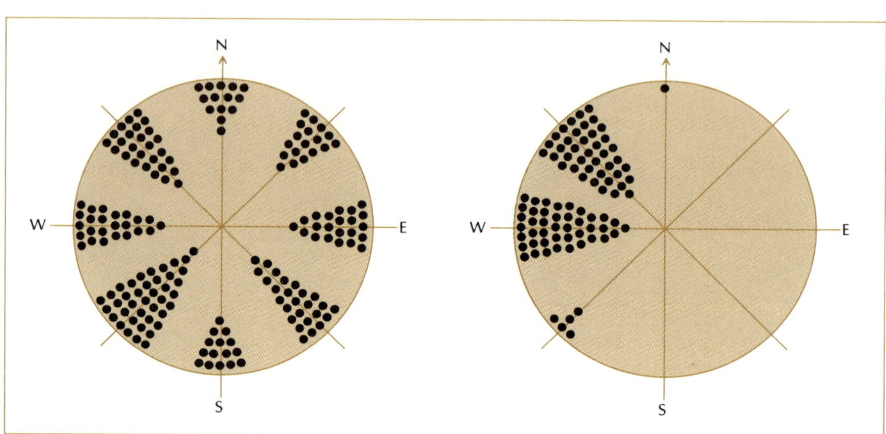

15

Sun is compass for migrating starlings. Orientation of caged starlings was recorded during their migratory season. Each dot represents ten seconds of activity directed toward a particular heading. When the daytime sky was overcast, starlings oriented randomly (left). When sky was clear, the birds tended to fly away from the sun: westward early in the day and northwest later (right).

In theory, birds can reach their destinations by dead reckoning, a combination of determining direction and then flying a certain distance to reach the goal (Fig. 15). But some birds are capable of true navigation: they can determine their changing position, be aware of the position of some distant goal, and fly directly to it (Fig. 16). This ability demonstrates itself in experiments in which birds are taken from nests containing their offspring and shipped to unfamiliar territory where they are released. The swiftness of their return to the nest is evidence of remarkable navigation. A classic case involves a manx shearwater, a seabird taken from its nest in Wales, flown to Boston, and released. The bird returned the 4800 kilometers to its nest in 12½ days. At an estimated flight speed of 48 kilometers per hour little time was left for the bird to wander from a direct, navigated flight path. Its method of navigation is not yet known, but may involve some mechanism for sensing the earth's magnetic field.

The displacement of birds captured in the midst of migration provides evidence of a map sense, or piloting ability. When inexperienced juvenile birds that had never been to their winter quarters were captured during migration and displaced several hundred miles off course before their release, they completed the migration, but with an error equivalent to the

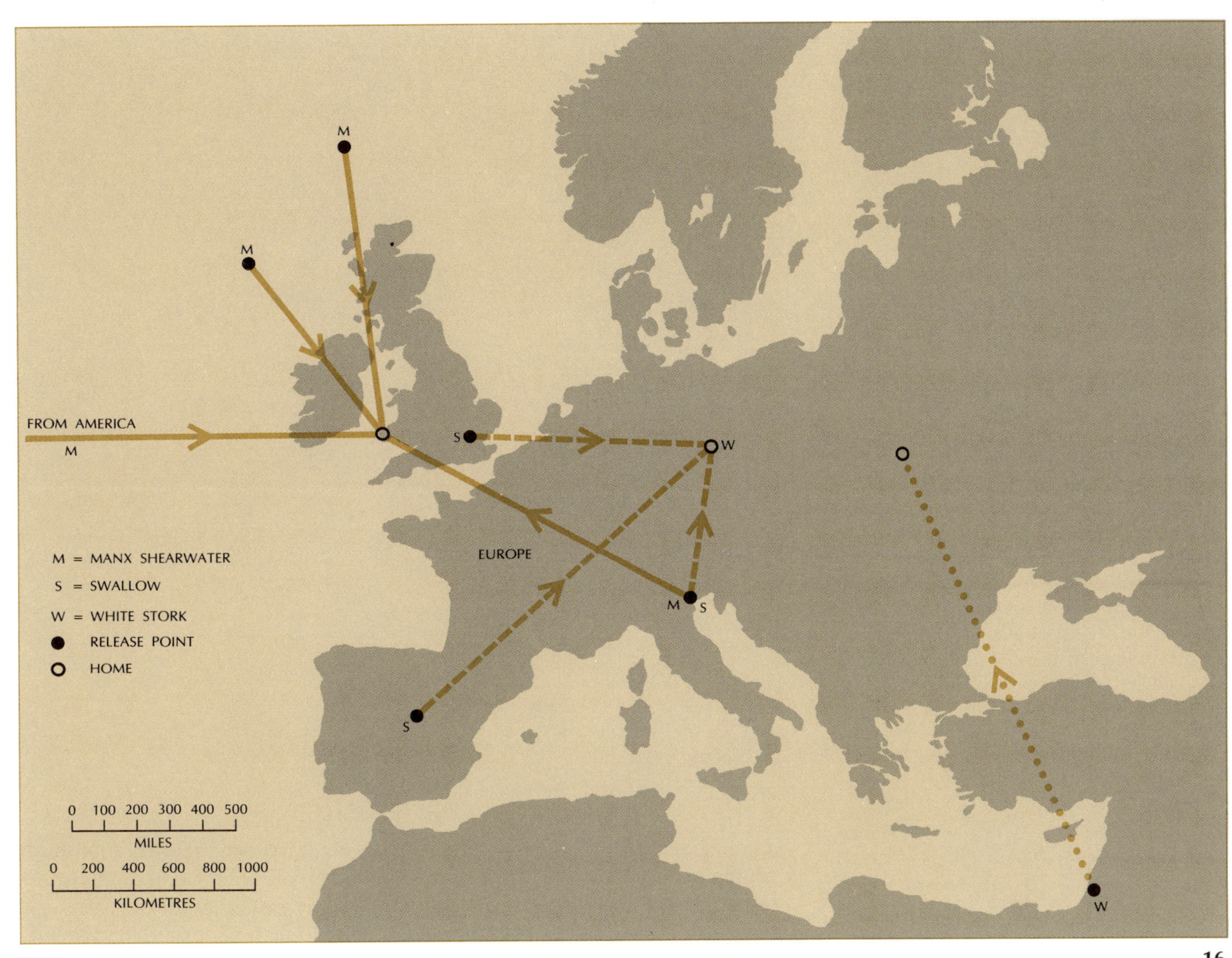

16

True navigation in animals is best demonstrated by a swift return to home after being released in distant and unfamiliar terrain. Map shows a few well-documented returns of banded birds to their home sites in Europe.

17

Piloting ability of birds is demonstrated by displacement experiments. Migrating starlings captured in the Netherlands (A) were transported to Switzerland (B), banded and released. Adult birds returned to their normal winter quarters, as indicated by the sites at which some of them were later recaptured (colored triangles). But young birds migrated to new winter quarters several hundred miles away (colored circles), following a flight path perpendicular to the normal one.

amount of the displacement (Fig. 17). In contrast, experienced adult birds captured and then released an equal distance off course corrected for the displacement and flew directly to their familiar winter quarters. This result suggests that such migrants inherit the ability to fly an appropriate direction and distance to reach the general vicinity of their winter quarters, but once the winter quarters have become familiar to them they can pinpoint their later return, perhaps piloting themselves by recognition of landmarks along the way.

SMELL

The long-range migrations of the Pacific salmon are guided, at least in part, by the sense of smell. Spawned in streams emptying into coastal waters from Alaska to California, the salmon fry grow for a while in the fresh waters, then gradually drift down to the sea, and finally mature in the open waters of the north Pacific. Upon reaching sexual maturity they begin their migratory run back to their birthplace. These salmon probably use a sun compass to guide themselves back to the coastline near the mouths of their native rivers. But when they enter the fresh waters of their native stream they depend on their chemical sense to guide them often hundreds of miles upstream to the exact branch where they were hatched. Their sense of smell is so precise that salmon can detect minor differences in chemical composition of adjacent stream branches and can correctly choose the one in which they were spawned.

There is also evidence that smell guides the green turtle on its migration from its pastures of turtle grass off the coast of Brazil to its nesting beaches on Ascension Island, a tiny speck thousands of miles away in the open Atlantic. Swimming near the surface, the turtles reach the general vicinity of the island, perhaps within 150 kilometers, by some compass mechanism, most likely a sun compass, then according to one hypothesis, they home in on the island by smell. Chemicals associated with the island may establish an olfactory gradient which guides the turtles to the very beaches where they were hatched. Much more evidence is required to confirm this hypothesis, but so far it stands as the most likely explanation of an astonishing navigational feat.

AGGRESSION

Aggression between animals of the same species usually involves competition for mates, for food, or for dominance in a group. Although aggression can be defined as the attempt or the threat to cause physical injury, prolonged fighting and serious injury are rare among animals under natural conditions. Usually the weaker animal quickly recognizes defeat and retreats or surrenders. Natural selection favors individuals who live to fight another day, no matter how; and in many species combat has become a ritualized contest in which the winner is decided merely by threatening displays or harmless tests of strength.

Threat postures of the black-headed gull emphasize the dark face mask and the beak. Such postures warn rivals of intention to attack.

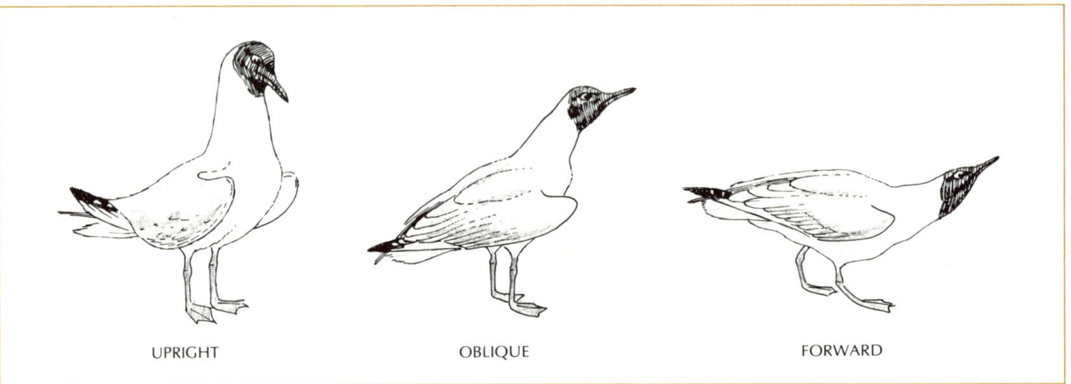

Animals equipped with deadly weapons such as beaks, claws, horns, and fangs have evolved effective surrender signals that automatically halt an attack. Among birds, the signal is often turning away the beak; among wolves it is offering the nape of the neck.

Aggression is most evident at the beginning of the breeding season, when the males of many species occupy a territory that they defend against other males of the species. Territoriality is common among vertebrates, particularly birds and mammals. Among birds of prey, the territory is the source of food, and may be as large as several square kilometers; among other birds, such as the herring gull, the territory may encompass only a few square meters around the nest.

Behaviorists disagree on the evolutionary history of territoriality. In many species territorial behavior seems to be a compromise between the advantages of spacing out and those of proximity. Spacing out diminishes population density, reduces disturbances to mating, keeps the nest site undisturbed, discourages other adults from preying on the broods, and in some species with large territories, provides an exclusive feeding area. Proximity provides a community alarm system and social defense against predators.

When a male intrudes upon another's territory, the resident male attacks or threatens the invader. The threat display is a signal between

Courtship displays of the black-headed gull begin with the male's long call (left). Female alights nearby, and then both birds strike ritualized threat postures (center) followed by an appeasement posture (right).

animals. The black-headed gull has three threat postures: the upright, the forward, and the oblique (Fig. 18). When a flying stranger approaches within a yard or so, the male strikes the oblique posture and emits a long call, a signal that conveys the message, ''This is my territory. Keep out!'' In general, males avoid other long-calling males. But if the stranger alights near the center of the territory, he is likely to be attacked. If he alights near the edge of the territory and approaches the center, the resident male may threaten him with the forward or the upright posture. Both of these threat postures emphasize the menacing beak and the ominous dark face mask. The intruder may retreat promptly, or he may return the threatening posture, possibly escalating the threat into an actual attack. Or he may surrender, acknowledging the rights of the resident male by assuming an appeasement posture.

The courtship displays of many species are rituals that serve to de-escalate aggression. Unmated females of black-headed gulls are attracted to calling males and alight near them. Male and female then adopt the forward posture, except that they stand parallel instead of facing each other. After a few seconds first one, then the other, jerks into the upright posture, then faces away (Fig. 19). This sequence of displays clearly indicates a conflict in both birds between three impulses: to attack, to escape, and to remain together. As two gulls become attracted to each other they repeat the same ceremony each time the female alights near the male: forward, upright, facing away. But the aggression of the male and the fear in the female gradually diminish; they spend more time together and the ritual becomes perfunctory. In a few hours or a few days the female begins to perform a head-bobbing movement that stimulates the male to re-gurgitate food and feed her. This is the final step of the mating ritual; soon afterwards, the gulls mate.

In the decade ahead, behavior will be studied more intensively than at any time in the past, not only for its own sake, but also for potential applications in areas of human concern. Pheromones, for example, might supplement toxic insecticides in mankind's perpetual war with the insect. Such com-pounds might be used to lure insects to traps or to confuse their search for mates. There will be further attempts to trace the animal origins of human behavior and the evolution of inherited behavior. Work with animals should also yield powerful insights into the nature of the chemical and neurological mechanisms that regulate behavior. Finally, the next five or ten years should narrow the rift between European ethologists, who favor the observation of animals in the wild, and American animal psychologists, who prefer to carry out controlled experiments under laboratory conditions.

READINGS

Burkhardt, D., W. Schleidt, and H. Altner, *Signals in the Animal World.* McGraw-Hill, New York, 1967

   A splendidly illustrated series of capsule essays on sense organs, environ-mental adaptations, and animal signal systems.

Dethier, V. G., and E. Stellar, *Animal Behavior.* Prentice-Hall, Englewood Cliffs, N.J., 1964 (paper)

   Surveys the topic of animal behavior at an intermediate level; particularly good account of the neurological basis of behavior.

Lorenz, K., *King Solomon's Ring.* Thomas Y. Crowell, New York, 1961 (paper)

   A popular account of an eminent ethologist's personal experiences with ani-mals and the phenomenon of imprinting.

Lorenz, K., *On Aggression*. Harcourt, Brace & World, New York, 1966
> Knowledge gained from a lifetime of studying animals, particularly their aggressive behavior, is applied in an attempt to understand and control human aggression.

Manning, A., *An Introduction to Animal Behavior*. Addison-Wesley, Reading, Mass., 1967 (paper)
> A professional and readable review of the entire field of animal behavior.

Marler, P., and W. J. Hamilton, III, *Mechanisms of Animal Behavior*. Wiley, New York, 1966
> Probably the most complete text available, but written primarily for advanced students.

Matthews, G. V. T., *Bird Navigation*. Cambridge University, New York, 1968 (paper)
> A thorough review of the literature on bird orientation and navigation by a respected worker in this field.

McGaugh, J. L., N. M. Weinberger, and R. E. Whalen (eds.), *Psychobiology*. Freeman, San Francisco, 1967 (paper)
> A collection of reprints from *Scientific American*. Reports of original work in fields ranging from the evolution of behavior to brain stimulation, emphasizing physiological and psychological studies.

McGill, T. E. (ed.), *Readings in Animal Behavior*. Holt, Rinehart & Winston, New York, 1965 (paper)
> A broadly based collection of original research papers.

Tinbergen, N., *Animal Behavior*. Life Science Library, Time-Life Books, New York, 1965
> A well-illustrated, popular treatment by a pioneer in the field of animal behavior.

von Frisch, K., *Bees: Their Vision, Chemical Senses, and Language*. Cornell University, Ithaca, N.Y., 1964 (paper)
> An easy-to-read classical account of the unraveling of the language of bees by the man who did much of the work.

15. ECOLOGY AND MAN

A disgruntled student once defined ecology as the study of everything, all at the same time. The field is so comprehensive that some ecologists regard biology as a subdivision of ecology, instead of the other way around. Ecology integrates data from various sciences, including biology, chemistry, climatology, and oceanography; it crosses the boundaries of geology, agronomy, genetics, and physics; it combines points of view as diverse as those of natural history and computerized mathematics. Its ultimate aim is to understand the intricate relationships of living things to one another and to their environment.

The basic concerns of ecology are energy, matter, space, and time. How do living things capture and use energy? How is matter recycled through the living world? How does geography influence the distribution of organisms? How do the plant and animal populations of a region change with the passage of time?

The interaction of these factors creates a dynamic, delicate balance between living things and their environment. The balance of nature is a complicated self-adjusting mechanism, but one that is highly vulnerable to tampering. For at least 50,000 years, human beings have seriously jeopardized this balance. In the last 200 years, the uncontrolled growth of both human population and human technology have not only upset the balance of nature, but have depleted and polluted the natural resources on

◀ **Twentieth-century still life.** Floating garbage includes dead fish, wooden board, and beach grass which will eventually decay and be recycled. Plastic cup and aluminum beer can are virtually indestructible by natural processes. They might prove to be the most enduring monuments to our time.

which that balance depends. The most important theme of 20th-century ecology is the impact of one blundering animal—man—on the living world.

HUMAN ECOLOGY

Will man eventually create an environment that he can no longer dominate? In the long run, the answer depends on whether or not we can control the "population bomb." We must use the powerful tools of ecological analysis to evaluate and understand man's place in nature. His dominance, even his survival on the planet, depends on his management of its limited resources. The alarming depletion of the earth's minerals, the pollution of its air and water, and the shortage of food are all basically the result of overpopulation. Almost four billion people live on earth today. By the year 2000 there will be nearly twice this number. If this rate of increase continues for about 200 years, the planet will be overrun by a human horde numbering 100 billion.

Man is no longer in harmony with his environment. Paying only lip service to conservation, he is living from day to day. The pious suggestion is often made that we must conserve our resources. But how? Many of these resources must be used now if we are to survive. We need more and more land for housing, highways, shopping centers, and factories. Increasing amounts of fuel must be burned to keep the expanding population warm in winter and cool in summer, and to produce the consumer products upon which we all depend. We must dispose of our industrial wastes and sewage even though it often kills entire populations of fish and other aquatic organisms. The need for lumber and paper products is increasing rapidly, even though lumbering often removes the natural erosion control of the forest cover.

No longer can we move out into the country to avoid industrially polluted air. We can never build enough swimming pools to replace the facilities originally provided by the now-poisoned lakes and rivers of the world. And to add to the miseries of mankind, we are beginning to starve.

Where will the populations of the future get their food? Will the energy resources available to man be adequate? No longer do the agriculturists reassure us that we can always make enough food to feed mankind just by intensifying our farming efforts and increasing our harvests. The untapped resources of the sea are often mentioned as the salvation for starving humanity. Using the antiquated and not overly ambitious fishing techniques of today, man is presently harvesting only about 50 million tons of fish per year. The oceans are so large and deep that they are considered by many to contain an unlimited supply of food. However, Dr. Wolf Vishniac of the University of Rochester has made some rather astounding calculations concerning this "vast storehouse of food." Using mathematical calculations based on the amount of light energy striking the surface of the oceans, the total surface area of the oceans, and the amount of energy required by the plants in the sea to produce organic compounds, he concluded that the oceans are producing just about all they can from the amount of light available from the sun. Furthermore, they are producing food at about the same rate as we are harvesting it. It appears that the oceans are barely holding their own. It is true that the total quantity of fish in the ocean is unknown and that there must be enough to feed the world for many years, but it will be a losing battle. We shall soon be removing food from the oceans faster than they can produce it. We must no longer look to the organisms of the sea as the solution to our future food crisis.

COMMUNITY AND ECOSYSTEMS

To obtain the knowledge necessary to proceed intelligently with the task of using our limited resources, man must study organisms in their natural habitats. He must discover the reasons that specific plants and animals are found in a given region. From this he must learn to infer the types and the extent of both the beneficial and the detrimental effects that might follow a change in that environment. Then he must have the wisdom to use this knowledge.

To the ecologist the basic unit of the living world is usually a population, a group of organisms of the same species. Collectively, the populations of all the species living in a particular area comprise a community. The biological community, coupled with the physical and chemical aspects of its environment, comprise an ecosystem (Fig. 1). A forest is an ecosystem. So is an aquarium, a city park, or an ocean. Spread in a thin film across the surface of the earth, all of the ecosystems taken together make up the biosphere.

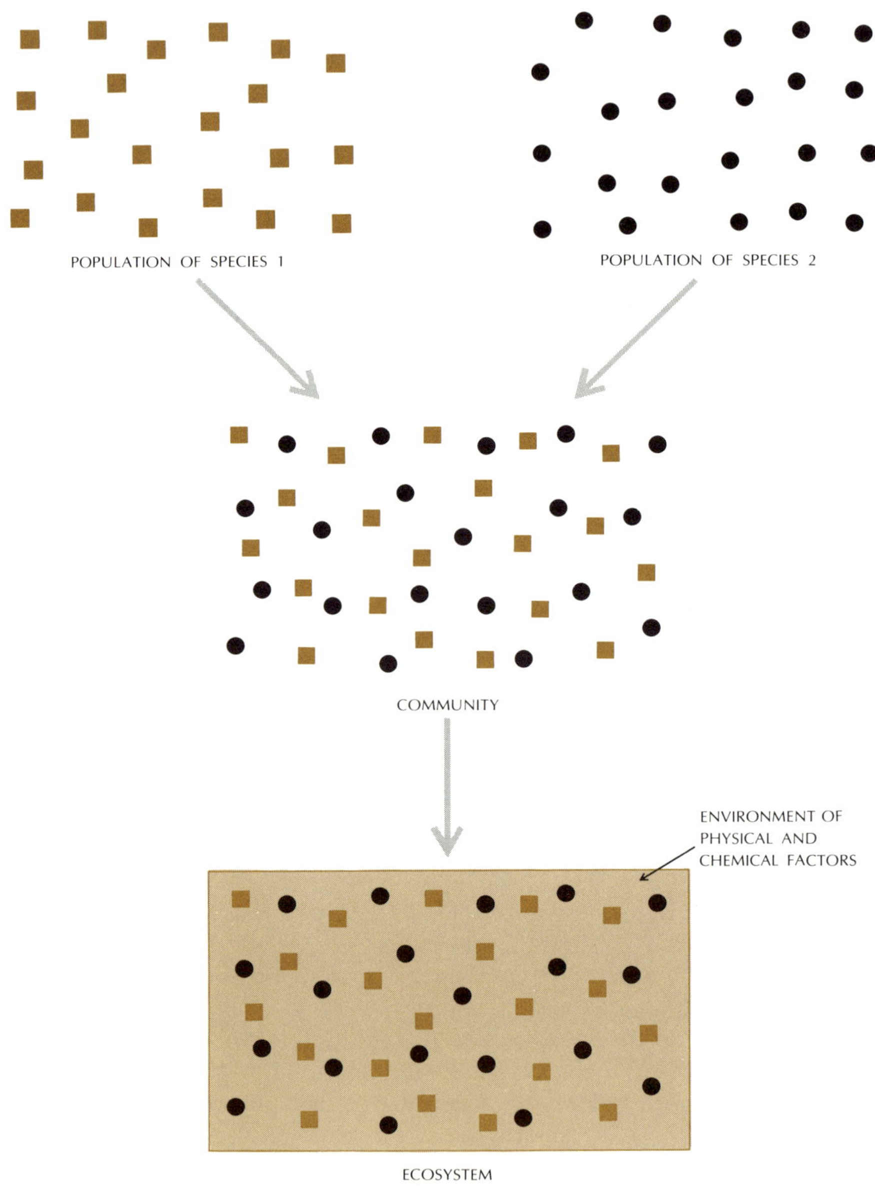

POPULATION OF SPECIES 1
POPULATION OF SPECIES 2
COMMUNITY
ENVIRONMENT OF
PHYSICAL AND
CHEMICAL FACTORS
ECOSYSTEM

On an ecological map the earth's two main habitats, land and water, are divided into distinct habitat zones called biomes: ocean, fresh water, desert, grassland, rain forest, deciduous forest, taiga, and tundra (Fig. 2). Five thousand square meters of desert, an area roughly the size of a football field, produce less than 2.5 kilograms of new biomass per day; the same area of grassland or deciduous forest produces from 2.5 to 15 kilograms per day; rain forests, 15 to 25 kilograms, about the same yield as cultivated farmland. The same surface area of coastal waters yields an average of about 7 kilograms a day, and the ocean depths less than 2.5 kilograms (Fig. 3). The different requirements for survival in each zone profoundly shape the energy economy and the makeup of the communities that inhabit them.

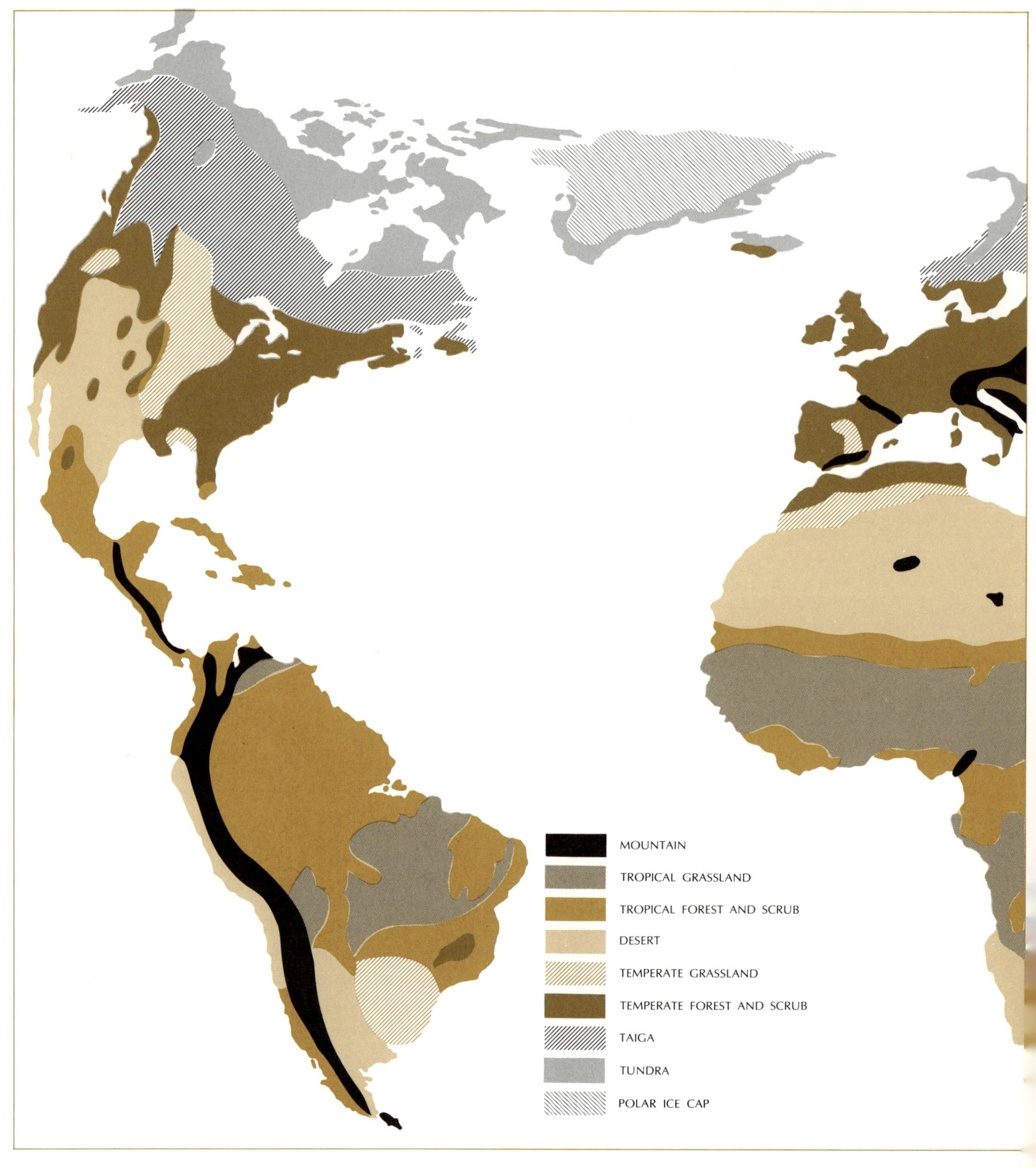

MOUNTAIN
TROPICAL GRASSLAND
TROPICAL FOREST AND SCRUB
DESERT
TEMPERATE GRASSLAND
TEMPERATE FOREST AND SCRUB
TAIGA
TUNDRA
POLAR ICE CAP

Major land biomes are shown in color, greys, and black. Highly productive temperate forest and grassland areas comprise only a small proportion of the world's land surface.

Phytoplankton in this photomicrograph include chains of diatoms and ▶
anchor-shaped dinoflagellates. Magnification about 70 ×.

3
Productivity of biomes. Deserts and deep oceans produce far less new living matter (biomass)
per square meter per day than intensive agriculture, estuaries, and coral reefs. Bars show range
of yield: lower values of each ecosystem are in dark color, upper limits in light color.

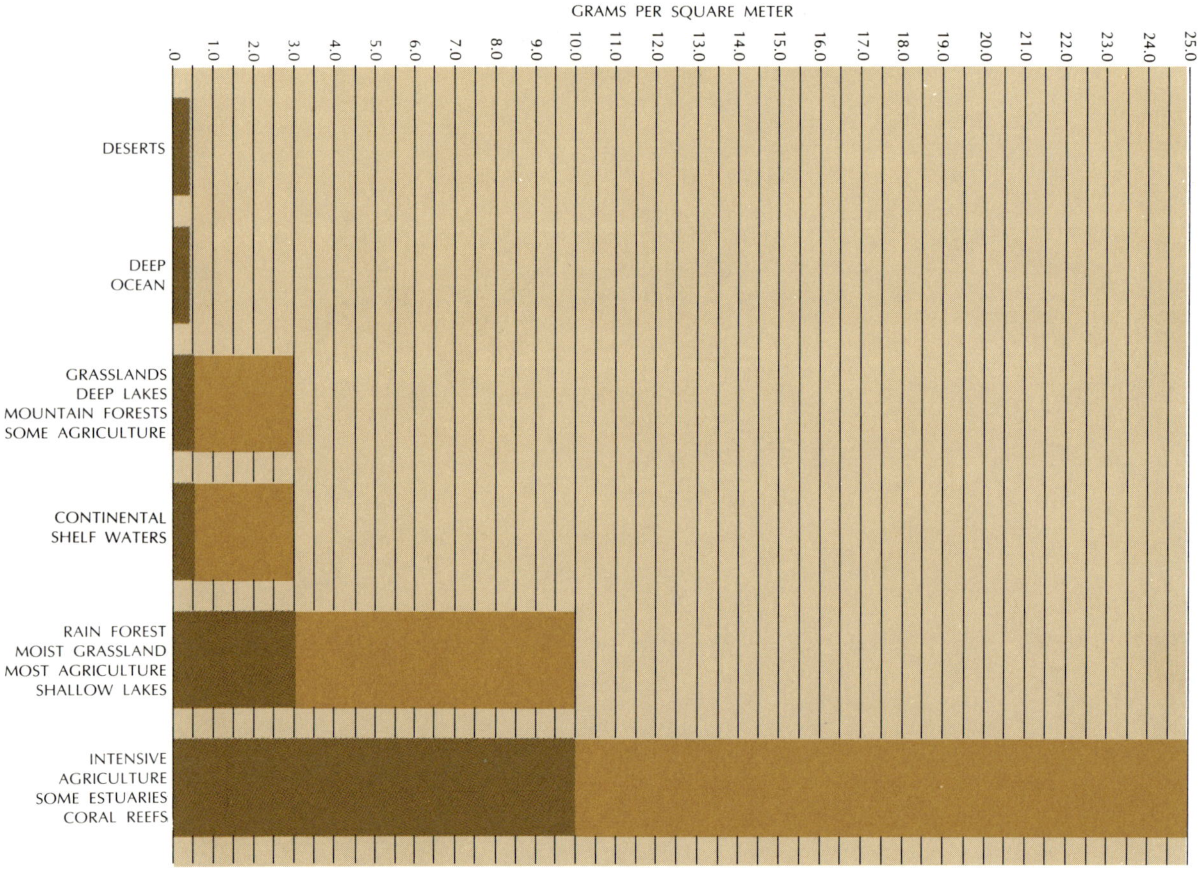

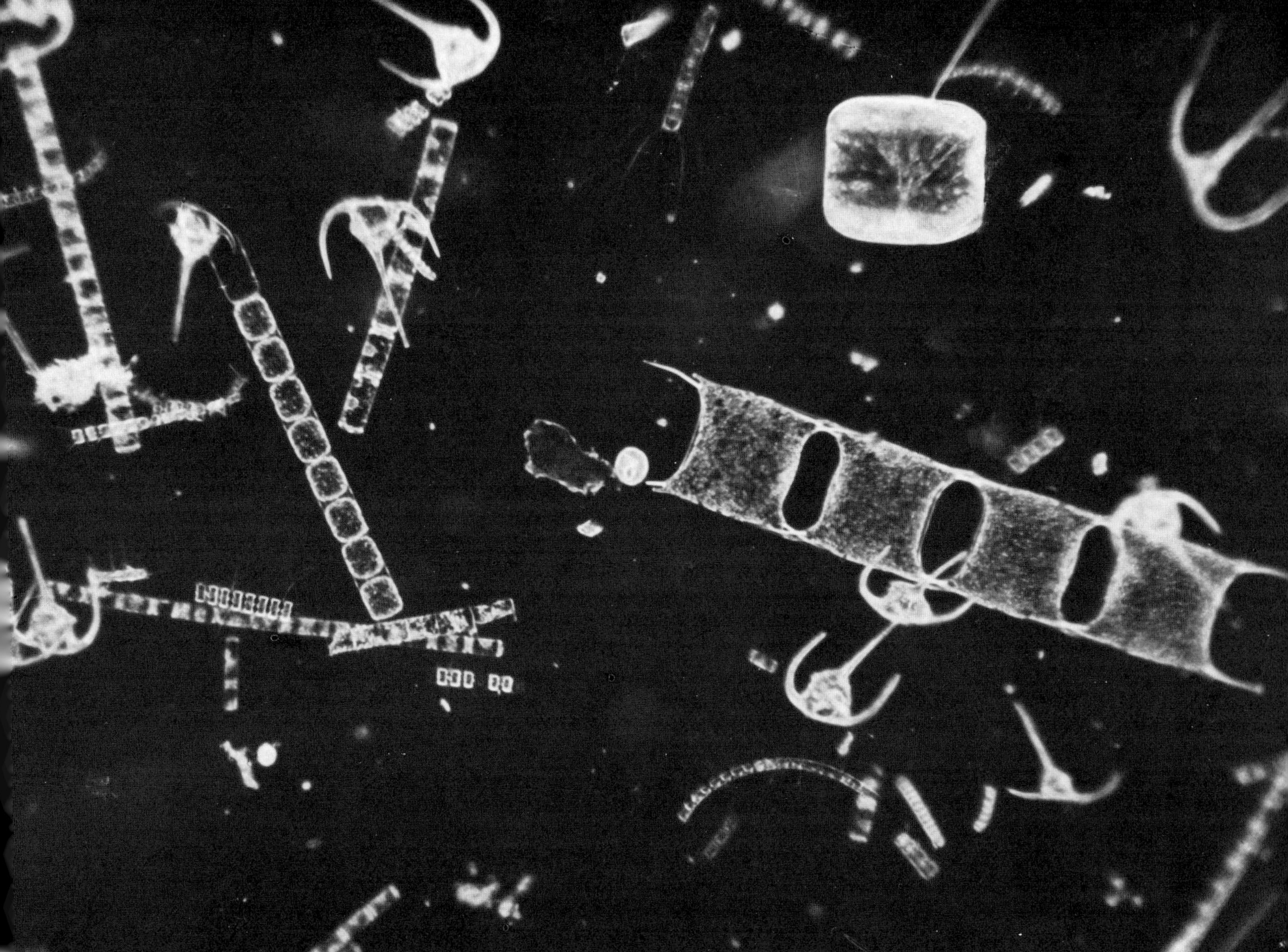

In terms of numbers and sheer tonnage of biomass, the most abundant form of oceanic life is plankton, a collective term for all floating or passively drifting organisms, most of them microscopic. Sometimes called the grass of the sea, the phytoplankton consist largely of bacteria and algae, particularly diatoms, which are single-celled algae encased in transparent silicon shells (Fig. 4). Many types of marine algae and bacteria when disturbed emit flashes of light, creating a luminescent wake behind boats and swimmers at night. Some surface waters harbor large populations of

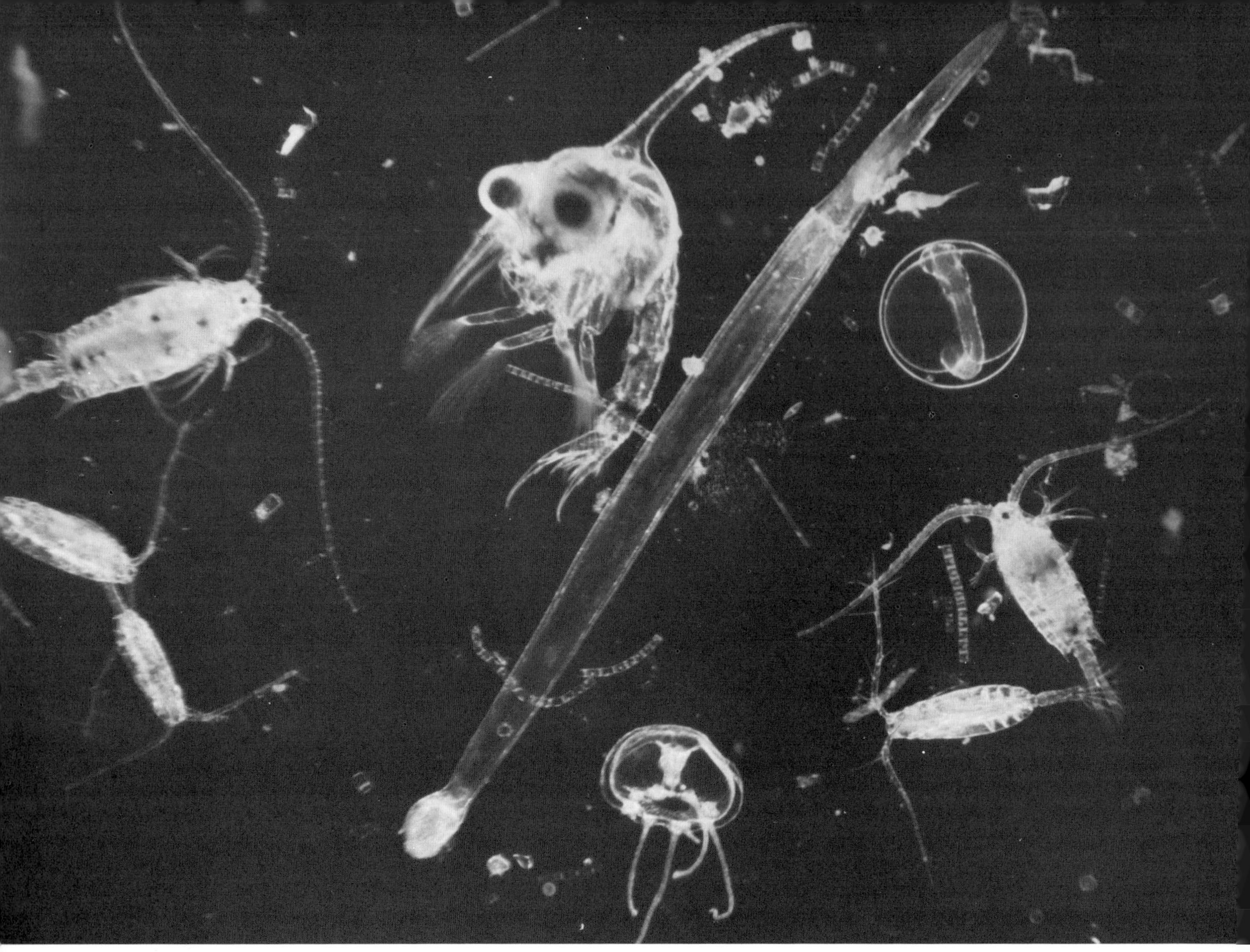

dinoflagellates, the reddish "fire algae" that sometimes multiply explosively, creating "red tides." The zooplankton are tiny animals including protozoa, krill (tiny shrimp), miniature crustaceans, the larvae of many kinds of organisms, and innumerable other microorganisms (Fig. 5).

The size and distribution of each plankton population depends largely on water temperature, light, and the abundance of nutrients. Because all chemical reactions, including those in living organisms, proceed faster at

◀ **Zooplankton.** A long thin arrowworm (*Sagitta setosa*) crosses the picture diagonally. Above it swims the larva of a crab. A fish egg drifts at upper right, and below it swim two copepods. Hydroid medusa is at bottom center. Magnification about 25 ×.

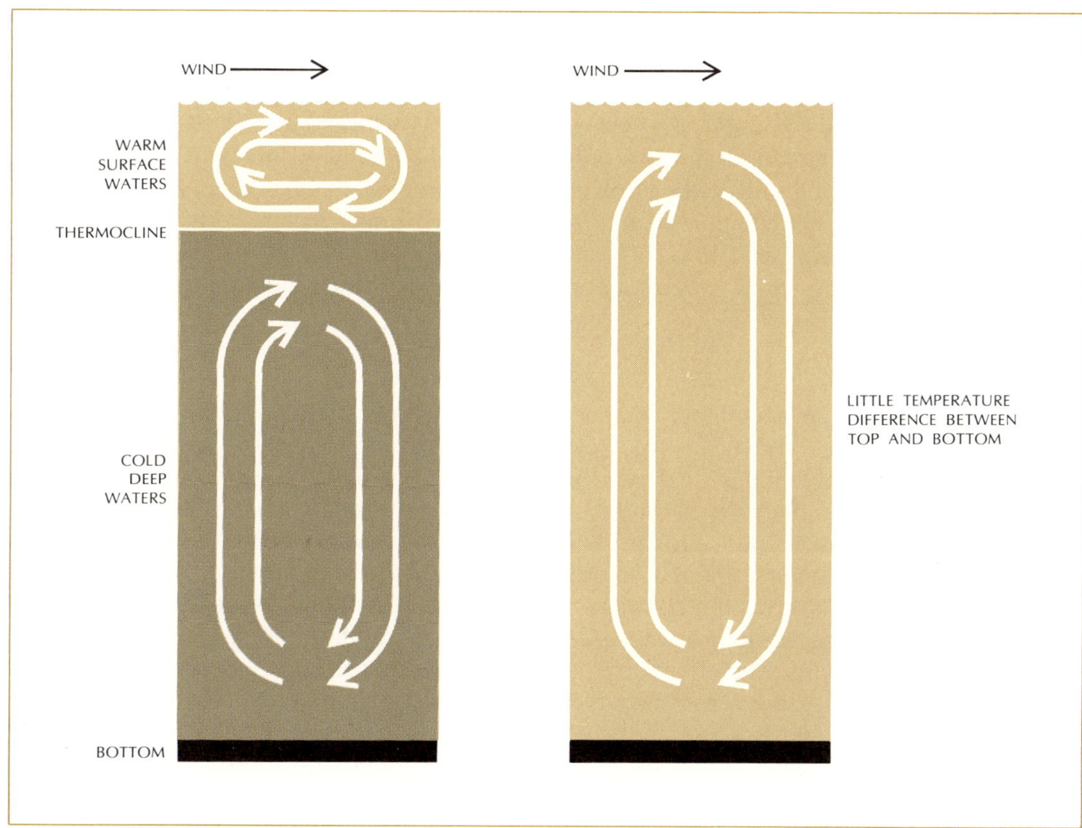

6
Temperature layering interferes with the mixing of deep and surface waters in warm seas (left). In cold seas there is considerable vertical mixing (right).

higher temperatures, we might expect warm seas to generate a much larger tonnage of biomass than cold seas. Actually the reverse is true because of a physical effect called temperature layering. Heated by the summer sun, the surface water in temperate and tropical regions warms and expands slightly and literally floats on the colder, denser layer below. The boundary between the two layers, known as the thermocline, inhibits vertical mixing of the two layers (Fig. 6). In cold seas this type of stable layering is rare.

In these cold areas, although a steady "rain" of dead organic matter sinks slowly to the bottom, some of it is returned to the surface by vertical mixing, providing an important source of nutrients to the organisms in the sunlit surface waters. The spectacular clarity of tropical seas, like those around the Bahama Islands, is largely due to the relatively low density of plankton. By inhibiting vertical mixing, the thermocline in the tropical areas decreases the recirculation of nutrients into the surface layers. This in turn helps explain why tropical waters, despite their diversity of species and high temperature, support a comparatively small total biomass.

ENERGY TRANSFER

Within a community, energy from the sun is captured by photosynthetic plants and flows from organism to organism in a series of steps called a food chain. In a typical food chain the basic energy converters are the green plants. The herbivores, or grazers, obtain energy by eating the green plants. These primary consumers in turn supply energy to the secondary consumers, the carnivores. Decomposers, usually bacteria and fungi, break down the complex compounds of the bodies of dead organisms and of the waste products of living ones. In addition, all of the organisms that comprise the chain are in turn fed upon by parasites. To complicate things further, a given organism can occupy more than one link in the chain at the same time: a human eating a forkful of potatoes and steak is simultaneously a primary and a secondary consumer. Because food chains tend to be tangled and complicated, most ecologists now prefer the term food web. Figure 7 is a highly simplified diagram of part of the food web of a western chaparral community. Such complexity explains why no food web for a natural ecosystem has been completely worked out.

Each level of nourishment in a food chain is called a trophic level. Green plants, the primary producers, comprise the first level, herbivores the second, carnivores the third, and so on. The flow of energy along a food chain from trophic level to trophic level is highly inefficient. Between each level, about 90 percent of the available chemical energy is lost to the environment. The existence of such a loss can be predicted from the Second Law of Thermodynamics (Chapter 3). The magnitude of the loss is more difficult to estimate. Howard Odum attempted this formidable task in a classic study of the ecology of Silver Springs, Florida. Odum set himself

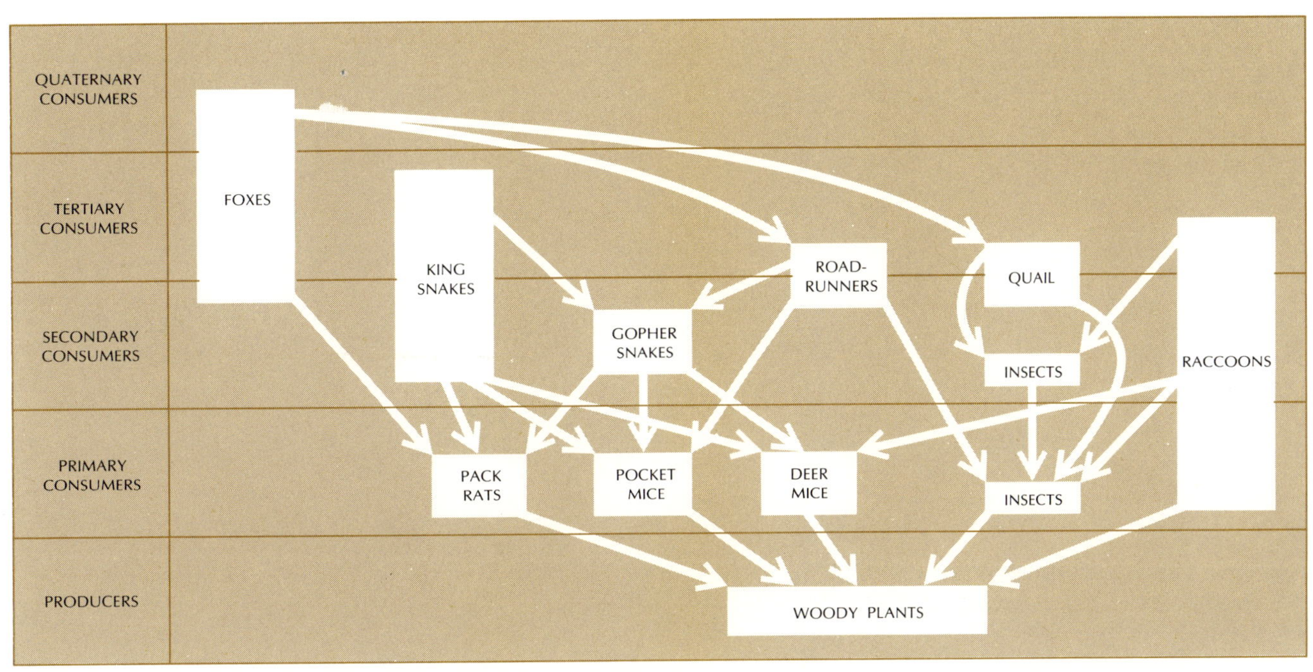

7

Food web of chaparral community. Arrows indicate food sources of each population. Deer mice feed on woody plants; king snakes feed on gopher snakes, deer mice, pocket mice, and pack rats.

the monumental task of deducing the food web of this stream and then of estimating (1) the amount of energy entering each trophic level, (2) the amount of energy converted to heat by respiration in each level, and (3) the amount of energy retained in a form available for use in the next trophic level. He calculated the ecological efficiency of each level by dividing the amount of energy available for use by members of one level into the amount of energy available for use by members of the next level. The efficiency turned out to be about 16 percent for the plants, 11 percent for the herbivores, and 6 percent for the first level of carnivores. Due to the difficulties

Energy transfer   447

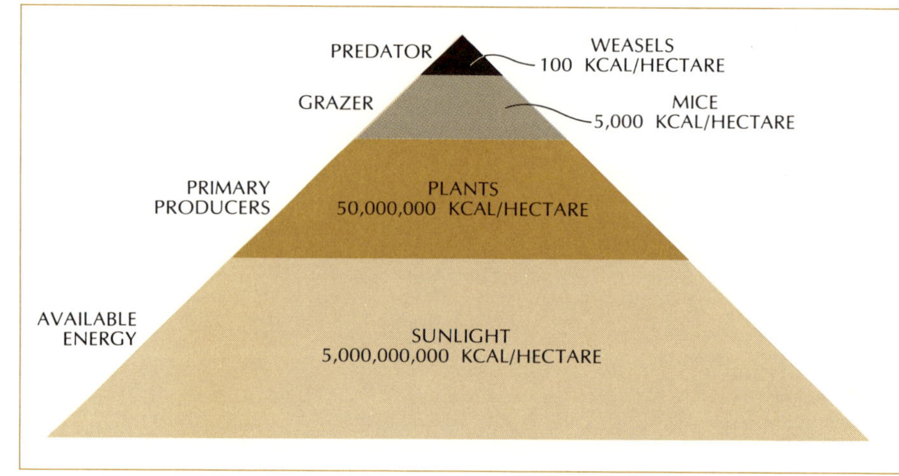

8
Energy pyramid depicts the energy value of the populations of organisms in three trophic levels as compared to the energy supplied to the ecosystem by the sun.

of collecting data in a complex natural environment, these figures were obviously only rough approximations. But they led to the design of laboratory experiments under more controlled conditions. The ecological efficiencies found in these studies ranged from 6 to 13 percent; ecologists now often use a figure of about 10 percent for the average ecological efficiency of each trophic level until they can make a more accurate estimate.

The flow of energy through a food chain can be represented as a pyramid (Fig. 8). Because of the low efficiency of energy conversion at each link, the flow is less in each higher layer. Moreover, the total reservoir of energy in each trophic level is generally much greater than at the next higher level. In other words, the green plants of a community typically contain much more total chemical energy than the herbivores, and the herbivores much more than the carnivores.

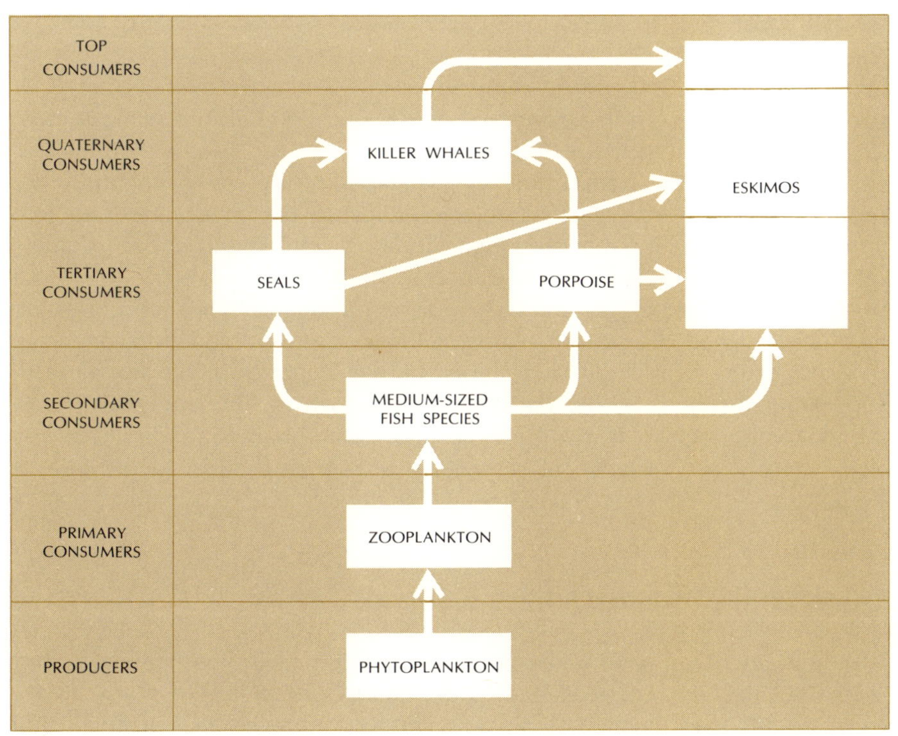

9
Arctic food web illustrated here consists of six trophic levels. Eskimos actually occupy at least three trophic levels. The shorter the food chain, the more efficient the utilization of energy trapped by the primary producers.

AN ARCTIC ENERGY PYRAMID

It is possible to calculate the amount of energy which must be captured by plants to support a food chain consisting of a given number of trophic levels. The usual three-level food chain consists of primary producer, herbivore, and carnivore. But some food chains are more complex, including one from the Arctic that consists of six trophic levels. In the Arctic Ocean, microscopic phytoplankton is fed upon by larger zooplanktonic organisms, which are eaten by small fish, which are eaten by seals and porpoises, which serve as food for killer whales, which are killed and eaten by Eskimos (Fig. 9). An Eskimo must ingest about 3000 calories a day to survive. If we assume only a 10 percent efficiency of energy transfer for each trophic level, this means that about 30,000 calories must enter the trophic level occupied by the killer whale; 300,000 calories must enter the secondary carnivore level of seals and porpoises; and an average of 300,000,000 calories per day must be absorbed by the phytoplankton.

By reducing the number of trophic levels in any energy pyramid, the proportion of the total amount of energy lost to the environment as heat can be decreased. For man, the ideal situation would be to utilize the sun's rays directly and to make his own food by photosynthesis. Since this is not possible, the next best alternative is for him to occupy the second trophic level and to feed directly on plants. This suggestion has led to extensive research on cultivating aquatic plants such as algae for direct human consumption. At present, most of the algae grown in this way are not too palatable although their caloric and protein content are high. In the future, faced with the alternative of starvation, man might adjust to the role of herbivore.

THE RECYCLING OF MATTER

Although energy is continually being degraded into useless heat in its one-way flow through an ecosystem, the chemical elements in nutrients are recycled again and again. In any specific trophic level, the consumer that feeds on the organisms of the previous trophic level breaks down the giant molecules in its food into simpler chemical fragments, then reorganizes these fragments into the complex macromolecules of its own cells. Not all of the matter from the lower trophic level is converted into the biomass of the next. Each step creates waste products: soluble inorganic and organic compounds and particles of detrital material. In short, the nutritional activity of each trophic level produces a quantity of new biomass, a pool of small molecules, and a pool of detritus. The consumer biomass is directly available to the organisms in the next trophic level. The soluble inorganic molecules can be reused by the plants in the first trophic level. The soluble organic molecules may serve as food for populations of fungi and bacteria. The detritus is broken down by these organisms into additional soluble compounds. Some of the detritus, as well as the new bacterial cells, serves as food for small animals. Thus, by various pathways, the chemical elements within an ecosystem are cycled back into the food web where they may be reused innumerable times (Fig. 10).

10

Matter is recycled through a food web. Small molecules that enter the web (upper left) are eventually converted either to biomass and passed to a higher level (right) or they become detritus that is broken down by decay bacteria and recycled (bottom).

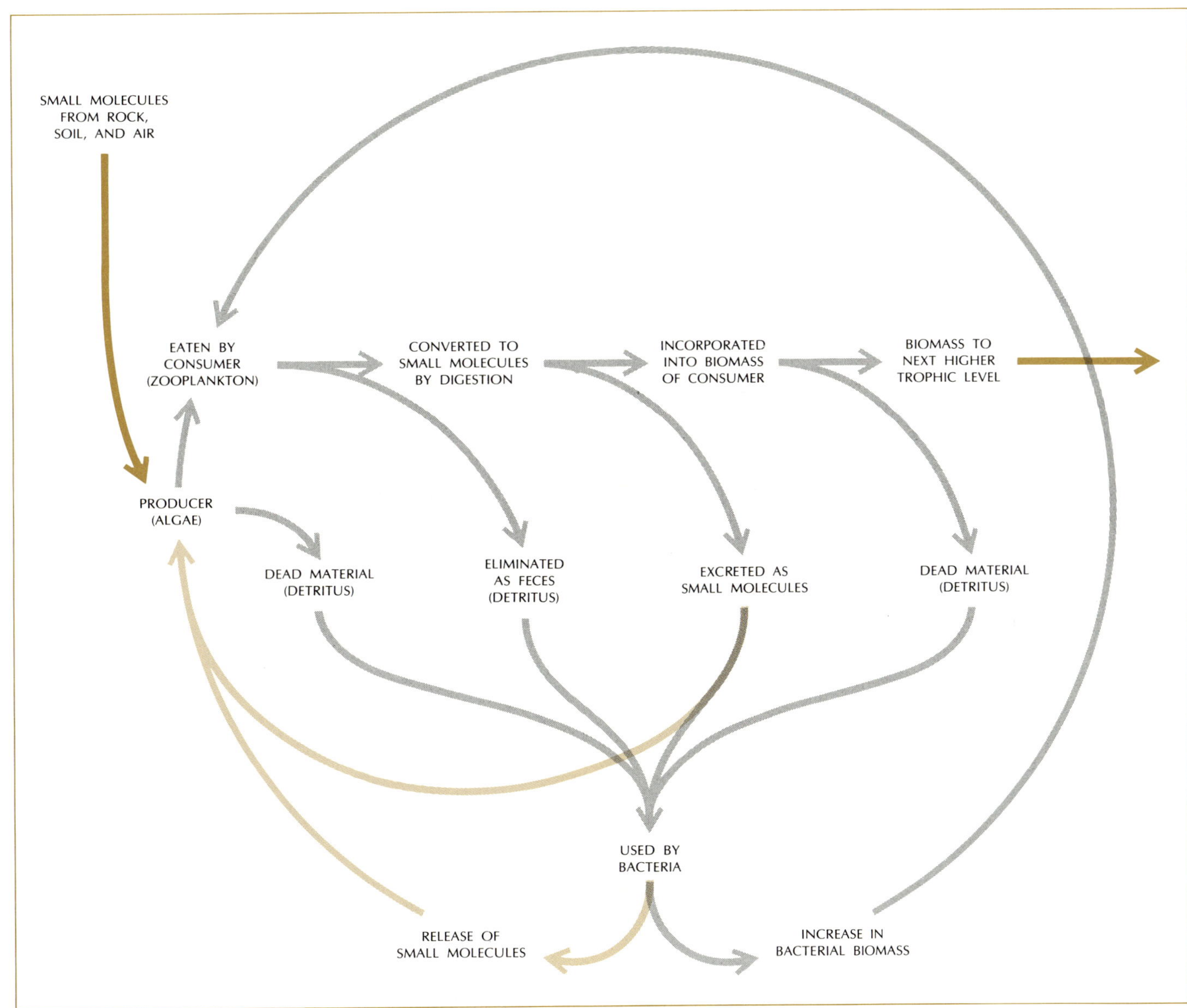

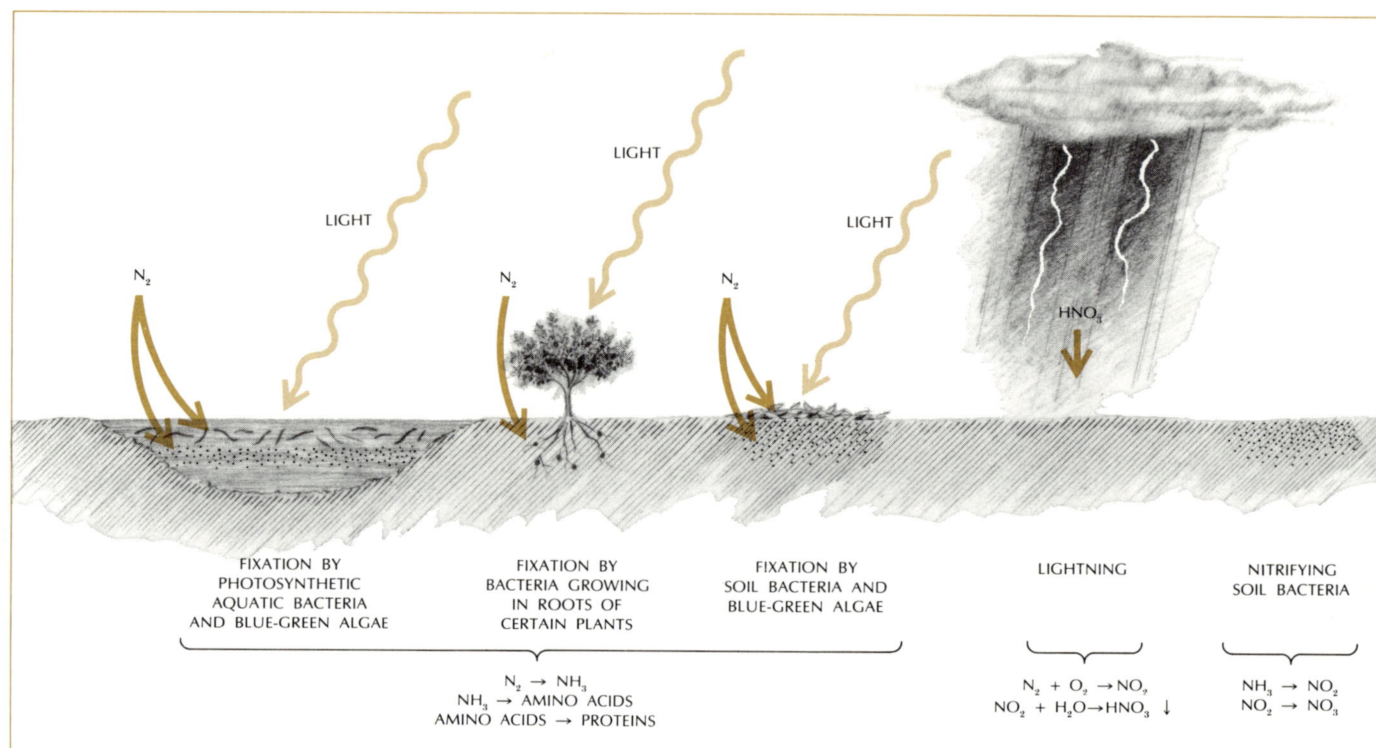

For example, nitrogen is abundant in the atmosphere as molecular N_2. But only a few bacteria and blue-green algae can "fix" atmospheric nitrogen; that is, convert it into more usable forms like nitrate and nitrite, which can be utilized by most plants. The plants incorporate the inorganic nitrogen into organic nitrogen-containing compounds such as amino acids, proteins, and nucleic acids. These compounds are then usable by members of other trophic levels in the food web. The detrital material from all of the trophic levels is attacked by decomposers which utilize the organic compounds in the detrital mass to make their own proteins and nucleic acids. The nitrogen in these organisms is then available for use by the

Nitrogen cycle. Molecular nitrogen in the atmosphere cannot be used by most organisms. Once converted to more usable forms by microorganisms, nitrogen is cycled through the biosphere.

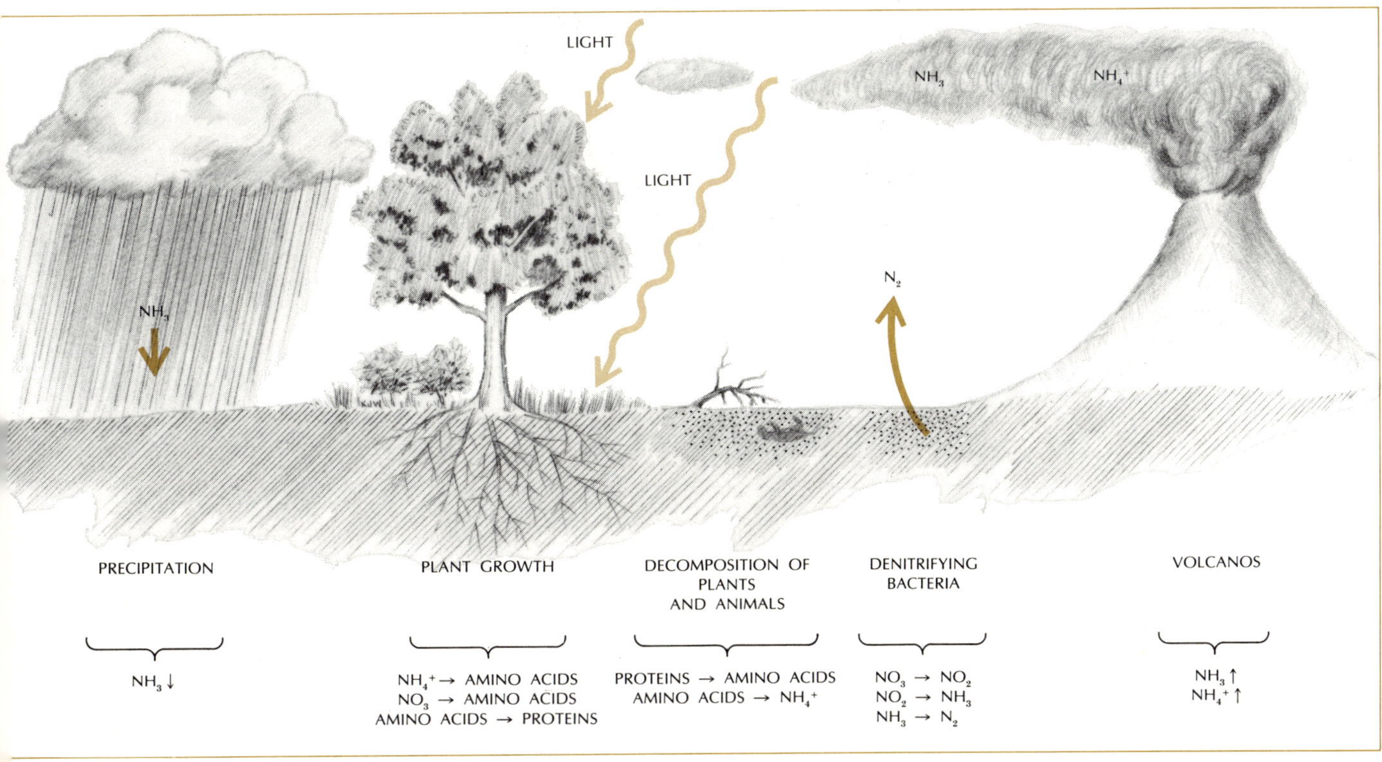

members of the primary consumer level and in turn by the subsequent steps in the food chain. Bacteria also convert some organic compounds back into inorganic compounds such as nitrate, nitrite, and ammonia which are then available for uptake by the plant populations (Fig. 11).

Thus, once the molecular nitrogen from the air is fixed by special microbial populations, it may recycle through various levels of the food web, first as a component of inorganic compounds, then as part of organic compounds, then perhaps again in inorganic form, but ultimately as part of the biologically important molecules essential to the life of every organism.

The recycling of matter   453

12
Mycorrhizae growing in association with a root of a hickory tree. Fungal hyphae surround and penetrate the tissues of the root. Nutrients from the soil can be taken up by the mycorrhizae and transferred to the conducting tissue of the tree root.

THE HUMAN IMPACT

Human tampering with even one trophic level of a delicately balanced ecosystem often leads to surprises and sometimes to disaster. The widespread and often indiscriminate use of pesticides has repeatedly produced more problems than it has solved. An excellent case history involves a recent effort to improve the health of the human population of the Malaysian island of Borneo by reducing the mosquito population. It seemed like a simple, straightforward attack on an ordinary, uncomplicated problem, involving massive spraying with the insecticide DDT. This was followed almost immediately by a dramatic decline in the mosquito population. The control program appeared to be a complete success. But DDT is a broad-spectrum insecticide; in the Borneo progam it killed many kinds of insects, including large populations of a special predatory wasp that normally fed on the native caterpillar population. When the wasps were killed by the DDT the caterpillar population exploded, feeding gluttonously on the leaves of living trees and on the leaves used on the roofs of the native houses. To the dismay of the inhabitants, the roofs began to collapse. But this was a minor inconvenience compared to another side effect of the spray program. The DDT also killed large numbers of houseflies, which then were eaten by small lizards called geckos. The lizards soon accumulated lethal doses of the DDT; the sick and dying geckos became easy prey for domestic house cats which in turn sickened and died. The decline in the cat population led to an increase in the number of rats, the carriers of many human maladies including the dread disease called the Plague. The situation had gotten out of hand. The quickest solution was to import more cats. And the most efficient method available was to ship them by air and drop them by parachute. Who would have though that spraying with DDT to kill a few mosquitoes would result in the invasion of Borneo by airborne cats? The ramifications of man's activities are often strange and wondrous to behold. The problem has been considered only in terms of mosquitoes versus chemicals. An ecologist could have predicted the retention and concentration of the DDT molecules in each trophic level, and might have suggested a more efficient alternative.

FARMING THE JUNGLE

Another example of human tampering is the conversion of forest to farmland. Such efforts are based on the simple logic that, whenever possible, we should remove all native plants and animals that have no immediate economic importance to man and replace them with highly productive domesticated organisms. This approach has been tried in the tropics with spectacularly poor results. The natural communities that occupy an area on earth have evolved over millions of years. The evolution of any individual population is intricately linked to the evolutionary trends among the other organisms in that ecosystem. The community develops as a whole, not as a cluster of isolated compartments; when its complex equilibrium is disturbed, its biological productivity suffers.

In the tropics lush native forests have been cut down and replaced by agricultural crops. The harvests are usually good for the first year or two, but suddenly the yield drops precipitously. The soil has been depleted of essential nutrients and is virtually worthless for farming. It seems unbelievable that farms in one of the richest and most complex ecosystem on earth can become so unproductive so fast.

Why did the crop plants deplete the soils of minerals while the forest did not? In the tropical rain forest there is an efficient recycling of nutrients. The vegetation and dead animal matter that fall to the forest floor are transformed by bacteria and fungi into simpler soluble compounds. These soluble materials are not washed away by the torrential rains, nor do they accumulate in thick layers of humus; instead, they are rapidly reabsorbed into the roots of the trees. The activity of special fungi known as mycorrhizae is important in this process (Fig. 12). These fungi live in intimate association with the cells of the roots of the forest trees. They derive organic food from the host tree and at the same time provide the tree with nutrients which they have absorbed from the soil.

When man cuts down the trees and replants the region with domestic crops, he disturbs this dynamic equilibrium. The crop plants lack mycorrhizae, and the soluble nutrients produced on the forest floor are leached out of the soil and lost. Moreover, crop plants are harvested regularly, a practice that prevents the accumulation of dead plant material—an important source of stored nutrients—on the ground. When the soil's reserves of essential elements are depleted, agriculture quickly becomes unprofitable and the area must be abandoned. The damage has been done; it will take many years for the natural community of organisms to repopulate the area and restore the fertility of the land.

THE WAR WITH THE INSECT

For political or religious reasons many people still refuse to admit that there is a real conflict between the human population explosion and the ability of the earth to support us. Some quote C. T. deWit, a Dutch biologist who has calculated that the earth, if properly managed, could theoretically support at least 146 billion people. This optimistic estimate is based on the development of sound management practices that would result in maximum crop yields from all the land available for agriculture, including arctic and tropic regions. Nevertheless, all biologists agree that there is a limit to the number of people the earth can support; they differ only in their estimate of how much time we have before we reach this limit.

Although the only rational long-range solution is a program of human population control, today man must exploit stop-gap measures to increase food production without jeopardizing the resources of the future. The war with famine almost always involves man in a battle with the insect, and all too often man and his insecticides have won the battle and lost the war. Preliminary successes have led to the wholesale application of chemicals in situations where their potential ecological effects are completely unknown. The results are illustrated by two additional case histories.

THE GYPSY MOTH

The first concerns the gypsy moth, a pest which during the 1930's defoliated about 500,000 acres of coniferous forests in the United States each year.

Control measures were apparently warranted. In 1945 a spraying program was initiated, and at first it seemed spectacularly effective. By 1947 the number of acres defoliated was reduced to less than 200,000. But despite continued spraying, in 1953 the number of acres defoliated had jumped to 1,500,000. The spraying was increased and by 1955 the number of acres defoliated had again dropped below 200,000. Apparently the spray program had paid off. Recently this conclusion was questioned by Kenneth Watt of the University of California at Davis. After a close examination of the records of forest defoliation for each year between 1925 and 1960, he began to wonder whether the reduction in defoliation during specific years was a result of the spraying program or merely the result of insect control by natural factors. He noted that in 1942 and 1943 when there was no spray program, the number of acres defoliated suddenly dropped below the 200,000 acre level. After extensive statistical analyses he concluded that weather was about twice as effective as the spray program in reducing defoliation by the gypsy moth. He deduced that spraying could save foliage if it was applied when the moth larvae were small, but that it did not significantly reduce the size of the moth population from generation to generation. Any reduction in one year merely reduced the competition for food among the survivors and resulted in a much larger population the following year. If the spray program were progressively increased year after year, the population might be effectively controlled, but at formidable cost. Eventually the insect population would build up a genetic resistance to the insecticide, making it necessary to substitute a new one and to begin the cycle again. The economic benefits of this type of program are highly questionable.

THE RING BARK BORER

The second case history concerns another insect control program in Malaysia. In 1956 cocoa was introduced into the state of Sabah as a commercial cash crop. The rich tropical forest was cleared to make way for the cocoa plantations. By 1966 there were 6000 acres in cultivation. But almost as soon as the cocoa trees were planted they were attacked by insect pests. The first was a ring bark borer that drills into the trunk of the cocoa tree and eventually kills it. In 1959 the broad-spectrum insecticides dieldrin and DDT were applied in high concentrations, but they were ineffective because the borer lives deep inside the plant.

In the same year additional pests appeared: caterpillars, aphids, and mealybugs. The spray program was intensified, but by 1961 caterpillars, loopers, leaf hoppers, bark borers, and bagworms had become so abundant that they threatened the very survival of the cocoa industry. Spraying was again intensified, but with little effect. Then a daring solution was suggested by the entomologists at the local Cocoa Research Station. They decided that the best hope of control was to stop the spraying. They realized that all of these insect pests were native to the region and had become a problem only after the spraying began. They reasoned that the broad-spectrum insecticides were killing off the natural enemies of the pests instead of the pests themselves.

The spraying was stopped, with immediate and dramatic results. A population of parasitic wasps attacked the loopers and reduced their numbers drastically. The plant hoppers were decimated by a parasitic fungus and by natural predators. The branch borers were parasitized by another species of wasp and rapidly declined in abundance. A special, highly selective pesticide was used for a year to reduce the bagworm population, which then succumbed to natural control by a parasitic fly. Within two years natural control had righted the havoc wrought by the dieldrin and DDT. Since that time only minor, highly selective control measures have been necessary to control a few low-density species of pest.

CONTROL BY STERILIZATION

A possible breakthrough in insect control has come from the laboratory of the Czechoslovakian biologist Karel Slama. He and his coworkers were searching for the "perfect" insecticide, a chemical that would somehow reduce the reproductive ability of individual insects without actually killing them. This would allow the affected individuals to remain in the population and to compete with their neighbors for food and for mates. Chemical sterilants have been used in the past but most have been nonselective; they sterilize both beneficial and harmful insects. Slama's technique is based on the activity of the insect's juvenile hormone, which promotes the growth of the larva and the development of the pupa, but inhibits the metamorphosis of the pupa into a sexually mature adult. Slama and his colleagues have synthesized an artificial juvenile hormone effective against the linden bug, a European insect pest. The chemical, which they call

DMF, is so potent that one microgram (one-millionth of a gram) will sterilize a female linden bug for life.

In their experiments they collected male linden bugs and gave them doses of DMF. When the males mated with untreated females, they passed along enough of the chemisterilizant to render the females sterile. In nature the results could be spectacular. A few males could infect many females. The females would continue to live, feed, and compete with normal linden bugs but they would produce no offspring. DMF appears to be highly specific; it affects only the linden bug. Because the natural form of the hormone is essential to the life process of the linden bug, the species cannot easily develop a genetic resistance to the artificial form. Thus with a small, inexpensive, periodic release of treated male bugs into the natural population, a serious pest can be controlled. The next task is to find other forms of the juvenile hormone that will be effective on other species of pest. Probably no single approach to selective control will provide a solution that applies to all insect pests; the habits and life cycles of insects are simply too diverse for that.

MAN VERSUS NATURE

Man today assumes himself to be master of the earth. He can create lakes or destroy mountains at will, and can disrupt entire biological communities with very little effort or thought. Such "progress" poses a problem which Aldous Huxley, author of *Brave New World,* expressed succinctly in an essay entitled "The Politics of Population":

> How does the human race propose to survive and, if possible, improve the lot and the intrinsic quality of its individual members? Do we propose to live on this planet in symbiotic harmony with our environment? Or, preferring to be wantonly stupid, shall we choose to live like murderous and suicidal parasites that kill their host and so destroy themselves?*

We do not have much more time to decide.

---

* "The Politics of Population," *The Center Magazine,* II, 2. March 1969. Published by the Center for the Study of Democratic Institutions. Reprinted by permission.

READINGS

Bates, M., *The Forest and the Sea*. Vintage Books, Random House, New York, 1965 (paper) .
> Beautifully written descriptions of the interrelationships of organisms in the oceans and the tropical rain forests.

Carson, R., *Silent Spring*. Fawcett, Greenwich, Conn., 1962 (paper)
> The classic bestseller that alerted the public to the devastating effects of pesticides on our environment and on us.

Dumont, R. and B. Rosier, *The Hungry Future*. Praeger, New York, 1969
> Both authors are French agronomists and they argue that by 1980 billions of people may begin to die of malnutrition unless agricultural technology is overhauled and population growth limited.

Ehrlich, P., *The Population Bomb*. Ballantine Books, New York, 1969 (paper)
> This bestseller, by a Stanford biologist, shattered public complacency about the short- and long-term consequences of irresponsible population growth.

Fuller, R. B., *Operating Manual for Spaceship Earth*. Simon & Schuster, New York, 1969 (paper)
> The designer of the geodesic dome discusses a variety of topics from his vast experience, and comes to a cautiously optimistic conclusion about the future.

Grobstein, C., *The Strategy of Life*. Freeman, San Francisco, 1965
> An overview of the whole science of biology in 113 pages. Clearly and forcefully written, it is strongly recommended as a supplement to this or any other introductory course. If you read only one other book in connection with this course, this should be the one.

Marx, W., *The Frail Ocean*. Ballantine Books, New York, 1965 (paper)
> A well-written examination of the pollution of the oceans. An eye opener.

McHarg, I. L., *Design with Nature*. Natural History Press, Garden City, N.Y., 1969
> A truly modern blend of city planning and fundamental ecology. Not only a plea, but a plan to stop the desecration of our landscapes. For the involved layman and the professional.

Murdoch, W. M. (ed.), *Environment: Resources, Pollution, and Society.* Sinauer Associates, Stamford, Conn., 1971 (paper)

> A collection of essays by experts in fields ranging from ecology to economics. Not easy reading, but one of the few works that distinguishes between situations which are truly alarming and those which simply require careful decisions.

Odum, E. P., *Ecology.* Holt, Rinehart & Winston, New York, 1966 (paper)

> A brief and turgid discussion of the dynamics of ecosystems. Stresses energy relations as well as the organisms that comprise ecosystems.

Richards, P. W., *The Tropical Rain Forest.* Cambridge University, N.Y., 1952

> A treatise on one of the richest and most complex types of ecosystem on earth.

Rienow, R., and L. T. Rienow, *Moment in the Sun.* Ballantine Books, New York, 1967 (paper)

> An exceptional description of man's impact on his environment. Highly recommended.

Rosebury, T., *Life on Man.* Berkley Publishing, New York, 1970 (paper)

> The many microorganisms that inhabit our skin, mouth, and gut aren't all bad. The author is very witty in demolishing the prudish human attitudes toward microbes. He is skeptical of our concern with deodorant, mouthwash, and disinfectant, and he includes a scholarly discourse on scatology and obscenity.

Rudd, R. L., *Pesticides and the Living Landscape.* University of Wisconsin, Madison, 1966 (paper)

> A well-documented description of the economic and ecological aspects of modern pesticides.

Sanderson, I. T., *The Continent We Live On.* Random House, New York, 1961

> Outstanding photographs in color and black and white. An introduction to the plant and animal communities of North America. Informative text.

Shepard, P., and D. McKinley, (eds.), *The Subversive Science: Essays toward an Ecology of Man,* Houghton Mifflin, Boston, 1969 (paper)

> An intriguing collection of articles running the gamut of biological relationships. Lends perspective and insight into the complex problems of the ecological approach to nature.

16. TOWARD THE TWENTY-FIRST CENTURY

"We have met the enemy and he is us."
Pogo

The scientists of recent decades have been sharply criticized for their moral detachment and lack of social commitment. The new scientist tends to be more political, more concerned with the impact of technology on the environment. Biologists in particular have become involved. They have alarmed the public with predictions of disaster, and in the process left themselves vulnerable to the criticism that their views are sometimes more than biology and less than science.

How valid are the scientists' prophecies? Reliable predictions are based on the ability to recognize significant trends and to project them into the future, preferably the near future. Prophecy is a risky task, but the risk seems worthwhile to men who have gazed into the future and been appalled by what they have seen. As authors, given a choice between scientific objectivity and social commitment, we have chosen commitment, and have written this chapter on sociobiology in the hope that in your hands it might help to change the shape of the future.

◄ **Power plants** will become a major source of atmospheric and thermal pollution in the immediate future. Photograph shows coal-fired plant on the shore of Lake Michigan near Gary, Indiana.

We are children of the wildest boom in human history. All the graphs slash upward: population, crops, energy consumption, economic growth. Citizens of a nation thousands of times richer than imperial Rome, we have tasted the bitter paradox of growth: that production and wealth can increase while our standard of living declines. In the words of economist E. J. Mishan, ''The spreading suburban wilderness, the near traffic paralysis, the mixture of pandemonium and desolation in the cities, a sense of spiritual despair scarcely concealed by the frantic pace of life'' are part of the price we pay for an expanding economy.

The resources of the planet are scarcely adequate to sustain another century of reckless expansion. The boom will destroy itself and perhaps us as well. The most dangerous threat appears to be the annihilation of all life by thermonuclear weapons or by chemical and biological warfare. Each of the almost yearly confrontations between major powers—in Korea, Berlin, Hungary, Czechoslovakia, Cuba, Laos, Vietnam, the Middle East—carries the risk that a strategic escalation of threat and counterthreat will trigger a war that would destroy hundreds of millions of people and contaminate the entire planet. The US has destroyed its stockpile of biological weapons (no doubt guided by self-interest, because bacteria and viruses are notoriously indiscriminate about whom they infect). But the major powers retain enormous stockpiles of poison gas and, as the US demonstrated in Vietnam, the will to use them. Five or ten more confrontations between great powers might leave us only a 50-50 chance of surviving to 1990. The establishment of some stable international peacekeeping mechanism is unquestionably the most crucial problem of our time. This is basically a political problem and relates to biology only in the sense that the functioning of the entire biosphere is now in jeopardy due to the power of one species . . . Man. We shall not discuss it further, except to point out that no solution is in sight, and the long-stagnant disarmament talks suggest that at the highest political levels the great powers seem to find it most convenient simply to ignore the problem.

The other problems that will become critical in the decades ahead include population, energy, pollution, and economic growth. Each is a facet of the environmental crisis created by an increasing population whose demands for an ever higher standard of living impose drastic demands on technology and on the limited resources of the planet. Solutions exist for most of these problems, at least in broad outline, but to many businessmen

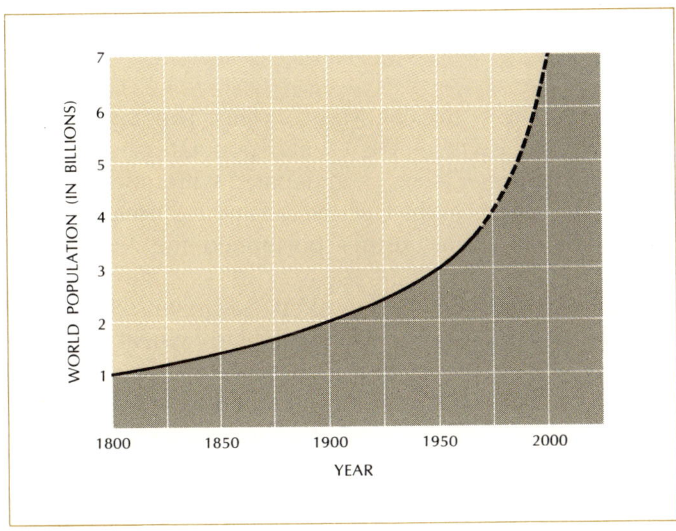

1

World population increases each year by about 70 million people, roughly the combined population of France and Canada. According to United Nations estimates, population of earth will reach seven billion by the year 2000.

and politicians the solutions seem unthinkable because they involve a halt to economic growth. The speed at which the crises sweep toward us leaves very little time to adjust our thinking. Historians have observed that democratic societies are not noted for acting promptly or for taking the long view. It is far easier in the short run to do nothing or to debate endlessly.

Regardless of what man does, the boom will end, and if we manage to avoid nuclear war, postindustrial man will revert to the more or less steady-state society characterized by a spaceship economy in which resources are continually recycled. Our only option is to decide whether we shall control the inevitable transition or whether it will simply be imposed upon us as the result of a series of ecological disasters.

POPULATION AND THE GREEN REVOLUTION

The population explosion presents a problem of both numbers and distribution. Not only are there too many people on earth but they tend to accumulate in metropolitan areas. Throughout most of his history on earth man has existed as a typical biological population, one whose high death rate was offset by a high birth rate. In the last two centuries, advances in medicine and public health have sharply reduced the death rate. As a result, the population of the earth increases by about 70 million people every year, and will probably reach seven billion before the year 2000 (Fig. 1).

At the end of the 18th century, the English economist Thomas Malthus predicted that eventually the growth of the human population would be limited by famine. Malthus' prophecy has been debated with more passion than reason for a century and a half. In the last decade another breakthrough in the technology of agriculture has further postponed the Malthusian catastrophe.

The Green Revolution, as it is sometimes called, began with the development of high-yield varieties of cereal grains, particularly wheat and rice, bred to respond to massive applications of nitrate fertilizers. Investigators also developed powerful techniques to cope with plant diseases and insect pests. Grants from the Ford, Rockefeller, and other foundations helped to put a "package" of seeds and techniques into the hands of farmers in poor countries like India and Mexico, often doubling local yields. The Green Revolution is not the answer to overpopulation but it will buy time, perhaps a decade or two, to work out a long-term solution.

The Green Revolution also has its dark side. It depends on enormous increases in the use of two serious types of pollutant: pesticides and fertilizers. American experience indicates what may happen in other countries. We have learned that many pesticides are resistant to breakdown by natural processes and tend to accumulate in the environment. DDT and other hard pesticides accumulate in certain plant and animal tissues and become concentrated as they pass upward through the food chain. DDT has found its way into human tissues; the breast milk of some American mothers was found to contain too much of it to meet Federal standards for milk shipped in interstate commerce.

At present, the global use of fertilizers totals about 56 million metric tons per year, almost all of it in advanced countries. US farmers use nitrates at the rate of 70 kilograms per arable hectare, compared with eight kilograms in India and five in Africa. In 30 years the world consumption of fertilizers could reach 700 million metric tons, with devastating impact on the inshore waterways of the overpopulated nations of the Third World. The excess nutrients in streams and lakes could choke them with blooms of foul-smelling, fish-killing algae.

Increased harvests have brought sudden prosperity to millions of farmers in underdeveloped nations. Their abrupt good fortune has aroused the aspirations of millions of others, particularly landless rural laborers, and has aggravated the classical political tensions between the haves and

and have-nots. The thrust of the Green Revolution is toward efficient large-scale farming, and this means the replacement of laborers by machines. In the Third World as in the US, more and more food is being produced by fewer and fewer hands. And throughout the world the unemployed rural laborers are migrating to cities.

CITY AND SUBURB

Because the city must bear the first shocks of the population explosion, the quality of life in urban areas has deteriorated appallingly in the last decade. According to a projection made by a United Nations team, by the end of the century 2.4 billion people will live in urban areas of 20,000 or more inhabitants. The UN report does not refer to "cities" but to "urban agglomerations." Growth is greatest around the most crowded metropolises. One third of all Argentinians now live in greater Buenos Aires, and the latest Soviet census indicated that Moscow has 400,000 more residents than its planners thought were there. Despite the efforts of many governments to encourage the pioneering of undeveloped regions, the migration to the city continues. Perhaps the most likely explanation is that the city is where the action is. It holds the promise of wealth and glamor, or at least of a job and relief from loneliness—promises that for the unskilled worker all too often wither to despair in a decaying slum.

Like a beehive, the city can be considered a kind of superorganism comprised of individuals with a variety of special skills. The essence of city life is high population density, which creates a lively diversity of activity in a compact area. Although humans can live virtually anywhere on earth, only in cities does the cross-fertilization of knowledge, talent, and wealth provide man with the opportunity to rise from mere subsistence to the potential grace of civilization.

Most American cities are plunging toward bankruptcy, and their fate is a portent of the fate of all cities. Other nations are repeating our mistakes with urban renewal and suburban sprawl just as they repeated our earlier mistakes with the automobile. Americans have long detested cities. Jefferson equated civic virtue with ownership of land, and the Constitution entrusted the fate of cities to state legislatures dominated by nonurban voters. Even today, when more than two-thirds of Americans live in urban agglomerations, we cherish the nostalgic fantasy that we are a nation of

2

Old-world city is designed for people. Aerial view of Marrakesh shows tightly packed houses. Each is built around a central courtyard that brings the outdoors inside. Winding streets, many of them dead-end, are too narrow for cars.

small town and country folk. Actually we are fast becoming a nation of suburbanites. The rich still maintain their city penthouses and rural estates but the middle class, long the backbone of urban life, has flocked to the suburbs.

The shift to the suburbs has aggravated the tyranny of the automobile. Between 1960 and 1970 the automobile population of the US increased twice as fast as the human population, and the growth was concentrated in the suburbs. Highways, parking lots, and gas stations now cover more of the US than do homes, stores, and schools. Dwelling between the city and what is left of the countryside, suburbanites are too thinly spread to be served by mass transit; a family must have one or two automobiles just to buy groceries and commute to work. Automobiles guzzle more than 35 percent of the total annual energy budget of the nation, and they spew out microscopic soot and noxious fumes at mind-boggling rates. Despite tight Federal emission laws, the internal combustion engine and its fuel simply cannot be cleaned up enough to make it acceptable in city traffic. It probably should be banned from the city, and our planning should be focused

on mass transit and on the restoration of much of the area of the inner city to homes and people instead of offices and cars. Even in death the car is a polluter: unprofitable as scrap, it is abandoned on the street or ends up in hideous junkyards. It is probably the greatest single contributor to the trashing of America.

The suburbs that ring every major American city can be envisioned as old-world cities turned inside out. Compare Marrakesh, Morocco, and Middletown, Rhode Island (Figs. 2 and 3). The old city of Marrakesh, built for people, is compact. Its streets are too narrow for cars, but shops and friends and the open desert are only a short walk away. The population density is very high, but residents live intimately with the outdoors because each house is built around a garden or courtyard. Middletown, a typical suburb, is a cluster of shopping centers connected by roads and dotted with houses and tiny patches of grass, an ironic reminder of the farmland that was bulldozed to make room for them. The per capita income is vastly higher in Middletown than in Marrakesh, but one wonders which of the towns is richer in the quality of life.

3
American suburb is designed for automobiles. There are no sidewalks. Streets, driveways, and garages occupy about as much space as houses. Lawns consume much of the land, but do not offer the privacy of the Islamic courtyard.

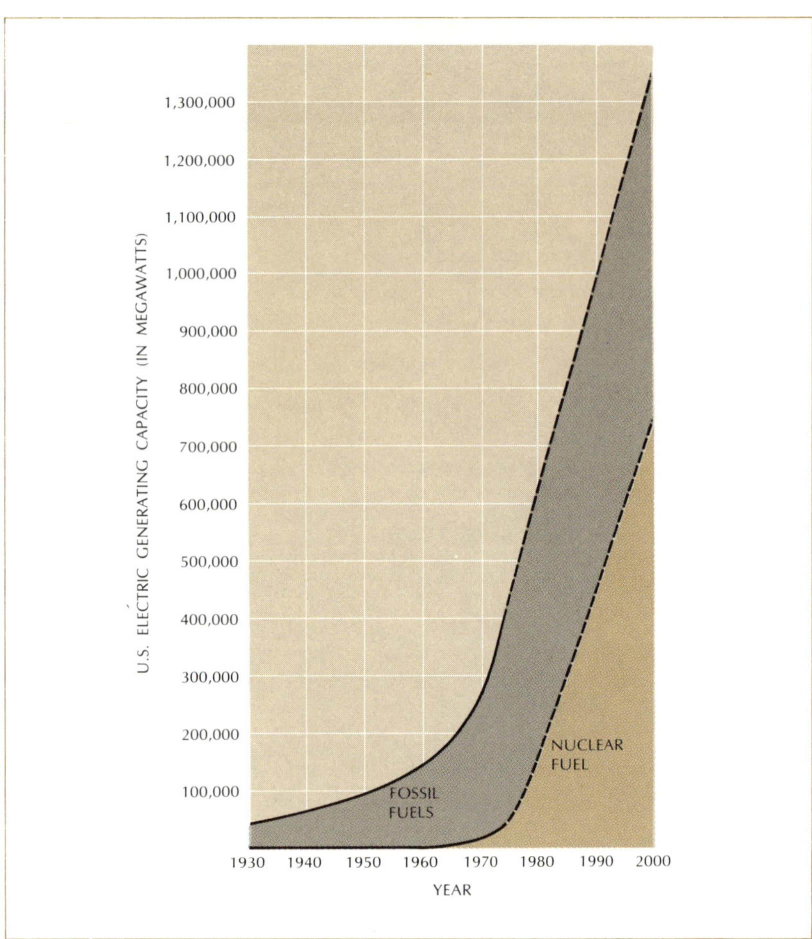

4
◄**Demand for electric power** in the US doubles every decade. Graph projects generating capacity required to meet demand in the next three decades. By the year 2000, according to the Atomic Energy Commission, nuclear plants will generate more than half of our electricity. Curve for total demand is based on estimate of Westinghouse Electric Corporation.

ENERGY AND ELECTRICITY

Hellbent for economic growth, the nations of the world are racing toward environmental bankruptcy, with the industrial countries in the lead. The US, with six percent of the world's population, consumes 35 percent of its energy. Our energy consumption is increasing four times faster than our population. The demand for electric power doubles every decade (Fig. 4).

At present, most of the electric power in the US is generated by plants that burn fossil fuels. Coal is the cheapest fuel, particularly if it is strip mined, a process that has devastated thousands of square kilometers in West Virginia, Kentucky, and southern Illinois. In the next three decades, power shovels and scrapers will tear great tracts of western states into a scarred wasteland.

Thermal pollution. Hot water discharged by a power plant on the Connecticut River appears as a bright cloud in this aerial photograph made with an infrared scanner. Thin line at center of photo is an artifact.

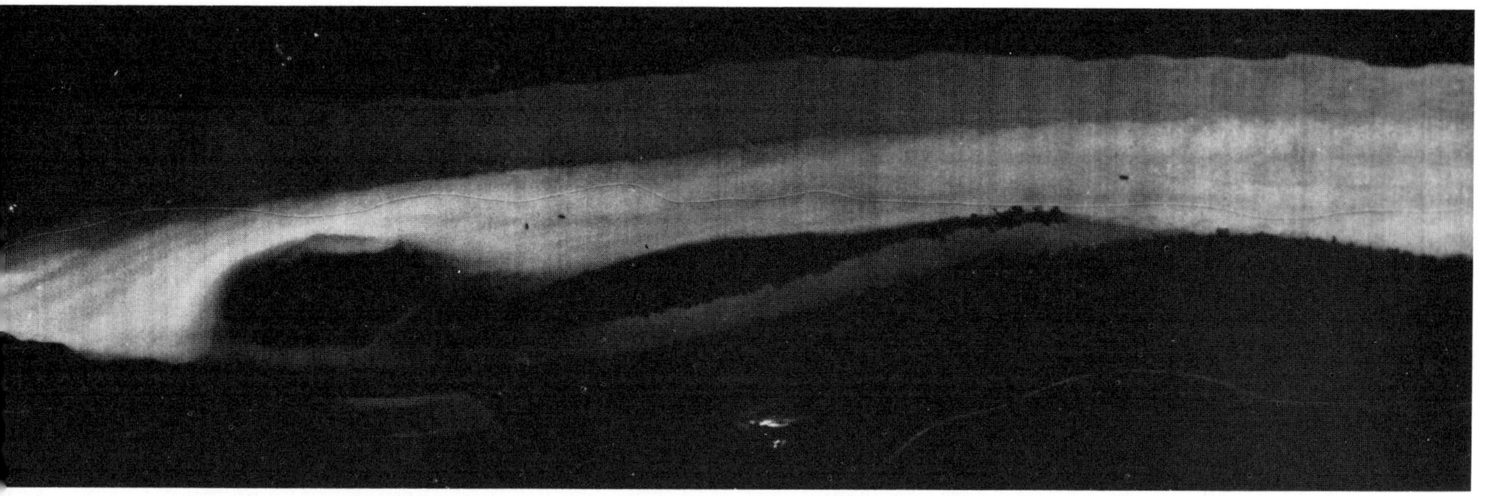

Uranium fuel has become even cheaper than coal, and most of the generating plants of the future will be nuclear. The technology of nuclear power was developed in the nick of time, because the future demand for power would have depleted the earth's reserves of fossil fuels in short order. Protests over the soot and sulfur dioxide emitted by coal-fired plants increase the attractiveness of nuclear reactors to utility companies. But nuclear reactors pollute waterways with awesome amounts of waste heat, and the chemical plants that recycle nuclear fuel elements may release dangerous radioactive waste into the environment.

Thermal pollution is a problem with both conventional and nuclear power plants (Fig. 5). In the boilers of a coal-fired plant, heat converts water into pressurized steam that spins the blades of a turbogenerator. The spent steam is cooled, condensed to water, and recycled through the boiler. The condenser is cooled by water from a nearby river or lake. In a nuclear plant the heat is produced by the fission of uranium fuel elements in the reactor core, but otherwise the generating cycle is similar. Nuclear plants require about 50 percent more cooling water than comparable coal-fired plants for two reasons: first, coal-fired plants vent some of their waste heat into the atmosphere via their smokestacks; second, for safety, nuclear plants are designed to operate at a lower temperature than conventional plants.

Right now, US power plants require almost 60 trillion gallons of cooling water per year, more than 13 percent of the total freshwater runoff. By the year 2000, waste heat from power plants could require at least 40 percent of the total runoff and perhaps as much as 70 percent. Clearly this is unacceptable, and the utilities are already searching for alternatives. Some new plants employ cooling towers to decrease aquatic thermal pollution while increasing atmospheric thermal pollution. Heated water from the plant's condensers percolates through giant silos and transfers some of its heat to the atmosphere before being pumped back into the river or lake. These towers are enormous; each of the four at the Keystone Steam Electric Station in western Pennsylvania covers the area of a football field (Fig. 6). Such plants lose perhaps 30 to 40 million gallons of water per day by evaporation, and the loss must be made up by drawing more water from the river. Cooling towers won't work everywhere. In humid areas the addition of so much heat and hot water to the atmosphere could cause recurrent fog. And many inland cities cannot afford to sacrifice more than a billion gallons of their water supply each year to cool power plants. The trend is to locate plants along the shores of lakes, estuaries, and oceans, but fishermen and conservationists are rightly concerned about the possible effects of dumping heat into shallow coastal waters. A nuclear plant being built at Calvert Cliffs on Chesapeake Bay will suck in 3.5 billion gallons of water every day and return it to the bay ten degrees warmer. The bay is an estuary where salt and freshwater meet, and massive quantities of warm water might affect the food supply of clams and oysters or disrupt the seasonal migrations of the native crabs. Citizens who opposed the plant argued in vain that the ecology of these organisms was not well enough understood to predict that thermal pollution would not harm the

6
Cooling towers dump waste heat from a power plant into the atmosphere. These towers at the Keystone Steam Electric Station in western Pennsylvania are almost 100 meters high and more than 75 meters in diameter at the base. Strong draft of air moving from bottom to top of tower cools water from plant's condensers.

shellfish harvest. Maryland may find that it has forever traded part of a gourmet crop for a few million kilowatts.

The plants that will generate our electricity in the year 2000 are now on the drawing boards or under construction. The trend is toward giant installations that must dump into the environment massive amounts of heat and, in the case of coal-fired plants, smoke, soot, and noxious gases. These plants will play a powerful role in regional ecosystems, in some cases with dismal results. The crust of the earth contains enough uranium to fulfill mankind's energy needs for another millenium. The ultimate limit on the use of energy will not be imposed by a scarcity of fuels but by the capacity of the environment to absorb huge doses of waste heat and other pollutants.

A mighty battle continues between utility companies and local residents over the locations of the new plants. If one includes the cost of transmission lines and distribution networks, the plants represent a capital investment that approaches $250 billion. That means a lot of profitable contracts, jobs, and heavy tax revenues for the communities in which the plants are eventually built. With so much money at stake, one can only be pessimistic about the fate of the environment.

CHEMICAL POLLUTION

Agriculture and industry vent thousands of different waste compounds into the air and slop about half a million kinds of pollutants into the earth's waterways. Many of these pollutants are exotic compounds, by-products of the synthesis of pesticides, plastics, and pharmaceuticals. Huge doses of these compounds are dumped into the environment with little or no study of the possible long-term effects. Microorganisms can adapt their enzyme systems to break down some of these pollutants as food, particularly the organic wastes from the petrochemical industry. There are even bacteria that can digest a tankerload of spilled crude oil, given enough time, and recycle it through the food chain. Pollutants that resist microbial attack, such as some detergents, simply accumulate in local ecosystems. A few suds in drinking water may be disgusting but not dangerous to humans. The real danger comes from the tons of lethal substances that industry flushes into the environment every day.

Mercury is a case in point. In the past 100 years, about 75 million kilograms of mercury have been used in the US, chiefly in the production of chlorine, caustic soda, plastics, and electrical equipment. In agriculture, mercury fungicides and bacteriocides are commonly sprayed on seeds and plants. Mercury compounds are also used to retard the growth of fungi and bacteria on wood, plastics, paint, paper, sewage plant filters, and hospital uniforms. The very effectiveness of mercury as a killer of micro-organisms should have given us second thoughts. The chemistry of bacteria and fungi is not basically different from the chemistry of humans. Soluble salts such as bichloride of mercury have long been recognized as being among the most toxic compounds on earth. Each year several deaths in the US are attributed to mercury poisoning, although it is difficult to diagnose. The health of an unknown number of citizens has been impaired by sublethal doses. In the 1930's, many workers in hat factories were afflicted by a malady marked by inflammation of the gums, diarrhea, chronic headaches, uncontrollable tremors, mental instability, blindness, coma, and in some cases death. The "Mad Hatter" syndrome was eventually attributed to the workers' exposure to mercury.

In 1969 alone, almost three million kilograms of mercury were used by American industry and agriculture. About half of this was recovered for reuse. The rest is still in the environment, either vented into the air by chemical plants and by furnaces that burned mercury-containing fuel oil, coal, and paper, or washed away as liquid industrial waste or runoff from farms. Mercury in aquatic ecosystems poses serious danger to man. It is progressively concentrated by plankton, fish, and marine birds. Many fish-eating birds have died of mercury poisoning, and so have some fish-eating humans. In Japan between 1953 and 1960, more than 100 people died or suffered drastic nerve and brain damage after eating seafood from Minamata Bay, which had been contaminated by mercuric oxide waste from a plastics factory.

The menace of mercury poisoning was brought home to Americans by the tragedy of the Huckleby family of Alamogordo, New Mexico. Ernest Huckleby, a laborer at the nearby Air Force base, discovered torn sacks of seeds, apparently discarded, in a shed owned by a seed company. The seeds had been coated with Panogen, a methylmercury fungicide, and dyed pink to indicate that the coating was poisonous. Huckleby loaded about five tons of the seeds onto his truck and fed them to a penful of hogs that he raised for the family table. Some of the hogs died of "the blind staggers," and later three of Huckleby's children were stricken. By Christmas of 1970, nine-year-old Ernestine Huckleby lay in a coma, blind and unable to speak. Her brother Amos was in a wheelchair, his vision impaired, his head bobbing uncontrollably. Their 21-year-old sister could walk only with crutches, and an infant boy who was in the womb when Mrs. Huckleby ate pork from the poisoned hogs was born blind.

By now we have recognized the danger of mercury and begun to adopt controls. The use of mercury in agriculture is declining and discharges of mercury from industrial plants are being sharply curtailed. Air, water, and food are monitored for traces of mercury compounds. But what about the lead in gasoline and paint? Or other heavy metals such as arsenic, cadmium, chromium, and nickel, which are also dangerous to all forms of life? Most likely business, government, and the public will pay no serious attention to them until a tragedy forces us to act.

CORPORATION VERSUS PUBLIC

It is an axiom of classical economics that the primary goal of a business venture is to maximize profit, but Harvard economist John Kenneth Galbraith argues that the primary aim of the modern corporation is growth. Skeptics might well ask themselves what corporation president would boast of a decrease in sales. The present trend toward the formation of mammoth conglomerate corporations further supports Galbraith's contention. With so powerful a commitment to growth the corporation stands as the most serious obstacle to the attainment of a steady-state economy. Corporations, as such, have no social consciences and their managers are not necessarily accountable to the public or to the government. Even the vast majority of stockholders has no significant voice in choosing the officers of a giant corporation or in influencing its policy decisions. A corporation like General Motors or International Telephone and Telegraph, with an annual cash flow higher than most nations' gross national product, can reshape the environment of a continent solely on the basis of costs and benefits to the company. The role of regulating the activities of corporations falls to the government, but in the words of ecologist William W. Murdoch, ''. . . large corporations not only have the power to pollute, they have the economic and political power to prevent, delay, and water down regulatory legislation. They also have the power and connections to ensure that the regulatory agencies don't regulate as they ought to.'' In the face of

Greasy smoke erupts from Firefighting School of Newport Naval Base. "Training fires" pollute Newport skies several days a week.

overwhelming evidence that cigarettes cause cancer, the tobacco industry has managed to keep them on the market. Similarly Detroit, confronted with irrefutable evidence that badly designed cars kill thousands of people and are major contributors to air pollution, has stalled or weakened regulatory legislation for a decade. It might be pointed out that governments set a dismal example in regulating themselves. Municipal incinerators are heavy polluters in New York City, and in Rhode Island the large Navy base at Newport pollutes the Narragansett Bay area with discharges of raw sewage, garbage, oil, and filthy smoke (Fig. 7).

Although some corporate advertising campaigns have proclaimed a fashionable interest in the environment, it usually turns out to be either a marketing gimmick or an attempt to mislead the public about the damage inflicted by the company's factories and products. After high-phosphate detergents were banned in a few cities, manufacturers rushed into the supermarkets a shelf-full of "nonpolluting" detergents with catchy names. Many of the "nonpolluters" contained caustic compounds that can burn skin and eyes and poison children. A 15-month-old infant in Connecticut died of internal burns after inhaling nonpolluting detergent.

Because of the danger that the corporate system poses to the environment, several political theorists have maintained that the corporation as we now know it is obsolete. Unfortunately the performance of the socialist nations does not promise a viable alternative. The industrial-minded bureaucracy that governs the Soviet Union appears to be as unresponsive to the welfare of the biosphere as any giant corporation. The protests of Soviet biologists failed to halt the pollution of Lake Baikal, and the Caspian Sea has become a sink for industrial waste, threatening the caviar-producing sturgeon with extinction. Pollution in the socialist countries is not yet so severe as in the West, but with expanding industrialization it is rapidly worsening.

To men born in a century shaped by the assumption that growth is good, the idea of an equilibrium population and a spaceship economy in which resources are continually recycled may seem revolutionary, and indeed it is. It seems unlikely that either the advanced industrial nations or the underdeveloped ones will voluntarily abandon their drive for growth and progress, the advanced countries because their political structure is dominated by industry, and the poor ones because of the fear that a halt to expansion might lock them into perpetual poverty.

AN END TO INDIVIDUALISM

As a political goal, the shutoff of economic growth appears to be unattainable. But Harvard psychologist B. F. Skinner looks at the problem in a different way. In his book *Beyond Freedom and Dignity,* he argues that human survival depends on a complete reshaping of the emotions of mankind, particularly the substitution of altruism for individualism. Aggressive individualism, an effective sort of behavior for taming a frontier or resisting the domination of tyrants, becomes dangerous to the species on a crowded and civilized planet. Man's freedom to be selfish exacts a hideous price in violence, overpopulation, and the plundering of resources.

Skinner maintains that human behavior is a product not of free will but of long and subtle social conditioning. He advocates the development of a technology of behavior: the engineering of psychological tools to manipulate human behavior toward socially desirable patterns. Such techniques have already been used successfully in a limited way with mental patients and juvenile criminals. But whether behavioral engineering can control the conduct of all mankind is open to doubt. The prospect stirs uneasy

"Excuse me, sir. I am prepared to make you a rather attractive offer for your square."

Drawing by Weber © 1971 The New Yorker Magazine, Inc.

An end to individualism   479

thoughts of George Orwell's *1984,* and fears about how such a powerful tool might be used by political leaders such as the chairman of the People's Republic of China, a Latin American dictator, or the president of the United States. Can man be programmed to act with the cooperation characteristic of an insect society? Should he be? And to what ends?

Biology now enters a schizophrenic future. The one path leads down the well-trod trail of rigorous scientific experimentation dealing with biological systems under controlled conditions; the other leads down the more tortuous path of interactions with the social and economic realities of human existence on earth. We must walk both trails with a sure tread. If we will but learn from the past, question the facts of the present, and innovate for the future, we humans may yet become a harmonious part of the community of life in our biosphere. In fact, you can bet your life on it.

READINGS

American Chemical Society, *Cleaning Our Environment: The Chemical Basis for Action.* ACS, Washington, D. C., 1969 (paper)

> Chemists discuss pollution: air, water, solid wastes, and pesticides. Presents 73 recommendations for further research. Nothing about thermal pollution, radioactive contamination, or aircraft exhaust.

Byerton, G., *Nuclear Dilemma.* Ballantine Books, New York, 1970 (paper)

> A journalist describes the problems of nuclear power plants. An indictment of the AEC, public utilities, and politicians.

DeBell, G. (ed.), *The Voter's Guide to Environmental Politics before, during and after the Election.* Ballantine Books, New York, 1970 (paper)

> An activist's guide on how to influence legislators to vote for ecologically sound programs.

Ehrlich, P. R., and R. L. Harriman, *How to Be a Survivor.* Ballantine Books, New York, 1971 (paper)

> An ecologist and a political scientist tell what man must do, in both over- and underdeveloped countries, to survive on Spaceship Earth.

Esposito, John C., *Vanishing Air.* Grossman Publishers, New York, 1970 (paper)

> A Ralph Nader study group report on air pollution. Facts about the sociology, politics, and enforcement of air pollution controls.

Handler, P. (ed.), *Biology and the Future of Man.* Oxford University, New York, 1970

> A look ahead by a team assembled by the National Academy of Sciences.

Helfrich, H. W., Jr. (ed.), *The Environmental Crisis: Man's Struggle to Live with Himself.* Yale University Press, New Haven, 1970 (paper)

> Twelve symposium lectures on topics such as the modification of weather, the Green Revolution, gross national product, and the role of the courts in the drive for a quality environment.

Jacobs, J., *The Death and Life of Great American Cities.* Random House, New York, 1961 (paper)

> A former editor with *Architectural Forum* attacks city planning and urban renewal. Neither city nor suburb will ever look the same to you once you've read it.

Mishan, E. J., *The Costs of Economic Growth.* Penguin Books, Baltimore, 1969 (paper)

> An increase in collective wealth does not necessarily improve the quality of life for the individual: "As the carpet of increased choice is being unrolled before us by the foot, it is simultaneously being rolled up behind us by the yard."

Murdoch, W. M. (ed.), *Environment: Resources, Pollution and Society.* Sinauer Associates, Stamford, Conn. (paper)

> Today amateurs spin a lot of woolly talk about the environment. This is a sourcebook by a score of professionals.

Rudofsky, B., *Streets for People.* Doubleday, New York, 1970

> If wheeling your Chevy down to the drive-in is your idea of a groove, this book isn't for you. Rudofsky celebrates the lively life of the streets and illustrates his ideas with photographs of old-world towns.

Portola Institute, *The Last Whole Earth Catalog.* Random House, New York, 1971 (paper)

> It's more than a catalog. If you haven't seen one, where have you been?

SCEP (Report of the Study of Critical Environmental Problems), *Man's Impact on the Global Environment.* MIT Press, Cambridge, Mass., 1970 (paper)

> Studies by six work groups of man's effect on climate, industrial pollutants, domestic and agricultural waste, energy resources, and ecology in general.

Skinner, B. F., *Walden Two.* Macmillan, New York, 1948 (paper)

> A novel about a commune organized according to the principles of behavioral engineering. Decades ahead of its time, it was as controversial in the 1950's as Skinner's *Beyond Freedom and Dignity* is today.

ACKNOWLEDGMENTS

Chapter 1

Opening photo, NASA. Fig. 1, Lowell Observatory. Figs. 2 and 3, Jet Propulsion Laboratory. Fig. 4, NASA. Fig. 5, Official U.S. Air Force photo. Fig. 7, B. J. Bok. Fig. 8, National Radio Astronomy Observatory. Fig. 10, N.Y. Academy of Medicine. Fig. 11, A. M. Siegelman. Fig. 12, after drawing from Carolina Biological Supply Co. Fig. 13, C. F. Robinow and J. Marak. Fig. 14, John R. Raper. Fig. 15, A. M. Siegelman.

Chapter 2

Fig. 2 (top), S. Mudd and D. B. Lackman, *J. Bact.,* **41,** 416, © 1941 by Williams and Wilkins Co.; (bottom), S. Mudd and T. F. Anderson, *JAMA,* **126,** 561–571 (1944). Fig. 3, Norma J. Lang. Figs. 4 and 5, Martin Deckart. Fig. 6, Grant Heilman. Fig. 7, Fleischmann Laboratories, Standard Brands Incorporated. Fig. 8, J. T. Bonner. Fig. 11 (left and right), Douglas P. Wilson. Fig. 12 (left), Robert Hermes, from National Audubon Society; (right), A. W. Ambler, from National Audubon Society. Fig. 13, Grant Heilman. Fig. 14 (left and right), American Museum of Natural History. Fig. 15 (top), Hugh Spencer; (bottom), Robert Hermes, from National Audubon Society. Fig. 16 (top), Hugh Spencer; (bottom), Robert H. Wright, from National Audubon Society. Fig. 19, Janet R. Stein. Fig. 20, Mercedes R. Edwards. Fig. 21, David Muench. Fig. 22, Bob Leatherman, from National Audubon Society. Fig. 23 (left), Woods Hole Oceanographic Institution; (right), U.S. Naval Oceanographic Office. Fig. 24, E. S. Boatman. Fig. 25, David Muench.

Chapter 3

Opening photo, F. C. Steward and K. Mühlethaler, *Ann. Botany,* **N. S. 17,** 295–325 (1953). Fig. 9, A. M. Siegelman.

Chapter 4

Opening photo, Helene N. Guttman and R. C. Styskal. Fig. 2, Fairchild Semiconductor. Fig. 5, Mercedes R. Edwards. Fig. 7, A. M. Siegelman. Fig. 8 photo, Keith R. Porter. Fig. 9, Keith R. Porter. Fig. 10 photo, Keith R. Porter. Fig. 11, A. M. Siegelman. Fig. 12 photo, Keith R. Porter. Fig. 13, W. P. Wergin. Fig. 14 (left), E. S. Boatman; (right), Norma J. Lang. Fig. 15, M. W. Steer. Fig. 16, A. M. Siegelman. Fig. 18 (A, B, C, D), G. G. Simpson and W. S. Beck, *Life, An Introduction to Biology,* Harcourt, Brace and World, 1965, p. 67; (E, F, G, H), V. A. Greulach and J. E. Adams, *Plants, An Introduction to Modern Botany,* Wiley, 1967, pp. 129, 131, 134.

Chapter 5

Opening photo, Keith R. Porter. Fig. 1, CCM: General Biological Supply House, Inc., Chicago. Fig. 3, A. M. Siegelman. Fig. 4, Brookhaven National Laboratory. Fig. 5, B. P. Kaufmann. Fig. 6, after E. Altenburg. Figs. 7 and 8, from E. D. P. DeRobertis, W. W. Nowinski, F. A. Saez, *Cell Biology,* 5th ed., W. B. Saunders, 1970. Fig. 9, diagrams adapted from F. W. Stahl, *The Mechanics of Inheritance,* Prentice-Hall, 1964, pp. 74–75; photos, A. M. Siegelman. Figs. 10, 11, and 12, A. M. Siegelman. Fig. 13, Robert Austrian, *J. exp. Med.,* **98,** 21–40 (July 1953). Fig. 15 photo, S. Brenner, R. W. Horne, et al., *J. molec. Biol.,* **1,** 281 (1959). Fig. 17 (left), Martin Deckart; (right), A. M. Siegelman.

Chapter 6

Opening photo, A. K. Kleinschmidt, *Biochim. biophys. Acta,* **61,** 861 (1962). Fig. 14, S. H. Wittwer, Michigan State University.

Chapter 7

Opening photo, Landrum B. Shettles. Fig. 1 (left), Werkfoto Carl Zeiss; (right), E. Kellenberger and C. Kellenberger, Basel Institute for Immunology. Fig. 5, Hugh Spencer. Fig. 6, T. F. Anderson. Figs. 7 and 8, Lynne Broderick. Fig. 9 photos, A. M. Siegelman. Fig. 10, Grant Heilman. Fig. 12, Triarch, Inc. Fig. 16, Grant Heilman. Fig. 20, U.S. Dept. of Agriculture.

Chapter 8

Opening photo, A. M. Siegelman. Fig. 2, DeKalb AgResearch, Inc. Fig. 5, Lynn Lamoreux.

Chapter 9

Opening photo, Lennart Heimer. Fig. 1, J. B. Gurdon, *J. Hered,* **53** (1), 5 (Jan.–Feb. 1962). Fig. 4, A. M. Siegelman. Fig. 6, adapted from Breuer and Pavan, *Chromosoma,* **7,** 371–386 (1955). Fig. 19, IBM.

Chapter 10

Opening photo, W. P. Wergin. Fig. 3, H. G. Schlegel and N. Pfennig, *Arch. Microbiol.,* **38** (1), 1961. Fig. 6, M. W. Steer. Fig. 13, Melvin Calvin.

Chapter 11

Opening photo, Keith R. Porter. Fig. 6, H. D. Agar and H. C. Douglas, *J. Bact.,* **70,** 427–434, © 1955 by Williams and Wilkins Co. Figs. 8 and 9 photos, Harry E. Morton through Society of American Bacteriologists. Fig. 15, Richard E. Dickerson and Irving Geis, *The Structure and Action of Proteins,* © 1969 by the authors. Reprinted by permission of the authors and Harper and Row.

Chapter 12

Opening photo, Elso S. Barghoorn, *Science,* **152,** 758–763 (1966). Fig. 2, Elso S. Barghoorn. Fig. 7, Institute of Molecular Evolution, University of Miami. Fig. 9, Walter Dawn, from National Audubon Society.

Chapter 13

Opening photo, American Museum of Natural History. Fig. 1, Stephen Karakashian. Fig. 2, J. William Schopf. Fig. 3, National Maritime Museum, Greenwich, England. Fig. 5 (left), Smith, Elder, and Co., London, through N.Y. Public Library; (right), Karl Weidmann, from National Audubon Society. Fig. 6, American Museum of Natural History. Fig. 10, from the experiments of Dr. H. B. D. Kettlewell. Fig. 14, N. W. Carkhuff, U.S. Geological Survey. Fig. 16, M. F. Glaessner, University of Adelaide, South Australia. Fig. 18, Milwaukee Public Museum. Fig. 22, American Museum of Natural History. Fig. 25, after reconstruction by British Museum (Natural History).

Chapter 14

Opening photo, Frederic A. Webster. Fig. 2, Paul Mundinger. Fig. 4, after W. C. Dilger. Fig. 5, Paul Mundinger. Figs. 6, 8, and 9, Yerkes Regional Primate Research Center of Emory University. Fig. 10, Paul Mundinger. Fig. 11, Katherine Ralls. Fig. 12 (top), Paul Mundinger; (bottom), Peter Marler.

Chapter 15

Fig. 2, adapted from R. H. Whittaker, *Communities and Ecosystems,* Macmillan, Collier-Macmillan, Ltd., 1970, pp. 154–155. Figs. 4 and 5, Douglas P. Wilson. Fig. 12, V. Greulach and J. E. Adams, *Plants,* Wiley, 1967.

Chapter 16

Opening photo, Charles E. Rotkin. Fig. 2, E. Vogel, Collection Musée de l'Homme. Fig. 3, John Hopf. Fig. 5, Hrb-Singer, Inc. Fig. 6, Pennsylvania Electric Co. Fig. 7, John Hopf.

INDEX

Acetabularia, 17–18, 130–133
Achondroplasia, 236
Acids: chemical, 92
    definition, 88
Actinospherium, 36
Active transport, 105–106
Adenine, 160, 164, 167, 172–174
Adenosine diphosphate, *see* ADP
Adenosine triphosphate, *see* ATP
ADP, 94–95, 297–298, 300, 322, 326
Aerobic organisms, 320, 334, 354
Aestivation, 62
Agapornis, 411
Aggression, 429–432
Albinism, human, 182, 236
Alcohol, ethyl, 323
Algae, 292, 294, 302, 442–444, 449–452, 466
    blue-green, 33–35
    fossil, 343
    life cycle of, 203
    marine, 17–18, 20, 105, 130–131
    symbiotic, 358–359
Alkaptonuria, 236
Alleles, 227, 240
    multiple, 243–244

Alloway, J. Lionel, 147
Alternation of phases, 205–206
Amino acids, 90–93, 169–176, 452–453
    biosynthesis of, 331–332
    essential, 311
    structural formulae, 91
Amoeba proteus, 112
Anaerobic organisms, 320, 324
Anemone, 46–47
Anonychia, 236
Antelope, 421
Archaeopteryx, 356–357, 395
Archeozoic era, 340–341, 385
Arctic food web, 449
Arteries, hardening of, 86
Arthropods, 48, 51–53
Ascorbic acid, 312
Asexual reproduction, 186–191
Asteroxylon, 387
Atherosclerosis, 86
Atmosphere, primitive, 343–344
Atoms, 78–81
ATP, 94–95, 169, 171, 297–298, 300–301, 303–304, 306, 310, 320–323, 325–332, 334, 336
Australopithecus africanus, 397
Automobile, 468
Autotroph, 310

Auxin, 267–268, 271
Axon, 272, 276

Bacteria, 31–33, 450–453
    cell growth of, 330–334
    electron micrographs of, 33
    fossil, 338–339
    genetics of, 144–157
    lactic acid, 324
    photosynthetic, 290–291, 297–300, 305
    reproduction of, 187–188, 192–194
    thermophilic, 59
Bacteria and DNA replication, 163–164
Bacteriochlorophyll, 290–291, 297, 299–300, 305
Bacteriophage, 148–153, 158–159
Bacterium, deep-sea, 64–65
Bahia Blanca, 362, 364
Bakanae disease of rice, 269
Barbarossa, Frederick, 23
Bark borer, ring, 457–458
Base, chemical definition of, 92–93
Beagle, H.M.S., 361–363
Behavior, 403–408, 410–413, 429–432
    human, 478
Beriberi, 312

Bilaterates, 44, 48
Biological control, 458–459
Biologist, activities of, 21–23
Biome, 439–441
Biosynthesis, 330–333
Bird calls, 421
Bird songs, 418–419
Bitter Springs chert, 359
Black Cloud, The, 11–13
Blending, 240–241
Blindness, night, 312
Blood: human, 181–182
  plasma of, 105
Blood sugar regulation, 264–265
Blood type, human, 244
Bonding potential of chemical element, 79–80
Bonner, James F., 372
Borneo, 454–455
Botulism, 32
Brain cells, 252–253
Branton, Daniel, 306

Calvin, Melvin, 37, 300, 302–304
Cambrian period, 340–341, 359, 389
Camel, 61
Capybara, 364–365, 368
Carnivores, 446–450
Carbohydrase, 317
Carbohydrates, 82–85
  biosynthesis of, 332
  digestion of, 316–317
Carbon, atomic structure of, 78–79
Carbon cycle of photosynthesis, 303
Carboxyl group, 87–88

Carotene, β, 311, 313
Carrier, genetic, 233–235, 241
Cause and effect, 68–69
Cell, 16–21
  chemical composition of, 74, 333
  eukaryotic, diagram of, 98–99
Cell division, *see* Cytokinesis
Cell theory, 16–17
Cells: human, 118–121
  plant, 119–121
Cellulose, 84–85
  electron micrograph of, 72–73
Cenozoic era, 396–397
Centromere, 137–138, 202, 220–221
Chase, Martha, 150
Chiasma, 202
Childhood's End, 13
Chimpanzee, 415
Chlamydomonas nivalis, 58
Chlamydomonas, reproduction of, 194–197
Chlorella, 37, 292, 294, 302, 358–359
Chlorophyll, 289–291, 296, 300, 340, 342
Chloroplast, 114–115, 116–117, 286–287, 294–296, 304–306, 328
Chordates, 48
Chromatids, 138–141, 199–202, 220–221, 230, 232, 246–247
Chromatium, 291, 305
Chromosome map, 136–137, 245, 248–250
Chromosome numbers, 139, 196–201

Chromosome puff, 266
Chromosomes, 107–108, 134–144
  diagrams of, 224–244
  human, 220–221
Citric acid, 326
Clarke, Arthur C., 13
Clock, atomic, 381
Clostridium botulinum, 32
Coacervates, 349–350
Colchicine, 216
Community, 437–439
Computers, 11, 281–284
  memory circuit, 100–101
Conifers, 392–393
Conjugation, 153–157, 192–194
Cork, 17
Corn, dwarf, 226–229
Corynebacterium, 64–65
Courtship, 411–413, 415–416, 431
Crick, Francis H., 159–160
Cristae, 308–309, 328
Crossing over, 199–202, 245–249
Cybernetics and Society, 21
Cycads, 392–393
Cytokinesis, 139, 142–143, 199–203
Cytokinin, 269
Cytoplasm, 111–112
Cytoplasmic inheritance, 153–157
Cytosine, 160, 164, 167, 172–174

Darwin, Charles, 267, 359–369, 372–373
DDT, 454–455, 457–458, 466
Deciduous forest, 439–441

Decomposers, 446, 450–453
Dendrite, 272
Deoxyribose nucleic acid, *see* DNA
Deoxyribose sugar, 92–93
Desert, 59–61, 439–442
Desmacidon, 45
Desmarestia, 105
Detritus, 450–452
Development, organismic, 254
deWit, C. T., 456
Diatoms, 442–443
Dickinsonia costata, 386
Dictyosome, 111
Dictyostelium discoideum, 40–42
Dieldrin, 457–458
Differentiation, cell, 254
Diffusion, 102–104
Digestion, 314–319, 336
Dihybrid, 237–240
Dimetrodon, 394–395
Dinoflagellates, 442–444
Dinosaurs, 393–395
Diplococcus pneumoniae, 145–148
Diploid number, 198, 203–206, 211–219
Diseases: genetic, 181–183
    human, 181–183
Displays, 411–413, 417, 431
DNA, 92–93, 158–160, 162–169, 173–174, 179–182, 260–262
Dobzhansky, Theodosius, 370
Dominance, incomplete, 240–241
Dominant gene, 225–231
Drosophila melanogaster, 54, 248–250
    chromosomes of, 135–137

Drug resistance in bacteria, 153–157
Dürer, Albrecht, 28
Dwarf plants, 177–181
Dwarfism: corn, 226–229
    human, 230–233

Earth, age of, 343
Earthworm, 48–49
Ecdysone, 266
Ecological efficiency, 447–448
Ecology, 435–436, 447, 454–458
Ecosystem, 437–439
Egg, human, 184–185
Eggs, 203, 205, 208–212, 214–215
Electricity, 470
Electromagnetic spectrum, 288
Electron transport system, 296–297, 326–327, 329
Electrons, 78–80, 295–297
Element, chemical, 80
Embryo: animal, 212
    plant, 209–211
Endocrine gland, 265
Endoplasmic reticulum, 110–111
Energy, 56–57, 75–78, 446–450, 470, 474
Entropy, 77–78, 104
Enzyme synthesis, control of, 258–259
Enzymes, 75, 90, 179–181, 316–317
Eobacterium isolatum, 338–339
Epiloia, 236
Erythrocyte, 18
Escherichia coli, 163, 257–259, 331–334

Eskimos, 449
Euglena gracilis, 96–97, 352–353
Eukaryote, 96–98
    reproduction of, 188–191, 194–215
Eukaryotic cells, 359
Evolution, 357, 360, 367–380, 399
    chemical, 345–347
    human, 396–399

Fats, 86–90
    biosynthesis of, 332
Fatty acids, 87–90
Fear grimace, 416–417
Fermentation, 320–321, 323–324, 336
Ferns, life cycle of, 203–205
Fertilization, 194–196, 209–211, 214–215
Fertilizers, 466
Fig-tree chert, 339, 342
Finches, 366–369, 406, 418–419, 421
Firefly, 416
Fish, three-spined stickleback, 415
Flatworm, 48–49
Flowering effect of day length, 23–24, 65, 67
Fluke, liver, 48
Food chain, *see* Food web
Food web, 446–453
Formula: chemical, 80–82
    structural, 81
Fossils, 338–339, 343, 356–357, 359, 364–365, 368, 381, 385–389, 393, 395–396
Fox, Sidney, 349

Fructose, chemical structure of, 82–83
Fruit fly, 54
  chromosomes of, 135–137
Fungi, 38–39
Funk, Casimir, 311

Galapagos Islands, 363, 366–369
Galaxy, 7–9
  age of, 342
Galbraith, John Kenneth, 476
Gametes, 194–196, 212–215
Gametogenesis, human, 212–215
Gametophyte, 203–209
Ganglion, 273, 275–276
Gene flow, 373, 377–378, 380
Gene interaction, 240
Gene pool, 371, 373–375
Genes, 138, 221–251, 257–262, 374–375, 377–379
Genetic code, 172–173
Genetic diseases, 181–183
Genetic engineering, 372
Genetic inheritance in bacteria, 144–157
Genetic traits, human, 236, 242, 244
Genotype, 224–225, 227–231, 233–235, 237–241, 244–245
Geologic calendar, 340–341
Germination, 269–271
Gibberellin, 178–181, 269–271
Ginkgo, 392–393
Glacier, continental, 378
Glucose, chemical structure of, 82–83

Glycogen, 84
Glycolysis, 321–322, 325, 328–329
Golgi bodies, 111
Gorilla, 415
Gout, 236
General Motors, 476
Grana, 286–287, 306
Grand Canyon, 381–383
Grassland, 439–442
Green revolution, 465, 466, 467
Griffith, Fred, 145
Grooming, 412–415
Growth, economic, 465, 476
Guanine, 160, 164, 167, 172–174
Gull, black-headed, 430–432
Gunflint chert, 342–343
Gurdon, J. B., 255–257

Habituation, 407
Hammerling, J., 130, 133
Haploid number, 198, 203–206, 208–209, 212–216, 218–219
Hardy, G. H., 374
Hardy-Weinberg law, 374
Head-bobbing, 411
Heat, 57–59
Heinlein, Robert A., 21
Helix, double, of DNA molecule, 160, 163–165, 168–169
Hemoglobin, 181–182, 241–242, 335
Herbivore, 446–449
Heredity, 222–223
Hershey, A. D., 150
Heterotroph, 310, 350, 353

Heterozygous condition, 234, 245–246
Homeostasis, 265–266
Homeotherms, 58
Hominids, 396–397
Homo africanus, 397–398
Homo sapiens, 398
Homologous chromosomes, 198–202, 230, 232
Homozygous condition, 234, 245–246
Honeybee, 422–424
Hooke, Robert, 17
Hormones, 263–271, 404–406
  insect, 458–459
Hoyle, Fred, 11–13
Huntington's chorea, 236
Huxley, Aldous, 459
Hybrids, 234, 241, 380, 410–411
Hyde, James Logan, 292
Hydrogen, atomic structure of, 78–79
Hydrolysis, chemical, 314–316
Hypochoeris tweedie, chromosomes of, 138
Hypothalmus, 404–405

Ice age, 396
Idiocy, amaurotic, 236
Imprinting, 408
Incubation patch, 406
Indole-3-acetic acid (IAA), 268, 271
Induction, gene, 257–261
Insecticides, 454–458, 466
Insects: body plan of, 52–53
  metamorphosis of, 266
Insulin, 264

International Telephone and Telegraph, 476
Islets of Langerhans, 264
Isotopes, 161, 292–293, 302–303
  definition, 150

Jacob, François, 257–259
Jellyfish, 46–47

Kamen, Martin, 292
Karyokinesis, 139
Kettlewell, H. D. B., 375–377
Knee jerk, 276–277
Krebs, Hans, 325
Krebs cycle, 325, 328–330
Kurosawa, E., 269

Lactic acid, 324
Lactobacillus acidophilus, 324
Lamarck, Jean Baptiste de Monet, chevalier de, 359–360
Latin name, 30–31
Learning, 406–408, 417–418
Lecithin, 89
Lederberg, Joshua, 151, 372
Lehninger, Albert L., 331–332
Lethal gene, 243, 250, 374
Leuconostoc mesenteroides, 310–311
Leucosolenia, 45
Life: boundary conditions, 56
  definition, 10–11
  distribution on earth, 55
  origin of, 339
Life cycles, 203–215
Life probes, 14–16
Life-span of organisms, 67–68
Light: effects on animals, 67
  visible, 288–289

Limulus, 389
Linkage, genetic, 227–229, 244–249
Linnaeus, Carl, 27–29
Lipase, 317
Lipids, 86–90
  digestion of, 316–317
Lovebirds, 411, 413
Lysosomes, 113, 115

Mad Hatter syndrome, 475
Magnification, microscopic, 33
Malaria, 241
Malaysia, 454, 457
Malthus, Thomas, 466
Mapping, chromosome, 245, 248–250
Margulis, Lynn, 357
Mars, 1–3, 14–16
Mass number, 79–80
Mating types in *Chlamydomonas,* 196
Megaspores, 207–209
Meiosis, 197–203, 205–209, 212–215, 230, 232, 247, 249
Membrane: cell, 101–102, 104–107, 110–111
  mitochondrial, 113
Membrane synthesis, 335–336
Mendel, Gregor Johann, 222–224, 373
Mercury, 474–476
Merman's shaving brush, 18, 20
Meselson, Matthew, 160, 162–163
Mesozoic era, 390–396
Metamorphosis, insect, 266
Metaphyta, 31, 42–43, 389

Metazoa, 31, 43–55, 385
Meteors, 5
Methane, molecular models of, 81
Methemoglobinemia, 181–182
Mice, coat color, 243
Microhyla carolinensis, 380
Microhyla olivacea, 380
Microspheres, 349–350
Microspores, 206–207
Migration, 424–426, 429
Milky Way, 7–9
Miller, Stanley, 345
Millipede, 51
Mishan, E. J., 464
Mitochondria, 113, 308–309, 328–330
Mitosis, 139–143, 188–191
Model, thought, 56
Molecule, definition of, 80–81
Mollusks, 50
Monera, 30–35, 385
Monod, Jacques, 257–259
Morgan, Thomas Hunt, 244–245
Moth: gypsy, 456–457
  peppered, 375–377
Mühlethaler, K., 306
Muller, H. J., 370
Multicellularity, 121–122
Murdoch, William W., 476
Mutagen, 250–251
Mutation, 165, 174, 250–251, 374–376, 379
Mycorrhizae, 454–456
Myopia, 234–235

Natural selection, 367–373, 376, 381, 389

Navigation, 423–429
Nearsightedness, 234–235
Nerve cell, *see* Neuron
Nerve impulse, 274–276
Nervous system, 272, 408–409
Neumann, John von, 284
Neuron, 252–253, 272–279
Neutrons, 78–80
Nitrogen, 452–453
Nitrogen fixation, 452–453
Nobel Prize, 37, 245, 304, 311
Nucleic acids, 75, 92–94, 144, 148–150, 159–171
Nucleoli, 109
Nucleotides, 92–94
Nucleus: atomic, 78–80
  cell, 107–109, 128–130, 133

Ocean, 439, 442–446
  as food source, 437
Ocean depths, 62–65
Odum, Howard, 446–447
Oparin, A. I., 339, 345
Operator gene, 258–262
Operon, 258–262
Opsin, 313
Organic chemicals, 73–74
Organisms, first true, 350
Organization, levels of, 123–125
Organizer, tissue, 263
Ortelius, Abraham, 27
Orwell, George, 480
Ostracoderm, 389
Ovary, human, 214–215
Ovum, 214–215
Oxaloacetic acid, 326
Oxygen revolution, 353–354

Paleontology, 381
Paleorivularia ontarica, 343
Paleozoic era, 340–341, 359, 385–391
Pancreatic fibrocystosis, 236
Pantothenic acid, 312, 326–327
Paramecia, 37, 154–155
Paramecium bursaria, 358–359
Park, R. B., 306
Parrot, African, 410–412
Pavlov, Ivan, 407
Pecking, 408–409
Penicillus, 18, 20
Peptide bond, 92–93, 170–171, 175
Permeability of cell membrane, 103–104
Pesticides, 454–458, 466
Phage, 148–153
Phalangeal synostosis, 236
Phenotype, 227–231, 234–236, 239–240
Phenylketonuria, 182–183, 236
Pheromones, 419–420, 432
Phospholipids, 88–90, 336
Photoperiod, 65, 67
Photosynthesis, 287, 290, 292, 294–304, 353–354
Photosynthetic unit, 305–306
Photosystems, 300–301
Phytoplankton, *see* Plankton
Pine, bristlecone, 66–67
Pistil, 209–211
Pituitary, 404–405
Plankton, 443–444, 446, 449
Plantlet, reproduction by, 189
Pneumonia, 145–148
Poikilotherms, 58

Poinsettia, 23–24
Polar bodies, 184–185
Pollen grain, 206–211, 249
Pollution, 468–469, 471
  air, 477
  chemical, 474
  thermal, 471–474
Polydactyly, 236
Polymers, 92, 347–349
Polymmia, 48–49
Polyploidy, 217–218
Polysaccharides, 82–85
Population, 437–439
Population explosion, 436, 456, 465, 467
Porphyria, 236
Precambrian era, 386
Preening, 412–413
Pressure, 62–63
Probability, genetic, 231, 233–235
Producer, primary, 446–450
Production, biological, 439, 442
Project Viking, 14–16
Prokaryote, 97, 116, 385
  reproduction of, 187–188
Protease, 317
Proteins, 90–93, 452–453
  biosynthesis of, 331–332
  digestion of, 316–317
  structure of, 335
  3-dimensional structure of, 176–177, 181
Protein synthesis, 166–177
Proterozoic era, 382, 385–386
Protista, 30–31, 35–42, 385
Protons, 78–80
Psilopsid, 386–387, 389
Pterosaurs, 393–395

Puberty, 212–213
Puffball, 190–191
Pure line, 234
Purkinje cell, 272
Pyridoxine, 312
Pyruvic acid, 321–325, 327

Radiates, 44, 46–47
Radioisotopes, 302, 381
Ragweed, 65, 67
Rain forest, 439–442
Randall, Merle, 292
Rat, kangaroo, 61
Recessive gene, 225–231
Reflex arc, 276–280
Regeneration of tissues, 191–192
Regulator gene, 258–262
Repression, gene, 257–262
Repressor, 258–262
Reproduction, 185–192, 194–197, 203–205, 212–215
  human asexual, 372
Resistance factor, 155–157
Resources, natural, 436, 464
Respiration, 320–321, 325–329, 336, 354
Respiratory chain, 326–327, 329
Retinene, 313
Reuben, Samuel, 292–294
Rhizomes, 189
Rhodopsin, 313
Ribose nucleic acid, *see* RNA
Ribose sugar, 92–93
Ribosomes, 109–110, 116, 166, 169–171
Ritualization, 412
Rivularia, 343

RNA, 92–93, 166–171, 173–174, 179–181, 260–262

Saccharomyces cerevisiae, 38–39, 323
Salmon, 53–54
  migration, 429
Saprolegnia, 38
Saturated fatty acids, 87–89
Schleiden, Matthias, 16–17
Schwann, Theodore, 16–17
Science fiction, 11–14, 21–22
Scientific method, 22–24
Scientific name, 29–31
Scorpion, 51
Scurvy, 312
Sea urchin, 52–53
Seed, structure of, 270
Seed germination, 269–271
Sexual reproduction, 192–194, 203–215
Sickle cell anemia, 182, 241
Sirens of Titan, 14
Skinner, B. F., 478
Skoog, Folke, 269
Skunk cabbage, 378–379
Slama, Karel, 458
Slime mold, 40–42
Smell, 429
Snail, 50
Snow algae, 58
Sociobiology, 463
Solar radiation, 288–289
Sonagram, 418–419
Sonar, animal, 423
Speciation, 377–380
Species, origin of, 377–380
Sperm, 203, 205, 207, 209–215
  salmon, 53–54

Sponges, 44–45
Sporangium, 203–205
Spore, 203–209
Sporophyte, 204–206, 209
Spriggina floundersi, 386
Squid, 50
Stahl, Franklin, 160, 162–163
Stamen, 207–211
Starch, 83–84
  digestion of, 319
Starch grain, 294
Starlings, 426–428
Starter gene, 260–262
Starvation, 318, 320
Stentor, 154–155
Steward, F. C., 42, 255
Stranger in a Strange Land, 21–22
Streptococcus pyogenes, 32
Stretch reflex, 277–279
Suburbs, 468–469
Sucrose, synthesis of, 82–83
Sugars, chemistry of, 82–85
Suture zone, 379–380
Symbiosis, 359
Symmetry, planes, 44
Symplocarpus foetidus, 378
Swanscombe man, 398
Swanson, Carroll A., 43
Synapses in neurons, 276, 279
Synapsis of chromosomes, 199–202, 232

Taiga, 439–441
Tail-wagging, 412
Teleology, 69
Telescope, radio, 7–9
Temperature, 57–59
Termites, 84–85
Territoriality, 430–431

Testis, human, 213
Theatri Orbis Terrarum, 26–27
Thermocline, 445
Thermodynamics, first law of, 76–77
Thermodynamics, second law of, 77–78, 104, 446
Thiamine, 312
Thymine, 160, 164, 167, 172–174
Tissue, regeneration of, 191–192
Toad, narrow-mouthed, 380
    South African clawed, 255–257
Totipotency, 255–257
Transduction, genetic, 151–153
Transformation, genetic, 144–148
Tribrachidium heraldicum, 386
Trihybrid, 240
Trilobite, 389–390
Triumphal Arch, 28

Trophic level, 446–450, 452, 454
Tropics, 454–458
Tryconympha campanulla, 85
Tubeworm, 48–49
Tundra, 439–441
Turtle, green, 429
2,4-D, 268
Tylosis, 236

Ultracentrifuge, 161–163
Uracil, 167, 172–174
Urbanization, 467
Urey, Harold, 345

Vacuole, 315
    contractile, 37–38
Virus, 148–153
Virus reproduction, 186–187
Vishniac, Wolf, 437
Visual cycle, 313–314
Vital force, 73
Vitamin A, 311–314
Vitamin B_1, 312

Vitamin B_6, 312
Vitamin C, 312
Volvox, 36
von Frisch, Karl, 422–424
Vonnegut, Kurt, Jr., 14

Wall, cell, of plants, 116–117
Wallace, Alfred Russell, 367
Wald, George, 311
Water, 59–62
Watson, James D., 159–160
Watt, Kenneth, 457
Weinberg, G., 374
Went, Fritz, 267
Wiener, Norbert, 21
Windflower, 378
Wöhler, Friedrich, 73

Yeast, 38–39, 323

Zinder, Norton, 151
Zooplankton, *see* Plankton
Zygote, 194–197, 203–205, 209–212

This book was set in Optima, a contemporary typeface designed by Hermann Zapf.

ABCDEFGH798765432